PROMISING
DRUG MOLECULES OF
NATURAL ORIGIN

PROMISING DRUG MOLECULES OF NATURAL ORIGIN

Edited by
Rohit Dutt, PhD
Anil K. Sharma, PhD, MPharm
Raj K. Keservani, MPharm
Vandana Garg, PhD

First edition published 2021

Apple Academic Press Inc.
1265 Goldenrod Circle, NE,
Palm Bay, FL 32905 USA

4164 Lakeshore Road, Burlington,
ON, L7L 1A4 Canada

CRC Press
6000 Broken Sound Parkway NW,
Suite 300, Boca Raton, FL 33487-2742 USA

2 Park Square, Milton Park,
Abingdon, Oxon, OX14 4RN UK

First issued in paperback 2021

Library and Archives Canada Cataloguing in Publication

Title: Promising drug molecules of natural origin / edited by Rohit Dutt, PhD, Anil K. Sharma, PhD, MPharm, Raj K. Keservani, MPharm, Vandana Garg, PhD.

Names: Dutt, Rohit, editor. | Sharma, Anil K., 1980- editor. | Keservani, Raj K., 1981- editor. | Garg, Vandana, (Assistant professor of pharmaceutical sciences), editor.

Description: Includes bibliographical references and index.

Identifiers: Canadiana (print) 20200305085 | Canadiana (ebook) 20200305190 | ISBN 9781771888868 (hardcover) | ISBN 9781003010395 (PDF)

Subjects: LCSH: Biological products. | LCSH: Antineoplastic agents—Therapeutic use. | LCSH: Materia medica, Vegetable. | LCSH: Herbs—Therapeutic use.

Classification: LCC RS431.A64 P76 2021 | DDC 615.7/98—dc23

Library of Congress Cataloging-in-Publication Data

Names: Dutt, Rohit, editor. | Sharma, Anil K., 1980- editor. | Keservani, Raj K., 1981- editor. | Garg, Vandana, (Assistant professor of pharmaceutical sciences), editor.

Title: Promising drug molecules of natural origin / edited by Rohit Dutt, Anil K. Sharma, Raj K. Keservani, Vandana Garg.

Description: Burlington, ON ; Palm Bay, Florida : Apple Academic Press, [2021] | Includes bibliographical references and index. | Summary: "This new volume, Promising Drug Molecules of Natural Origin, explores potential beneficial drug substances derived from nature. It presents the general principles, characteristics, evaluation techniques, and applications involved in drug molecules from natural sources, such as plants and marine life. With chapters from renowned experts from around the world, the chapters in this volume address the challenges of standardization of herbal medicines, methods of characterization of natural medicines and phyto-constituents, and quality control methods for herbal medicines. Several chapters in the book focus on the evolution of phyto-constituents in cancer therapeutics, while others deal with applications for other diseases, such as diabetes and neuroinflammatory disorders. The volume also specifically reviews heterocyclic drugs from plants. Key features: Provides an exhaustive overview of drugs originating from nature Covers scientific study protocols for characterization Presents diverse applications and uses for herbal medicines Looks at the use of herbal medicines for specific diseases and health conditions, such as diabetes and neuroinflammatory disorders This volume will be a valuable resource for faculty and advanced students in pharmaceutics as well as researchers, scientists, and industry professionals in medicine and drug development"-- Provided by publisher.

Identifiers: LCCN 2020035005 (print) | LCCN 2020035006 (ebook) | ISBN 9781771888868 (hardcover) | ISBN 9781003010395 (ebook)

Subjects: MESH: Biological Products | Antineoplastic Agents--therapeutic use | Phytotherapy--methods | Herbal Medicine--standards

Classification: LCC RS431.A64 (print) | LCC RS431.A64 (ebook) | NLM QV 241 | DDC 615.7/98--dc23

LC record available at https://lccn.loc.gov/2020035005
LC ebook record available at https://lccn.loc.gov/2020035006

ISBN: 978-1-77188-886-8 (hbk)
ISBN: 978-1-77463-894-1 (pbk)
ISBN: 978-1-00301-039-5 (ebk)

About the Editors

Rohit Dutt, PhD
Professor and Associate Dean, School of Medical and Allied Sciences,
GD Goenka University, Gurugram, Haryana, India

Rohit Dutt, PhD, is Professor and Associate Dean at the School of Medical and Allied Sciences of GD Goenka University, Gurugram, Haryana, India. He is a medicinal chemist with expertise in phytochemical screening and herbal drug development. Presently, he is pursuing research in the field of lead identification and optimization from natural sources. He is a referee and reviewer of manuscripts for many international journals. Several of his drug design tools were incorporated into leading CADD software being marketed by various multinational companies. He has more than 40 research/review publications to his credit, most of them in international journals. In 2015, he was honored with an "APTI Best Young Pharmacy Teacher of the Year Award" for his excellent contribution to the field of pharmacy education.

Anil K. Sharma, PhD, MPharm
Assistant Professor, School of Medical and Allied Sciences,
GD Goenka University, Gurugram, India

Anil K. Sharma, PhD, MPharm, is Assistant Professor at the School of Medical and Allied Sciences, GD Goenka University, Gurugram, India. He has more than 10 years of experience in academics. He has published 28 peer-reviewed papers in the field of pharmaceutical sciences in national and international reputed journals as well as 15 book chapters and 12 edited books. His research interests encompass nutraceutical and functional foods, novel drug delivery systems (NDDS), nanotechnology, health science / life science, and biology / cancer biology / neurobiology.

Raj K. Keservani, MPharm
Faculty Member of B. Pharmacy, CSM Group of Institutions,
Prayagraj, India

Raj K. Keservani, MPharm, is Faculty Member of B.Pharmacy, CSM Group
of Institutions, Prayagraj, India. He has more than 10 years of academic
(teaching) experience at various institutes in India within pharmaceutical
education. He has published 30 peer-reviewed papers in the field of pharmaceutical sciences, in national and international journals, 15 book chapters,
three co-authored books, and 12 edited books. He is also active as a reviewer
for several international scientific journals. His research interests include
nutraceutical and functional foods, novel drug delivery systems (NDDS),
transdermal drug delivery, health science, cancer biology, and neurobiology.

Vandana Garg, PhD
Assistant Professor, Department of Pharmaceutical Sciences at
M.D. University, Rohtak, India

Vandana Garg, PhD, is Assistant Professor in the Department of Pharmaceutical Sciences at M.D. University, Rohtak, India. Her field of specialization
is phytochemical screening of natural resources. Her expertise is in tissue
culture. Isolation and extraction of phytoconstituents enabled her to conduct
research in the areas of herbal drug design and development. Her present area
of research is the identification and development of anti-cancer agents from
natural sources. She is well-versed in various analytical techniques used for
the isolation and characterization of phytochemicals. She has more than 14
years of experience in academia. She has published 29 peer-reviewed papers
in the field of pharmaceutical sciences in national and international reputed
journals as well as six book chapters and one edited book.

Contents

Contributors

Mohammad Mumtaz Alam
Associate Professor, Department of Pharmaceutical Chemistry,
School of Pharmaceutical Education and Research (SPER), Jamia Hamdard, New Delhi – 110062, India
Tel.: +91-7004101839, E-mail: drmmalam@gmail.com

Anjali
Delhi Institute of Pharmaceutical Sciences and Research, Department of Pharmaceutical Chemistry,
Sector-3, MB Road, Pushp Vihar, Delhi – 110017, India

Debasish Bandyopadhyay
Department of Chemistry, School of Earth Environment and Marine Sciences (SEEMS),
The University of Texas Rio Grande Valley, 1201 West University Drive, Edinburg,
Texas – 78539, USA, Tel.: +1(956)5789414; +1(956)6653824, Fax: +1(956)6655006,
E-mail: debasish.bandyopadhyay@utrgv.edu

Mahima Chauhan
Department of Pharmaceutical Sciences, M.D. University, Rohtak, India

Harish Dureja
Department of Pharmaceutical Sciences, Maharshi Dayanand University, Rohtak, Haryana – 124001, India

Rohit Dutt
School of Medical and Allied Sciences, G.D. Goenka University, Gurugram, India

Faizana Fayaz
Delhi Institute of Pharmaceutical Sciences and Research, Department of Pharmaceutical Chemistry,
Sector-3, MB Road, Pushp Vihar, Delhi – 110017, India

Valeria Garcia
Department of Chemistry, The University of Texas Rio Grande Valley, 1201 West University Drive,
Edinburg, Texas – 78539, USA

Vandana Garg
Department of Pharmaceutical Sciences, Maharshi Dayanand University, Rohtak, Haryana – 124001,
India, E-mail: vandugarg@rediffmail.com

Felipe Gonzalez
Department of Chemistry, School of Earth Environment and Marine Sciences (SEEMS),
The University of Texas Rio Grande Valley, 1201 West University Drive, Edinburg,
Texas – 78539, USA

Vishakante Gowda
Department of Pharmaceutics, JSS College of Pharmacy, JSS Academy of Higher Education and Research,
Mysuru – 570015, Karnataka, India

Himangini
Delhi Institute of Pharmaceutical Sciences and Research, Department of Pharmaceutical Chemistry,
Sector-3, MB Road, Pushp Vihar, Delhi – 110017, India, Tel.: +91-9810074460,
E-mail: himanginibansal@gmail.com

Johurul Islam
Department of Pharmaceutics, JSS College of Pharmacy, JSS Academy of Higher Education and Research, Mysuru – 570015, Karnataka, India

Monika Elżbieta Jach
Department of Molecular Biology, The John Paul II Catholic University of Lublin, Konstantynów Street 1I, Lublin – 20-708, Poland, E-mail: monijach@kul.lublin.pl

Suresh Kumar
School of Medical and Allied Sciences, G.D. Goenka University, Gurugram, India

Bhaskaran Mahendran
Department of Pharmaceutics, JSS College of Pharmacy, JSS Academy of Higher Education and Research, Mysuru – 570015, Karnataka, India

Arunachalam Muthuraman
Department of Pharmacology, JSS College of Pharmacy, JSS Academy of Higher Education and Research, Mysuru – 570015, Karnataka, India; Pharmacology Unit, Faculty of Pharmacy, AIMST University, Semeling, Bedong – 08100, Kedah, Darul Aman, Malaysia, E-mail: arunachalammu@gmail.com

Nallapilai Paramakrishnan
Department of Pharmacognosy, JSS College of Pharmacy, JSS Academy of Higher Education and Research, Mysuru – 570015, Karnataka, India

Muthusamy Ramesh
Department of Pharmaceutical Analysis, Omega College of Pharmacy, Edulabad – 501 301, Hyderabad, Telangana, India

Narahari Rishitha
Department of Pharmacology, JSS College of Pharmacy, JSS Academy of Higher Education and Research, Mysuru – 570015, Karnataka, India

Anna Serefko
Department of Applied Pharmacy, Medical University of Lublin, Chodźki Street 4a, Lublin – 20-093, Poland

Mohammad Shaquiquzzaman
Department of Pharmaceutical Chemistry, School of Pharmaceutical Education and Research, Jamia Hamdard, New Delhi – 110062, India

Deepak Sharma
School of Medical and Allied Sciences, G.D. Goenka University, Gurugram, India, E-mail: deepak.sharma@gdgoenka.ac.in

Sandeep K. Thakur
School of Medical and Allied Sciences, G.D. Goenka University, Gurugram, India

Thirumalaraju Vaishnavi
Department of Pharmacy Practice, JSS College of Pharmacy, JSS Academy of Higher Education and Research, Mysuru – 570015, Karnataka, India

Rajavel Varatharajan
Pharmacology Unit, Faculty of Pharmacy, AIMST University, Semeling, Bedong – 08100, Kedah Darul Aman, Malaysia

Garima Verma
Department of Pharmaceutical Chemistry, School of Pharmaceutical Education and Research, Jamia Hamdard, New Delhi – 110062, India

Balasubramanyam I. Vishwanathan
Department of Pharmaceutical Chemistry, JSS College of Pharmacy,
JSS Academy of Higher Education and Research, Mysuru – 570015, Karnataka, India

Kripi Vohra
Department of Pharmaceutical Sciences, Maharshi Dayanand University, Rohtak,
Haryana – 124001, India

Ghazala Zia
Department of Pharmaceutical Sciences, M.D. University, Rohtak, India

Abbreviations

AAS	atomic absorption spectrophotometry
ABC	ATP binding cassette
AC	*Androctonus crassicauda*
AChE	acetylcholinesterase
ACS	American Cancer Society
AD	Alzheimer's disease
ADA	American Diabetes Association
AFB1	activity against aflatoxin B1
AGE	advanced glycation end products
AHP	American Herbal Pharmacopeia
ALL	acute lymphoblastic leukemia
ALS	amyotrophic lateral sclerosis
AML	acute myeloid leukemia
AMPK	AMP-activated protein kinase
APCI	atmospheric pressure chemical ionization
API	Ayurvedic Pharmacopoeia of India
APS	Astragalus Polysaccharides
AR	aldose reductase
ASIV	astragaloside IV
BBB	blood-brain barrier
BP	blood pressure
CAM	chick chorioallantoic membrane
CAPE	caffeic acid phenethyl ester
CAT	catalase
CDC	control and prevention
CDTs	cytolethal distending toxins
CE	capillary electrophoresis
CEDAD	capillary electrophoresis diode array detection
CF	crataegus flavonoids
CGE	capillary gel electrophoresis
CHD	coronary heart disease
CIEF	capillary isoelectric focusing
CLL	chronic lymphocytic leukemia

CML	chronic myeloid leukemia
CMV	cytomegalovirus
CNS	central nervous system
CPT	camptothecin
Cr	chromium
CR	complete response
CVD	cardiovascular diseases
CZE	capillary zone electrophoresis
DA	dihydroaustrasulfone
DASH	dietary approaches to stop hypertension
DHA	dihydroartemisinin
DIMS	direct infusion mass spectrometer
DNA	deoxyribonucleic acid
DPP4	dipeptidyl peptidase IV
DPPT	deoxypodophyllotoxin
DSC	differential scanning calorimetry
DT	diphtheria toxins
DTA	differential thermal analysis
EA	ellagic acid
ECI	electron capture ionization
EGCG	epigallocatechin-3-gallate
EGF	epidermal growth factors
EGFR	epidermal growth factor receptor
EI	electron impact
EIAs	enzyme-inducing anticonvulsants
ELSD	evaporative light scattering detection
EMA	European Medicines Agency
EMT	epithelial-mesenchymal transition
ESI	electrospray ionization
ETA	exotoxin A
FAO	Food and Agriculture Organization
FBF	flat bottom flask
FBG	fasting blood glucose
FDA	Federal Drug Administration
FE	fluorescence emission
FEP	first-episode psychosis
FPG	fasting plasma glucose
FRAP	ferric reducing antioxidant power

FTIR	Fourier-transform infrared
GABA	γ-aminobutyric acid
GACP	good agriculture and collection practices
GADPH	glyceraldehyde-3-phosphate dehydrogenase
GC	gas chromatography
GDM	gestational diabetes mellitus
GI	glycemic index
GIST	gastrointestinal stromal tumor
GIT	gastrointestinal tract
GKB	ginkgolide B
GLA	gamma-linolenic acid
GLC	gas-liquid chromatography
GMP	good manufacturing practices
GOT	glutamate oxaloacetate transaminase
GPx	glutathione peroxidase
HB-EGF	heparin-binding EGF-like growth factor
HCL	hair cell leukemia
HD	Huntington's disease
HDL	*high*-density lipoprotein cholesterol
HILIC	hydrophilic interaction chromatography
HIV	human immunodeficiency virus
HMVEC	human microvascular endothelial cell
HOMA	homeostatic model assessment
HOMA-IR	homeostatic model assessment of insulin resistance
HPLC	high-performance liquid chromatography
HPLC–DAD	high-performance liquid chromatography diode array detection
HPTLC	high-performance thin-layer chromatography
HRM	high resolution melting analysis
HRT	hormone replacement therapy
HSCCC	high speed counter-current chromatography
HSV	herpes simplex virus
ICD	implantable cardioverter defibrillator
ICP	inductively coupled plasma
IDDM	insulin-dependent diabetic mellitus
IEC	intestinal epithelial cells
IV	inserted intravenously
IκB	inhibitory kappa B

KF	kahalalide F
LA	linoleic acid
LAM-O	lamellarin O
LC-MS	liquid chromatography-mass spectrometry
LC-NMR	liquid chromatography-nuclear magnetic resonance spectroscopy
LC–QTOF	LC with quadrupole time-of-flight
LDH	lactate dehydrogenase
LDL-	low-density lipoprotein
LFn	lethal factor
LLC	Lewis lung cancer
LPO	lipid peroxides
LQ	*Leiurus quinquestriatus*
LS	longitudinally
MDA	malondialdehyde
MDR	multidrug-resistance proteins
MMP	matrix metalloproteinase
MND	motor neuron disease
MOMP	mitochondrial outer membrane permeabilization
MPF	maturation promoting factor
MS	mass spectrometry
MS	multiple sclerosis
MTD	maxima tolerated dose
NAA	neutron activation analysis
NADPH	Nicotinamide Adenine Dinucleotide Phosphate
NCI	National Cancer Institute
NEFA	non-esterified fatty acids
NMR	nuclear magnetic resonance
NOD	non-obese diabetic
NPLC	normal phase isolation chromatography
NSAIDs	non-steroidal anti-inflammatory drugs
NSCLC	non-small cell lung cancer
OGTT	oral *glucose tolerance test*
PAL	phenylalanine ammonia-lyase
PARP	poly (ADP-ribose) polymerase
PC	protein carbonyl
P-CABs	potassium-competitive acid blocker
PCNA	proliferating cell nuclear antigen
PD	Parkinson's disease

PICs	protease inhibitor concentrates
PNS	peripheral nervous system
PPARγ	peroxisome proliferator-activated receptorγ
PPB	pyrogallol-phloroglucinol-6,6-bieckol
PPG	postprandial plasma glucose
PTWI	provisional tolerable weekly intake values
RCC	renal cell carcinoma
R_f	retardation factor
ROS	reactive oxygen species
RP IPC-HPLC	reversed-phase ion-pairing HPLC
RP	reversed-phase
RT	reverse transcriptase
SARM's	selective androgen receptor modulator
SAX-HPLC	strong anion-exchange HPLC
SCA	spinocerebellar ataxia
SCC	squamous cell carcinoma
SCFA	short-chain fatty acids
SCLC	small cell lung cancer
SEC	size exclusion chromatography
SGPT	serum glutamate pyruvate transaminase
SNc	substantia nigra pars compacta
SOD	superoxide dismutase
SU	sulfonylureas
T1DM	type 1 diabetes mellitus
T2DM	type 2 diabetes mellitus
TBARS	thiobarbituric acid reactive substance
TC	total cholesterol
TGA	thermogravimetric analysis
TGF-β	tumor growth factor-beta
THC	tetrahydrocannabinol
TLC	thin layer chromatography
TRAIL	TNF-alpha-related-apoptosis-inducing-ligand
TS	transversely
TTL	tubulin-tyrosine ligase
UHPLC	ultra-high performance liquid chromatography
UV/VIS	ultra-visible spectroscopy
VaD	vascular dementia
VBL	vinblastine
VCL	vincristine

VDS	vindesine
VEGF	vascular endothelial growth factor
VFL	vinflunine
VRL	vinorelbine
VTA	ventral tegmental area
WHO	World Health Organization
XRD	x-ray powder diffraction

Preface

The substances derived from nature have been accredited for several years as a vivacious source of drugs as therapeutic agents against diverse diseases. Especially, the utility of drug substances of natural origin is documented in microbial products well before the emergence of modern synthetic drugs. The key interrelated specializations such as chemistry, biochemistry, and synthetic science have apparently helped to assess the efficacy of natural products from the beginning of human civilizations. Therefore, with scientific progression in current molecular as well as cellular biology, analytical chemistry, and pharmacology, the inimitable features of such drug molecules derived from nature are being pooled so as to explore the diversity of natural medicines relevant to their safety and efficacy.

There are plenty of evidences where natural products have convincingly demonstrated success in a variety of therapeutic domains. Yet adequate screening of phytopharmaceuticals poses a huge challenge to researchers. Endeavors are being made globally to harmonize the authentication procedures, process validation, and regulatory requirements pertaining to medicines originated from natural sources.

The present book strives to provide requisite information to its readers relevant to potential drug substances derived from nature. The text of this book is written by highly skilled, experienced, and renowned scientists and researchers around the globe with up-to-date information to offer insights on herbal medicines to readers, researchers, academicians, scientists, and industrialists worldwide.

This book *Promising Drug Molecules of Natural Origin*, comprises 11 chapters with four sections that describe general principles, characteristics, evaluation techniques, and applications. One section of the present text is devoted to herbal drug interventions in cancer therapeutics.

The first section titled *Drugs from Nature and Their Evaluation* consists of four chapters.

Chapter 1, *Facts about the Standardization of Herbal Medicines*, written by *Vandana Garg* and *colleagues*, gives a general introduction of drugs of natural origin. Further, it encompasses the validation and authentication procedure followed by the role of global regulatory agencies.

The details of the latest trends in natural products have been presented in Chapter 2, *Current Perspectives and Methods for the Characterization of Natural Medicines,* written by *Muthusamy Ramesh* and *associates.* The authors have given an introduction to marine drugs followed by different techniques of characterization. The key methods of evaluation have been presented in a tabulated manner.

Chapter 3, *Characterization of Phytoconstituents,* written by *Himangini Bansal* and *Sakshi Bajaj,* provides an overview of different techniques used to evaluate pharmaceuticals originated from nature.

The quality control aspects of herbal medicines are described in Chapter 4, *Quality Control of Herbal Medicines,* written by *Deepak Sharma* and *colleagues.* The chapter highlights several standards laid down by the World Health Organization to characterize natural products. There is a brief mention of harmonization initiatives across the globe to establish the veracity of herbal products.

The second section titled *Herbal Medicines in Cancer Therapeutics* consists of three chapters.

Chapter 5, *Anti-Cancer Agents from Natural Sources,* written by *Debasish Bandyopadhyay* and *Felipe Gonzalez* focuses on cancer therapeutics through drugs derived from nature. The authors have described the anticancer molecules from a variety of phytoconstituents in a wholesome manner. The striking moieties have been listed as a table mentioning the type of cancer for which they have shown promise.

A focused discussion on the role of marine drugs in curbing the menace of cancer has been provided by Chapter 6, *Current and Future Perspectives of Marine Drugs for Cancer Disorders: A Critical Review,* written by *Bhaskaran Mahendran* and *colleagues.* The authors have provided a preamble on cancers followed by a specific description of marine drugs as potential anticancer tools. In brief, the authors have attempted to address the clinical status of marine drugs as well.

Chapter 7, *Edible Pulses: A Part of Balanced Diet to Manage Cancer,* written by *Vandana Garg* and *associates,* gives particular uses of different pulses in treating cancer beyond their nutritional applications in households. In addition, the nutritional value of pulses and other health benefits associated with the consumption of pulses has also been discussed.

The third section titled *Natural Drugs with Heteroatoms* consists of two chapters.

The multi-faceted applications of drugs from nature having heteroatoms have been discussed in Chapter 8, *Heterocyclic Drugs from Plants,* written by *Debasish Bandyopadhyay* and *colleagues*. The authors have strived to compile all the chemical structures with appropriate figures. The text is elaborate and meticulously presented.

Chapter 9, *Heterocyclic Drugs Design and Development,* written by *Garima Verma* and *associates,* has attempted to provide a broad picture of natural medicines having heteroatoms to its readers. The authors have touched upon nearly all chemical entities used therapeutically having derived from nature.

The fourth section titled *Diverse Applications of Herbal Medicines* consists of two chapters.

The specialized applications of herbal medicines in control of diabetes are given by Chapter 10, *Role of Natural Agents in the Management of Diabetes,* written by *Monika Elżbieta Jach* and *Anna Serefko*. The chapter begins with the proposed mechanisms of herbal medicines in regulating carbohydrate disorder followed by a description of plants used for check of glycaemic levels.

Chapter 11, *Marine Drugs: A Source of Medicines for Neuroinflammatory Disorders,* written by *Arunachalam Muthuraman* and *colleagues,* describes the role of marine phytomedicines in the treatment of a variety of inflammations. This book chapter explores the marine drugs category and their molecular mechanism for neuroinflammatory disorders such as Alzheimer's disorder (AD), Parkinson's disorders (PD), multiple sclerosis (MS), stroke, vascular dementia (VaD), and neuropathic pain.

Part I

Drugs from Nature and Their Evaluation

CHAPTER 1

Facts about Standardization of Herbal Medicines

VANDANA GARG,[1] GHAZALA ZIA,[1] MAHIMA CHAUHAN,[1] and ROHIT DUTT[2]

[1]Department of Pharmaceutical Sciences, M.D. University, Rohtak, India

[2]School of Medical and Allied Sciences, G.D. Goenka University, Gurugram, India

ABSTRACT

Medicinal plants and their products have been used in people's life since ancient times. A large number of scientific reports highlight the benefits of using medicinal plants and herbs as an alternative to modern synthetic drugs. This increases the incidence of self-consumption and self-prescription of herbal plants as a source of medicines and as nutraceuticals. Misidentification of the constituent plants may lead to the inclusion of undesirable, unrelated species, with a potential health risk to the end-users. Substitution of the product's ingredients either intentionally or inadvertently can have a negative effect on both consumers and producers. To avoid such types of incidences, WHO and many regulatory agencies have specified the guidelines for the safety, efficacy, and quality assessment of herbal plants. Regulatory authorities of different countries have set different criteria of standardization as per their suitability. To avoid the variation, common standardized parameters and limits should be approved which should be followed globally. This overview covers the different procedures involved in the process of standardization of crude herbs and herbal formulations.

1.1   INTRODUCTION

Nowadays, renewed interests in medicinal plants have been increased due to the wide range of highly complex secondary metabolites. Plants play a beneficial role as a folk medicine, food supplements, and nutraceuticals, in modern medicines, synthesis of chemical entities and pharmaceutical intermediates (Pandey et al., 2014). Though there are numerous advantages of herbal medicines being easily available, cost-effective, range-wide range of medicinal uses, this leads to the growing demand in the International market as well. Now most of the medicinal herbs arc commercialized in markets as processed or modified forms such as tablets, capsules, powders, and in dried material (Veldman et al., 2014; Kool et al., 2012). To meet this growing demand and to decrease the non-availability of the herbal drugs they are being adulterated with substandard material, that's the main reason for the withdrawal of various herbal drug formulations from the International market. However, spurious, substituted, and adulterated medicines put the life of consumers in danger (Ize Ludlow et al., 2004; Skalli et al., 2002; Ouarghidi 2012). Adulteration is a common practice of addition, substitution, sophistication, spoilage, and inferiority either done intentionally or by mistake. Seasonal (time of collection) and chemotypes (composition of the secondary metabolites) variation, of the plants used in the commercialized herbal formulation is another important issue (Chanda et al., 2014). It is found that the long-term usage of herbal medicines shows some side effects which refute the safety consideration of herbal medicines. Aloe has attained the seventh position as the most widely used herb in the world in the treatment of minor bruises. Now its internal usage has been increased which in the long term causes potassium deficiency. Anthraquinones glycosides containing drugs should not be taken for more than 2 weeks as these may cause an electrolyte imbalance. This review highlights different standardization parameters and measures for the quality control of herbal medicine.

The term "standardization" may be defined as a set of information, which is necessary to control the reasonable consistency of products. Inherent variation attained by different agriculture and manufacturing process can also be assured qualitatively by the process of standardization (Chanchal et al., 2016). Methods of standardization include all parameters that contribute to their identity, efficacy, and quality of herbal drugs. Correct identification of the sample, morphological, and pharmacognostic evaluation such as the presence of volatile matter, quantitative evaluation (ash values, extractive values, moisture content, etc.), phytochemical evaluation, test for the presence of xenobiotics, microbial load testing, toxicity testing, and biological activity are the various parameters taking into considerations.

Amongst these, the phytochemical profile has a special significance since, it bears a direct effect on the activity of the herbal drugs. Preliminary steps for standardization of crude drugs include pharmacognostical study, which provides all the valuable information regarding the physical appearance and chemical constituent of the crude drug. It also provides various parameters, which help in identifying adulteration in dry powder, because when a plant gets dried and powdered it loses its morphological identity and easily prone to adulteration (Chanda et al., 2014). Current quality and authentication assessment methods mainly based on morphology and analytical phyto-chemistry based methods are detailed in different pharmacopeias (Raclariu et al., 2018). An error in the identification of vernacular name, quantification organoleptic, physical, chemical, and microbiological parameters will lead to sub standardization of herbal medicines and formulation which eventually affect the health of consumers. Recently advances are made on molecular techniques to detect substitution and adulteration in medicinal plants. Detection of adulteration is even possible in processed forms (Coghlan et al., 2012; Newmaster et al., 2013).

The following are the steps involved in the standardization of raw materials, and their formulations if we want to check its identity, purity, and quality.

1.2 AUTHENTICATION OF PLANTS

It involves the taxonomical botanical identification of crude drugs as prescribed in the official pharmacopeias. Plant name should be written in Latin words, where the first letter represents the Genus of the plant while the second one is a species (both in italics font) sometimes followed by varieties of the plant. Like in the case of broccoli and cabbage, both these vegetables have the same botanical name, i.e., *Brassica oleracea* but are of different varieties as the former one is of *italica* and later is *capitata*. So, the plant which is to be identified should be fully authenticated with the help of references given in the Pharmacopoeia or on the basis of their morphological and microscopical characters.

If the morphology of the two plants is quite similar and it is hard to distinguish or identify them even for an expert then DNA bar-coding is found to be helpful. DNA bar-coding in combination with High Resolution Melting analysis (Bar HRM) is an effectual, way to precisely identification of species in plants. Chloroplast DNA (cpDNA) is another suitable method used in the identification of plants (Kress et al., 2005). There is a region in the chloroplast genome of each plant species which show sufficient variation that will be convenient for identification. The CBOL Plant Working Group have studied about 907 plant species by using a variety of different gene and non-gene regions in the cpDNA and found that the two regions, rbcL, and matK, exhibit a promising efficient role in the discrimination in plant species (CBOL Plant Working Group, 2009). The greatest advantage of DNA metabarcoding is its ability to identify every single species within complex multi-ingredient and processed mixtures simultaneously (Raclariu et al., 2018). Only 15% of investigated Veronica herbal products contained the target species *Veronica officinalis* L., whereas the main known adulterant, *Veronica chamaedrys* L., was detected up to 62% (Raclariu et al., 2017).

DNA barcoding is found to be useful in the identification of plants at species-level, but there are some disadvantages to the technique. Because it is expensive and time-consuming, not easy to apply in developing countries due to financial constraints and limited availability of chemicals and consumables. There is a need of an hour to develop and validate various methods that should be reliable, more economical, and rapid than DNA bar-coding. Now a day's Bar-HRM is currently in use for taxonomic identification and the detection of adulteration in food and agriculture products (Ganopoulos et al., 2013; Madesis et al., 2012; Madesis et al., 2013).

1.2.1 MACROSCOPICAL /MORPHOLOGICAL/ORGANOLEPTIC EVALUATION

Wherever we have to authenticate a specimen of the plant material, pharmacopeial standards are available to provide a reference. Visual inspection is the quickest and simplest method for identification. Further, if a sample is found to be extensively different, in terms of color, odor, or taste in comparison with the reference, it is considered as not satisfying the parameters of standards or it may be a new species/variety (Bijauliya et al., 2017).

Differentiation of crude drugs may be done on the basis of organoleptic characters such as color, odor, taste, size, and shapes surface characteristics particularly textures and fractures. Examples include a bitter taste of *nux vomica*, the aromatic odor of Umbelliferae fruits, fractured surfaces of cascara, quillia bark, compound squill of cinchona, and brown color of cinnamon. The shape of the drug may be conical, cylindrical, subcylindrical, and fusiform. Taste of crude drugs may sweet, sour, saline, bitter, and tasteless. The difference in the shape of leaves considered on the basis of types of arrangements of leaves, margin, apex, and its base.

1.3   MICROSCOPICAL EVALUATION

The microscopic evaluation provides a detailed examination of organized drugs by studying their histological characters. Microscopy can be done by both qualitative and quantitative purposes.

1.3.1   QUALITATIVE MICROSCOPY

Qualitative microscopy of the leaf, root, and powdered drug can be done by sectioning transversely (leaf, bark, and root) and longitudinally (root, bark) by staining with phloroglucinol-HCl and methyl orange as per requirement. Photomicrography of stained sections can be done by using a camera. Different histological characters that may be seen in the microscope are cell walls, cell contents, calcium oxalate crystals, starch grains, trichomes, fibers, vessels, etc.

Different parts of the crude drugs may be easily differentiated with the help of microscopic evaluation such as powdered clove fruit contain starch grain while it is absent in the bud of clove. Interspecies variations are detected microscopically like in the case of *Hypericum perforatum* which is commonly substituted with *H. Patulum* due to limited availability of the former one. Transverse section of *H. perforatum* stem consists of compressed thin phloem, hollow pith, and absence of calcium oxalate crystals but in the case of *H. patulum* there is a broader phloem, partially hollow pith and presence of calcium oxalate crystals (Sarin et al., 1996).

1.3.2   QUANTITATIVE MICROSCOPY

The stomatal index is one of the broadly used parameters in microscopic characters of a crude drug, which points out the individual species within a genus. The stomatal index of an individual species may be different from others except in few cases like in the comparative microscopic study of four different species of Solanaceae. Lowest (13%) stomatal index is found in *Capsicum annum* whereas the highest (25%) stomatal index is reported in *Datura innoxia* (Ajayan et al., 2015).

1.4 PHYSICAL EVALUATION

Medicinal plants should be free from insects or molds and their excreta, including visible contaminants such as earthy matter harmful and poisonous foreign matter and chemical residues. To maintain the quality of crude drugs different parameters such as moisture content, viscosity, melting point, solubility, optical rotation, refractive index, ash value, extractive value and foreign organic matter should be evaluated. The presence of foreign organic matter should not be more than 2% w/w in crude drugs. Total ash signifies the amount of silica, which adheres to the plant surface externally. Ash value is determined by firstly weighing the crude drug minus weighing after incineration of the drug at a high temperature of more than 400°C. Exhausted ginger can be easily detected if its ash value varies from 6.0% w/w total ash and 1.7%w/w water-soluble ash (Bijauliya et al., 2017). High acid insoluble ash in drugs such as clove, senna, and licorice confirmed that these drugs may be contaminated with earthy material (Bele et al., 2011). The moisture content of crude drugs should be controlled. Moisture can decompose active constituents and decreases its potency, like in the case of ergot and digitalis it should not be more than 8 and 5% w/w, respectively. Distillation and gas-chromatographic methods are mainly used for the presence of moisture content of the drug (Evans et al., 2009). Another parameter that magistrate the purity of the crude drug is the melting point. Pure phytochemicals and chemicals have a constant and very sharp melting point (Gautam et al., 2010). The melting point of pure wool fat and cocoa butters are 34–44°C and 30–33°C, respectively (Kokate et al., 2015). The existence of the contaminant in a drug can be easily concluded by solubility studies. Glycosides are extracted by water and alcohol while their aglycones moieties are soluble in ether and benzene. Pure asafoetida is soluble in carbon disulfide (Kokate et al., 2015).

The optical rotation of oil is one of the important parameters as it affects the biological activity of oils. Some oils are optically active compounds either laevorotatory (+) or dextrorotatory (-), e.g., caraway oil, +74° to +80°and lemon oil, +57° to + 70°. The purity of the volatile oil containing drugs can be determined by the amount of oil present in the drug. In the case of dill and cardamom seeds volatile oil should not be less than 2.5% and 4.0%, respectively. Extractive value signifies the amount of active constituents extracted with solvent from a given amount of medicinal plant (Evans et al., 2009). Depending upon the nature of the constituent's extractive value of drug can be determined. Likewise, the drug-containing constituents like sugars, carbohydrates, mucilage water extractive value are usually determined. For tannins and resins containing drugs ethanol-soluble extractive value is calculated. Ether soluble extractive value is calculated for volatile and nonvolatile containing drugs. Other parameters include bitterness value (presence of bitter constituents, e.g., nux vomica), foaming index(in saponins containing drugs), and crude fiber (rhizome ginger contain a large amount of oleoresin and starch) should be determined. Thermogravimetric analysis (TGA), differential thermal analysis (DTA), and differential scanning calorimetry (DSC) are generally utilized to study physical or chemical changes that occur in various herbal drugs products. These instruments may also be used in preformulation studies and also used to check the compatibility between herbs and other ingredients (Silva Juinior et al., 2011).

1.5 CHEMICAL EVALUATION

The presence of chemical constituents present in the crude drugs can be estimated qualitatively and quantitatively. Qualitative estimation can be done by the confirmation of different tests in which change in color confirms the presence of the active constituents. Drangendroff's, Mayer's, Hager's, Wagner's test confirms the presence of alkaloids in herbal drugs. The presence of glycosides may be confirmed by the performing modified Bontrager's tests. Million's test confirms the presence of amino acids. Drugs containing fixed oil, volatile oil, waxes, and balsams, generally acid value, iodine value, ester value, saponification value, and weight per ml are calculated. These tests confirm the presence and different classes of secondary metabolites.

Different analytical methods and chemical markers are employed to quantify the presence of secondary metabolites present in the crude drug. Chemical markers can be classified into four groups: -active principle, active marker, analytical marker, and negative marker. Most of the plants are standardized by using active markers though they have therapeutic potentials and efficient. Although analytical markers have no therapeutic potential but they are having a defined range of a particular set of compounds. The disadvantage of analytical markers is, encloses the whole set of constituents rather than a specific constituent, e.g., total alkaloid content determines the presence of alkaloids in the drug but which type of alkaloids are present in the sample active markers is not signified. To ensure the presence of a particular one should go for the active markers. Quantitative chemical evaluation of different crystalline materials, minerals, and metallic based herbal drugs and formulations may be done with X-ray powder diffraction (XRD). XRD data confirms the development of a phospholipid complex with emodin (Singh et al., 2011) and naringenin (Semalt et al., 2010). The high crystalline nature of tin-based herbal formulation of *Vangaparpam* is determined by XRD by studying their intense sharp diffraction peaks (Sudhaparimala et al., 2011).

The chromatographic fingerprinting technique is a cost-effective primary analysis, gives a unique fingerprint which comprises of a set of peaks, which are characteristic of the herbal ingredients present in them. HPTLC, HPLC, and other chromatographic techniques help in both qualitative and quantitative analysis and also useful in the detection of adulteration. Recently these chromatographic techniques are used in collaboration with other analytical techniques which increase its efficiency and decrease the analytical time such as Gas Chromatography–Mass Spectroscopy (GC–MS), Mass Spectroscopy (HPLC–MS), Nuclear Magnetic Resonance Spectroscopy (HPLC–NMR), Capillary Electrophoresis-Diode Array Detection (CEDAD), High-Performance Liquid Chromatography Diode Array Detection (HPLC–DAD). Gas-liquid chromatography (GLC) is used for generating the fingerprint profiles of volatile oils and fixed oils of herbal drugs. Gas chromatography is used quantitatively for the identification and isolation of fixed and volatile oil in herbal medicines. When gas chromatography is used in combination with mass spectrometry it becomes cost-effective and less time-consuming. HPTLC is a handy tool and commonly used in pharmaceutical industries for the identification process development and detection of adulterants in herbal medicines. It also used for the determination of mycotoxin and pesticide contents. Polyherbal herbal formulation containing *Syzygiumjam bolanum, Bacopamon nieria, Cannabis sativa, Withania somnifera* are quantitatively evaluated by HPTLC in terms of stability, repeatability, accuracy, and phyto-constituents. HPLC techniques (Preparative and analytical) are commonly used for isolation and purification of herbal components. Two polyherbal drug formulations—Shereeshadi Kashaya and Yastyadivati, Vasicine is the major bioactive alkaloid of *Adhatoda vasica*, is estimated by HPLC. The vasicine content was found to be 18.1 mg/100 g in Shereeshadi Kashaya and 0.7 mg/100g in Yastyadivati (Anupam et al., 1992).

Liquid Chromatography-Mass Spectroscopy: (LC-MS) LC-MS is a useful tool in drug development and detection of sensitivity and specificity. Laser mass spectroscopy with 600 MHz is the best instrument for the determination of molecular weight peptides, proteins, and isotopes pattern (Zhang et al., 2009). For separation and structural elucidation of unknown compounds, mixtures, light, and oxygen-sensitive substances, the combination of liquid chromatography with NMR spectroscopy is one of the most powerful and time-saving methods. Supercritical fluid extraction is the most advanced technology in which compressed carbon dioxide gas-liquid is used. The main advantage of using SCFE is less amount of plant material is required but it is very expensive. Supercritical fluid chromatography is a fusion of gas and liquid chromatography. This technique is used for the analysis of nonvolatile and thermolabile compounds where other analytical techniques are not applicable (Pandey et al., 2014).

Capillary electrophoresis (CE) is another précised analytical technique which increases the sensitivity and specificity. It is commonly used for the quality control of flavonoids and alkaloids. It also requires less solvent and decreases the analytical time as compared to HPLC. *Radix scutellariae* showed a decrease in analysis time from 40 minutes to 12 minutes when analyzed by CE (Pietta et al., 1991; Wang et al., 2005).

1.6 TOXICOLOGICAL TESTING AND MICROBIAL ASSAY

To ensure the toxicity of drugs, one must go through the microarray, cell lines, *in vivo*, and *in vitro* techniques. The promising nature of drugs may be confirmed by doing it's *in vitro* study, which is carried outside the body in laboratories, e.g., growing cancer cells independently in a test tube in laboratories and confirming the anticancer activity of drug by doing MTT assay and NCI-60 Tumor Cell Line Screen. Further, *in vivo* studies are needed to check the response of the body and validate the therapeutic potential of the drug. Likewise, for evaluation of anti-inflammatory drugs – paw edema model and croton oil ear edema in rats and mice, oxazolone induced ear edema in mice are generally studied. As the toxicity of herbal medicines is based on the dose chosen, *in vivo* studies also confirm the minimum effective and maximum toxic dose of the drug.

Pathogenic microorganisms such as Enterococcus, Enterobacter, Clostridium, Shigella, Pseudomonas, and Streptococcus should be not present in the drug sample. It has been observed that herbal products particularly contain high starch content, are more easily prone to microbial growth. Factors responsible for the increased growth of microbes are the negligence occurs during growth stages like harvesting, processing, and in the storehouse. Plant material contains a marked amount of pesticide residue which is inevitable. Commonly used pesticides are organophosphates, carbamates, and chlorinated hydrocarbons. Thus, different pharmacopeias have prescribed limits for the presence of microbes and amount of pesticide residue. It also includes detailed test methods for determining 34 potential pesticides with their mandatory limits. The presence of heavy metals either accidental or intentional ultimately causes danger to the user's health. Limits tests for toxic heavy metals like mercury, lead, cadmium, and arsenic are mandatory to check whether it is crude drug or herbal formulation. Limit tests are defined as a simple and straightforward method of determination of amount of heavy metal present on the basis of change in the color reaction by adding specific chemical reagents. The amount of heavy metal present (in parts per million) can be estimated by comparing with the standard value in the guidelines (WHO, 1988a). Atomic absorption spectrophotometry (AAS), inductively coupled plasma (ICP), and neutron activation analysis (NAA) are employed to analyze the amount of metal quantitatively (Watson et al., 1999). Potential intake of the toxic metal can be estimated by calculating the recommended dose with the amount present in the product. This calculated intake is further compared with the Provisional Tolerable Weekly Intake values (PTWI) for toxic metals. This process of estimating the toxicological perspective of herbal preparation has been validated by the Food and Agriculture Organization of the World Health Organization (FAO-WHO) (De Smet et al., 1997).

The risk of contamination by radionuclides is rare, but if it occurs may be hazardous. WHO has collaborated with other international organizations, developed guidelines to overcome the events spread due to the contamination by radionuclides. The main emphasis has been given to the health risk arising due to the nuclear accidents popularly happened in Chernobyl and Fukushima. Radionuclide's contamination depends upon the specific radionuclide and quantity of contaminated plants consumed. Thus, contaminated herbal medicine consumed by an individual is likely to be a health risk. No permissible limits have been proposed for radioactive contamination their fore should be carried out.

1.7   ROLE OF THE WORLD HEALTH ORGANIZATION (WHO)

WHO has recommended good agriculture and collection practices (GACP) and good cultivation practices for medicinal plants. Their main motive is to ensure the quality of medicinal plants by applying standards and using modern analytical methods. Guidelines have prescribed the reference material to assess the quality of medicinal plants on account of the presence of contaminant and pesticide residue (Geneva, Switzerland: World Health

Organization, 2007). When there is a need of herbal formulations, the WHO has set parameters for the preparation of standardized herbal formulations which includes quality control of crude drug and finished products followed by the stability testing and safety assessment or toxicological studies. Also, in 2007, there was an implementation of WHO Guidelines on good manufacturing practices (GMP) for herbal medicines and for assessing the quality of herbal medicines with reference to contaminants and residues.

1.8 REGULATORY AUTHORITIES OF RESPECTIVE COUNTRIES

Several pharmacopeias for herbal medicines available are Ayurvedic Pharmacopoeia of India (API), British Herbal Pharmacopoeia, Chinese Herbal Pharmacopoeia, and Japanese Standards for Herbal Medicine, United States Herbal Pharmacopoeia. These Pharmacopeias consists of monographs describe analytical, physical, and structural standards for herbal drugs (Pandey et al., 2014). API consists of monographs of eighty common Ayurvedic herbal drugs. In 2009, the AYUSH department get collaborated with the Quality Council of India and introduced a certification scheme for AYUSH drug products. Various guidelines from different authorities such as United States Pharmacopeia, US Food and Drug Administration, and International Conference on Harmonization have provided a framework of validation process which includes accuracy, detection, linearity, precision, range, and specificity (De Smet et al., 1997). The label of the product is a primary source, which gives identification and all necessary information related to the herbal product, currently, there is no regulatory authority that certifies herb and herbal formulation as being labeled correctly. It has been observed that herbal formulation labels usually cannot be trusted to reveal what is in the package. Analysis of herbal formulation has shown that consumers have less than a 50% chance of actually getting what is enlisted above the container. Certain herbal supplements have been found significant differences between what is listed on the label and what is in the container.

1.9 CONCLUSION

As per the WHO report, more than 80% population of the world is dependent on herbal medicines. Since all the countries do not have uniform standardization parameters. So, it is a big irony, people of the one country are not fully benefitted from the product of another country on account of dissimilar regulatory guidelines. It may be said that a sizable majority of the world is deprived of the benefits of herbal medicines. So, there is a need of an hour that a common guideline should be formulated at an international level. Herbal medicines are required to be validating on each and every step. The process of standardization is a major concern in developing countries due to the selling of adulterated herbs by the fake seller for their own profit. There is no crystal-clear definition of the term "standardization" this causes variation according to intellectual. Also marketed herbal products should be validated on account of their therapeutic activity and safety profile. The regulatory authority should include guidelines related to the proper labeling of the herbal medicinal product. This reduces the risk of improper use and adverse reactions.

KEYWORDS

- **Adulteration**
- **Microbial assay**
- **Atomic absorption spectrophotometry**
- **Capillary electrophoresis**
- **Medicinal plants**

REFERENCES

Ajayan, K. V., & Babu, R. L., (2015). Variability of stomatal index and chlorophyll content in four species of solanaceae members. *Int. Res. J. Biol. Sci, 4*(2), 16–20.

Anupam, S., Krishan, L., & Handa, S. S., (1992). Standardization: HPLC determination of vasicine in polyherbal formulations. *Pharm. Biol., 30*(3), 205–208.

AOAC, (2005). *AOAC International* (18th edn.). AOAC International, Gaithersburg, MD.

Bele, A. A., & Khale, A., (2011). Standardization of herbal drugs: An overview. *Int. Res. J. Pharm, 2*(12), 56–60.

Bijauliya, R. K., Alok, S., Chanchal, D. K., & Kumar, M., (2017). A comprehensive review on standardization of herbal drugs. *Int. J. Pharm. Sci., 8*(9), 3663–3677.

CBOL Plant Working Group, (2009). A DNA barcode for land plants. *PNAS, 106*, 12794–12797.

Chanchal, D. K., Niranjan, P., Alok, S., Kulshreshtha, S., Dongray, A., & Dwivedi, S., (2016). A brief review on medicinal plant and screening method of antilithiatic activity. *Int. J. Pharmacogn., 3*(1), 1–9.

Chanda, S., (2014). Importance of pharmacognostic study of medicinal plants: An overview. *J. Pharmacogn. Phytochem., 2* (5), 69–73.

Chawla, R., Thakur, P., & Chowdhry, A., (2013). Evidence based herbal drug standardization approach in coping with challenges of holistic management of diabetes: A dreadful lifestyle disorder of 21st century. *J Diabetes Metab Disord, 12*(1), 35.

Coghlan, M., Haile, J., Houston, J., Murray, D., White, N., & Moolhuijzen, P., (2012). Deep sequencing of plant and animal DNA contained within traditional Chinese medicines reveals legality issues and health safety concerns. *PLoS Genet., 8*(4), e1002657.

De Smet, P. A. G. M., Keller, K., Hansel, R., & Chandler, R. F., (1997). *Adverse Effects of Herbal Drugs* (Vol. 3, pp. 137–145). Springer-Verlag, Heidelberg.

Evans, W. T., (2009). *Trease and Evans Pharmacognsoy* (16th edn., pp. 121–132), Elsevier Limited. Ganopoulos, I., Aravanopoulos, F., Madesis, P., Pasentsis, K., Bosmali, I., & Ouzounis, C., (2013). Taxonomic identification of Mediterranean pines and their hybrids based on the high resolution melting (HRM) and TRNL approaches: From cytoplasmic inheritance to timber tracing. *PLoS One., 8*(4), e60945.

Gautam, A., Kashyap, S. J., Sharma, P. K., Garg, V. P., Visht, S., & Kumar, N., (2010). Identification, evaluation, and standardization of herbal drugs: An overview. *Der Pharmacia. Lettre., 2*(6), 302–315. https://www.cabdirect.org/cabdirect/abstract/20113033868 (accessed on Nov 20, 2018).

Ize Ludlow, D., Ragone, S., Bruck, I. S., Bernstein, J. N., Duchowny, M., & Peña, B. M., (2004). Neurotoxicities in infants seen with the consumption of staranise tea. *Pediatrics, 114*, 653–656.

Kokate, C. K., Purohit, A. P., & Gokhale, S. B., (2015). *Pharmacognosy* (51st edn., Vol. 7, pp. 19–26). Nirali Prakashan.

Kool, A., De Boer, H. J., Krüger, Å., Rydberg, A., Abbad, A., & Björk, L., (2012). Molecular identification of commercialized medicinal plants in Southern Morocco. *PLoS One., 7*(6), e39459.

Kress, W. J., Wurdack, K. J., Zimmer, E. A., Weigt, L. A., & Janzen, D. H., (2005). Use of DNA barcodes to identify flowering plants. *PNAS, 102*, 8369–8374.

Madesis, P., Ganopoulos, I., Anagnostis, A., & Tsaftaris, A., (2012). The application of Bar-HRM (barcode DNA-High resolution melting) analysis for authenticity testing and quantitative detection of bean crops (Leguminosae) without prior DNA purification. *Food Control, 25*, 576–582.

Madesis, P., Ganopoulos, I., Bosmali, I., & Tsaftaris, A., (2013). Barcode high resolution melting analysis for forensic uses in nuts: A case study on allergenic hazelnuts (*Corylusavellana*). *Food. Res. Int., 50*, 351–360.

Newmaster, S. G., Grguric, M., Shanmughanandhan, D., Ramalingam, S., & Ragupathy, S., (2013). DNA barcoding detects contamination and substitution in North American herbal products. *Biomed. Central, 11*, 222.

Ouarghidi, A., Powell, B., Martin, G. J., De Boer, H. J., & Abbad, A., (2012). Species substitution in medicinal roots and possible implications for toxicity in Morocco. *Econ. Bot., 66*, 370–382.

Pandey, A., & Tripathi, S., (2014). Concept of standardization, extraction, and pre phytochemical screening strategies for herbal drug. *J Pharmacogn Phytochem, 2*(5), 115–119.

Patil, S. G., Wagh, A. S., Pawara, R. C., & Ambore, S. M., (2013). Standard tools for evaluation of herbal drugs: An overview. *Pharma Innovation, 2*(9), 60–65.

Pietta, P., Mauri, P., Rava, A., & Sabbatini, G., (1991). Application of micellarelectrokinetic capillary chromatography to the determination of flavonoid drugs. *J Chromatogr A, 549*, 367–373.

Raclariu, A. C., Heinrich, M., Ichim, M. C., & Boer, H. D., (2018). Benefits and limitations of DNA barcoding and metabarcoding in herbal product authentication. *Phytochem Anal., 29*(2), 123–128.

Raclariu, A. C., Mocan, A., Popa, M. O., Vlase, L., Ichim, M. C., Crisan, G., Brysting, A. K., & De Boer, H., (2017). *Veronica officinalis* product authentication using DNA metabarcoding and HPLC-MS reveals widespread adulteration with *Veronica chamaedrys*. *Front Pharmacol, 8*, 378.

Sarin, Y. K., (1996). *Illustrated Manual of Herbal Drugs Used in Ayurveda* (pp. 1–422). *CSIR*, New Delhi.

Semalt, A., Semalty, M., Singh, D., & Rawat, M. S. M., (2010). Preparation and characterization of phospholipid complexes of naringenin for effective drug delivery. *J Incl Phen Macro, 63*(3), 253–260.

Silva, J. J. O. C., Costa, R. M. R., Teixeira, F. M., & Barbosa, W. L. R., (2011). Processing and quality control of herbal drugs and their derivatives. *J Herb Med, 14*(1), 115– 114.

Singh, D., Rawat, M. S. M., Semalty, A., & Semalty, M., (2011). Emodin-phospholipid complex: A potential of herbal drug in the novel drug delivery system. *J Therm Anal Calorim, 8*(3), 284–291.

Skalli, S., Alaoui, I., Pineau, A., Zaid, A., & Soulaymani, R., (2002). *Atractylis gummifera* L. poisoning: A case report. *Bull. Soc. Pathol. Exot., 95*(4), 284–286.

Sudhaparimala, S., Kodi, C. M., Gnanamani, A., & Mandal, A. B., (2011). Quality assessment of commercial formulations of tin-based herbal drug by physicochemical fingerprints. *Indian J Sci Technol, 4*(12), 1710–1714.

Veldman, S., Otieno, J., Gravendeel, B., Andel, T. V., & Boer, H. D., (2014). Conservation of endangered wild harvested medicinal plants: Use of DNA bar-coding. *Novel Plant Bioresources: Applications in Food, Medicine and Cosmetics*, 81–88.

Wang, L. C., Cao, Y. H., Xing, X. P., & Ye, J. N., (2005). Fingerprint studies of *Radix Scutellariae* by capillary electrophoresis and high-performance liquid chromatography. *Chromatographia, 62*, 283–288.

Watson, D. G., (1999). *Pharm Anal*. Churchill Livingstone, Edinburgh.

Zhang, Q., & Ye, M., (2009). Chemical analysis of the Chinese herbal medicine Gan-Cao (licorice). *J Chromatogr A, 1216*(11), 1954–1969.

CHAPTER 2

Current Perspectives and Methods for the Characterization of Natural Medicines

MUTHUSAMY RAMESH,[1] ARUNACHALAM MUTHURAMAN,[2,3]
NALLAPILAI PARAMAKRISHNAN,[4] and
BALASUBRAMANYAM I. VISHWANATHAN[5]

[1]*Department of Pharmaceutical Analysis, Omega College of Pharmacy, Edulabad – 501 301, Hyderabad, Telangana, India*

[2]*Department of Pharmacology, JSS College of Pharmacy, JSS Academy of Higher Education and Research, Mysuru – 570015, Karnataka, India*

[3]*Pharmacology Unit, Faculty of Pharmacy, AIMST University, Semeling, Bedong – 08100, Kedah, Darul Aman, Malaysia, E-mail: arunachalammu@gmail.com*

[4]*Department of Pharmacognosy, JSS College of Pharmacy, JSS Academy of Higher Education and Research, Mysuru – 570015, Karnataka, India*

[5]*Department of Pharmaceutical Chemistry, JSS College of Pharmacy, JSS Academy of Higher Education and Research, Mysuru – 570015, Karnataka, India*

ABSTRACT

Plants, animals, marine, minerals, and microorganisms are the promising natural sources of therapeutic agents. Natural medicines are widely used for the ailments of various disorders including the cardiovascular, respiratory, and central nervous systems. Phytochemical constituents of natural origin have shown multiple pharmacological actions that include anti-inflammatory, anticancer, antioxidant, and anti-infective. Natural resources constitute a diverse class of phytochemical constituents from simple to complex molecular structures like taxol and vincristine. The natural compounds play a pivotal role in the design and development of newer drugs. The natural resources have been a productive origin by providing a novel hit/lead molecules for emerging therapeutic targets. In recent days, the isolation of marine products and characterization of natural constituents have been increasing continuously in literature. However, the characterization of phytochemical constituents has been found as a critical process in many cases. The current chapter describes computational methods, chromatography techniques, and spectrometry-based characterization of natural compounds. In addition, the recent developments in the structural elucidation and characterization of the phytochemical structural entity would be discussed. The current study may provide a newer vision for the characterization of complex phytochemical moieties of natural origin.

2.1   INTRODUCTION

Chemicals present naturally in the plants are called phytochemicals (Baxter et al., 1998; Molyneux et al., 2007). Alkaloids, glycosides, essential oils, oleoresin, flavonoids, and sterols are major secondary metabolites of plants included in phytochemicals. The bioactive phytochemicals have a protective role against several illnesses (Craig, 1997). Phytochemicals contribute defense action of plants (Craig, 1997). It also serves as pigments and non-essential nutrients. The phytochemicals are used in the various ailments. The eight major actions &molecular mechanisms are identified for their properties. Such actions are (i) help the cells to keep healthy from free radicals associated cell death, e.g., lycopene from tomato, and watermelon and allium from onion (Naik et al., 2006); (ii) prevent the heart diseases, e.g., ellagic acid from grapes and strawberries (Liu, 2007); (iii) protect from cancer, e.g., monoterpenes from citrus fruits (Weng and Yen, 2012); (iv) turn up the body's defense mechanism, e.g., flavonoids from green tea (Hughes, 1988); (v) acts as antioxidants, e.g., carotenoids from carrots (Kim et al., 2003); (vi) show antibacterial activity, e.g., allicin from garlic (Nascimento et al., 2000); (vii) interfere with cell DNA, e.g., saponin from beans (Rajendran et al., 2011); and viii) function as detoxifying agent, e.g., non-starch polysaccharides (Percival, 1997).

2.1.1   MARINE CONSTITUENTS

Ocean has rich and largest biomass of marine products. The marine resources are used for numerous ailments of human disease. A plethora of bioactive molecules was reported from the marine species. The marine sources are the primary reservoir of diverse therapeutic molecules (Scheuer, 2013). Marine drugs possess novel drugs for the treatment of human diseases. Many of the novel marine species are under pipelines for the drug screening process. Algae, marine bacteria, cyanobacteria, sponges, invertebrates, gorgonians, bryophytes, fungi are the inevitable sources of marine drugs (Asakawa et al., 2013; Beck et al., 2012; Lira et al., 2011; Osako and Teixeira, 2013; Pauletti et al., 2010; Rahman et al., 2010; Rateb and Ebel, 2011; Sheu et al., 2014). Marine sources have provided insights into the diverse biological activities. Therefore, marine sources have great potential to be used for therapeutic benefits as part of pharmaceuticals in the design and development of lead discovery.

2.2 METHODS TO CHARACTERIZE THE PHYTO AND MARINE CONSTITUENTS

A numerous number of analytical techniques are used for the structural illustration and characterization of natural medicines. The analytical technologies need to be reliable and robust for high-throughput routine analyses in the characterization. Therefore, an integrated approach of analytical techniques is preferred over a single methodology to characterize the phytochemicals. It includes *in-silico* tools, computational methods, high-performance liquid chromatography, Nuclear magnetic resonance spectroscopy, mass spectrometry, and liquid chromatography-mass spectrometry. Different types of analytical techniques used for structural illustration and characterization of phytochemicals and marine constituents are illustrated in Figure 2.1.

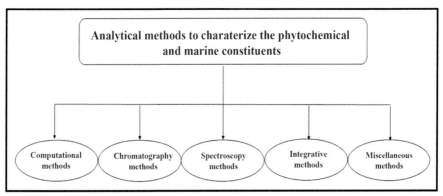

FIGURE 2.1 Different types of analytical techniques used in the characterization of phytochemical and marine constituents.

2.3   CHARACTERIZATION OF PHYTOCONSTITUENTS AND MARINE DRUGS BASED ON *IN-SILICO* TOOLS AND COMPUTATIONAL APPROACHES

The development of *in-silico* tools has allowed sophisticated system modeling to visualize natural products in the graphical representation. Computational methods like pharmacophore modeling (DISCOVERY STUDIO), molecular docking (Auto Dock), molecular dynamics (AMBER), and quantum chemical analysis (GAUSSIAN) resulted in many theoretical ideas to assess the therapeutic and metabolic profiles of phytochemical constituents, marine constituents, and drug molecules. These computational models are aimed at unveiling the systemic function of drug metabolism. Computational methods/models help to cover the research gap between drug and biological systems in the establishment of newer drugs. *In-silico* methods to link genomic space to chemical, space has been proposed as an important strategy in the genomics-driven products identification of newer natural products. The computational approach uses "genes to metabolites approach" for analyzing the secondary metabolite. Khater et al., described the available *in-silico* tools and computational approaches to analyze the secondary metabolites in biosynthetic pathways. Further, the study revealed the computational algorithms of retro-biosynthetic based transformations of biological chemical agents for the treatment of chronic disorders. The approach has been proposed to identify the genes associated with the biosynthesis of polyketide or non-ribosomal peptide (Khater et al., 2016). Hirte et al., utilized the homology modeling strategy to characterize the amino acids of the active site for structural characterizing and function of aphidicolan-16-ß-ol synthase. Raptor X and MODELLER software packages were used to run a homology modeling. The study identified the catalytically active amino acids like I^{626}, T^{657}, Y^{658}, A^{786}, F^{789}, and Y^{923} based on mutation. The study confirmed the role G-loop of diterpene synthases in the opening and closing mechanism of the enzyme (Hirte et al., 2018).

2.4   ISOLATION AND CHARACTERIZATION OF NATURAL MEDICINES BASED ON CHROMATOGRAPHY TECHNIQUES

Chromatography is a major method for the separation of bioactive products from natural resources. This technique works based on the distribution of molecules in different phases. The natural compounds are distributed into two different phases, i.e. stationary phase and mobile phase. Based on the relative distribution of the chemical constituents, the constituents are separated. Chromatography is functioning by different methods: (i) column/adsorption chromatography; (ii) partition chromatography; (iii) paper chromatography; (iv) thin-layer chromatography; (v) gas-liquid chromatography; (vi) gas-solid chromatography; and (vii) ion-exchange chromatography. The parameters such as retention factor, selectivity, efficiency, retention time, and peak area are investigated for the structural characterization of marine products based on chromatography. Different types of chromatography techniques employed in the isolation and characterization of phytochemicals and marine constituents are tabulated in Table 2.1.

TABLE 2.1 Different Types of Chromatography Techniques Employed in the Isolation and Characterization of Phytochemical and Marine Constituents

No.	Types	Stationary Phase	Mobile Phase
1.	Column chromatography	Solid	Liquid
2.	Partition chromatography	Liquid	Liquid
3.	Paper chromatography	Liquid	Liquid
4.	Thin-layer chromatography	Solid	Liquid
5.	Gas-liquid chromatography	Liquid	Gas
6.	Gas-solid chromatography	Solid	Gas
7.	Ion-exchange chromatography	Solid	Liquid

2.4.1 THIN-LAYER CHROMATOGRAPHY

Thin-layer chromatography is used for the characterization of natural compounds by a qualitative and quantitative manner. The major principles of thin-layer chromatography are adsorption or partition or both depending upon the adsorbent (stationary phase) and solvent (mobile phase). Silica gel, alumina, cellulose powder, kieselguhr, and Sephadex gel are used as the stationary phase. The mobile phase is carbon tetrachloride, benzene, ethyl acetate, dichloromethane, hexane, chloroform, ethanol, and water. The chemical components having an efficient affinity to stationary phase travel slower, whereas the components with a lower affinity toward the stationary phase travel faster. A stationary phase is used as a silica gel coated strip and organic solvents are play as mobile phases. The adsorbent is loaded on thin glass/sheet/plastic and it acts as a stable phase. About ~25 mm thickness coating on the glass plate was used for thin-layer chromatography. Plaster of Paris is used as a binder to makes coating of adsorbent. After the preparation of plates, all the plates are activated by keeping it in 100°C for 3 hr. The organic solvent are spotted at the bottom of the stationary phase. The solvent or mixture of solvent is allowed to move up to the plate by capillary action. The compounds are separated with respect to its affinity, size, and charge. Retardation factor (R_f) value (distance traveled by solute/distance traveled by solvent front) is used as a characteristic parameter in thin-layer chromatography. However, the nature of adsorbent, mobile phase, the thickness of layer, temperature, and sample loading are the critical factors and it can alter the R_f value. Thin-layer chromatography separates the components into the individual by using a finely divided adsorbent (solid/liquid) spread over a plate. The technique is employed under temperature and pressure-controlled laboratory. Thin-layer chromatography provides many advantages over other chromatography, i.e., (i) non-volatile/low volatile chemical compounds able to analyzed; (ii).

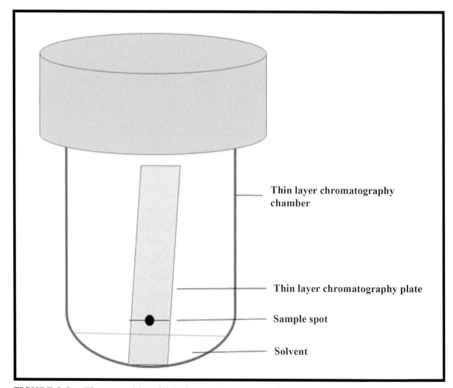

FIGURE 2.2   The assembly of thin-layer chromatography.

The constituents with diverse polarity (polar/non-polar/low polar) can be studied; and (iii) multiple and a large quantity of samples can be handled (Kumar et al., 2013; Ciura et al., 2017). Roh et al. employed thin-layer chromatography for the separation of lycopene from *Lycopersicon esculentum*. Silica Gel 60 F254 (0.25mm) activated by warming of plates at 110°C for the period of 10 min to ascertain the purity of lycopene. The method was suggested to be useful in large-scale purification of lycopene (Kyun et al., 2013). The assembly of thin-layer chromatography is illustrated in Figure 2.2.

2.4.2 COLUMN CHROMATOGRAPHY

Column chromatography is employed for the separation and purification of solids and liquids. This is a solid-liquid technique, where the solid is a gel loaded glass plate and organic liquid is the mobile phase. The basic principle involved in the column chromatography is adsorption. The adsorbent is packed in the glass column and organic solvent is carry the soluble compounds allowed through the column. The dissolved compounds is separated via organic solvent based mobile phase movement and introduced into the stationary phase. The components are separated by mobile phase with relative affinities of stationary phase. The compounds which have more attraction toward the stationary phase move slowly and eluted later, whereas the compound with a lower affinity toward stationary phase moves faster and eluted first. Alumina, starch, calcium carbonate, calcium phosphate, silica, and magnesia are the adsorbent used as stationary phase for column chromatography. Silica gel is widely used for polar compounds isolation and alumina is used for less polar compounds separation. The variable particle sizes are used as adsorbent *i. e.,* from 50–200 μM for column chromatography. The spherical shape of particles, uniform in size, chemically inert, inexpensive, and easily available with mechanical stability properties enhances the efficiency of compound separation. Petroleum ether, ester, carbon tetrachloride, cyclohexane, water, ether, benzene, toluene, and acetone are used for the solvent phase. The solvents with lower boiling points enable the easy recovery of constituents. The columns are made with glass and sometimes, the burette is used as a column. The columns are available the ratio of 10:1, 30:1 and 100:1 (length: width). The compound separation is achieved with the narrow and long column. The preparation of the column is by packing the bottom of the glass with cotton/wool followed by the adsorbent. On top of the column, a paper disk is kept in order to avoid the disturbance of column material. Then, the dissolved samples introduced at the top of the column. The mobile phase is passed into the column for carrying the components. It leads to isolates the compounds from the complex mixture and the eluted compounds are analyzed. Column dimensions, particle size, temperature, nature of the solvents are the critical factors that may affect the efficiency of column chromatography. The assembly of the column chromatography is illustrated in Figure 2.3.

Column chromatography is used to separate, isolate, identify, and estimate a wide range of phytochemicals and marine constituents (Bajpai et al., 2016). The isolation of alkaloids explained by Lim et al. from the extract of leaves of *Kopsia arborea* using column chromatography (Lim and Kam, 2008).

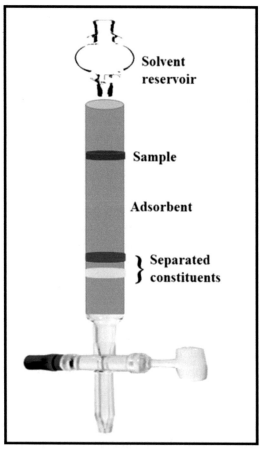

FIGURE 2.3   A simplistic model of column chromatography.

The column was prepared by silica gel and methanol was used as a solvent. The study isolated six alkaloids of the family of the methyl chano-fruticosinate group (Lim and Kam, 2008). The column chromatography based characterized phytochemical and marine constituents are tabulated in Table 2.2.

TABLE 2.2   The Analytical Techniques in the Characterization of Phytochemical and Marine Constituents

No.	Analytical Methods	Analytical Techniques	Compounds Analyzed	References
1.	Computational methods	*In-silico* tools and computational methods/models (pharmacophore modeling, molecular docking, QSAR, and molecular dynamics)	Polyketide or nonribosomal peptide.	Khater et al., 2016
		In-silico tools and computational methods/models (homology modeling and molecular docking)	Bifunctional Aphidicolan-16-ß-ol Synthase	Hirte et al., 2018
2.	Chromatography methods	Thin-layer chromatography	Lycopene	KyunRoh et al., 2013
		Paper chromatography	Enteramine, octopamine, tyramine, and histamine	Erspamer and Boretti, 1951
		Column chromatography	methyl chanofruticosinate	Lim and Kam, 2008
		High-performance liquid chromatography	Prenylflavonoids and bitter acids	Bertelli et al., 2018
		High-performance thin-layer chromatography	Carvone from spearmint essential oil (Mentha spicata L.) and resveratrol from Fallopia multiflora	Do et al., 2014
		Ion-exchange chromatography	Saxitoxins	Papageorgiou et al., 2005

TABLE 2.2 *(Continued)*

No.	Analytical Methods	Analytical Techniques	Compounds Analyzed	References
3.	Spectrometry methods	UV-Vis spectroscopy	Diarylheptanoids	Alberti et al., 2018
		Infrared spectroscopy	Caffeine and catechins	Lee et al., 2014
		Nuclear magnetic resonance spectroscopy	α-Galactosidase	Bakunina et al., 2018
		Mass spectrometry	Carrageenan	Li et al., 2018
4.	Integrative methods	Liquid chromatography-Nuclear magnetic resonance spectroscopy	Marine constituents	Pérez-Victoria et al., 2016
		Liquid chromatography-mass spectrometry	Steviol and its glycosides	Molina-Calle et al., 2017

2.4.3 PAPER CHROMATOGRAPHY

Paper chromatography is the simplest and widely used chromatography technique for the isolation, identification, and quantification of natural compounds. In paper chromatography, solvent flow is occurring on a filter paper. Two major principles are involved in paper chromatography, i.e., (i) paper adsorption chromatography: A paper impregnated in silica media, and it acts as a stationary phase. The organic solvent acts as the mobile phase; and (ii) paper partition chromatography-moisture present in the pores of filter paper acts as a stable phase and organic solvent acts as a mobile phase. The separation of particles in paper chromatography is based on partition than adsorption. What man filter papers No. 1, 2, 3, 4, 20, and 40 are used for stationary medium. It consists of 98–99% α-cellulose and 0.3–1% β-cellulose. In addition, the modified papers (acid/base washed papers), hydrophilic papers, and hydrophobic papers are also used as the stationary phase. To run a paper chromatogram, the sample is dissolved in a solvent and spotted at the base of the paper. The paper is dried and allows developing the spots on a paper. Two types of mobile phase composition are used: (i) hydrophilic mobile phase, i.e., isopropanol:ammonia:water; methanol:water; and n-butanol:glacial acetic acid:water; and (ii) hydrophobic mobile phase, i.e., dimethyl ether:cyclohexane. While running paper chromatography, the atmosphere must be saturated with the solvent vapor. The chromatogram is developed in different ways: (i) ascending; (ii) descending; (iii) ascending and descending; (iv) circular/radial; and (v) two dimensional. After developing the chromatogram, the components are detected visually if it is colored. Alternatively, iodine chamber, UV chamber, reagents are used for the detection of a specific class of components. The R_f (distance traveled by sample/distance traveled by the solvent) and R_x values (distance traveled by sample/distance traveled by the standard) are used as an indicator of sample class with the reference sample. Paper chromatography is employed for the separation of compounds from mixtures, carbohydrates, proteins, amino acid sand impurities (Block et al., 2016). A systematic assembly of paper chromatography is illustrated in Figure 2.4.

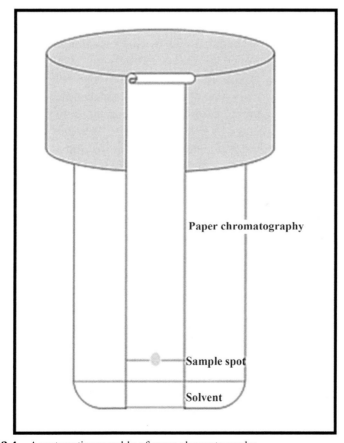

FIGURE 2.4 A systematic assembly of paper chromatography.

Octopamine, tyramine, enteramine histamine, and other allied substances were characterized by paper chromatography. These compounds are separated from extracts of posterior salivary glands of octopods and other tissue extracts of vertebrates as well as invertebrates (Erspamer and Boretti, 1951). The paper chromatography based characterized phytochemical and marine constituents are listed in Table 2.2.

2.4.4   HIGH-PERFORMANCE LIQUID CHROMATOGRAPHY

High-performance liquid chromatography is also called HPLC. It is an advanced analytical technique over column chromatography. In this technique, the mobile phase is pumped with pressure through a stationary phase of the column which is packed irregularly shaped particles to speed up the analytical procedure. The solvent phase is pumped through a column with a pressure of, 1000–3000 psi. HPLC is effective techniques in the analysis of phytochemical and marine constituents. The approach speeds up the analysis over the traditional column. HPLC is carry small volume of liquid as a sample is injected into the column comprising porous particles (stationary phase). The particle size varies from 5–10 µM. When mixtures of compounds are passed into the column, components are isolated based on their interaction property between the sample and stationary phase by the movement of liquid. The separated components are detected by part of the HPLC device is called detector. Generally HPLC has two method of operation, based on their working principle employed on separation i) normal phase chromatography; and ii) reverse phase chromatography. Normally, chromatography techniques are working based on polar (hydrophilic) stationary phase and non-polar (hydrophobic)mobile phase. In HPLC, it is reverse phase functions, i.e., non-polar stationary phase and polar mobile phase. As many of the pharmaceuticals/drugs are polar in nature, reverse phase HPLC is mainly used in pharmaceutical industries. Based on the principle employed on elution, HPLC is classified into two different categories: (i) isocratic elution; and (ii) gradient elution. The polar component of the mobile phase is constant for isocratic elution. In contrast, the gradient elution pattern, the mobile phase is the reversed process of HPLC analysis. Gradient elution is reduced the retention time and therefore, the components are eluted faster. The approach improves the shape and height of the peak. Further, HPLC is classified into two different modes of operation: (i) analytical HPLC (the compounds are not recovered; (ii) preparative HPLC (the compounds are recovered). The instrumentation of HPLC includes a solvent reservoir, mixing vessel, pressure pump, guard column, a sample injector, column, detector, and collector. Retention time, retention volume, separation factor, resolution, height equivalent theoretical plate, efficiency, and asymmetry factor are the HPLC are employed to detects the components from the test sample. A schematic diagram of the instrumentation of high-performance liquid chromatography is illustrated in Figure 2.5.

HPLC is employed in analytical chemistry and bioanalytical chemistry to isolate, separate, quantify, and characterize the constituents from the mixture. The advantages of HPLC include faster separation, higher resolution, and reproducibility. Bertelli et al. employed HPLC techniques in association with nuclear-magnetic-resonance spectroscopy (NMR) to characterize the bioactive constituents from *Humuluslupulus* L. (hop). The method quantitatively estimated prenylflavonoids and bitter acids in the commercially available hop. The accuracy of the HPLC method was compared with the NMR technique. The method developed by Bertelli et al. was suggested to be applicable in the quality control of hop plant material and hop-based products (Bertelli et al., 2018). The HPLC based characterized marine constituents are listed in Table 2.2.

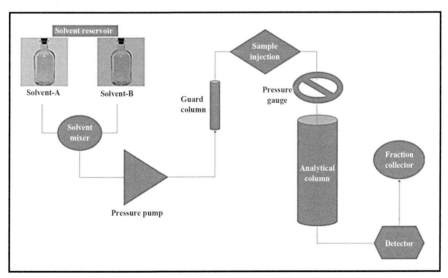

FIGURE 2.5 A schematic diagram of the instrumentation of high-performance liquid chromatography.

2.4.5   *HIGH-PERFORMANCE THIN-LAYER CHROMATOGRAPHY*

High-performance thin-layer chromatography (HPTLC) is an advanced version of thin-layer chromatography. It is a sophisticated, and automated analytical technique to conduct qualitative and quantitative analysis. It is also known as flatbed chromatography. It enables multiple sample handling and complicated separation. The chromatogram of HPTLC is easily visible and the samples are detected using a UV/visible or fluorescence detector. It employs a pre-coated plate in the analysis of samples. The layer thickness is ~100–200 µM. The sample volume is low and the efficiency is high in comparison to TLC. The basic principle of HPTLC based separation of components is adsorption. HPTLC used in the pharmaceutical industry, food analysis, clinical laboratory, natural product characterization, etc. In some cases, HPTLC found to superior over the HPLC technique. Tien Do et al. compared the HPLC and HPTLC for the efficient isolation of marine secondary-metabolites from the extracts of *Mentha spicata* L. and *Fallopia multiflora*. The parameters like chromatographic separation, purity, and quantity of isolated compounds, the solvent consumption, the duration and cost of operations were considered for the comparison. The results have shown HPTLC as a superior method over the HPLC method (Do et al., 2014). The HPTLC-based characterized phytoconstituents and marine compounds are listed in Table 2.2.

2.4.6   *ION-EXCHANGE CHROMATOGRAPHY*

The principle involved in ion-exchange chromatography is the interaction between ions of the sample and the charged site of the stationary phase. The column of ion-exchange chromatography contains charged functional groups for the stationary phase. Resins, gels, and inorganic ions are used as ion-exchangers. The resin is used to separate the small molecules. Gels are employed to separates the macromolecules like nucleic acids, and proteins. Inorganic ions are used at high temperatures and strong basic conditions. Based on the charged functional groups of ion-exchange chromatography, it is further classified into two types: (i) cation-exchange chromatography (negatively charged groups bound to stationary phase); and (ii) anion-exchange chromatography (positively charged groups bound to stationary phase). Cation-exchange chromatography, the positive charge of sample attracted to the anion of the stationary phase that may result in the exchange of cation. It is widely used for the separation of protonated bases. Anion-exchange chromatography has a negative charged sample and attracted to the cation of a stationary phase that may result in the exchange of anion so it is employed to the separation of acidic samples. The usage of ion-exchange chromatography is: (i) to convert one salt into another; (ii) to prepare deionized water; (iii) to separate ions; (iv) to purify the hard water; and (v) to measure the drugs and metabolites from biological fluids (Willard et al., 1988; Skoog et al., 2017). Cyanobacterial-derived saxitoxins were analyzed by using high-performance based ion-exchange chromatography via chemical oxidation and fluorescence detection as described by Papageorgiou et al. (2005). Anion-exchange column (Source 15Q PE 4.6/100) and two Source 15S PE 4.6/100 cation-exchange columns were employed in a series. The quantitative analysis of paralytic shellfish poisons produced by *Anabaena circinalis* and other cyano-bacteria are also tested by the same these techniques (Papageorgiou et al., 2005). The ion-exchange chromatography based characterized phytochemical and marine constituents are listed in Table 2.2. A schematic diagram of ion-exchange chromatography is illustrated in Figure 2.6.

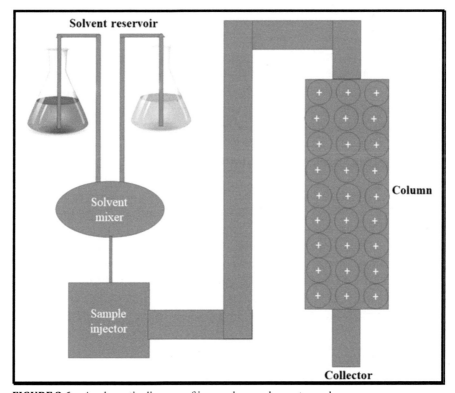

FIGURE 2.6 A schematic diagram of ion-exchange chromatography.

2.5   STRUCTURAL CHARACTERIZATION OF PHYTOCONSTITUENTS AND MARINE DRUGS BASED ON SPECTROMETRY METHODS

The spectroscopical method of compound detection and isolation is due to the interaction between matter with light/electromagnetic radiation. UV-Vis spectroscopy, infrared spectroscopy, NMR, and mass spectrometry are employed to characterize the bioactive constituents of nature. The spectrum of compounds has atoms/molecules and can fragment under certain conditions. Spectral graph of absorbed and emitted radiation and frequency and wavelength helps to detect the compounds. A spectrophotometer is an instrument employed to obtain the spectrum.

2.5.1   UV-VISIBLE SPECTROSCOPY

UV spectroscopy is another type of spectroscopy. The basic principles of UV spectroscopy are absorption of light and make the changes of the incident after passing to samples. It lies between the wavelength of 200–400 nm and visible spectroscopy lie at the wavelength of 400–800 nm. The instrument used for obtaining the spectrum is UV-Vis spectrophotometer. Ethyl alcohol and hexane are the solvents widely used to prepare the sample for UV-Vis spectroscopy. UV-Vis spectrum assists to characterize the aromatic group of compounds and conjugated dienes in qualitative analysis. In quantitative analysis, UV-Vis spectroscopy also helps to determine the molar concentration of constituents present in a given sample. In addition, it is also used to detect impurities, isomers, and molecular weight (Perkampus, 2013). UV-Vis spectroscopy was employed to characterize diarylheptanoids in association with other spectral techniques (Alberti et al., 2018). Diarylheptanoids have a specific absorption range, i.e., 250–290 nm. Acetonitrile was used as a solvent to get the UV-Vis spectrum. Further, a wider absorption band observed for curcumin, i.e., 410–430 nm. Keto-enol tautomerism of curcumin was characterized from the intra- and intermolecular hydrogen bonding. UV-Vis spectroscopy method is also used for the quantification of the curcuminoid content of *Curcuma Longa* extract (Alberti et al., 2018). The UV-Visible spectroscopy-based characterized phytoconstituents and marine compounds are listed in Table 2.2.

2.5.2   INFRARED SPECTROSCOPY

Infrared spectroscopy is working based on the interaction of infrared light radiation on a matter. Infrared radiation is a part of the electromagnetic spectrum and it covers by visible and microwave region waves. Infrared spectroscopy related to the rotational and vibrational energy of the molecule. When infrared frequency and natural frequency of molecules are matches, then a molecule absorbs the infrared radiation. The absorption rate of infrared radiation is indicated by the excitation of molecules from the ground state to a higher energy state. Absorption of radiation occurs only at the bonds that are accompanied via changes of dipole moment. In infrared spectroscopy, the change in vibration depends upon three factors, i.e., (i) mass of atoms; (ii) bond strength; and (iii) arrangement of atoms. No two compounds have the same absorption except enantiomers. Therefore, the absorption of radiation is detected by the absorption of radiation by molecule and alteration in the infrared spectrum. The spectrum consists of several peaks which are characteristic for functional groups/bonds present in a molecule. The sample molecule is characterized by analyzing the infrared spectrum. The presence of caffeine and catechins in the leaves of green tea (*Camellia sinensis L.*) was characterized by Lee et al. (2014). The study suggested that near-infrared reflectance spectroscopy is also a useful tool for the rapid detection of chemical constituents present in tea leaves (Lee et al., 2014). The infrared spectroscopy based characterized phytoconstituents and marine compounds are listed in Table 2.2.

2.5.3 NUCLEAR-MAGNETIC-RESONANCE SPECTROSCOPY (NMR)

NMR plays a vital role for the characterization and quantification of phyto-chemical constituents. The chemical constituents are characterized based on δ-values of spectral data and the quantification is calculated from the peak areas of NMR spectra. Further, the structure of phytochemical constituents determined by isotopic labeling techniques. The basic principle of NMR is the absorption as well as the re-emission of radiofrequency energy by atomic nuclei. To structural characterizing of metabolites, the nucleus of hydrogen (^{1}H-NMR), carbon (^{13}C-NMR), nitrogen (^{14}N-NMR), oxygen (^{17}O-NMR), and phosphorous (^{31}P-NMR) are widely studied. Bakunina et al. characterized α-Galactosidase from Pseudoalteromonas KMM 701 and its C494N and D451A mutants. The NMR characterization of molecules covers the structural property, physicochemical property, and catalytic properties. NMR technique was employed to identify the transglycosylation products from reaction mixtures (Bakunina et al., 2018). The NMR spectroscopy-based characterized phytoconstituents and marine compounds are listed in Table 2.2.

2.5.4 MASS SPECTROMETRY

Mass spectrometry records the spectral data by plotting the mass-to-charge ratio (m/z) against the intensity of ions. The spectrum is generated from the gas phase ions. Mass spectrometer performs three basic fundamental functions like (i) The metabolites are converted into ions by the bombardment by high-energy electrons; (ii) The ions are accelerated and separated by the changes of mass-to-charge ratios in an electric or magnetic field; and (iii) detection of ions. Mass spectrometry is classed into different categories based on its working principle, i.e., (1) Direct infusion mass spectrometer (DIMS) - in DIMS, the sample is directly injected into the spectrometer; and (2) Tandem mass spectrometer (MS/MS) - the mass analyzer is combined with the additional mass analyzer. The ions of the first mass analyzer were ionized further in the second mass analyzer. Car-19 is the most thermostable κ-carrageenase. The κ-carrageenase producing thermophilic bacterial strains were obtained from sediment samples in Indonesia. carrageenase was characterized by electrospray ionization method of mass spectrometry from hot spring bacterium by Li et al. (2019). Car-19 is used for the preparation of carrageenan oligosaccharides with plant protection activity (Li et al., 2018). The mass spectroscopy-based characterized phytoconstituents and marine compounds are listed in Table 2.2.

2.6 INTEGRATIVE TECHNIQUES IN THE CHARACTERIZATION OF PHYTOCONSTITUENTS AND MARINE DRUGS

2.6.1 LIQUID CHROMATOGRAPHY-NUCLEAR MAGNETIC RESONANCE SPECTROSCOPY

Liquid chromatography-nuclear magnetic resonance spectroscopy (LC-NMR), in which HPLC is integrated with NMR. Adsorption, ion-exchange, gel permeation are the principle implemented in LC- for sensitive and unstable compounds. The LC-NMR analysis is automated and faster to characterize the phytochemical, marine, and other constituents. De-replication of marine constituents is the driving force for this study. Pérez-Victoria et al. employed the combined strategies of LC-UV/NMR/MS for the de-replication of marine constituents. Analytical instrumentation and spectral database were found as an essential requirement to achieve the de-replication of marine constituents. The integrative methods provided more structural information, characterization of regioisomers, and stereoisomers. The use of databases reduces the cost and money (Pérez-Victoria et al., 2016). The LC-NMR spectroscopy-based characterized phytoconstituents and marine compounds are listed in Table 2.2.

2.6.2   *LIQUID CHROMATOGRAPHY-MASS SPECTROMETRY*

Liquid chromatography-mass spectrometry (LC-MS) identifies the drug metabolites of polar to non-polar. LC-MS separates metabolites based on polarity, molecular weight, and hydrophobicity. Two different types of basic principles were employed in LC, i.e., (i) normal liquid phase chromatography where the polar stable phase is employed; and (ii) reverse-phase liquid chromatography where non-polar stable phase is employed. Mass spectrometry serves as a detector to identify the metabolites. Molina-Calle et al. demonstrated the characterization of Stevia leaves constituents by LC with quadrupole—a time of flight (LC-QTOF) mass spectrometry (Molina-Calle et al., 2017). Mediterranea Sea (C18) column has 5 µ m, 15 × 0.46 cm dimension and it is used in LC-MS device. The source of electrospray ionization was used in mass spectrometry. Steviol and its glycosides were determined from the analysis of polar and non-polar components of Stevia leaves. A total of eighty-one compounds of different chemical classes of flavonoids, quinic acids, caffeic acids, diterpenoids, sesquiterpenoids, amino acids, fatty acids, oligosaccharides, glycerolipids, fatty amides purines, and retinoids were identified. The study was proven to benefit the production of commercial products from Stevia (*Steviarebaudiana bertoni*) leaves (Molina-Calle et al., 2017). The LC-MS spectroscopy-based characterized phytoconstituents and marine compounds are listed in Table 2.2.

2.7   CONCLUSIONS

Phytochemical and marine constituents are characterized by computational, chromatography, and spectroscopy techniques. These natural constituents serve as a useful lead compound in the search for new therapeutic agents for the treatment f several diseases. It could be potential therapeutic drugs for neuropathic pain syndromes, anticoagulants, and antioxidants. However, additional studies are necessary in order to investigate the therapeutic activities of marine sources in human subjects.

ACKNOWLEDGMENTS

The authors acknowledge the faculty members of Omega College of Pharmacy, Hyderabad for their support. The authors are also thankful to JSS College of Pharmacy, JSS Academy of Higher Education and Research, Mysuru, India; and Faculty of Pharmacy, AIMST University, Semeling, Malaysia for their support to make this book chapter.

KEYWORDS

- **chromatography**
- **computational chemistry**
- **isolation**
- **marine drugs**
- **saxitoxins**
- **spectroscopy**

REFERENCES

Alberti, A., Riethmüller, E., & Béni, S., (2018). Characterization of diarylheptanoids: An emerging class of bioactive natural products. *J. Pharm Biomed. Anal., 147*, 13–34.

Asakawa, Y., Ludwiczuk, A., & Nagashima, F., (2013). Phytochemical and biological studies of bryophytes. *Phytochemistry., 91*, 52–80.

Bajpai, V. K., Majumder, R., & Park, J. G., (2016). Isolation and purification of plant secondary metabolites using column-chromatographic technique. *Bangladesh J. Pharmacol., 11*(4), 844–848.

Bakunina, I., Slepchenko, L., Anastyuk, S., Isakov, V., Likhatskaya, G., Kim, N., Tekutyeva, L., et al., (2018). Characterization of properties and transglycosylation abilities of recombinant α-galactosidase from cold-adapted marine bacterium *Pseudoalteromonas* KMM 701 and its C494N and D451A mutants. *Mar. Drugs., 16*(10), 349.

Baxter, H., Harborne, J. B., & Moss, G. P., (1998). *Phytochemical Dictionary: A Handbook of Bioactive Compounds from Plants.* CRC Press: United States.

Beck, C., Knoop, H., Axmann, I. M., & Steuer, R., (2012). The diversity of cyanobacterial metabolism: Genome analysis of multiple phototrophic microorganisms. *BMC Genomics, 13*(1), 56.

Bertelli, D., Brighenti, V., Marchetti, L., Reik, A., & Pellati, F., (2018). Nuclear magnetic resonance and high-performance liquid chromatography techniques for the characterization of bioactive compounds from *Humulus lupulus* L. (hop). *Anal Bioanal. Chem., 410*(15), 3521–3531.

Block, R. J., Durrum, E. L., & Zweig, G., (2016). *A Manual of Paper Chromatography and Paper Electrophoresis*. Elsevier: Amsterdam, Netherlands.

Ciura, K., Dziomba, S., Nowakowska, J., & Markuszewski, M. J., (2017). Thin-layer chromatography in drug discovery process. *J. Chromatogr A., 1520*, 9–22.

Craig, W. J., (1997). Phytochemicals: Guardians of our health. *J. Am. Diet Assoc., 97*(10): S199–S204.

Do, T. K. T., Hadji-Minaglou, F., Antoniotti, S., & Fernandez, X., (2014). Secondary metabolites isolation in natural products chemistry: Comparison of two semi preparative chromatographic techniques (high pressure liquid chromatography and high performance thin-layer chromatography). *J. Chromatogr. A., 1325*, 256–260.

Erspamer, V., & Boretti, G., (1951). Identification and characterization, by paper chromatography, of enteramine, octopamine, tyramine, histamine and allied substances in extracts of posterior salivary glands of octopoda and in other tissue extracts of vertebrates and invertebrates. *Arch. Int. Pharmacodyn. Ther., 88*(3), 296–332.

Hirte, M., Meese, N., Mertz, M., Fuchs, M., & Brück, T. B., (2018). Insights into the bifunctional aphidicolan-16-ß-ol synthase through rapid biomolecular modeling approaches. *Front Chem., 6*, 101.

Hughes, Jr. C. L., (1988). Phytochemical mimicry of reproductive hormones and modulation of herbivore fertility by phytoestrogens. *Environ Health Perspect., 78*, 171–174.

Khater, S., Anand, S., & Mohanty, D., (2016). *In silico* methods for linking genes and secondary metabolites: The way forward. *Synth. Syst. Biotechnol., 1*(2), 80–88.

Kim, D. O., Jeong, S. W., & Lee, C. Y., (2003). Antioxidant capacity of phenolic phytochemicals from various cultivars of plums. *Food Chem., 81*(3), 321–326.

Kumar, S., Jyotirmayee, K., & Sarangi, M., (2013). Thin-layer chromatography: A tool of biotechnology for isolation of bioactive compounds from medicinal plants. *Int. J. Pharm. Sci. Res., 18*(1), 126–132.

KyunRoh, M., Hee, J. M., Nam, M. J., Sook, M. W., Mee, P. S., & Suk, C. J., (2013). A simple method for the isolation of lycopene from *Lycopersicon esculentum*. *Bot. Sci., 91*(2), 187–192.

Lee, M. S., Hwang, Y. S., Lee, J., & Choung, M. G., (2014). The characterization of caffeine and nine individual catechins in the leaves of green tea (*Camellia sinensis* L.) by near-infrared reflectance spectroscopy. *Food Chem., 158*, 351–357.

Li, J., Pan, A., Xie, M., Zhang, P., & Gu, X., (2019). Characterization of a thermostable κ-carrageenase from a hot spring bacterium and plant protection activity of the oligosaccharide enzymolysis product. *J. Sci. Food Agric., 99*(4), 1812–1819.

Lim, K. H., & Kam, T. S., (2008). Methyl chanofruticosinate alkaloids from *Kopsia arborea*. *Phytochemistry, 69*(2), 558–561.

Lira, N. S., Montes, R. C., Tavares, J. F., Silva, M. S. D., Da Cunha, E. V., Athayde-Filho, P. F. D., Rodrigues, L. C., Dias, C. D. S., & Barbosa-Filho, J. M., (2011). Brominated compounds from marine sponges of the genus Aplysina and a compilation of their 13C NMR spectral data. *Mar. Drugs., 9*(11), 2316–2368.

Liu, R. H., (2007). Whole grain phytochemicals and health. *J. Cereal. Sci., 46*(3), 207–219.

Molina-Calle, M., Priego-Capote, F., & De Castro, M. L., (2017). Characterization of Stevia leaves by LC-QTOF MS/MS analysis of polar and non-polar extracts. *Food Chem., 219*, 329–338.

Molyneux, R. J., Lee, S. T., Gardner, D. R., Panter, K. E., & James, L. F., (2007). Phytochemicals: The good, the bad, and the ugly? *Phytochemistry, 68*(22–24), 2973–2985.

Naik, G., Priyadarsini, K., & Mohan, H., (2006). Free radical scavenging reactions and phytochemical analysis of triphala, an Ayurvedic formulation. *Curr. Sci., 90*(8), 1100–1105.

Nascimento, G. G., Locatelli, J., Freitas, P. C., & Silva, G. L., (2000). Antibacterial activity of plant extracts and phytochemicals on antibiotic-resistant bacteria. *Braz. J. Microbiol., 31*(4), 247–256.

Osako, K., & Teixeira, V. L., (2013). Natural products from marine algae of the genus *Osmundaria* (Rhodophyceae, Ceramiales). *Nat. Prod. Commun., 8*(4), 533–538.

Papageorgiou, J., Nicholson, B. C., Linke, T. A., & Kapralos, C., (2005). Analysis of cyanobacterial-derived saxitoxins using high-performance ion-exchange chromatography with chemical oxidation/fluorescence detection. *Environ. Toxicol., 20*(6), 549–559.

Pauletti, P. M., Cintra, L. S., Braguine, C. G., Silva, M. L. A., Cunha, W. R., & Januário, A. H., (2010). Halogenated indole alkaloids from marine invertebrates. *Mar Drugs, 8*(5), 1526–1549.

Percival, M., (1997). Phytonutrients and detoxification. *Int. J. Nutr., 5*(2), 1–4.

Pérez-Victoria, I., Martín, J., & Reyes, F., (2016). Combined LC/UV/MS and NMR strategies for the dereplication of marine natural products. *Planta Med., 82*(9–10), 857–871.

Perkampus, H. H., (2013). *UV-VIS Spectroscopy and its Applications*. Springer Science and Business Media: Berlin, Germany.

Rahman, H., Austin, B., Mitchell, W. J., Morris, P. C., Jamieson, D. J., Adams, D. R., Spragg, A. M., & Schweizer, M., (2010). Novel anti-infective compounds from marine bacteria. *Mar, Drugs, 8*(3), 498–518.

Rajendran, P., Ho, E., Williams, D. E., & Dashwood, R. H., (2011). Dietary phytochemicals, HDAC inhibition, and DNA damage/repair defects in cancer cells. *Clin. Epigenetics., 3*(1), 4.

Rateb, M. E., & Ebel, R., (2011). Secondary metabolites of fungi from marine habitats. *Nat. Prod. Rep., 28*(2), 290–344.

Scheuer, P. J., (2013). *Marine Natural Products: Chemical and Biological Perspectives*. Academic Press: United States of America.

Sheu, J. H., Chen, Y. H., Chen, Y. H., Su, Y. D., Chang, Y. C., Su, J. H., Weng, C. F., et al., (2014). Briarane diterpenoids isolated from gorgonian corals between, 2011 and, 2013. *Mar. Drugs., 12*(4), 2164–2181.

Skoog, D. A., Holler, F. J., & Crouch, S. R., (2017). *Principles of Instrumental Analysis*. Cengage learning: Massachusetts, United States.

Weng, C. J., & Yen, G. C., (2012). Chemo preventive effects of dietary phytochemicals against cancer invasion and metastasis: Phenolic acids, monophenol, polyphenol, and their derivatives. *Cancer Treat Rev., 38*(1), 76–87.

Willard, H. H., Merritt, Jr. L. L., Dean, J. A., & Settle, Jr. F. A., (1988). *Instrumental Methods of Analysis*.

CHAPTER 3

Characterization of Phyto-Constituents

HIMANGINI, FAIZANA FAYAZ, and ANJALI

Delhi Institute of Pharmaceutical Sciences and Research, Department of Pharmaceutical Chemistry, Sector-3, MB Road, Pushp Vihar, Delhi – 110017, India, Tel.: +91–9810074460, E-mail: himanginibansal@gmail.com (Himangini)

ABSTRACT

Phyto-pharmaceuticals materials are obtained from the plants or the plant materials in the form of fresh or dried which is procured after extraction of plants. Once the compounds are separated, their chemical composition is confirmed by various techniques like thin-layer chromatography (TLC), high-performance thin-layer chromatography (HPTLC), high-pressure liquid chromatography (HPLC), gas chromatography (GC), etc. These chemical constituents have been extensively indicated in various types of diseases with high therapeutic potency, which can lead to their vast developments as clinical agents.

3.1 INTRODUCTION

Herbal medicine and their preparations have been widely used since long because of their natural origin and lesser side effects as compared to modern medicines. Traditional herbal medicines are obtained from the medicinal plants especially in the rural developing areas. In earlier days, Veda used to take care for each and every patient separately; they used to prepare the medicines as per the condition of the individual, by considering their sex, body mass, age, etc., on a small scale. But now a day's herbal medicines are being prepared on large scale in industries and manufacturers have to obtain genuine raw materials, follow proper standardized operating procedures with quality control, etc. Medically active chemical components can be present in any part of plant such as in leaf, root, stem, seeds, flowers, bark, etc., and are recognized as Phytoconstituents. Phytoconstituents have been considered as plant medications and have being considered for their antirheumatic, anticancer, antidiabetic, antimicrobial, antiprotozoal, and other biological activities.

3.2   TYPES OF PHYTOCONSTITUENTS

3.2.1   *PHENOLICS*

Phenolics are plant metabolites and are typically unfold throughout the plant state. Phenolics otherwise called phenols or polyphenolics (or polyphenol removes) are substance parts that happen all around as common shading colors and are in charge of the shading. They are obtained naturally from phenylalanine by means of the biochemical reaction of phenylalanine ammonia lyase (PAL). They are characterized into three categories (i) phenolic acids, (ii) flavonoid polyphenolics (flavonones, flavones, xanthones, and catechins), and (iii) non-flavonoid polyphenolies. Caffeic acid as appeared in Figure 3.1 is the most widely recognized phenolic compound in the plant verdure pursued by chlorogenic acid which is known to cause allergic dermatitis among. They have been the subject of an incredible number of substances, organic, agricultural, and therapeutic investigations. Some of them are polymerized into bigger particles, for example, the proanthocyanidins and lignins. Besides, phenolic acid additionally present in plants as esters or glycosides hybridized with other constituents, for example, flavonoids, alcohols, hydroxyl fatty acids, sterols, and glucosides (Hung, 2010).

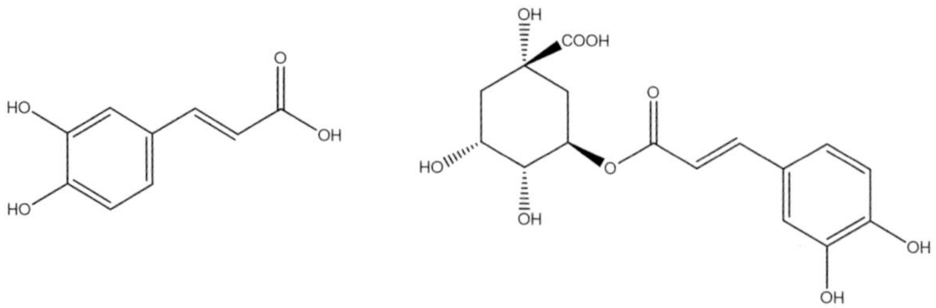

Caffeic acid   Chlorgenic acid

FIGURE 3.1   Structures of common phenolic compounds.

3.2.2 ALKALOIDS

Alkaloids are essential, nitrogen-containing constituents that happen for the most part in plants kingdom. They are considered as subordinates of pyridines, quinoline, or isoquinoline. The metabolism of the amino acid gives the nitrogen in the alkaloidal structure. The level of basicity, relies upon the chemical moiety, nature, and its presence in the moiety (Sarker et al., 2007). They are highly soluble in alcohol, sparingly soluble in aqueous solvent. They are soluble in the salt form but they are bitter in taste. They form crystalline salt when reacted with salts without delivering water (Firn, 2010). Alkaloids and concentrates of alkaloids have been utilized all through mankind's history in many forms such as poison, as drugs, and psychoactive medications. Numerous alkaloids are noxious, yet they have physiological impact which makes them profitable as meds. For instance, curarine, is used as muscle relaxant; atropine is utilized in the dilation of pupils and cocaine is used as local anesthetics. Certain opioid alkaloids are also having medicinal value, for example, morphine (Figure 3.2) and codeine are indicated in the intense pain and some of the alkaloids are also used as antibiotics (Wink, 2015).

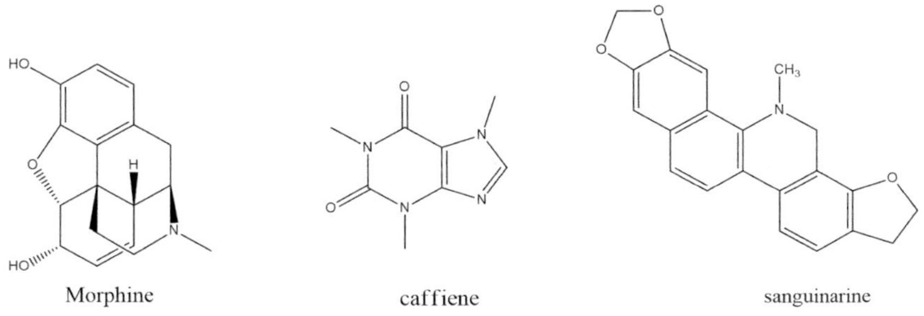

Morphine   caffiene   sanguinarine

FIGURE 3.2 Structures of morphine, caffeine, and sanguinarine.

3.2.3 GLYCOSIDES

Glycosides are blends containing starch and a non-sugar development in a comparable molecule. Glycosides are characterized as the buildup results of sugars (counting polysaccharides) with a large group of various assortments of natural hydroxy (once in a while thiol) compounds (constantly monohydrate in character), in such a way, that the hemiacetal moiety of the starch must participate in the buildup. The carbohydrate or glycone is appended by an acetal linkage at carbon particle 1 to a nonsugar buildup or aglycone. On the basis of its pharmacological activity, sugar component and chemical property of aglycon component, glycosides are classified. Examples include cardiac glycosides (like digitalis acts on the heart), anthracene glycosides (like aloe and rhubarb used as purgative, and for treatment of skin diseases), chalcone glycoside (anticancer), alcoholic glycosides (salicin used as analgesic), cyanogenic glycosides (like amygdalin, prunasin) are used as flavoring agents in many pharmaceutical preparations. Amygdalin as shown in Figure 3.3 has been also utilized as antimalignant agent (HCN which is evolved in gastro kills cancer cells), and also as a cough suppressant in various preparations (Abraham et al., 2016). Overdose of cyanogenic glycosides can be lethal.

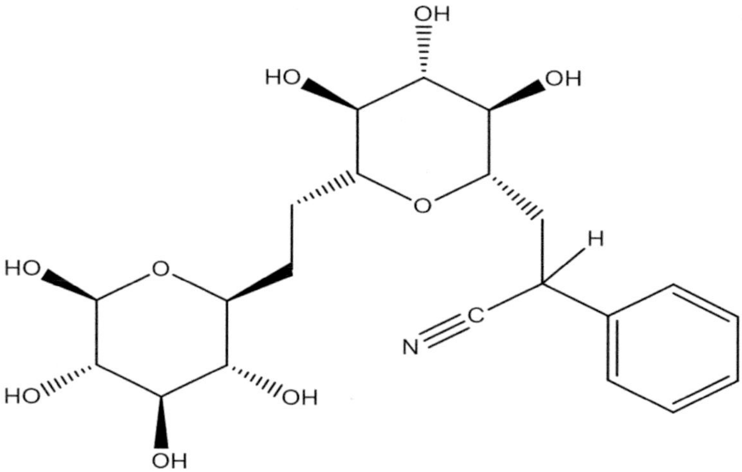

Amygdalin

FIGURE 3.3   Structure of amygdalin.

3.2.4   SAPONINS

From the plant *Saponaria vaccaria,* the word saponin is taken, which was once utilized as soap. They comprise of polycyclic aglycones joined to at least one sugar side chains. The frothing capacity of saponins is because of the blend of a hydrophobic (fat solvent) sapogenin and a hydrophilic (water dissolvable) sugar part, which upon hydrolyzis produces aglycone, "sapogenin." There are two kinds of sapogenin: steroidal and triterpenoidal. Saponins are highly poisonous, causing hemolysis of blood and are very well known as cattle poisoning agents (Akinpelu et al., 2012). They have a severe and bitter taste and it also causes irritation on the mucous membrane. The most significant saponin drugs are quillaia and senega. Commercial saponins are obtained essentially from the plant *Yucca schidigera* and *Quillaja saponaria*. Yucca and Quillaja Saponins are utilized in beer industries, to create stable froth. A few saponins are dangerous and are known as sapotoxin. Studies have outlined the helpful impacts of saponins on blood cholesterol levels, bone conditioning, tumor, and as an immune booster. They have been widely utilized in cosmetic industries like cream, shampoos, and facial cleansers as a result of their cleanser properties.

3.2.5 TERPENES

Terpenes are a wide and diverse faction of compounds which are isolated from terpentine, which can be isolated from the pine trees. Terpenes are the hydrocarbons of plant root of the general equation $(C_5H_8)^n$ as well as their oxygenated, hydrogenated, and dehydrogenated subsidiaries. On prolong heating of terpenes gives isoprene as the result. A molecule of terpene consists of two or greater isoprene units coupled in a 'head to tail' style (Yadav et al., 2014). Examples of commonly important monterpenes include eugenol, limonene, camphor, terpenolen, carvone, α-cubebene, taxadiene, etc. included in Figure 3.4. Table 3.1 incorporates the classification of terpenoids as indicated by the quantity of isopropene units

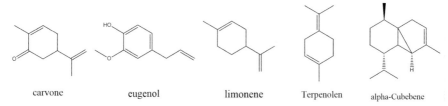

| carvone | eugenol | limonene | Terpenolen | alpha-Cubebene |

FIGURE 3.4   Structures of some common terpenes.

TABLE 3.1 Types of Terpenoids According to the Number of Isopropene Units

Type of Terpenoid	Number of Carbon Atoms	No. of Isoprene Units	Examples
Monoterpene	10	2	Limonene, menthol, pinene, citral, camphor, etc
Sesquiterpene	15	3	Artemisinin, nerolidol, farnesol etc
Diterpene	20	4	Forskolin, phytol, cafestol etc
Triterpene	30	6	α-amyrin, squalene, lanosterol etc
Tetraterpene	40	8	β-carotene, lycopene etc
Polymeric terpenoid	Several	Several	Ruber

3.2.6 TANNINS

Tannins are complicated, nonnitrogenous derivatives of polyhydroxy benzoic acid having a special quality to tan, i.e. to change things into the leather. The word tannin basically refers to the use of wooden tannins from oak in tanning animal hides into leather; hence the words "tan" and "tanning" for the remedy of leather. They have acidic nature which is present because of the phenolic or carboxylic group present in its moiety. They create complex structure with gelatin, carbohydrates, proteins, and alkaloids. They are mainly obtained in aerial parts of plants like fruits, barks of immature plants, leaves, and disappear on ripening. Tannins have molecular weights ranging from 500 to over 3,000 and up to 20,000. The astringent nature of tannins reasons dry and puckery feeling inside the mouth when fed on unripe (Ashok et al., 2012).

The types of tannins are:

1. *Gallotannins:* Gallotannins refers to the class of hydrolyzable tannins which have polyphenolic and a polyol residue. They undergo esterfication and forms Gallic acid (Figure 3.5), binds with hydroxyl group of polyol carbohydrate as glucose. Examples cloves, bearberry leaves, maple, chestnut, etc.

2. *Ellagitannins:* It is one of the varied class of hydrolyzable tannins, formed after the oxidative linkage of galloyl groups in 1,2,3,4,6-penta-galloyl glucose. Examples Castalagin, castalin, grandinin, terflavin B, etc.

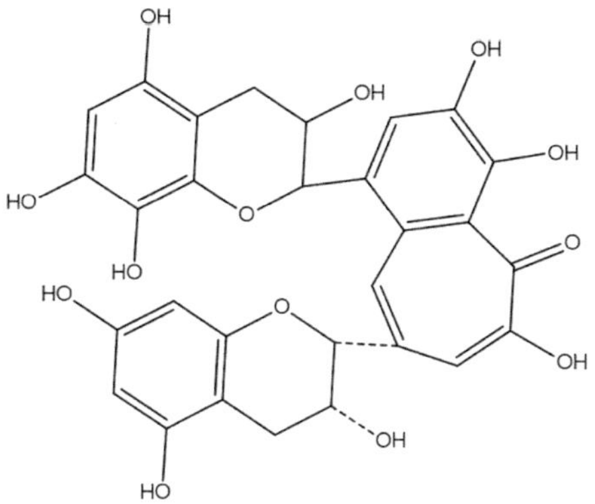

Theaflavin 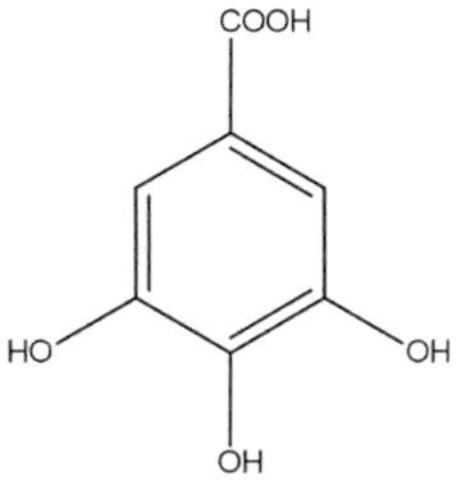   Gallic acid

FIGURE 3.5   Structures of gallic acid.

3.   *Condensed tannins:* Oligomeric and polymeric proanthocyanidins
     compounds are known as Condensed tannins which are opposed
     to hydrolyzis and are obtained from flavanols, flavan-3,4 diols,
     catechins. By the successive condensation of single building
     blocks, condensed tannins are formed. When treated with acid
     they decompose into phlobaphenes. They are present in roots and
     rhizomes of krameria and male fern, bark (cinnamon, cinchona,
     oak), in the seeds of cacoa, kola, areca, and in the leaves of
     hamamelis, etc.

3.2.7   ANTHRAQUINONES

Anthraquinones bearing 9,10-dioxoanthracene moiety is in the category of
aromatic compounds, till date 79 naturally occurring anthraquinones have
been distinguished which consist of chrysophanol, aleo-emodin, dianthrone,
emodin, dianthranol, rhein, catenarin, cascarin, and physcion. They are
oranges in color, highly soluble in dilute alcohol and water. Naturally occurring anthraquinones shows a wide spectrum of biological activities such as
anti-inflammatory, antimicrobial, cathartic, phytoestrogen, anticancer, vaso-
relaxing, and diuretic activities (Chien et al., 2015). Some of the examples
are illustrating in Figure 3.6.

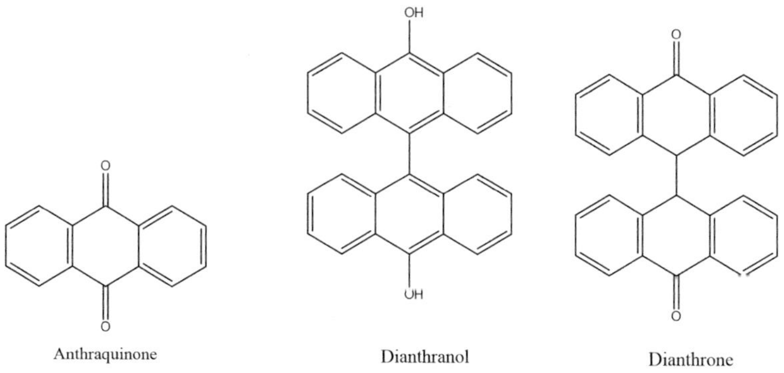

FIGURE 3.6   Structures of anthraquinones.

Seroids assume a significant job in the organic procedures of plants, for example, in development and growth, cell division, and protection from harm from natural burdens like chilly climate and so on. Some plant steroids are likewise helpful for their belongings when administer by human on the grounds that their quality decreases the cholesterol level in the blood. Chemical structure of all the steroids has carbon atoms connected by single or double bonds and organized into four interconnected rings. Extra functional groups are present on various carbon atom which change one steroid to another. Various steroids have various properties that differ as indicated by the quantity of double bonds in the carbon rings and the composition of the attached functional groups. One of the most biologically potent plant steroids is Brassinolide ($C_{28}H_{48}O_6$), which is significant for the improvement of plant cells and advancing the plant's growth. Brassinolide is prepared from Campesterol ($C_{28}H_{48}O$), another plant steroid that is component of a group of related steroid components known as phytosterols. Different examples of steroids include ergosterol, lupeol, β-sitosterol, cholesterol, and campesterol, are illustrated in Figure 3.7 (Ogbel et al., 2015).

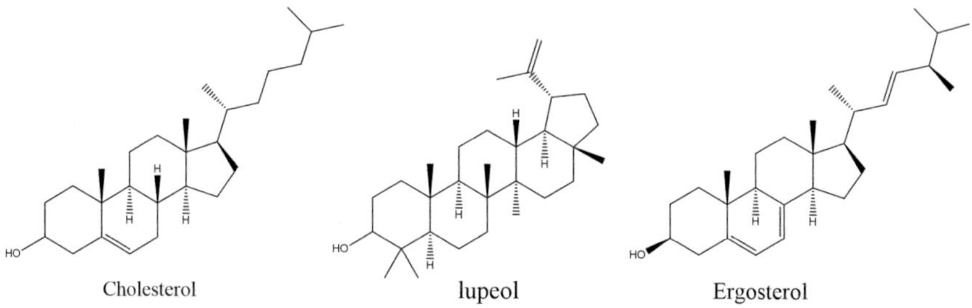

FIGURE 3.7   Structures of common steroids.

3.2.8 ESSENTIAL OIL

Essential oils also called volatile oils, aetherolea, and ethereal oils (Figure 3.8) contain the characteristic smell of the plant from which they are extracted such as eucalyptus oil, rosemary oil, clove oil, etc. They are available in different pieces of plants as in fruits (fennel), petals (saffron), seeds (caraway), strip (lemon), bark (cinnamon), and so forth and are large extricated by refining, by utilizing steam. Different procedures incorporate articulation, solvent extraction, absolute oil extraction, resin tapping, and cold squeezing. They are colorless liquids, marginally dissolvable in water yet exceedingly soluble in natural solvents like ether, alcohol, and so on. They are utilized in different drugs attributable to their therapeutic properties; they additionally discovered incredible use in scents, cleansers, flavoring agents, and so on.

They are colorless liquids, slightly soluble in water but highly soluble in organic solvents like ether, alcohol, etc. They are used in various medications owing to their medicinal properties; they also found great use in perfumes, soaps, flavoring agents, etc., (Rassem et al., 2016).

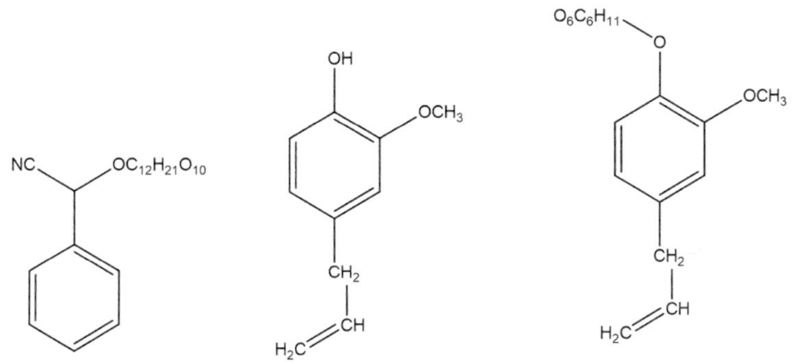

Amygdalin   Gein   Eugenol

FIGURE 3.8   Important structures of essential oils.

3.3   VARIOUS EXTRACTION TECHNIQUES FOR MEDICINALLY SIGNIFICANT PLANTS

3.3.1 MACERATION

It is the best approach of extraction, where whole or coarsely powdered drug is located in a closed vessel with a suitable solvent or menstruum and is allowed to stand for at least five days at room temperature with occasional shaking. The liquid is then strained off and the solid residue (Marc) is tough-pressed to extract the solution. The strained and expressed liquids are mixed and clarified through filtration or decantation (Azwanida, 2015).

3.3.2 INFUSION

Infusion is utilized when plant or rough medication extricate contains constituents which are volatile in nature or which dissolves or discharges active ingredients effectively in water. In this strategy, the appropriate fluid is bubbled and afterward poured over the plant or unrefined medication, which is then permitted to soak in the fluid for a specific timeframe. The fluid may then be stressed or the herbs generally expelled from the solvent. The amount of time the herbs are left in the liquid depends on the purpose for which the infusion is being prepared. Typically soaking for not more than 15 to 30 minutes or until the blend cools, will make a refreshment with ideal flavor.

3.3.3 DIGESTION

It is the type of maceration in which delicate warmth is utilized during the extraction procedure, if warmth ought not modify the active elements of plant material. By and large temperature in the scope of 35° and 40°C is utilized.

3.3.4 DECOCTION

The decoction is utilized for active compounds that doesn't alter with temperature. In this procedure, the drug is bubbled in water for 15 to an hour, it is then cooled, stressed, and sifted. Decoction time fluctuates relying upon the ingredients of plant to extract leafy stems, roots, blossoms, and leaves are boiled in water for around 20 minutes, while the branches and other hard parts can require as long as 60 minutes. When the decoction is done it is important to filter the fluid through a material. The decoctions should be used fresh and used within 24 hours. The procedure is mainly used to extract water solvable, thermostable compounds.

3.3.5 PERCOLATION

It is an extraction procedure that includes the moderate drop of a dissolvable through a fine substance until it retains certain constituents and comes out through the narrow base of the compartment. For the preparation of tinctures and fluid extracts this system is generally used. A typical percolator is utilized, which is a limited cone-shaped molded vessel, open at both the finishes, entire medication or coarsely powdered medication is saturated with the suitable menstrum for about 5 to 6 hr, mass is stuffed and after the packing container is closed form the top. After that extra menstrum is introduced in an effort to create the shallow layer on the top of the mass, and the blend is permitted to remain in the percolator for a time of 24 h. At that point the opening of the percolator at the bottom side is opened and the fluid is permitted to trickle gradually. The marc is the squeezed so as to evacuate the additional fluid, The stressed and expressed fluids are blended and cleared up by filtration or decantation.

3.3.6 SOXHLET EXTRACTION

Soxhlet extraction is employed once active elements of the drug don't seem to be soluble within the solvent, then it turns into essential to extract the crude drug via the movement of warm menstruum for a large period of time. The fixed oils from leaves and seed from the drug are extracted via non-stop hot percolation manner using different solvent with different polarity. The equipment used for this is known as soxhlet apparatus that encompass three components:

1. Boiling solvent containing flask;
2. Extractor in which the medication to be separated is packed; and
3. A condenser wherein the solvent vapors are again condensed once more into solvent.

3.4 CHARACTERIZATION OF PHYTOCONSTITUENTS

Characterization of compounds obtained by means of chromatographic combined strategies and are strongly endorsed with the aim to become aware of the purity of natural drugs, in view that they might signify the "chemical integrities" of the natural drugs and consequently used for quality control of the natural products (Kamboj, 2012).

Purity is closely associated with secure usage of medicines including the factors like contaminants (e.g., foreign particulate present as other herbs), ash values, and heavy metals, etc. But, within the current generation of analytical strategies, improved purity test includes radioactivity, bacterial, and fungal infection, pesticide residues, and aflatoxins (Swatantra et al., 2010). Analysis such as spectroscopic techniques (UV, IR, MS, and NMR), gas chromatography (GC), thin-layer chromatography (TLC), high-performance thin-layer chromatography (HPTLC), high-performance liquid chromatography (HPLC), volumetric, and gravimetric analysis are important for quantitative as well as qualitative characterization of small amounts of foreign matter used for quality control and standardization (Booksh et al., 1994).

Setting of measure is the most significant region of interest for the assessment in natural medications with unknown variety of phytoconstituents. At times, pharmacopeia's is ways to deal with distinguish the active ingredients or marker with percentage extractable matter in the form of assay. Another technique is used for the determination of essential oil i.e. steam distillation. An immense cluster of modernized chemical analytical strategies can be useful for identification of active components, for example, ultra-visible spectroscopy (UV/VIS), HPLC, GD, TLC, mass spectrometry (MS), or hyphenated GC and MS (GC/MS) (Farooqui et al., 2014).

3.4.1 THIN-LAYER CHROMATOGRAPHY

Thin-layer Chromatography in Phytochemistry is the primary source to provide the information on TLC as it implies to the separation, identification, quantification, and isolation properties of component of herbal plant. (Lalla et al., 2000). Further, it is used widely for the assessment of phytochemical of natural drugs, because:

1. It required a minimum amount of sample;
2. It provides semi qualitative and quantitative confirmations; and
3. Fingerprinting strategies utilizing HPLC and GLC is likewise completed in measurement of a synthetic constituent.

Identification should be possible by examination between the spots of indistinguishable Rf values about an equivalent size acquired, with a test sample and a standard on a similar TLC plate (Figure 3.9).

Right now, TLC is as yet observed to be most normally utilizing the system for the investigation of analysis of herbal drugs since different pharmacopeias like Chinese medication monographs and analysis, American Herbal Pharmacopeia (AHP), and so forth employed TLC as a first characteristic observation of constituent.

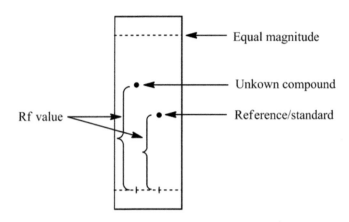

FIGURE 3.9 TLC plate.

TLC is the least difficult strategy for beginning screening method together with other chromatographic procedures qualitative estimation of natural compounds. (Habib et al., 2000; Lee et al., 2002) (Tables 3.2 and 3.3).

TABLE 3.2 Example of Solvent System Used in TLC

Sl. No.	Phytoconstituents	Stationary Phase	Solvent System	
			Mobile Phase	Ratio
1.	*Harhra (Terminaliachebula and Gallic acid)*	Silica gel	Toluene-ethyl acetate-formic acid,	5:5:1
2.	*Azadirachta indica, Catharanthus roseus,* and *Momordica charntia*	Silica gel	Dichloro methane-methanol	2:8
3.	*Mushroom extracts*	Silica gel	Dichloromethane-ethyl acetate-methanol	3:1:1
4.	*Strychnos nux vomica*	Silica gel	Chloroform-ethyl acetate-diethyl amine	0.5:8.5:1
5.	*Quinones*	Silica gel 60	dichloromethane-n-hexane	8:2

TABLE 3.3   Comparison Between TLC and HPTLC

Parameter	TLC	HPTLC
Technique	Manual	Automatic
Efficiency	Less	High (Due to smaller particle size)
Sample volume	1–5 µl	0.1–0.5 µl
Shape of sample	Circular (2–4 nm Dia)	Rectangular (6 mm L × 1 mm W)
Separation time	20–200 min	3–20 min
Sample tray per plate	≤ 9	≤ 37 (72)
Detection limits (Absorption)	1–5 pg	100–500 pg
Layer	Lab Made/Pre-Coated	Pre-coated
Mean particle size	8–12 µm	5–6 µm
Thickness of layer	240 µm	100 µm
Plate Height	30 µm	12 µm
Solid Support	Silica Gel and Kiesulguhr	Normal Phase-Silica Gel Reverse phase- C8 & C18
Sample Spotting	Manual Spotting (Capillary/pipette)	AUTO sampler (Syringe)
Detection limits (Fluorescence)	50–100 pg	5–10 pg
PC connectivity, method storage	No	Yes
Validation, Quantitative Analysis, Spectrum Analysis	No	Yes
Analysis Time	No	Yes
Analysis Time	Lesser	Shortage Migration Distance
Wavelength Range	255 or 360 nm	200 or 800 nm
Scanning	Not possible	UV/ visible/fluorescence scanner scans the chromatogram qualitatively and quantitatively

3.4.2 *HIGH-PERFORMANCE THIN-LAYER CHROMATOGRAPHY (HPTLC)*

HPTLC is a modernized model of TLC with higher performance and detection limits for separation of active constituents. HPTLC technique in particular used for identity of pesticide content material, mycotoxins, and in quality manage of medicinal plant life and broadly hired in process improvement, identity and detection of adulterants in pharmaceutical enterprise. The benefits of HPTLC are that it requires smaller quantity of mobile section to run numerous samples concurrently than in HPLC. Some other gain of HPTLC, Chromatogram screening may be done with the equal or unique situations. Consequently, software of HPTLC can be finished by using simultaneous assay of numerous additives in a multicomponent formulation. Validation assessment of stability and consistency of various parts can be possible with this technique (Table 3.4).

TABLE 3.4 Example of Mobile Phase Used in HPTLC for Phytoconstituents

Sl. No	Phytoconstituents	Eluent System (v/v)
1.	Alkaloids	Toluene: Ethyl Acetate: Diethyl Amine [7.1:2.0:1.0]
2.	Flavonoids	Ethyl Acetate: Water: Formic Acid: Acetic Acid [10.0: 2.6:1.1:1.1]
3.	*Polar Compounds* Anthraglycosides, Alkaloids, Cardiac Glycosides, Flavonoids, Saponin	Methanol: Ethyl Acetate: Water [35:10:3. 1.0]
4.	*Lipophilic Compounds* Coumarin, Napthoquinones, Velpotriate	Ethyl Acetate: Toluene [9.2:0.8]
5.	Lignans	Chloroform: Methanol: Water [7.1:3.0:0.4]
		Chloroform: Methanol [8.9:1.1]
		Toluene: Ethyl Acetate [6.9:3.1]
6.	Essential Oil	Toluene: Ethyl Acetate [9.2:0.8]
7.	Saponin	Chloroform: Acetic Acid: Methanol: Water [6.5:3.2:0.1:0.8]
8.	Triterpenes	Ethyl Acetate: Toluene: Formic Acid[5.0:5.0:1.5]
		Chloroform: Toluene: Ethanol [4.0:4.0:1.0]
9.	Terpenes	Chloroform: Methanol: Water [6.5:2.4:0.5]
10.	Cardiac Glycosides	Ethyl Acetate: Methanol: Water [8.2:1.0:0.8]

Source: Wagner (1996).

HPTLC is the effective analytical tool based on the similar talents of TLC. The benefits of HPTLC over the TLC are to perceive the fingerprints of additives with accuracy, versatility, high velocity, particular sensitivity and simple sample guidance.

Several advantages of the usage of HPTLC for the characterization of compounds at the same time as evaluating the comparing techniques, like HPLC, spectrophotometry, titrimetry, and so forth.

The essential benefits of the HPTLC are:

- Most adaptable, reliable, and cost-efficient separation technique;
- This method is selective detection principle, full of automation, screening, validation, least sample preparation, full optimization;
- It is subtle tool used for natural action affirmation of mixtures of prescribed drugs, phytoconstituents, and foodstuffs;
- Smooth separation technique with colored compounds;
- type of solvents as mobile stages is completely vaporated in advance than the detection step;
- Maintain Stability during the Two-dimensional separations;
- Reagents are certain to detect the specific individual spots;
- HPTLC can be used as a combined approach and permitting to detect the unique light-absorption characteristics of compounds HPTLC; and
- This technique minimizes the exposure of toxic organic effluents and notably decreases the environmental pollution (Patel et al., 2008; Patel et al., 2012; Shuijun et al., 2006).

3.4.2.1 CLASSIFICATION OF HPTLC

The HPTLC can be classified into four categories:

1. Classical thin-layer chromatography;
2. High-performance TLC;
3. Ultra thin-layer chromatography; and
4. Preparative TLC.

They basically differ in particle size distribution and width of the sorbent coats.

3.4.2.2 METHODOLOGY FOR HPTLC ANALYSIS

For a qualitative and quantitative analysis method development generally starts with planar thin-layer chromatography. There are few literary works clarifying the physicochemical properties and nature of the sample like stability, structure, solubility, volatility, and polarity that's required new analytical procedure during establishing (Koll et al., 2003). This method includes hit and trial-error process. Steps concerned in HPTLC method are (Nicoletti, 2012).

Initial Steps: The initial steps are as follows:

- Selection of stationary phase;
- Mobile phase selection and its optimization;
- Application and preparation of sample;
- Chromatogram development; and
- Process of detection.

Quantization: The validation of HPTLC method is done by the following factors:

- Linearity and specificity;
- Accuracy and precision;
- Range, detection limit; and
- Quantization limit and robustness

3.4.3 HIGH-PERFORMANCE LIQUID CHROMATOGRAPHY (HPLC)

Over the previous years, high-pressure fluid chromatography (HPLC) has accomplished the most broad application in the characterization of natural drugs which is otherwise called high-performance liquid chromatography, is somewhat column chromatography in which column is stuffed stationary stage (molecule estimate 3–50 μm) with a little bore (2–5mm), another finish of which is appended to a mobile phase (pressurized fluid eluent). The three basic standards of HPLC regularly utilized are: partition, adsorption, and ion exchange.

HPLC is a prevalent apparatus for the characterization of phytoconstituents. The significant preferred standpoint of HPLC over different systems that it is not constrained to the volatile compounds. Reversed-phase (RP) columns are usually utilized in the HPLC for the separation of natural drugs. Less tedious tasks, resolution, and sensitivity are the basic factors in HPLC.

There are two kinds of preparative HPLC:

- high-pressure HPLC (pressure less than 20 bar).
- low-pressure HPLC (typically less than 5 bar)

The perfect separation condition for the HPLC relies on pump pressures, different compositions of the mobile phases, pH modification, and so on. So as to accomplish the great division, a few new methods have been found in the research field of chromatography. Presently, recently created logical techniques like high-speed counter-current chromatography (HSCCC), electrokinetic capillary chromatography (MECC), low-pressure size exclusion chromatography (SEC), reversed-phase ion-pairing HPLC (RP IPC-HPLC), and strong anion-exchange HPLC (SAX-HPLC) are being used. In some examples, the examination of non-chromophoric mixes now shown by utilizing HPLC combined with evaporative light scattering detection (ELSD), is an extraordinary recognition strategy. This new detector focused just on the size, shape, and number of eluate particles rather ultraviolet detector helps in examination of chromophore or structure of constituent. Moreover, the subjective examination or structure illustration of the compound segments might be shown by utilizing hyphenated HPLC systems, for example, HPLC-IR, HPLC–MS, HPLC-NMR (Lazarowych et al., 1998; Li et al., 1999; Li et al., 2003)

3.4.4   ULTRA-HIGH-PERFORMANCE LIQUID CHROMATOGRAPHY (UHPLC)

Presently, UHPLC has been most important strategy for the quality control of therapeutic plants. UHPLC conveys liquid chromatographic investigation to another dimension by equipment changes of the regular HPLC hardware. The real favorable position of UHPLC over HPLC is that it can face up a pressure of at maximum 8000 psi and makes it feasible to perform at high resolution to gain advanced sensitivity and resolution. Smaller molecule size and shorter columns size estimate prompts accomplish higher division effectiveness in lesser time utilization (Kong et al., 2010). In the most recent couple of decades, UHPLC fingerprints of natural items were set up instead of regular HPLC methods (Jiang et al., 2010).

3.4.5   HYDROPHILIC INTERACTION CHROMATOGRAPHY (HILIC)

HILIC has pointed on great division nature of hydrophilic mixes in natural fingerprinting. It is basically utilized for extraction of polar drugs utilizing aqueous solvents, which may be better segregated by this technique (Liu et al., 1992). This chromatography may be employed as another choice rather than ordinary normal phase isolation chromatography (NPLC) and permits the isolation of polar substance on polar stationary phases with water-advanced mobile phases. Its rule dependent on the apportioning between a aqueous layer in the hydrophilic stationary phase and hydrophobic mobile phase having five-forty% water in organic solvent. HILIC is eco-friendlier methodology than HPLC in light of the fact that it includes water and polar organic solvents as mobile phase.

3.4.6 ELECTROPHORETIC METHODS

In the early 1980s, capillary electrophoresis (CE) was developed as a powerful analytical and separation device. It detects the purity/complexity of a sample and can deal with every kind of charged components of sample from simple inorganic ions to DNA. Thusly, the utilization of fine electrophoretic techniques expanded in the investigation of natural drugs in last past years. The working of CE examination can be performed by electric field worked in tight cylinders which prompts division of numerous mixes. The separation of different charged components caused due to applied voltage in between buffer filled capillaries which generates the production of ions depending on their mass and charge ratio. Frequently used electrophoresis techniques are capillary zone electrophoresis (CZE), capillary gel electrophoresis (CGE), and capillary isoelectric focusing (CIEF). CE is the most proficient strategy utilized for the division and investigation of modest number of analytes with excellent partition capacity. In the meantime it has comparative specialized qualities as that of liquid chromatography; anyway it is a superior technique for building up the chemical fingerprints of the natural medications.

The several advantages of CE are:

- versatile and powerful separation tool;
- high separation efficiency;
- analyze mixtures of low-molecular-mass components;
- improved resolution and throughput; and
- improve the reproducibility of both mobility and integral data based on internal standards.

Capillary electrochromatography and CE carry better comprehension with nature and properties of arrangement of natural drugs, particularly when united with the incredible spectrometric detectors (Stuppner et al., 1992; Yang et al., 1995). The hyphenated strategies, for example, CE-DAD,

CE-MS, and CE-NMR, have additionally immediately been utilized for the investigation of the samples from natural drugs.

3.4.7 GAS CHROMATOGRAPHY (GC)

GC is another modern investigative apparatus utilized for the characterization, quantization, and distinguishing proof of volatile drugs. It very well may be utilized in cosmetics, pharmaceuticals, and furthermore in natural toxins. The use of GC is to investigate human breath, blood, salivation, and other emission containing a lot of organic volatiles.

The benefits of GC are:

- Efficient controlling separation.
- Sensitive recognition for the examination of essential oils (Zhu et al., 2010).

The restrictions of GC are:

- Analysis of phytoconstituents is constrained to the essentials oils as there are possibilities of degradation of thermo-labile constituents (David et al., 1999).

GC might be utilized for the characterization of herbal drugs having complicated mixtures of same compounds. The GC–MS evaluation of essentials oils results in lowering analysis instances (40–100 sec) in addition to decreased detection limits, makes quicker evaluation with high efficiency (Wang et al., 2001) (Table 3.5).

TABLE 3.5 Specification of Gas Chromatography

Sample	Column Type	Carrier	Detector	Injection
0.25cc natural gas sample	Helium (8.6ml/ min at 60°C)	HP-PLOT Q, (30 m x 0.53 mm x 40 μm)	TCD 250°C	Split mode (100 ml/min)

3.5 HYPHENATION TECHNIQUES

Within the ongoing advances, joining a chromatographic separation with a spectroscopic detector with a view to get better structural identifications has turned out to be the most vital approach a few of the analytical equipment.

The preferred technique for the analysis of phytoconstituents is the combination of column liquid chromatography or capillary GD with a UV-VIS or a mass spectrometer (HPLC-DAD, CE-DAD, GC-MS, and LC-MS). The extra data can be acquired by nuclear magnetic resonance (NMR), atomic emission, fluorescence emission (FE) or Fourier-transform infrared (FTIR) spectrometry. In addition, the data obtained by these hyphenated techniques are called two-way data which justifies the data collecting from one way chromatogram and the other way spectrum.

3.5.1 LC-IR, LC-MS, AND LC-NMR

The hyphenated technique has become revolutionary change in the identification, characterization, and quantization of herbal drugs. One of the best methods developed by combining liquid chromatography and infrared spectroscopy is known as LC-IR. LC-IR predicts the peaks of functional groups in mid-IR region which makes structure elucidation helpful. But detection technique of IR is much slower than MS or NMR. Two big tools in series may be used in these techniques are flow cell method and solvent elimination methods. Liquid chromatography-mass spectrometry (LC-MS) combines liquid chromatography with the capabilities of MS. LC-MS has very high sensitivity and selectivity for identification of mixture of separated components.

There are two fundamental parts of LC/MS process based on delicate ionization methods: atmospheric pressure chemical ionization (APCI) and electrospray ionization (ESI). The coupling of abilities of fluid chromatography (LC) with NMR is regularly an incredible procedure in all viewpoints in discovering compound data from complex mixtures of analytes. Several more hyphenated NMR techniques have been designed like LC-SPE-NMR with the modification by utilizing a solid-phase extraction device. The integration of capillary LC with NMR also provides benefits by lowering the detection limit to a nanogram range (Gong et al., 2003).

3.5.2 GC-MS

GC-MS is the aggregate of two effective device Gas spectrometry and Mass chromatography which is the most delicate and particular method and gives data on the molecular weight as well as the compound structure. In MS, it depends upon ionization of molecule and then separate the ions. The methods of ionization integrated with GD are electron impact (EI) and electron capture ionization (ECI) method. Electron Ionization specifically pulls in towards positive ions, while ECI is utilized for negative ions (ECNI).

GC–MS was the first successful online tool used in the analysis of essential oil in herbal products. It provides information regarding the qualitative and quantitative features of the herbal drugs. Significant advantages for GC–MS, that is: (1) combination of GC–MS with the capillary column, enhances the ability of separation, which can provide high-quality chemical fingerprint; (2) coupled GC–MS gives qualitative and relatively quantitative information which will be helpful in research for structure elucidating the relationship between chemical constituents in herbal medicine and its therapeutic activity. Therefore, GC–MS should be the most promising technique for the characterization of the volatile phytoconstituents. (Li et al., 2003; Maillard et al., 1993).

3.5.3   HPLC–DAD, HPLC–MS, AND OTHERS

HPLC–DAD is an overall procedure utilized in most analytical research centers. UV spectroscopy brings the subjective examination of complexed structures a lot simpler than before by distinguishing the peak and after that looking at the standard spectra of the known compound with test sample. Particularly, with the advent of electrospray MS with liquid chromatography has changed the angles towards the investigation of phytoconstituents. Different examinations additionally revealed for the presentation and utilization of LC–MS in the characterization of Phytoconstituents. Inside several beyond years, the expanding utilization of LC–MS and HPLC–DAD become most significant instrument in the characterization of plant constituents. Reviews have been reports that utilization of HPLC is fruitful methodology for the investigation of the dynamic synthetic constituents in plants, particularly, the hyphenated strategies. Besides, hyphenated HPLC–DAD–MS techniques accept advantages of chromatography as a superior identification with a better partition method. DAD and MS are currently giving on-line UV and MS data for separate peaks in a chromatogram and one could identify the chromatographic peaks effectively on-line, which made hyphenated procedures all the more powerful, helpful, and less tedious technique. At present, the tendency of the hyphenation or multi-hyphenation are spread all through the four spectroscopic detectors, say FTIR, NMR, MS, and UV is in advancement. A "complete examination gadget" has been set up, i.e., on-line HPLC– UV (DAD)-FT-IR-NMR-MS analyzer. These methods make a multi-channel transformation in characterization apparatuses and give a superior peak purity check and peak identification of complicating Phyto-constituents (Rajani et al., 2001; Revilla et al., 2001; Mellon et al., 1987; Wolfender et al., 1995; Wolfender et al., 1993; Wolfender et al., 1994).

KEYWORDS

- **atmospheric pressure chemical ionization**
- **capillary electrophoresis**
- **chromatography**
- **herbal medicines**
- **phenols**
- **terpenes**

REFERENCES

Abraham, K., Buhrke, T., & Lampen, A., (2016). Bioavailability of cyanide after consumption of a single meal of foods containing high levels of cyanogenic glycosides: A crossover study in humans. *Arch. Toxicol., 90,* 559–574.

Akinpelu, A. B., Oyedapo, O. O., Iwalewa, O. E., & Shode, F., (2012). Biochemical and histopathological profile of toxicity induced by saponin fraction of *Erythrophleum suaveolens* bark extract. *Phytopharmacology., 3*(1), 38–53.

Ashok, P. K., & Upadhayaya, K., (2012). Tannins are astringent. *J. Pharmacogn. Phytochem., 1*(3), 45–50.

Azwanida, N. N., (2015). A review on the extraction methods uses in medicinal plants, principle, strength, and limitation. *Med. Aromat. Plants., 4,* 196.

Booksh, K. S., & Kowalski, B. R., (1994). Theory of analytical chemistry. *Anal. Chem., 66,* 782–791.

Chien, S. C., Wu, Y. C., Chen, Z. W., & Yang, W. C., (2015). Naturally occurring anthraquinones: Chemistry and therapeutic potential in autoimmune diabetes. *Evid. Based Complement. Alternat. Med., 2015,* 357357.

David, F., Gere, D., & Scanlan, F., (1999). Instrumentation and applications of fast high-resolution capillary gas chromatography. *J. Chromatogr. A., 842,* 309–319.

Farooqui, N. A., Dey, A., Singh, G. N., Easwari, T. S., & Pandey, M. K., (2014). Analytical techniques in quality evaluation of herbal drugs. *Asian J. Sci. Res., 4*(3), 112–117.

Firn, R., (2010). *Nature's Chemicals: The Natural Products that Shaped Our World.* Ann. Bot. Oxford University Press, Oxford.

Gong, F., Liang, Y. Z., Pei-S, X., & Chau, F. T., (2003). Information theory applied to chromatographic fingerprint of herbal medicine for quality control. *J. Chromatogr. A., 1002*, 25–30.

Habib, M. Y., Islam, M. S., Awal, M. A., & Khan, M. A., (2005). Herbal products: A novel approach dor diabetic patients. *Pak. J. Nutr., 4*(1), 17–21.

Hung, V., (2016). Phenolic compounds of cereals and their antioxidant capacity. *Crit. Rev. Food Sci. Nutr., 56*(1), 25–35.

Jiang, Y., David, B., & Tu, P., (2010). Recent analytical approaches in quality control of traditional Chinese medicines: A review. *Anal. Chim. Acta., 657*, 9–18.

Kamboj, A., (2012). Analytical evaluation of herbal drugs. *Drug Discovery Res. Pharmacogn.* Published by Intech. open 10.5772/26109.

Koll, K., Reich, E., Blatter, A., & Veit, M., (2003). Validation of standardized high-performance thin-layer chromatographic methods for quality control and stability testing of herbals. *J. AOAC Int., 86*, 909–915.

Kong, W., Jin, C., Liu, W., Xiao, X. H., Zhao, Y. L., Zhang, P., & Li, X. F., (2010). Development and validation of a UPLC-ELSD method for fast simultaneous determination of five bile acid derivatives in calculus bovis and its medicinal preparations. *Food Chem., 120*, 1193–1200.

Kushwaha, S. K. S., Kushwaha, N., & Maurya, N. A. K., (2010). Rai role of markers in the standardization of herbal drugs: A review. *Arch. Appl. Sci. Res., 2*(1), 225–229.

Lalla, J. K., Hamrapurkar, P. D., & Mamania, H. M., (2000). *J. Planar. Chromatogr., 13*, 390.

Lazarowych, N. J., & Pekos, P., (1998). Use of fingerprinting and marker compounds for identification and standardization of botanical drugs: Strategies for applying pharmaceutical HPLC analysis to herbal products. *Drug Inf. J., 32*, 497–512.

Lee, J., Min, B., Lee, S., Na, M., Kwon, B., Lee, C., Kim, Y., & Bae, K., (2002). Cytotoxic sesquiterpene lactones from carpesium abrotanoides. *Planta Med., 68*, 745.

Li, N., Lin, G., Kwan, Y. W., & Min, Z. D., (1999). Simultaneous quantification of five major biologically active ingredients of saffron by high-performance liquid chromatography. *J. Chromatogr. A, 849*(2), 349–355.

Li, X. N., Cui, H., Song, Y. Q., Liang, Y. Z., & Chau, F. T., (2003). Analysis of volatile fractions of *Schisandra chinensis* (Turcz.) baill. Using GC-MS and chemometric resolution, *Phytochem. Anal., 14*(1), 23–33.

Liu, C. L., Zhu, P. L., & Liu, M. C., (1999). Computer-aided development of a high-performance liquid chromatographic method for the determination of hydroxyanthraquinone derivatives in Chinese herb medicine rhubarb. *J. Chromatogr. A., 857*, 167–174.

Liu, Y. M., & Sheu, S. J., (1992). Determination of quaternary alkaloids from coptidis rhizoma by capillary electrophoresis. *J. Chromatogr., 623*(1), 196–199.

Maillard, M. P., Wolfender, J. L., & Hostettmann, K., (1993). Use of liquid chromatography thermospray mass spectrometry in phytochemical analysis of crude plant extract *J. Chromatogr., 647*, 147–154.

Nicoletti, M., (2012). Phytochemical techniques in complex botanicals. The XXI century. Analytical challenge. *J. Chromatogr. Sep. Tech.*, p. S10.

Ogbe, R. J., Ochalefu, D. O., Mafulul, S. G., & Olaniru, O. B., (2015). A review on dietary phytosterols: Their occurrence, metabolism, and health benefits. *Asian J. Plant Sci. Res., 5*(4),10–21.

Patel, R. B., & Patel, M. R., (2008). An introduction to analytical method development for pharmaceutical formulations. *Pharmacol. Rev., 6*(4), 1–5.

Patel, R., Patel, M., Dubey, N., Dubey, N., & Patel, B., (2012). HPTLC method development and validation: Strategy to minimize methodological failures. *J. Food Drug Anal., 20*(4), 794–804.

Rajani, M., Ravishankara, M. N., Shrivastava, N., & Padh, H., (2001). A sensitive high-performance thin-layer chromatography method of estimation of diospyrin, a tumor inhibiting agent from stem bark of *Diospyros Montana. J. Planar. Chromatogr., 14*, 34.

Rassem, H. H. A., Nour, A. H., & Yunus, R. M., (2016). Techniques for extraction of essential oils from plants: A review. *Aust. J. Basic Appl. Sci., 10*(16), 117–127.

Renger, B., (1999). Benchmarking, A. HPLC and HPTLC in pharmaceutical analysis. *J. Planar Chromatogr., 12*, 58–62.

Revilla, E., Beneytez, E. G., Cabello, F., Ortega, G. M., & Ryan, J. M., (2001). Value of high-performance liquid chromatographic analysis of anthocyanins in the differentiation of red grape cultivars and red wines made from them. *J. Chromatogr. A., 915*, 53–60.

Sarker, S. D., & Nahar, L., (2007). *Chemistry for Pharmacy Students: General, Organic and Natural Product Chemistry.* John Wiley and Sons Ltd, the Atrium, Southern Gate. Chichester, West Sussex, England.

Shuijun, L., Gangyi, L., Jingying, J., Xiaochuan, L., & Chen, Y., (2006). Liquid chromatography-negative ion electrospray tandem mass spectrometry method for the quantification of ezetimibe in human Plasma. *J. Pharm. Biomed. Anal., 40*(4), 987–992.

Stuppner, H., Sturm, S., & Konwalinka, G., (1992). Capillary electrophoresis analysis of oxindole alkaloids from *Uncaria tomentosa. J. Chromatogr., 609*, 375–380.

Wagner, H., (1996). Atlas *"Plant Drug Analysis: A Thin-layer Chromatography"* (2[nd]edn). Springer.

Wang, Y., Sheng, L. S., & Lou, F. C., (2001). Analysis and structure identification of trace constituent in the total ginkgolide by using LC/DAD/ESI/MS. *Yao Xue Xue Bao, 36*, 606–608.

Wink, M., (2015). Modes of action of herbal medicines and plant secondary metabolites. *Medicines (Basel), 2*(3), 251–286.

Wolfender, J. L., & Hostettmann, K., (1995). Phytochemistry of medicinal plants. *Recent Adv. Phytochem.* (Vol. 29, pp. 189). Plenum Press, New York.

Wolfender, J. L., Maillard, M. P., & Hostettmann, K., (1993). Liquid chromatographic thermospray mass spectrometric analysis of crude plant extracts containing phenolic and terpene glycosides. *J. Chromatogr., 647*, 183–190.

Wolfender, J. L., Maillard, M. P., & Hostettmann, K., (1994). Thermospray liquid chromatography mass spectrometry in phytochemical analysis. *Phytochem. Anal., 5*, 153.

Yadava, N., Yadava, R., & Goyal, A., (2014). Chemistry of terpenoids. *Int. J. Pharm. Sci. Rev. Res., 27*(2), 272–278.

Yang, S. S., & Smetena, I., (1995). Evaluation of capillary electrophoresis for the analysis of nicotine and selected minor alkaloids from tobacco. *Chromatographia, 40*(7–8), 375–378.

Zhu, H., Wang, Y., & Liang, H., (2010). Identification of *Portulaca oleracea* L. from different sources using GC-MS and FT-IR spectroscopy. *Talanta., 81*, 129–135.

CHAPTER 4

Quality Control Methods for Herbal Medicines

DEEPAK SHARMA, SURESH KUMAR, and SANDEEP K. THAKUR

School of Medical and Allied Sciences, G.D. Goenka University, Gurugram, India, E-mail: deepak.sharma@gdgoenka.ac.in (D. Sharma)

ABSTRACT

Herbal-based medicines are developing as an essential wellspring of elective therapeutics in the cutting edge time of human services rehearses. With the frequent usage of herbal medicines, issues pertaining to the quality control needs to be deliberated and regularized. The World Health Assembly, in its resolutions, affirms the necessity for standards of herbal medicine. In this chapter a series of parameters for assessing the standards for Herbal drugs are enlist and elaborate.

4.1 INTRODUCTION

Quality is the assurance to meet the listed and non-listed requirements of the end-user. The word "quality" springs from the Latin term 'qualis' meaning 'of what kind.' Quality is of preponderant importance especially when it is associated with pharmaceuticals or medication and most importantly when Herbal Medicines are in consideration (Shinde, 2009).

Protocol related to the formulation, granting a license, and sales of the herbal medicines, available in the market as OTC products, are entirely different, additionally, the safety and efficacy regulations are also not well defined. Also, the major constituents of herbal products are not defined effectively. Standardization requires not only the execution of GMP, additionally; a study related to pharmacodynamics, pharmacokinetics, dosage, stability, shelf-life, toxicity evaluation, and chemical profiling is vital. Different prerequisites according to WHO Rule resembles pesticide's buildup, aflatoxin content, substantial metals defilement, are similarly significant. DNA examination can be considered for the separation of phytochemically vague certifiable medication from substituted or defiled medication (Choudhary, 2011).

Standards for herbal drugs are kept on the priority, by the World Health Organization (WHO) since its beginning. The need for standards for Herbal medicines is confirmed by the World Health Assembly; vide its resolution, WHA31.33 (1978), WHA40.33 (1987), and WHA42.43 (1989). In this chapter, a progression of parameters for surveying the norms for herbal drugs are enrolled and explained (Shinde, 2009).

Antiquated drug treatment, according to the perspective on WHO, incorporates natural resources (plants, medicines from animals, non-common treatments, manual procedures, and exercise) alone or in the mix with routine therapeutic treatment to keep up the prosperity of the general public (Ajibesin, 2017).

Plants as a whole or in parts either processed or in an unprocessed manner are considered as herbal medicines. It may include leaves, blossoms, fruit, seed, stems, wood, bark, roots, rhizomes, or other plant parts. Natural preparations incorporate pulverized or powdered materials or concentrates of plant parts, tinctures, and fatty oils, produced by extraction, fractionation, purification, concentration, or alternative physical or biological processes (Mahomoodally, 2013).

4.1.1 PHARMACOPOEIAL (OFFICIAL) STANDARDS

The qualitative and quantitative standardization of herbal products are guaranteed in context to the norms stated in a pharmacopeia. The recommended Pharmacopoeial strategies, incorporate yet not constrained to Logical, physical, and basic measures for the herbal drugs. As a result of greater decent variety and changes in the synthetic constituents, powerful distinguishing proof and examination of crude drugs is important. To amplify the confirmation of quality, all official monographs have certain measures and explicit procedures for explicit herbal samples, a portion of the recorded ones are: Alkaloids content by Dragendorff test, Fatty substance by acid value, iodine value, saponification value, carbohydrate substance by molish test (Ekka, 2008).

4.2   WHO GUIDELINES FOR QUALITY CONTROL

Quality is sketched out as the aggregate of the considerable number of elements contributing straightforwardly or in a roundabout way to the well-being, adequacy, and worthiness of the medication. Standardization includes every one of the steps taken during production and quality control of the item that winds up in steady nature of unequivocal drug product (WHO Press, 2011).

Notwithstanding absence of typical quality control strategies, standardization of herbal resources/drugs still remains a major challenge for the scientific community. Keeping the perspective on the unpredictability of polyherbal formulations, it is need of an hour to draft stringent rules for the confirmation of the nature of such formulations. Additionally, the differed pharmaceutical dose forms incorporating herbal drugs medications require the advancement of standard conventions for natural drugs (WHO Press, 2011).

Some of the parameters to be considered during the assurance of the quality of Herbal medicines are discussed below.

4.2.1   GENERAL ADVICE ON SAMPLING

Sampling is a procedure for choosing sample members from a population, so that, the sample represents the complete populations. The normally followed sampling strategies are: Simple Random Sampling-where the examples are chosen from the populaces by an arbitrary way, every item has equivalent possibility of determination, Stratified Sampling-the populace is part into independent groups, known as strata. Then, sample is (often a simple random sample) drawn from each strata. Cluster Sampling -separate teams are prepared from the sample, known as clusters. Then, from the population, a cluster is chosen, in simple random fashion (MacKenzie, 2005).

4.2.2   MORPHOLOGICAL OR ORGANOLEPTIC EVALUATION

It is qualitative evaluation based on study of organoleptic characters (like color, taste, odor, size, shape) and some specific characters like feel-look, etc. Evaluation may be done like studying the surface fracture of cascara and quillia bark, etc. Aromatic odor is present in *Umbelliferous* fruits while, sweet taste can be found in liquorice (Gupta, 2007 and Ansari 2011).

4.2.3 POWDER FINENESS AND SIEVE SIZE

By going through a run of the mill set of strainers, fineness of a powder is frequently determined. For a molecule estimate over 75 micrometers, favored methods for deciding the powder fineness is sieving, while, for smaller particle size measure the system should be substantial. Systems utilized for sub-sifter particles incorporates: elutriation by methods for air or water, sedimentation, hydrometer readings, and the darkening of a light emission by dilute suspensions of the particles. It's classed by the ostensible gap of sieve mesh, communicated in micrometers, over that a nominative measure of the powder can pass and results are demonstrated as d90, d50, d10, where 90, 50 and 10 show the extent of the sample goes through the sifter (Table 4.1). Then again, the ensuing table can be utilized to point the outcomes (Heywood, 2018).

TABLE 4.1 Classification of Particle Size

Classification of Powder (Fineness)	D50 Sieve Opening (Micro-Meters) (USP29NF24 Page 2754)	Particle Size in Sieve no. (WHO Guidelines)
Very Coarse	Anything more than 1000	All the particles clear through a sieve no. 2000 and not less than 60% withstand a sieve no. 355.
Coarse	Max. 1000 min. 335	All the particles clear through a sieve no. 710 and not less than 60% withstand a sieve no. 250.
Moderately Fine	Max. 335 min. 180	All the particles pass through a sieve no. 355 and not less than 60% withstand a sieve no. 180.
Fine	Max. 180 min. 125	All the particles clear through a sieve no. 180.
Very Fine	Max. 125 min. 90	All the particles clear through a sieve no. 125.

Source: WHO (2011).

4.2.4 FOREIGN MATTER ESTIMATION

Plant samples are analyzed for the presence of extraneous matter. Herbal/Plant samples have to be completely free of noticeable indications of defilement (molds or bugs, animal contamination, together with animal squander). No unusual smell, staining, or indications of debasement should be identified. Any dirt, stones, sand, soil, and diverse remote inorganic matter ought to be evacuated before the size reduction of materials. Sort the extraneous staple either by visual examination, utilizing an amplifying focal point or with the help of a satisfactory strainer. Compute the substance of each group in grams per one hundred grams of air-dry example (Patwekar, 2015).

4.2.5 MACROSCOPIC AND MICROSCOPIC EXAMINATION

To build up the identity and immaculateness of herbal samples, macroscopic, and microscopic portrayal of the sample is essential. If conceivable, sample material ought to be compared with pharmacopoeial benchmarks. Visual technique is the least difficult and the most straightforward strategy to guarantee the identity and the immaculateness. On visual investigation (color, odour, consistency, and taste), if the sample isn't observed to be as determined, it can be rejected (Roy Upton, 2011).

Macroscopic identification depends on the shape, measure, color, odor, surface characteristics, surface appearance, fractures, if any, and the presence of the cuts. Macroscopic identification ought to be finished with due consideration as the substitutes or adulterants may intently take after the genuine sample. In such case, infinitesimal portrayal can be useful (WHO press, 2011).

For the investigation of the internal structure of plant cells, microscopy is advantageous. For the location of adulterants and contaminants of the herbal plant samples microscopy is an important tool, which provides resources not only for evaluating realness & nature of plant product but also, to measure different cells and tissues silhouette, the virtual locus, the biochemical-nature of cell walls, and nature of cell content of crude medication. Microscopic investigation of plant samples could be subjective and quantitative. Subjective research incorporates investigations of the transverse areas of leaf, root bark, additionally the underneath photo of the longitudinal segment of root bark with or without re-coloring. In case of powder microscopy, entirely diverse coloring agents like for location of calcium oxalate or starch grains iodine can be used though phloroglucinol meant for recognition of lignified segments. Reckonable research of selected pharmacognostic constraints like stomatal range, stomatal index, vein-islet no., vein termination no. furthermore, palisade ratio are utilized for recognizable proof, immaculateness assurance, and investigation of unrefined leafy samples. For quantitative microanalysis, morphological, and histological drawing of plant structures and several elective moment arrangements (e.g., trichomes, glands, stomata, calcium oxalate crystals) is additionally utilized. The characteristic microscopic characters incorporates trichomes, collenchyma, palisade, and spongy parenchyma, stomatal recurrence, their index, vein-islet, vein termination no., palisade ratio, shape, and size, besides as vascular bundles, vascular tissue and so on., including physical constants for verdant medications while cork cambium, primary cortex, phloem fibers, medullary rays, endodermis, pericycle, pericycle vascular bundles, and so forth., inside the slanting and longitudinal section including physical constants remain as distinguishing microscopic characters from root, stem, and so on. (WHO Press, 2011).

4.2.5.1   PRELIMINARY TREATMENT OF THE SAMPLE

Dried plant material is required to be mollified before microscopy. For this, it is generally be soaked in water from a couple of hours to medium-term or until the material is been mellowed, similar to, bark, and wood, or any other thick material, in order to be cut effectively (WHO Press, 2011).

4.2.5.2   PREPARATION OF SPECIMENS FOR POWDERED MATERIAL

A drop of water is to be placed over the slide and the powdered sample is to be exchanged to it with the assistance of a dampened needle. Once exchanged, it ought to be blended with the assistance of the needle tip and should be secured with the cover-slip. The overabundance liquid is to be cleared off utilizing a filter paper; care must be taken to keep the sample free from air bubbles (WHO Press, 2011).

4.2.5.3   SECTION CUTTING

The plant material under examination is to be cut into the appropriate length and mollified. The segment might be cut transversely (TS) or longitudinally (LS) utilizing a sharp disposable cutter. Thumb and index finger can be used to hold thick materials like roots, wood, stem, and so on for section cutting. Dainty materials like leaves or petals can be cut by putting in the middle of the two parts of pith. Seed or fruits which are flat and very small in size can be embedded into a hard paraffin wax square (WHO Press, 2011).

4.2.5.4   DETERMINATION OF STOMATAL INDEX

Observe the TS of the material below a 40X objective piece of a microscope having an eyepiece of 6X, outfitted with an illustration paper. Imprint (X) meant for every cell of epidermis and in lieu of a stoma a hover (O). *Stomatal Index* is dogged as:

$$\text{Stomatal Index} = (S \times 100) \div (E + S)$$

where, S is the stomata present in a unit area of leaf; E is the epidermal cells in a unit area of leaf.

   Note: At least 10 judgments is to be done and the normal incentive for the stomatal index is to be determined (WHO Press, 2011).

4.2.5.5   ESTIMATION OF PALISADE RATIO

The numbers of palisade cells present per epidermal cell for not less than ten estimations and compute as the average number (WHO Press, 2011).

$$palisade\ ratio = \frac{Total\ no.of\ Palisade\,Cells}{No.\ of\ Epidermal\ Cells}$$

4.2.5.6   VEIN-ISLETS NUMBER

The leaf's mesophyll gets separated into little bits of photosynthetic tissue by an astomosis of the veins and veinlet's; these little segments/regions termed as "Vein-Islets." Quantity of vein-islets per mm^2 is named as "Vein-Islet number." In certain species/ entirely developed leaves, this esteem appeared to stand consistent and is independent of the age or measure of leaf. Few species which are firmly related can be refined on the basis of vein-islet number (WHO Press, 2011).

4.2.5.7   HISTOCHEMICAL DETECTION

It incorporates the discovery of starch grains, aleurone grains, fats, fatty-oils, volatile oils, and resins. Calcium oxalate or calcium carbonate crystals, lignified cell wall, cellulose cell wall, mucilage, and tannin, utilizing different reagents over the TS and watching the equivalent under the microscope (WHO Press, 2011).

4.2.6   TLC-THIN-LAYER CHROMATOGRAPHY

Thin-layer chromatography is the least complex procedures to distinguish and evaluate different constituents of herbal medications, just as optional metabolite of pharmacological intrigue. Here, thin layer of adsorbent is used to separate compounds, glass plate bearing thin layer of silica gel (aluminum foil or plastic sheet may be used in place of glass plates), acknowledged as stationary phase and the segments been eluted with a fluid acknowledged as mobile phase, typically a blend of natural solvents of various extremity (Siddiqui, 2017).

The proportion of the separation gone by the solute front to the separation gone by the solvent front is called retention factor (Rf) (Siddiqui, 2017).

$$\text{Retention} = \frac{\text{Distance travelled by the solute front}}{\text{Distance travelled by the solvent front}}$$

4.2.7   ASH VALUE

The buildup staying after cremation is the slag substance of plant drug material. Total ash technique is utilized to gauge the aggregate sum of solid staying next to cremation. It is helpful to distinguish poor grade items, depleted items, abundance of sand and earthy issue. Total ash frequently comprises of carbonates, silica, phosphates, and silicates (Kaur, 2018).

Acid insoluble ash is the buildup gotten subsequent to boiling of total ash with dilute hydrochloric acid and afterward igniting the unsolvable substance (Kaur, 2018). Water-soluble ash is the distinction in weight among absolute (total) ash& buildup of total ash after treatment with water (Kaur, 2018).

4.2.8   ASCERTAINMENT OF EXTRACTABLE MATTER

Ascertainment of the potential components present in crude drug when extracted with specific solvent. There are various strategies (methods) for assurance of extractive matter–cold maceration, hot extraction, and hot continuous extraction (soxhlet) strategy (Azwanida, 2015).

- ***Cold Maceration:*** Transfer the predefined measure of the powdered herbal drug in a conical flask. Until otherwise specified in the official monograph, macerate thru 100ml of specified solvent for 6Hrs, shake, and permit representing 18hrs. After filtration, transfer the filtrate to a flat bottom flask (FBF) and keep the FBF over a water bath, at 100°C and dissipate to dryness, then cool and weigh. Extractable matter content is then determined, in mg/g of air-dried material (Azwanida, 2015).
- ***Hot Extraction:*** as specified in individual monograph, specified amount of the powdered herbal material is transferred to a conical flask. Weigh to acquire total weight, after including the determined amount of water then shake and allow standing for 1hr. after that, reflux for 1hr. Rearrange towards the first weight by solvent. Tremble, filter, and then transfer the filtrate to a FB-Flask then dissipate at 100°C, to constant weight, upon a water-bath for six-seven hours, cool the remaining and calculate the weigh right away. Finally, extractable matter content is calculated in mg per g of air-dried material (Azwanida, 2015).
- **Hot Continuous Extraction (Soxhlet)**: Soxhlet powdered plant material by placing it in a strainer sack, then heating the extracting solvent and condensing the vapors. The condensed solvent drips into the thimble having the crude material, and extracts it. When the liquid level increases in the chamber up to the uppermost level of siphon tube, the liquid contents siphon into FB-flask. The method continues until a drop of solvent from the siphon tube does not leave residue on dryness. With this method, large amounts of drugs are extracted less solvent is used (Azwanida, 2015).
- **Infusion:** Macerated with cold or hot water, the crude drug, for short period of time, to get dilute solutions of the freely-soluble components of crude plant drugs.
- **Decoction:** Extract the thermostable and water-soluble components of crude plant material by boiling it for a specified time, in predefined volume of water; cool and filter (Azwanida, 2015).

4.2.9 DETERMINATION OF BITTERNESS VALUE

Strong bitter taste compounds can be used as appetizing agents. Quinine hydrochloride concentration is compared with the threshold concentration of the extract to determine the bitterness value of an extract. Bitterness is communicated as parts proportional to the bitterness of a solution having

1 gm of quinine hydrochloride in 2000 ml. This is compared with drug after dilution (Bijauliya, 2017).

$$\text{Bitterness (unit/gm)} = \frac{200 \times C}{a \times b}$$

where, *a* is the concentration of quinine hydrochloride stock; *b* is the test solution volume in tube with threshold bitter concentration; and *C* is the quinine hydrochloride quantity in the tube with threshold bitter concentration.

4.2.10   DETERMINATION OF VOLATILE MATTER AND WATER

The water content of the crude plant drug ought to be limited to avoid deterioration of crude plant material because of change in the chemical nature or contamination of microbe. Water content can be estimated by heating a drug at 105°C in a hot air oven to a steady weight. It quantifies the total change in weight of a material when the sample is dried (Shulammithi, 2016).

$$LOD = \frac{final\ weight - initial\ weight}{initial\ weight} \times 100$$

The toluene distillation method can be used for the drugs containing volatile constituents.

4.2.11   DETERMINATIONS OF ESSENTIAL OILS

Essential oils are the volatile constituents of the crude plants drug and are a blend of terpenes, sesquiterpenes, and their oxygenated subsidiaries and sometimes a portion of the aromatic compounds, volatile at room temperature and are sleek in appearance. Usually, essential oils contain/or may contain the active constituents of the plants. Typically, volatile oils ban be quantified by water distillation of the plant part, gathering the distillate in a graduated cylinder, and then isolate the oil-phase from water phase upon standing (WHO Press, 2011).

4.2.12   DETERMINATION OF HEMOLYTIC ACTIVITY

Hemolytic activity can be estimated by comparison of plant sample with reference material. Hemolytic activity of 1000 units/g is seen in Saponin (WHO Press, 2011).

$$\text{Haemolytic activity} = \frac{1000 \times A}{B}$$

where, A is the amount of standard saponin used to produce total hemolysis (g); B is the amount of crude plant material used to produce total hemolysis (g).

4.2.13   DETERMINATION OF TANNINS

Tannins belong to the class of astringents that binds to proteins, amino acids, alkaloids, and certain other organic compounds. Typically, they are water-soluble polyphenolic biomolecules.

Estimation of tannins can be done by adding 150 ml of water to a specified amount of powdered drug or extract, in a beaker. After warming up for 30 min. the same is to be cooled and transferred to a volumetric flask of 250 ml and volume is made up to 250 ml with water. Allow to stand and filter the same when the solid settles down (Ashok, 2012).

Identification test: a solution of tannins, gets buff color precipitate on addition of an aqueous solution of gelatin and sodium chloride.

4.2.14   DETERMINATION OF SWELLING INDEX

Under specified conditions, the volume taken up by 1gm plant material on swelling is considered as its swelling index (WHO Press, 2011).

Swelling Index = Final Volume – Initial Volume

4.2.15   FOAMING INDEX

Due to the presence of saponins, plant material have a capability to foam; this capability of foaming is known as foaming index (WHO Press, 2011).

$$\text{Foaming Index} = \frac{1000}{A}$$

where, A is the amount of the preparation, in ml, used in the tube to observe a foam to a height of 1 cm.

4.2.16 PESTICIDE RESIDUES

Residues of chemicals (pesticides) may be present in the plant sample, which gets accumulated because of the various agricultural practices, like spraying of chemicals, soils treatments, and fumigation. Several pesticides contain chlorine and/or phosphate, can be analyzed by estimating total organic chlorine or total organic phosphorus. On the other hand, pesticides can be arranged based on their expected use, for instance: fungicides, rodenticides, bactericides, herbicides, bug sprays, and so on (Nawaz, 2003).

4.2.17 ARSENIC AND TOXIC METALS

Subjective investigation depends on shading responses with extraordinary reagents, such as thioacetamide estimation. Instrumental methods are available for quantitative determinations, even if, the impurities present in traces (Ahmed, 2011).

4.2.18   MICROORGANISMS CONTENT

Usually, because of the poor method of cultivation, handling, and storage, herbal drugs may contain a large number of bacteria and molds. The method of microorganism content determination is based upon determining the total aerobic microbial count, the total fungal count, together with tests for the presence of E-coli, Staphylococcus aureus, Shigella, and Pseudomonas aeruginosa and Salmonella spp (Kunle, 2012).

4.2.19   DETERMINATION OF AFLATOXINS

Aflatoxins are toxic, highly carcinogenic and naturally occurring mycotoxins produced by many species of Aspergillus (a fungus). On a prolonged exposure to a highly humid environment, or from stressful conditions like, drought, crops becomes susceptible to Aspergillus infection. High moisture content (at least 7%) and high temperature are the favorable conditions. Milk of the animals fed on the contaminated feed, may contain the toxins (WHO, 2018).

4.3 STABILITY TESTING OF HERBAL DRUGS

Since, the entire herb or herbal product is considered as the active matter, stability testing of the herbal drugs is a challenging task. Here, storage condition and shelf-life for the given product is judge by studying the variations in the quality and quantity of constituents with time under the influence of various environmental factors (temperature, light, oxygen, moisture). Typically, stability studies ought to be performed on at least three production batches for the proposed shelf-life, using modern analytical techniques (Kunle, 2012). For herbal drugs or herbal drug preparation having known therapeutic activity, the variation during proposed shelf-life should be within the limits of ±5% of the initial assay value, considering the factors like climate, harvesting, and biological variance, the natural variation of the marker (Nikhat, 2013).

4.3.1 IMPORTANCE OF STABILITY TESTING

Stability studies are used to ensure the quality, strength, purity and integrity of product during its entire shelf life. Typically, used for Shelf Life estimation (Sachan, 2013)

4.3.2 STRESS TESTING

To identify the degradation product and to establishing the degradation pathway is known as Stress Testing. As per official guidelines, most of the herbal products fail the conditions of Stress Test, therefore, it is normally considered as not necessary. If required, only 25°C/60% RH, condition is examined, in a manner similar to the chemical drug substances or products. (10)

4.3.3 CHALLENGES IN QUALITY CONTROL

Compared to chemically defined substances, Quality Control of Herbal Products is a challenging task, because of the following reasons (Nikhat, 2013):

1. Herbal drug and/or herbal drug preparations normally contain a large number of constituents and the constituents actually producing the desired effect are unknown, with combinational drug therapy, the condition becomes more sever;
2. Herbal constituents are usually unstable in nature; and
3. Any variations in factors related to cultivation (light, soil condition, etc.) affect the active constituent of the drug under cultivation, qualitatively, and quantitatively (Sindhura, 2018).

4.4   FUTURE TRENDS IN QUALITY CONTROL OF HERBAL DRUG

The traditional approach proves to be insufficient to control quality of herbal drugs, so to meet the daily developing needs of the market more advanced and developed technology of standardization are to be considered like, Chromatographic fingerprinting and DNA fingerprinting. The chromatographic technique involves the separation and identification of lead compound from other constituents. Irrespective of the part of the plant used, DNA fingerprint of genome identical although the quantity of phytochemical constituents may vary with the plant part used, physiology, and environment (Nikam, 2012).

KEYWORDS

- **herbal medicines**
- **morphological evaluation**
- **phytochemical constituents**
- **quality control**
- **stability studies**
- **World Health Organization**

REFERENCES

Ahmed, I. B. E., (2011). *Quality Control of Herbal Medicines and Related Areas*. In Tech Janeza Trdine, Croatia.
Ajibesin, K. K., (2017). Herbal medicine: Evidential, experiential, circumstantial. *Niger Delta Med. J., 1*(2), 40–62.
Ansari, S. H., (2011). *Essentials of Pharmacognosy*. In: Birla Publications Pvt Ltd.

Ashok, P. K., et al., (2012). Tannins are astringent. *J. Pharmacog Phytochem., 1*(3), 45–50.

Azwanida, N. N., (2015). A review on the extraction methods use in medicinal plants, principle, strength and limitation. *Med. Aromatic Plants, 4*(3), 1–6.

Basar, S. U., Rani, S., & Zaman, R., (2013). A review on stability studies of unani formulations. *J. Pharm. Sci. Inn, 2*(4), 1–8.

Bijauliya, R. K., et al., (2017). A comprehensive review on standardization of herbal drugs. *In. J. Pharm. Sc. Research*, 3663–3677

Choudhary, N., & Sekhon, B. S., (2011). An overview of advances in the standardization of herbal drugs. *J. Pharm. Educ. Res., 2*, 55–70.

Ekka, N. R., (2008). Standardization strategies for herbal drugs: An overview. *Research J. Pharm. and Tech., 1*(4), 310–312.

Fact Sheet on Mycotoxins Containing its Details Published by WHO. Available at: https://www.who.int/news-room/fact-sheets/detail/mycotoxins (accessed on 20 May 2020).

Gupta, M., & Sharma, P. K., (2007). Test book of pharmacognosy, ayurvedic formulations. In: Pragati Prakashan, Meerut (1st edn.). II.

Heywood, H.(1938). *Measurement of Fineness of Powdered Materials.* Available online on http://citeseerx.ist.psu.edu/viewdoc/download?doi=10.1.1.817.4178&rep=rep1&type=pdf (accessed on 20 May 2020).

Kaur, M. Determination of Ash Value. Available online on http://www.yourarticlelibrary.com/medicine/ayurvedic/determination-of-ash-values/49966 (accessed on 20 May 2020).

Kunle, O. F., et al., (2012). Standardization of herbal medicines: A review. *Int. J. Biodiversity Conservation., 4*(3), 101–112.

MacKenzie, D. I., & Royle, J. A., (2005). Methodological insights-designing occupancy studies: General advice and allocating survey effort. *J. App. Eco.*, 1105–1114.

Mahomoodally, M. F., (2013). Traditional medicines in Africa: An appraisal of ten potent African medicinal plants. *Evid. Based Complement Alternat. Med.*, 1–14.

Nawaz, S., (2003). Pesticides and herbicides—residue determination. *Encyclopedia of Food Sciences and Nutrition., 2*, 4487–93.

Nikam, P. H., Kareparamban, J., Jadhav, A., & Kadam, V., (2012). Future trends in standardization of herbal drugs. *J. Applied Pharm. Sc., 2*(6), 38–44.

Nikhat, S., & Fazil, M., (2013). Determination of the shelf life and expiry date of herbal compound drugs: A review *In. J. Sc. Res. Management, 1*(8), 415–420.

Patwekar, S. L., et al., (2015). Standardization of herbal drugs: An Overview. *Pharm. Inn. J., 4*(9), 100–104.

Quality Control Methods for Herbal Materials, (2011). WHO Press, World Health Organization, Geneva, (pp 14–187).

RoyUpton, R. H., et al., (2011). *American Herbal Pharmacopoeia Botanical Pharmacognosy— Microscopic Characterization of Botanical Medicines*. CRC Press Boca Raton.

Sachan, N. K., Pushkar, S., Sachan, A. K., & Ghosh, S. K., (2013). Thermal stability and drug-excipient compatibility studies of peppermint and caraway oils for formulation of chewable tablets. *Asian J. Chem., 25*(11), 5930–5934.

Shinde, V. M., et al., (2009). Application of quality control principles to herbal drugs. *Int. J. Phytomed., 1*, 4–8.

Shulammithi, R., et al., (2016). Standardization and quality evaluation of herbal drugs. *J. Pharm Bio Sc., 11*(5), 89–100.

Siddiqui, M. R., (2017). Analytical techniques in pharmaceutical analysis: A review. *Arabian J. Chem., 10*(1), S1409–S1421.

Sindhura, D. S., & Jain, V. K., (2018). Challenges in formulating herbal cosmetics. *In. J. App. Pharm., 10*(6), 47–53.

Srivastava, S., & Misra, A., (2018). Quality control of herbal drugs: Advancements and challenges. In: Singh, B., & Peter, K. V., (eds.), *New Age Herbals-Resource, Quality and Pharmacognosy* (1st edn, pp. 189–209) Springer Nature: Singapore.

Part II
Herbal Medicines in Cancer Therapeutics

CHAPTER 5

Anti-Cancer Agents from Natural Sources

DEBASISH BANDYOPADHYAY[1,2] and FELIPE GONZALEZ[1]

[1]*Department of Chemistry*

[2]*School of Earth Environment and Marine Sciences (SEEMS),*
University of Texas Rio Grande Valley, 1201 West University Drive,
Edinburg, Texas – 78539, USA, Tel.: +1(956)5789414,
Fax: +1(956)6655006, E-mail: debasish.bandyopadhyay@utrgv.edu

ABSTRACT

One-sixth of the total deaths worldwide are directly caused by cancer which estimated 9.8 million deaths only in 2018. Furthermore, cancer-related mortalities are projected to rise to over 13.1 million by 2030. Many cancers are drug-resistant and practically no treatment is available. Cancer has negative impacts on the world-economy as only in 2010 the total economic cost was estimated at approximately US$1.16 trillion although the extensive role of cancer research is opening new frontiers to solve this fatal problem. Although surgery, immunotherapy, radiotherapy etc. can be used in the treatment of various cancers, still chemotherapy, either individual or along with other therapeutic procedures, is used to treat cancer in majority cases. Consequently, there is a general demand to discover/develop new and selective anticancer agents to eliminate/remove both cancer incidence and cancer related mortalities. Alternatively, Mother Nature provides us a lucrative, apparently unlimited wear house of various natural compounds that exhibit excellent anticancer activities. The utilization of the nature-made compounds and/or their novel cores, to develop

---

This chapter is respectfully dedicated to Professor (Dr.) Bertha Connie Allen to honor her lifelong devotion to Chemical Sciences.

the final drug entity, is still an interesting and highly promising area of drug discovery research. Since, 1940s to the end of 2014, the US Federal Drug Administration (FDA) or its similar organizations have approved 175 small molecules for the treatment of various types of cancers. Out of these 175 small molecule anticancer drugs 131 (74.85%) are completely non-synthetic molecules. Interestingly, 85 anticancer drug molecules (48.57%) are either actually being natural compounds or semi-synthetic molecules. Many natural or semi-synthetic compounds are also being used to treat several other diseases, e.g., about 78% of drugs that are generally used to combat microbial infections, are either obtained from nature or chemically-modified natural compounds. This chapter describes the pertinent recent examples of anticancer nature-derived compounds with special emphasis to commercial drugs (both natural and semi-synthetic). Based on their source (terrestrial plants, marine organisms, microbes, and animals) and core structure nature-derived molecules have been classified in several categories. Discussion on pharmacokinetics of various recently reported anticancer molecules is an interesting section of this chapter. In addition, brief discussion on phytoceuticals as chemosensitizer and radiosensitizer has been covered in this chapter.

5.1   INTRODUCTION

Cancer is the second leading cause of the death worldwide, falling short of cardiovascular and its related diseases. In developed countries, cancer is the foremost reason of death. According to the World Health Organization (WHO, 2018), cancer accounted for 8.8 million mortalities in 2015. Low- and middle-income countries represent approximately 70% of the global cancer deaths (WHO, 2018). Five major classes of cells/tissues in humans that are known to become cancerous: carcinoma, lymphoma, sarcoma, leukemia, and myeloma. Besides these, there are brain and spinal cord cancers. Carcinomas begin in epithelium tissue cells, these cells cover in the exterior (skin) and interior (organs), about 80-90% of all types of cancers begin with malignant epithelium cells (Macmillan Cancer Support, 2017). There are diverse types of epithelial cells in human body, thus, various types of carcinoma are possible. The most common carcinomas are squamous cell carcinoma (SCC), adenocarcinoma, transitional cell carcinoma, and basal cell carcinoma. Sarcomas start in connective tissues, which include bones, tendons, cartilage, and other organ supportive tissues. Sarcoma has two major types: bone sarcomas (osteosarcomas) and soft tissue sarcomas. According to Cancer Research UK, sarcomas represent only 1% of all cancer diagnoses (Types of Cancer, 2017). Leukemias are cancers that begin in blood cells. Similarly, types of leukemia include acute myeloid leukemia (AML), ALL, chronic myeloid leukemia (CML), chronic lymphocytic leukemia (CLL), and the rare type of leukemia which is known as hair cell leukemia (HCL) (Leukaemia, 2017). The lymphatic system contains huge number of vessels and important lymph nodes. Its main function is to transport lymph, a clear fluid that comprises of numerous white blood cells. Due to this reason, cancerous lymphomas can be seen in anywhere in the body. Some white blood cells contain plasma cells, that produce antibodies to fight infections. They are primarily formed in the bone marrow and occasionally become cancerous. In the United States, on average every 3 minutes one person is diagnosed with a certain type of blood cancer (Facts and Statistics, 2018). The central nervous system (CNS) that consists of brain and spinal cord is known as the most complex organ in human body. There are many types of brain tumors and fortunately in major case they are benign (non-cancerous). The most common cancerous brain tumors are meningiomas, gliomas, and glioblastomas. According to the American Cancer Society (ACS) about 24,000 people are estimated to be diagnosed with a malignant form of brain tumor every year only in the US (Key Statistics for Brain and Spinal Cord Tumors, 2018) (Figure 5.1).

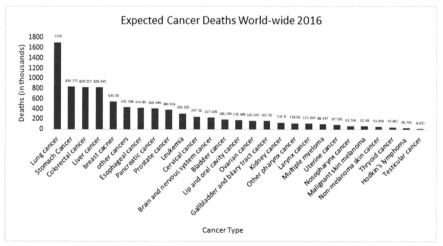

FIGURE 5.1   Expected number of deaths worldwide during 2016.

Source: Ritchie (2018). How many people in the world die from cancer? Our World in Data.

In general, the most common type of cancers with high death rates globally are lung, liver, colorectal, stomach, and breast cancer (see Figure 5.1). Currently, lung cancer ranks first having highest incidences of mortalities and morbidities than any type of cancers. It can broadly be categorized into two major types, non-small cell lung cancer (NSCLC) and small cell lung cancer (SCLC). Around 85% of lung cancers are of NSCLC type. The most common type of NSCLC is adenocarcinoma, followed by SCC, and large cell carcinoma. SCLC counts on the rest 15% of lung cancers (ACS, 2016). Tobacco smoke has been recognized as the leading risk factor in lung cancer and related diseases. In the US, 80% to 90% of all lung cancers are somehow linked to tobacco products (Centers for Disease Control and Prevention (CDC), 2017). Even if someone does not directly smoke tobacco products, secondary/indirect smoking which is the involuntary action of inhaling tobacco smoke from others, accounts for about 41,000 deaths in each year (American Lung Association, 2018). In modern world, cancer is a worse curse to the human civilization. The reason of 'why cancer is so difficult to treat' is because cancer is a multiple set of diseases that work in the same way. Initially it starts at a specific site and spread through invasion and metastasis. As stated by the National Cancer Institute (NCI), invasion occurs when cancerous cells migrate and penetrate neighboring tissues. After invasion, cancer cells eventually migrate to other organs through lymphatic and blood vessels, where they circulate around the body invading normal tissue (Rajabi et al., 2017). Currently, the most common methods of treating most cancers are surgery, radiation, and chemotherapy. Apart from these, immunotherapy, targeted therapy, and hormone therapy have accomplished encouraging results. Regardless, chemotherapy is still used in most of the cases which use cytotoxic (toward cancer cells) chemical substances. Most of the commercially available cancer drugs are organic molecules although a few organo metallic's are also known.

Cancers can be of drug-resistance and commercial anticancer drugs do not produce expected health effects. Several reasons have been identified for drug resistance. These reasons may act individually or in combination. These are: epigenetic, drug efflux, in-house repairing of damaged DNA, inhibition of cell death, epithelial-mesenchymal transformation, change of drug target, and most importantly drug inactivation. To counter this misfortune, scientists around the globe look for new, novel, and selective anticancer therapeutics with higher cytotoxicity toward malignant cells and reduced toxicity (ideally no toxicity) toward healthy cells, from the large biodiversity in the world, mother nature. Mother nature serves as a repository of many unknown anticancer compounds. Currently, 25% of all medications are directly being acquired from diverse natural sources. When it comes to anticancer drugs, 75% are either directly or indirectly related to a natural resource (Newman et al., 2016). Many natural compounds have been extracted from plant, marine, microbial, and organismal sources to determine their effectiveness against cancer cells. In general, 65–75% of natural drugs come from plants. Marine natural products represent 15–20%, microorganisms 10–15%, and animal natural products account for a sheer less than 3% (see Figure 5.2) (Newman et al., 2016). In nature, primary metabolites are produced by organisms to promote adequate growth, essential development, and successful reproduction. Organisms also produce secondary metabolites to ensure competitiveness, species to species interactions, and overall protection (Pichersky et al., 2000). Many secondary metabolites are very specific to an individual species and eventually recognized as bioactive compounds like phenolics, flavonoids, alkaloids, and terpenoids etc. Several nature-derived compounds have been known to contain immense anticancer activity (Avato et al., 2017; Joshi et al., 2017; Majumder et al., 2017). Currently, there are over one million natural products known and plenty of these compounds have been recognized as phytoceuticals that are capable of treating several ailments. Only in 2008, there were 225 natural product-based drugs in preclinical, clinical phases I, II, and III. Many compounds are tested on various types of cancer cells but very few compounds enter in the actual clinical phase. A possible reason of this might be the cancerous cells in human body express very different physiologies that are reflected by pharmacodynamics and cell lines (*in vitro*) are unable to recognize many proteins, transcription factors, metabolites at a given time.

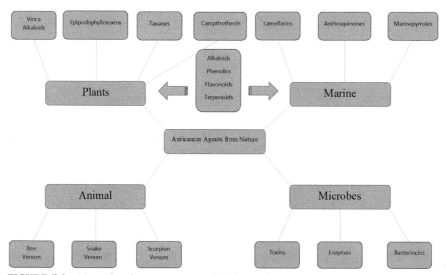

FIGURE 5.2   Natural anticancer agents and their main sources.

5.2 ANTICANCER AGENTS FROM PLANT (BOTANICAL/ TERRESTRIAL) SOURCES

The number of plant species in the world is expected to be approximately 374,000 which includes algae, liverworts, and eudicots (Christenhusz et al., 2016). Because of this wide range of diversity, plants are considered as important source for newly derived anticancer drugs. On average there are more than 100 botanical having anticancer or anti-infective activity are under active clinical trails(Harvey, 2008). Possible reason might be the natural products are generally less toxic than synthetic compounds and overall safer in a broad sense. According to WHO estimation more than 80% of various African and Asian countries rely on natural and traditional treatment techniques, learned from their ancestors hereditarily. This includes many varieties of spices such as turmeric as well. The global herbal market is expected to achieve $5 trillion worldwide by 2050 (Chaudhary et al., 2011). Mother nature is very selective and creates highly specific compounds. Many anticancer compounds are too complicated to synthesize, this means that the only way to get these compounds and use them for our benefit is through isolation and extraction techniques. These techniques are time consuming and sometimes very expensive. Regardless, many compounds were discovered, but the mother nature is too vast and diverse and many more are yet unknown. Commonly known anticancer compounds are categorized in four principal classes *viz.* Vinca alkaloids, epipodophyllotoxins, taxanes, and camptothecins (Desai et al., 2008).

5.2.1 VINCA ALKALOIDS

Anticancer activities of vinca alkaloids are well-known for the past many years. The first study detailing the anticancer activity of vinca alkaloids was reported in 1954 (The Canadian Medical Hall of Fame, 2018). Since then, extensive investigations have been conducted on these alkaloids to evaluate their cytotoxic effects (Gidding et al., 1999). The amazing cytotoxicity made some of the vinca alkaloids as chemotherapeutic drugs that are used extensively to treat a wide range of malignancies. Currently, vinblastine (VBL), vincristine (VCL), vinorelbine (VRL), and vindesine (VDS) remain the four most clinically used anticancer drugs. In Europe, VBL, VCL, VRL, VDS, and later a newer drug vinflunine are clinically approved anticancer drugs (Bennouna et al., 2008; Schutz et al., 2011). Vinca alkaloids bind to β-tubulin and consequently inhibit the microtubules formation in small concentrations. This operation is fast and reversible which occurs during the metaphase cell process. When the drug is administered in higher concentration, microtubule polymer mass is significantly reduced (Sisodiya, 2013) that in turn stops the development of a mitotic spindle. Vinca alkaloids, like many other anticancer drugs, also target healthy cells to some extent, to produce harmful side-effects. This is because microtubules are involved in many body processes. According to Correia et al. (2001), there is enough evidence to suggest two vinca alkaloid binding sites are present per mole of tubulin dimer. Studies conducted on vinca alkaloids (*in vitro*) have showed to inhibit malignant angiogenesis. Angiogenesis is the construction of new blood vessels and although this process is good for healthy cells, but angiogenesis caused by malignant cells provides a direct pathway for oxygen and nutrients to feed cancerous tumors. Nishida et al. determined that angiogenesis is regulated by an equal number of activators and inhibitors. However, for complete angiogenesis to occur a complete up-regulation of angiogenetic activators is not sufficient. An up-regulation of activators must be associated with a down-regulation of inhibitors to initiate angiogenesis (Nishida et al., 2006). In medical settings, vinca alkaloids specifically VBL, VCL, and VRL are being used extensively individually or with a combination of other chemotherapeutic agents. These are referred to as first-line chemotherapeutic regimens or drugs accepted by medical communities as the primary form of treatment for cancer types and stages.

The vinca alkaloid vinflunine (VFL) is a second-line drug that is frequently administered in treating metastatic bladder and urothelial cancer. It (VFL) received approval from the European Medicines Agency (EMA) in 2009 (Bellmunt et al., 2009). Up to now, the FDA has not approved any second-line chemotherapeutic drugs for patients who suffer from metastatic bladder cancer (Mamtani et al., 2014). In phase I clinical trials, VFL was administered to various patients to determine the drugs maximum tolerance and safest dosage (Bennouna et al., 2003). based on the results of phase one trails, it was determined that the safest dosage was 320 mg/m2 over 15–20 min in an interval of three weeks for most of the patients. In the following trials (phases II & III), VFL was tested to evaluate its effects on *cis*-platin and carboplatin-resistant (two organo-plainum drugs) urothelial cancer patients. VFL demonstrated promising results, and subsequently received approval and commercialized. VFL works similar to all vica alkaloids, it docks to tubulin at a specific binding site. This reduces microtubule networks through preventing microtubule polymerization. *In vitro* studies revealed induction of G2+M arrest, causing apoptosis to initiate through mitotic accumulation during the transition from metaphase to anaphase (Kruczynski et al., 2002; Pourroy et al., 2004). Further studies conducted by Etiévant et al. (2001) on P388 leukemia cells, concluded that VFL could successfully overcome drug resistance. Figure 5.3 provides stereo defined structures of some most common types of anticancer vinca alkaloids.

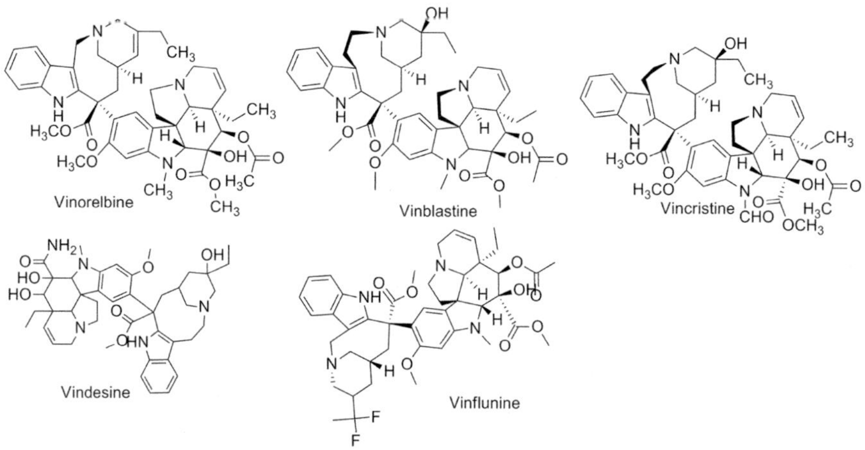

FIGURE 5.3   Anticancer vinca alkaloids.

5.2.2   PODOPHYLLOTOXINS

Podophyllotoxins are essential and naturally occurring compounds, found in the genus *Podophyllum*. Many podophyllotoxin derivatives have demonstrated anticancer potential. Out of these, etoposide, and teniposide are very significant anticancer agents. Etoposide (brand name Etopophos) in human forms a ternary complex with the topoisomerase II enzyme. Topoisomerases II is an enzyme present in cells to help in the DNA unwinding process. Because etoposide disrupts the activity of topoisomerase II. Consequently, DNA fails to rewind itself and ruins (Pommier et al., 2010). Etoposide is inserted intravenously (IV) and sometimes orally to treat testicular, glioblastoma, and certain lung cancers. To achieve maximum effect, etoposide is usually monitored with other drugs to produce synergistic effect (combined drug therapy). Teniposide (brand name Vumon), chemical name 4'-demethylepipodophyllotoxin, is a chemotherapeutic anticancer compound frequently executed to treat brain tumors, Hodgkin's lymphoma, and acute lymphocytic leukemia (ALL).

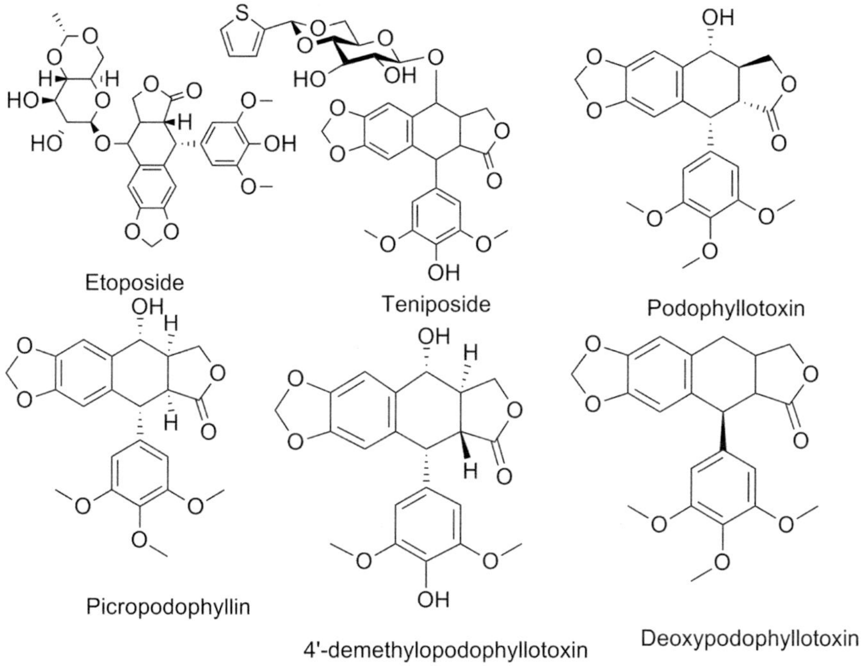

FIGURE 5.4 Structure and stereochemistry of frontline anticancer podophyllotoxins.

New studies have evaluated some other uncommon podophyllotoxin derivatives and determined comparatively low potency in various cancers. Few of these include picropodophyllin, 4'-demethylpodophyllotoxin, and deoxypodophyllotoxin (DPPT). Currently, a phase I clinical study is being conducted by Ekman et al. (2015) to evaluate picropodophyllin (AXL717) as a potential treatment for NSCLC. The patients treated with AXL717 demonstrated a decrease in regulation of IGF-1R, an important insulin-like *growth factor 1 gene*, on granulocytes. This indicates that picropodophyllin could potentially be used as a third-line form of treatment when it comes to NSCLC. In breast cancer *in vitro* research (Zilla et al., 2014), 4'-demethylpodophyllotoxin and few of its cognates demonstrated cytotoxic potential in several cell lines. The derivative 4'-demethylpodophyllotoxin itself demonstrated great potential on MCF-7 through nuclear condensation. Observations also revealed destruction of important micro-tubular proteins, protrusion of cellular membrane, and apoptosis. Apoptosis occurs due to docking of 4'-demethylpodophyllotoxin to kinase-2 checkpoint, an essential regulatory protein found in cells. By modifying the kinase-2-checkpoint G2/M arrest occurs causing a mitotic shutdown of cells to occur. DPPT is another significant podophyllotoxin derivative that induces apoptotic cell death in prostate cancer cell lines (DU-145). Hu et al. (2016) reported that DPPT, at low (50–100 nM) concentrations, induced apoptosis in a gradually increasing dose-dependent manner in DU-145 cell lines. Cellular viability was significantly reduced in DU-145 cells when treated with DPPT. Overall, elevation of casepase-3, p53, Bax, and PTEN proteins and a reduction of p-AKt proteins were observed, which are all key characteristics in apoptosis. Figure 5.4 provides chemical structures some common podophyllotoxins.

5.2.3 TAXANES

Taxanes are a class diterpenes belong to the family *Taxaceae*, specifically the Pacific yew. They have gotten much attention to the medical world because of two very common anticancer compounds: paclitaxel and docetaxel. Paclitaxel and docetaxel are administered in treating ovary, bladder, and esophagus carcinomas. Recent interest is to find out their effectiveness in treating early/very early stages of said cancers (Hagiwara et al., 2004; Rowinsky, 1997). The main mechanism of action that taxanes follow, is through the disruption of microtubule function. It works by stabilizing GDP-bound tubulin in cells, causing depolymerization to occur, which ultimately inhibits the completion of cell division (Nogales et al., 2006). Although very useful and promising anticancer agents, they bind to multidrug-resistance proteins (MDR), specifically P-glycoprotein (Pgp). This causes limitations on their effectiveness in treating some cancers efficiently (Goldstein, 1996). In 2010, a new semi-synthetic taxane, cabazitaxel, received approval from the FDA (Sebastian de Bono et al., 2010). Cabazitaxel has a low affinity to P-glycoprotein due to its methoxy groups located at C7 and C10. Preclinical studies conducted on docetaxel-effective cell lines HL60 (promyelocytic leukemia), Calc18 (breast carcinoma), P388 (lymphoblastic leykemia), and KB (cervical adenocarcinoma) determined cabazitaxel effectively promoted cytotoxicity on these cell lines (Bissery et al., 2000). Because of this reason, it is considered the drug of choice when tumors have become resistant to docetaxel (Kartner et al., 1983). Cabazitaxel contains a rather important and unique ability, it can freely penetrate the blood-brain barrier (BBB)

(Abidi, 2013). The BBB serves as a strainer, it prevents many substances from entering to brain. Clinical studies have not conducted to determine its interaction on BBB.

There are various other taxanes currently under clinical evaluation. Milataxel (TL139, phase II), larotaxel (XRP9881, phase III), ortataxel (IDN5109, phase II), and tesetaxel (DJ-927, phase II) are undergoing clinical trials (Figure 5.5). Phase II trials of milataxel on patients still suffering from previously treated advanced colorectal cancer was done. The clinical phase ended with the conclusion that no clinical activity was demonstrated (Ramanathan et al., 2007). Regardless, patients that are suffering from recurring or advance malignant mesothelioma could participate in another milataxel clinical trial to evaluate its response rate, effectiveness, and selectivity when taken orally by patients with malignancy mesothelioma (An Efficacy Study of Milataxel (TL139) Administered Orally for Malignant Mesothelioma (TL139), 2008). Larotaxel recently completed phase III clinical trials and its investigation to determine its response on advanced pancreatic cancer, when introduced intravenously along with 5-fluoruracil and capecitabine, has shown promising results (Larotaxel Compared to Continuous Administration of 5-FU in Advanced Pancreatic Cancer Patients Previously Treated with A Gemcitabine-Containing Regimen (PAPRIKA), 2016). Additional phase III clinical trial research conducted by Stemberg et al. (2013), on larotaxel with cisplatin for advanced bladder cancer and advanced metastatic urothelial tract cancer failed to demonstrate a higher rate of improvement when compared to cisplatin and gemcitabine (Sternberg et al., 2013). As of 2016, ortataxel has completed its phase II clinical trials on patients that are diagnosed with recurring glioblastoma. Current investigations are determining ortataxel's ability to treat lung cancer, lymphoma, kidney cancer, and glioblastomas (Ortataxel, 2018; An Efficacy Study of Ortataxel in Recurrent Glioblastoma (Ortataxel), 2016). In phase II trials tesetaxel failed to deliver promising results against metastatic colon cancer. Moreover, clinical trials have been performed to evaluate testaxel's efficiency to treat other solid tumors (Beeram et al., 2010; Moore et al., 2006; Roche et al., 2006). Additional research on tesetaxel was conducted against advanced-stage breast cancers. Tesetaxel was administered orally to evaluate its safety and therapeutic effect as a first-line therapeutic drug (Tesetaxel as First-line Therapy for Metastatic Breast Cancer, 2012). Notably, taxanes are excellent mitotic poisons that are very useful against many kinds of cancers. Detailed in-depth investigation is required to evaluate the cytotoxicity of these valuable taxanes.

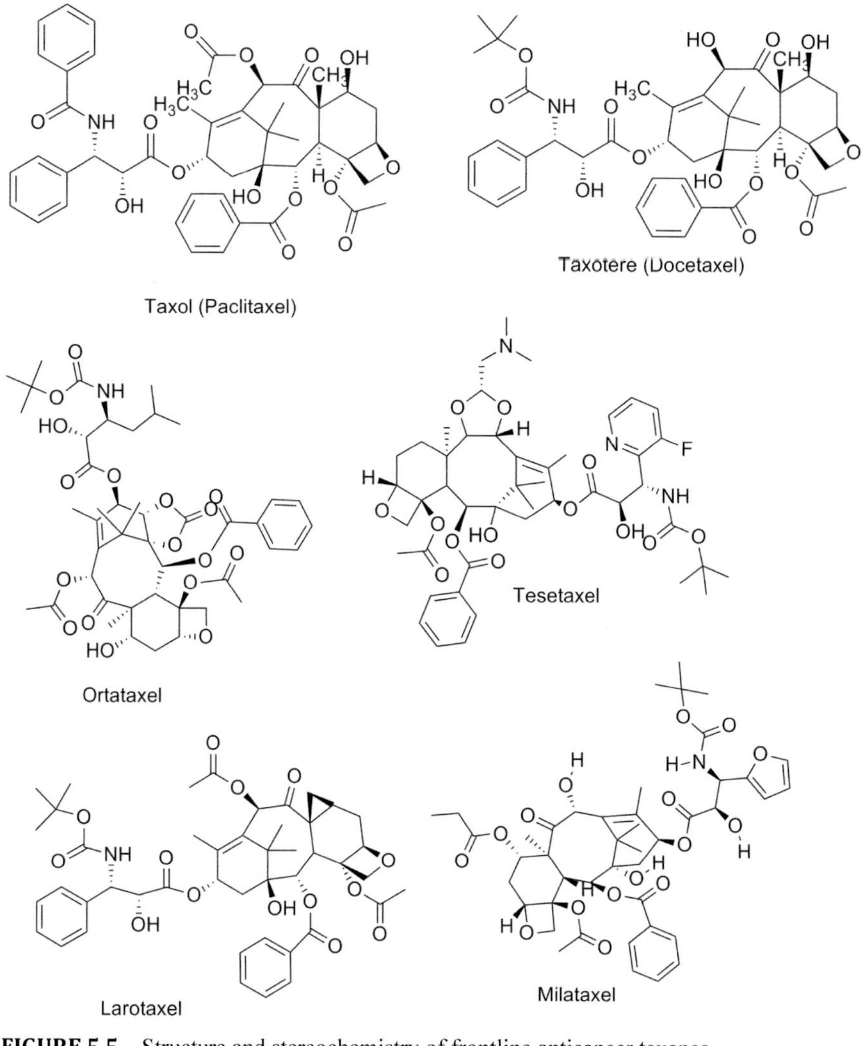

FIGURE 5.5 Structure and stereochemistry of frontline anticancer taxanes.

5.2.4 CAMPTHOTHECINS

Camptothecin (CPT) was first discovered in 1966 by M. E. Wall and M. C. Wani during their investigation of the bark and stem of the *Camptotheca acuminate* tree. CPT is a topoisomerase I (Topo-I) inhibitor as it can dock the enzyme Topo-I resulted in DNA re-ligation, subsequently DNA damage and ultimately apoptosis (NCI Dictionary of Cancer Terms,

2018). The *Camptotheca acuminate* tree is native in Asia (China) and people have been using this plant as folklore medicine for centuries (Basili et al., 2009). CPT works by targeting and inhibiting the nuclear enzyme DNA topoisomerase I (Camptothecin, 2018). Since CPT experiments concluded that CPT alone contains a low solubility (bioavailability), their effectiveness against cancer cells would are considered inefficient. Because of this reason, CPT analogs have been synthesized and tested against cancerous cells. CPT analogs such as topotecan and irinotecan, have demonstrated promising results against leukemia and many forms of solid tumors (Liu et al., 2015). Topotecan was approved as a second-line treatment for metastatic ovarian cancer in over 70 countries (Armstrong et al., 2005). Patients are usually treated with topotecan after failing first-line treatments. Since its discovery, many trials have been tested on topotecan. For instance, in phase II trials by Ardizzoni et al. (2005), topotecan was combined with *cis*-platin to quantify the synergistic therapeutic activity on patients with refractory SCLC (Ardizzoni et al., 2005). The results demonstrated that 62% of the patients developed myelosuppression, a decrease in bone marrow activity. Thus, five deaths were closely related to the treatment of topotecan and cisplatin. Regardless, the study concluded that topotecan and the *cis*-platin combination could be effective in refractory SCLC (Ardizzoni et al., 2005).

Although topotecan is an excellent anticancer compound, some clinical trials have presented low-yielding results. For example, Stathopoulos et al. (2005) evaluated in a (phase II) clinical trial the effectiveness of topotecan with paclitaxel on patients having recurrent SCLC. As in the trial mentioned before, myelosuppression was the main side effect associated with topotecan. The clinical trial only produced a 26.8% response rate on the 45 patients it was tested on (Stathopoulos et al., 2005). Additional research is required to determine its effectiveness on other cancer types. Another promising CPT analog is irinotecan, it is used to treat metastatic colon and rectal cancers. In another trial (phase I), conducted by Gilbert et al. (2003), to evaluate irinotecan effectiveness and maxima tolerated dose (MTD) in treating adults with recurrent malignant glioma and the MTD of irinotecan was 3.5 times greater in patients that were also taking enzyme-inducing anticonvulsants (EIAs). This opened the pathway for more clinical research on irinotecan for other types of brain cancers.

Three other campotothecin derivatives contain anticancer activities which include belotecan, lurtotecan, and exatecan. Belotecan has been tested by Choi et al. (2010) in a phase II trialalong with carboplatin (combination therapy) to assess its anticancer effect on patients that suffer from recurrent ovarian cancer. The clinical trial included 38 patients and became successful with an overall response rate of 57.1% and comprehensive response rate of 20%. Lurtotecan is highly water-soluble and due to this most research conducted on it focused on delivering the drug through liposomal formulation, NX211. Studies (Desjardins et al., 2001) on mice-bearing KB cancer cells determined that lurtotecan delivered in NX211 could improve the chances of drug-to-tumor contract. Exatecan, another analog of camptothecin, has a useful anticancer effect on advanced peritoneal, ovarian, and tubal cancer when novel anticancer agents such as topotecan, taxane, and *cis*-platin failed (Verschraegen et al., 2004). The phase II clinical study demonstrated its effectiveness when patients administered a daily dose in 5 times a day continuously every 3 weeks. Figure 5.6 provides the structure of anticancer campotothecin. Anticancer effect of a few significant natural and semi-synthetic drugs is summarized in Table 5.1.

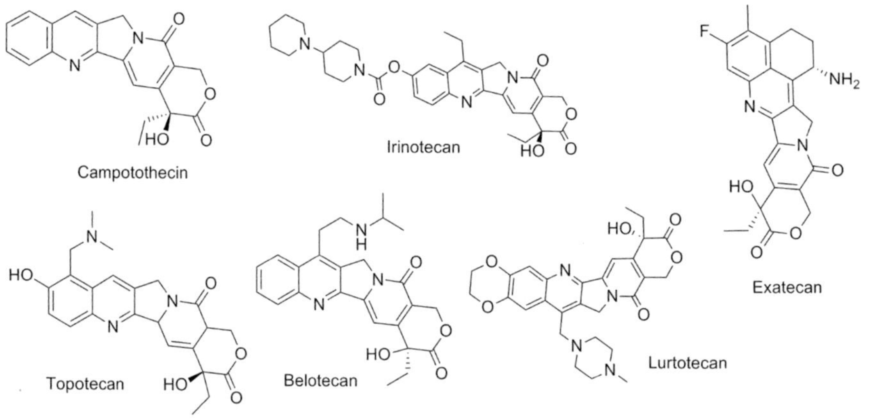

FIGURE 5.6   Structure and stereochemistry of anticancer campotothecins.

5.3   ANTICANCER AGENTS FROM MARINE ORIGINS

When it comes to discovering new anticancer agents, many people think of exploring the terrestrial plant kingdom only. Interestingly, the marine world is a gigantic and least explored area that contains enormous biodiversity. Most studies are conducted on terrestrial plants because such specimen is much easier to collect and harvest.

TABLE 5.1 Common Compounds from Plants that Contain Anticancer Properties

Compound	Species	Cancer Type	Type of Anticancer evaluation	Mechanistic Route	References
Vinblastine	*Catharanthus roseus*	Lung cancer (NSCLC)	*In vivo* human model	Tumor reduction. βIVb-tubulin regulation.	(Bennouna, et al., 2008; Correia et al., 2001; Gidding et al., 1999; Kruczynski et al., 2002)
Vincristine					
Vinorelbine		Breast cancer	*In vitro* human cell lines	Inhibition of microtubule formation	
Vindesine		Pancreatic cancer			
Vinflunine					
Podophyllotoxins (derivatives)	*Odophyllum peltatum*	Cervical cancer	*In vivo* murine models	Disruption of Topoisomerases II process	(Ekman et al., 2015; Hu et al., 2016; Zilla et al., 2014)
Picropodophyllin		Ovarian cancer	*In vitro* human cell lines		
4'demethylopodophyllotoxin		Lung cancer			
Deoxypodophyllotoxin		Prostate cancer			
Etoposide		Leukemia			
Teniposide					

TABLE 5.1 (Continued)

Compound	Species	Cancer Type	Type of Anticancer evaluation	Mechanistic Route	References
Paclitaxel	*Taxus brevifolia*	Myeloma	*In vivo* murine models	Apoptosis through caspase-3 activation.	(Abidi, 2013; Bissery et al., 2000; Goldstein, 1996; Hagiwara et al., 2004)
Docetaxel		Glioblastoma			
Cabazitaxel		Lung carcinoma	*In vitro* human cell lines	Tumor reduction.	
Milataxel		Colorectal cancer		Inhibition of cell growth.	
Larotaxel		Leukemia			
Ortataxel		Ovarian		Alternation of genes.	
Tesetaxel		Prostate cancer		Disruption of microtubule function.	
Camptothecin (derivatives)	*Camptotheca acuminata*	Colorectal	*In vivo* murine xenografts	Topoisomerase I inhibitor	(Ardizzoni et al., 2005; Choi et al., 2010; Desjardins et al., 2001)
Topotecan		Lung cancer (NSCLC)	*In vitro* human cell lines		
Irinotecan		Ovarian	*in vivo human models*		
Belotecan		Gliomas			
Lurtotecan		pancreatic			
Exatecan		leukemia			
		lymphoma			

Collecting marine compounds are more challenging to find and it is not easy to culture the specimen. Regardless of the challenges that presented, many previously unknown pharmaceuticals have been continuously reporting and many are yet to be discovered. Approximately 230,000 to nearly 250,000 species of organisms from oceans have been reported till now (Montaser et al., 2011). National Cancer Institute (NCI) reported that comparative preclinical screening on marine and terrestrial samples showed that 1% of the screened marine samples contain anti-tumor potential whereas only 0.1% of terrestrial samples demonstrated antineoplastic effect. Alternatively stated, probability of finding unrecognized anticancer agents in marine sources is about 10 times higher than terrestrial plant sources (Munro et al., 1999). One of the initially approved (in Europe) marine antitumor drugs is trabectedin (ET-743), found in the marine tunicate *Ecteinascidia turbinate*. The U.S. Food and Drug Administration (FDA) approved trabectedin for liposarcoma and leiomyosarcoma (trade name Yondelis ®; FDA Approves Trabectedin to Treat Two Types of Soft Tissue Sarcoma, 2015). Trabectedin damages the DNA by docking to the transcription factor FUS-CHOP in myxoid lipo-sarcoma. Trabectedin can also reverse the transcription program and cause apoptosis in myxoid liposarcoma (Grohar et al., 2011). Another famous marine-derived anticancer agent is halichondrin B, first found in the marine sponge *Halichondria okadai* in 1986. It chemotherapeutic active became known in 1992, when it demonstrated high cytotoxicity towards murine leukemia cells (Halichondrin B (NSC 609395) E7389 (NSC 707389), 1992). In 1992, the total chemical synthesis of Halichondrin B was reported (Aicher et al., 1992). This achievement led the formulation of eribulin mesylate (Halaven), a synthetic equivalent of halichondrin B, which was approved by FDA in November 2010 to treat advanced breast cancer (Huyck et al., 2001). Eribulin works by uniquely inhibiting the dynamics of microtubules. It binds to high-affinity sites located at the terminal of the microtubules, causing apoptosis to occur. This chapter does not discuss all the marine-derived anticancer agents instead highly significant anticancer agents are discussed.

5.3.1 LAMELLARINS

Lamellarins are class of marine alkaloids, first isolated from sea snails, mollusk *Lamellaria*, in 1985. Since their discovery, over 50 different types of laminarins have been identified. Their anticancer activity widely diverse, i.e., these compounds are active against different types of malignancies. Perhaps the most explored and well-studied lamellarin is lamellarin D (LAM-D) and

O (LAM-O). A study conducted by Facompre et al. (2003) on LAM-D and its nonplanar analog, lamellarin 501 (LAM-501), identified these molecules as outstanding Topo-I inhibitors. Using human leukemia cells, particularly CEM, CEMC2, P388, and P388CPT5, the mechanistic studies of LAM-D were conducted. The cell lines that were heavily affected by LAM-D & LAM-501 were exteinascidin-743 resistant ovarian cells (IGROV and IGROV-ET), melanoma (MEL-28), pancreatic (PANC-1), breast (SK-BR3), and lung (A-549) cancers. Extremely promising outcomes were shown in prostate cancer(LNCap) cells, but only LAM-D expressed high inhibitor growth but LAM-501 did not. LAM-D has a is similar effect as camptothecin (CPT). They both are Topo-I inhibitors. Initially it was expected that LAM-D alone could provide a higher inhibitory effect. This was not the case, when it was connected to the stabilization of Topo-I—DNA complex, LAM-D was about 5 times weaker than CPT. Regardless, LAM-D stabilizes Topo-I—DNA complex through a different and unique site. Many researchers have also conducted studies on other types of lamellarin-derivatives that belong to the lamellarin series.

Lamellarin O (LAM-O) is a pyrrole alkaloid, found in the Australian marine sponge, *Lanthella sp*, that is shown to express inhibition of ATP binding cassette (ABC) transporters and cytotoxicity. Haung et al. (2014) reported the antineoplastic activity of LAM-O on colon metastasized from lymph nodes SW-620 and SW-620 Ad300 and BCRP transporters. LAM-O expressed mild (low) cytotoxicity toward Sw-620 and SW-620 Ad300 having IC_{50} values of 22.0 μM and 22.3 μM, respectively. Using Calcein AM assay, LAM-O was found as a P-gp (an ATP transporter) inhibitor. Doxorubicin (*aka* adriamycin) is a frontline anticancer drug with diverse antineoplastic activities. Although very effective, doxorubicin becomes resistant in some cancers. The resistance arises from specific ECM proteins which usually bind to different cells, specifically pro-survival cells (Lovitt et al., 2018). In Huang's study, LAM-O had the potential to reverse survival cells, specifically P-gp mediated doxorubicin resistance in both cell lines. Furthermore, there was an inhibition of BCRP after LAM-O was used in a study.

5.3.2 *MARINOPYRROLES*

Marinopyrroles are marine-based compounds that were first discovered in California, shortly after the discovery of a new marine Streptomyces bacterium. The compound of interest is marinopyrrole A, which demonstrates promising antibacterial and anticancer potential. The marinopyrrole family is comprised of various substitutions around a characteristic bipyrrole struc-ture (Hughes et al., 2010). Actinomycete strain CNQ-418, a relatively and recently recognized new extant *Streptomyces* species was collected from a sediment sample. This sample was confirmed to possess 98.1% similarity with *S. sannurensis* by 16SrRNAgene sequencing methods. Upon culturing this variant, the extract demonstrated possible antibacterial, and later anti-cancer activity (Hughes et al., 2008). This began the interest in observing this strain to isolate biologically active novel pharmaceuticals with multi-dimensional medicinal properties. This discovery led to few first isolation and structural determination analysis of marinopyrroles A ($C_{22}H_{12}Cl_4N_2O_4$) and B ($C_{22}H_{11}BrCl_4N_2O_4$), and recognition of their likely bactericidal action on methicillin-resistant *Staphylococcus aureus* (MRSA) (Cheng et al., 2013). The seawater provided these compounds were essential for growth of this microbe and is likely the halogen source for these bacterial to produce marinopyrroles. CNQ-418 possessed an unusual 1,3'-bipyrrole core attached to two salicyloyl substituent's with marked antibiotic activity directed at MRSA. Good conformational stability at room temperature was noticed in the highly clustered *N, C*-biaryl bond which establishes an important axis of chirality for marinopyrroles A-E; natural product marinopyrroles are also strictly in an M-configuration (Yamanaka et al., 2012). Generation of these metabolites is possible synthetically through the Paal-Knorr pyrrole conden-sation technique (Yamanaka et al., 2012), which was used to synthesize 1,2'-bypyrrole cores, which is the backbone of marinopyrroles. Halogena-tion of this precursor was treated with *N*-bromosuccinimide to achieve the characteristic 4,4', 5', 5'-tetrahalogenation, seen in metabolites of marine organisms. The structure of these natural product compounds includes pres-ence of multiple halogens and of chiral centers at specific points, with a distinctive *bis*-pyrrole backbone. The marinopyrrole B has received further support to be an atropo-enantiomer with an M-configuration based on X-Ray analysis (Yamanaka et al., 2012). While this confirmation is more stable at room temperature, the M-(±)-marinopyrole A when subjected to elevated temperatures it could racemized to produce non-natural P-(+)-atropo-enan-tiomer (Yamanaka et al., 2012). This provides excellent antibacterial and anticancer activity (Haste et al., 2015). It was observed that a majority of the MRSA-active compounds demonstrated high cytotoxicity toward HCT-116, human colorectal malignant cells, exhibiting IC_{50} values between 1 to 5 µM; this finding inferred plausible effectiveness to be an anticancer agent as more ideal route of investigation.

The Bcl-2 proteins are a group of apoptosis suppressor genes, which include Bcl-2, Bcl-XL, Bim, and mcl-1 (Hata et al., 2015). These proteins serve to prolong the cellular life cycle, during a cancerous mutation however, overexpression of these genes may up survival of the metastatic cell due the cell cycle arrest at the G_0 phase. Prior studies have identified these genes as viable drug targets. Mérino et al. (2012) reported that ABT-737 &ABT-263 (navitoclax), are highly effectual inhibitors of Bcl-2 proteins; however, these drugs exhibit lower affinity to Mcl-1 protein which prevents apoptosis (Mérino et al., 2012). Thus, further research to search for a more selective agent of which can be more combatant in cancer types with elevated Mcl-1 levels is preferred. When cancer forms, oncogenes or tumor suppressor genes may specify for proteins that upregulate the survival of a mutant cell. This ultimately leads to cancer, in some scenarios the malignant cell is capable in forming vasculature to improve blood flow to itself and even perhaps use the body's own circulatory organ as a vector to spread (Science Daily, 2018). Therapeutic treatments with marinopyrroles are required for the initiation of apoptosis and counter these pro-survival proteins. The cellular process of apoptosis is caspase-dependent, which are cysteine proteases; their activation is facilitated through two pathways, extrinsic (Ashkenazi, 2015) or intrinsic (Elumalai et al., 2012). The extrinsic pathway is permitted by a death receptor, whereas the intrinsic pathway is derived from mitochondrial influence. Upon activation, caspases exert their effects *via* proteolytic cleavage; the resulting occurrences being DNA destruction and blebbing of the cellular membrane, both of which are characteristics of apoptosis in either circumstance of activation. Bcl-2 proteins are highly involved with the intrinsic mode of apoptotic activation by having regulatory possessions on the mitochondrial outer membrane permeabilization (MOMP); subsequently cytochrome c and Smac are released from the mitochondrial cytosol (Shamas-Din et al., 2013). The apoptogenic factors upon release increase the activation of caspases, inducing cell death.

Still, there is enough deficiency of small molecules, capable of inhibiting Mcl-1. In prior studies, marinopyrrole A has demonstrated selective binding to Mcl-1; the properties of this agent show activation of Mcl-1 dependent apoptosis; though not for Bcl-2 and Bcl-XL dependent, and decreased Mcl-1 protein levels attributed to proteasomal degradation. Results were similar in two separate studies dealing with different cell variants, leukemia, and melanoma cancer cells. However, further reports indicated that marinopyrrole A does not alter the Mcl-1 expression; follow up studies stated that marinopyrrole A itself does not cause the observed proteasomal degradation does not play any role in altering the Mcl-1 expression, either in observed treatments. Hughes et al. (2008) revealed actin as the main target for marinopyrrole A, not the expected Mcl-1 in HCT-116 cells. Their finding sparked interest in further selectivity of marinopyrrole A. It was reported that (Eichhorn et al., 2013) initial model centered on observing cellular consciousness to marinopyrrole A directed at Mcl-1 dependent HeLa cells with RS4;11 leukemia cells (human) for comparison. The reasoning for the latter was there have been prior indicative of RS4;11 being dependent on Blc-2 proteins for cell persistence. It was 72 h assessment of samples with varying concentrations (0.1-100 mM) of natural marinopyrrole A with interest being on expression. In brief, their experimental design showed that marinopyrrole A was more effective in combating the Bcl-2 dependent RS2;11 cells (IC_{50}: 2mM); in contrast Mcl-1 dependent HeLa cells exhibited IC_{50}: 20mM. Their immune blotting tests also specified that the Mcl-1 expression was unchanged in marinopyrrole A administration. Their study further analyzed if there would be any observed selective cytotoxicity exhibited by natural marinopyrrole A against two murine leukemia cell lines, with differing apoptotic dependencies, Mcl-1 versus Bcl-2. In brief, it concluded that marinopyrrole A had not much preference for either within the same cell line, demonstrating comparable effectiveness against either; Mcl-1 dependent leukemia cells (IC_{50}: 2.5 mM) and Bcl-2 dependent leukemia cells (IC_{50}: 2 mM). The activity of the Bcl-2 (B-cell lymphoma-2) gene family are act as apoptosis suppressors. Overexpression of these proteins may block or delay apoptosis, necessary to inhibit cell proliferation through terminating the cell cycle at the G_0 phase (Akl et al., 2014). Mcl-1 (myeloid cell leukemia) overexpression is common in many tumor types and, as other proteins in its family, favors cell survival; it is capable of granting further resistance to chemotherapeutic efforts and thus it is representing an ideal cancer treatment target (Ertel et al., 2013).

5.3.3 ANTHRAQUINONES

Anthraquinones are naturally occurring aromatic compounds, found in certain plant families *viz. Fabaceae, Liliaceae, Polygonaceae,* and *Rhamnaceae.* Their presence are seen in marine-derived fungi, for example, *Microsporum* sp. The major use of anthraquinones is as color pigment and they are used commercially to prepare natural dyes. Anthraquinones affect the cell cycle of malignant cells by disrupting it. Cancer cells may follow diverse pathways to avoid the influence of antineoplastic medicines which eventually may result in DNA damage. To make sure that DNA is not damaged, checkpoint proteins are able to control pathways by manipulating cell cycle. Tumorigenesis occurs when the tumor is formed because the cells lose checkpoint controls, this is the main reason why chemotherapy is required to regulate the cell cycle. Aside from being used commercially, anthraquinones are used extensively in pharma industries. For instance, *Rheum palmatum* or Rhubarb, a good source of anthraquinones, is still being used today. Traditionally, Rhubarb is used as a phytomedicine to treat constipation, jaundice, and some ulcers. But studies conducted on Rhubarb determined that its anthraquinones are able to prevent the proliferation of numerous cancer cells.

In rhubarb, the main anticancer anthraquinones were emodin, aloe-emodin, rhein, and chrysophanol (Figure 5.7). Early studies conducted on emodin showed that it could prevent cell proliferation in breast, cervical, colon, and prostate cancers (Chang et al., 1999; Pecere et al., 2000; Ren et al., 2018). Also, emodin has little or no cytotoxic effect in normal cells, suggesting that normal cells are comparatively safe than cancer cells when it comes to emodin-induced cytotoxicity. Aloe-emodin could inhibit the cell growth in many malignant tumors like hepatoma (liver) (Cha et al., 2005), human lung carcinoma (Chan et al., 1993; Jeon et al., 2012) and leukemia (Yeh et al., 2003). Jeon et al. (2012) treated hepatoma (HUH-7) cells with a dose-dependent concentration of aloe-emodin. He concluded that aloe-emodin could decrease CAPN2 and UBE3A, two essential proteins, which reduces cell growth and speeds up apoptosis. Since the mechanistic routes of aloe-emodin is unknown in H640 (lung cancer) cells, Yeh et al. (2003) conducted a study to evaluate the cytotoxicity of aloe-emodin in H640 cells. When introduced, the initial observation was apoptosis due to the modification of cAMP-dependent protein kinase. Other important protein expressions that were modified were BCL-3, protein kinase C, caspase-3, and p38.

The study identified the protein, responsible for apoptosis by aloe-emodin, was p38. Lee et al. (2001) conducted a study to investigate to evaluate the impact of aloe-emodin on the pharmacodynamics of CH27 (human lung squamous cell carcinoma) cells. The reported mechanism was like that of Yeh et al. CH27 experienced apoptosis through alteration of expression of the proteins BCLX(L), Bak, & Bag-1. These modifications are due to the translocation of Bax and Bak from the previous cytosolic to a particulate fraction. This could potentially cause an activation of capase-3, -8 and -9, which could conclude that apoptosis was initiated through the Bax pathway. In a recent report (Ren et al., 2018) showed chrysophanol-induced apoptosis in MCF-7 and MDA-MB-231 cells. Cell growth was inhibited by chrysophanol's potential to inhibit the NF-κB activity. The protein which was down-regulated was Bcl-2, and the crucial transcription factors p65 and IκB phosphorylation. An increase in paclitaxel potential (synergy) was observed with chrysophanol in both the cell lines. Moreover, the effects of chrysophanol-induced apoptosis could be notably reduced by NF-κB inhibition with ammonium pyrrolidine dithiocarbonate.

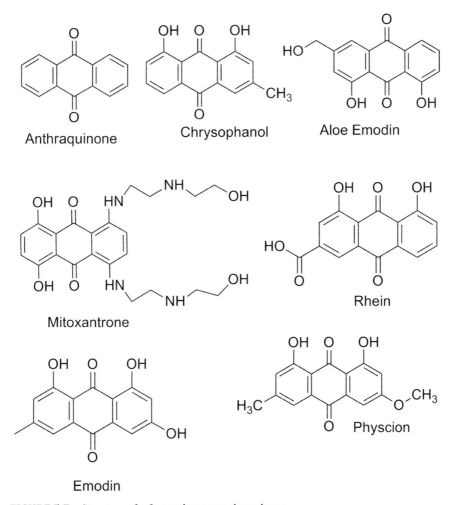

FIGURE 5.7 Structure of a few anticancer anthraquinones.

5.4   ANTICANCER AGENTS FROM MICROBES

The total number of unique microbial species on earth is estimated to be well over 1 trillion. Currently only about 0.01% are known. The small overall percentage has provided us great benefit in fighting with a few deadly diseases like cancer. Microbial organisms have three domains Eukarya, Prokarya, and Archaea. The eukaryotic and prokaryotic domains serve a rich reservoir filled with bioactive molecules waiting to be discovered (Chen et al., 2004). It might be a common curiosity, if so many microbial species exist, then why is there very little research conducted on microbes? The answer is the extraction and evaluation of anticancer properties of the microbes-originated compounds are significantly expensive. But few frontline drugs, for example rapamycin, have been isolated from this natural source. Rapamycin was reported in 1975 as an antifungal agent. When intense research was conducted on rapamycin, its anticancer potential was revealed which was previously unknown. This prompted an era and shift into discovering new anticancer agents in microbial species. Consequently, many anticancer drugs originated in microbial sources were validated. Temsirolimus (Torisel®) and ridaforolimus were chemically derived from rapamycin. Torisel® is administered to treat renal cell carcinoma (RCC) intravenously and works by inhibiting mTOR in cancerous cells (Ren et al., 2018). Protein mTOR plays a crucial role in normal cell development and division. Temsirolimus also inhibits angiogenesis by reducing the proliferation of vascular endothelial growth factor (VEGF) signaling (Giddings et al., 2013; Torisel (temsirolimus) injection, 2018; Ren et al., 2018). *Toxoplasma gondii*, a dangerous protozoon that causes the disease toxoplasmosis, has recently been known as anticancer antigens. At first, it was believed that the antigens were only useful to treat neurodegenerative diseases. Research concluded that lung, pancreatic, and ovarian cancer could be treated with T. *gondii*'s uracil auxotrophic carbamoyl phosphate synthase enzyme. This section deals with toxins, enzymes, proteins, and antibiotics, derived from microbes, that contain anticancer properties.

5.4.1   *TOXINS*

Bacteria-originated toxins are toxic substances that bacteria secrete to attack other bacteria or host cells. These toxins are classified into two simple forms: endotoxins and exotoxins. Endotoxins are toxic substances that bacteria secrete outside their cell wall. Endotoxins are toxic substances, usually lipids, that are located within the cell (Bacterial Toxins, 2018). Regardless of the mechanism used, damage to the host is negligible *viz.* toxins are often considered non-invasive and they cause either very limited damage or no damage on their host. A primary advantage of bacterial toxins is that they can attack the host from a distance, requiring them to recognize receptors to bind too. Most toxins produced by bacteria come as proteins that are encoded with the bacterial genes (Wan et al., 2006). Based on bacterial strain and the characteristics of the toxins, they may cause fever, hypotension, shock, hypoglycemia, and ultimately death. Some toxins have been investigated to validate their anticancer activity (Henkel et al., 2010). The few most studied toxins include diphtheria, cytolethal distending, and exotoxin a toxins.

Diphtheria toxins (DT) are secreted by the *Corynebacterium dipheriae*, a gram-positive bacterium that is known to cause an infectious disease known as diphtheria. This toxic substance is encoded by a *tox* gene, which is caused by a bacteriophage, that lives inside certain bacteria. To produce DT, a *C.diphtheriae* bacterium needs to contain the *tox+* phages. Recent studies have evaluated mutated DT to validate its anticancer activity. *In vivo* evaluation was reported (Lubran et al., 1988)to validate the effects of CRM197, a mutated DT carrier protein that is often used for vaccines, on human adrenocortical carcinoma. CRM197 could bind to heparin-binding EGF-like growth factor (HB-EGF), which is common in adrenocortical carcinomas. By binding to HB-EGF, CRM197 reduced angiogenesis and initiated apoptosis. Notably, CRM197 also inhibited cancerous cell migration (Lubran, 1988) that prevented metastasis. Another study (Zhang et al., 2010) carried out by using a DT mutant DT385, *in vivo* on chick chloroallantoic membrane (CAM), Lewis lung cancer (LLC) mouse model, and 18 different cancer cells. DT385 inhibited angiogenesis in CAM but were resistant to endothelial cells. In LLC cells, the subcutaneous growth was slightly inhibited. The tumor size reduced meaningfully. DT385 showed mild to strong sensitivity in 15 out of 18 tested cancer cell types. Human malignant cell lines U-87 MG (glioma), HEK293 (kidney), Hela (cervical), Calu-3 (lung), U251 (glioblastoma), and 293T (lung) showed high sensitivity to DT385. Moderate sensitivity was noted in MDA-MB-231 (breast), Colo201 (colon), Colo205 (colon), PC-3 (prostate), HT1080 (fibrosarcoma), and LNCap (prostate). Weakly sensitive cell lines were MCF7 (breast), HCT116 (colon/rectum), Hep3 (cervical), NB4 (promyelocytic leukemia), HL-60 (myeloid leukemia), and BT-20 (breast) (Zhang et al., 2010). DT385 prevents cell proliferation by inducing apoptosis.

Cytolethal distending toxins (CDTs) are a type of heterotrimeric toxins, produced by specific gram-negative mucocutaneous bacteria (Jinadasa et al., 2011). CDTs are AB-type toxins with DNase potential, which allows them to penetrate the targeted cell's DNA (Guerra et al., 2011). They are primarily composed of cdtA, cdtB, and cdtC subunits. Because of this reason, it has been studied constantly to fully understand the mechanistic pathways in cancer cells. Bachran et al. (2014) extracted CdtB from *Haemophilus ducreyi* and fused it to the N-terminal 255 amino acids of *Bacillus anthracis* toxin lethal factor (LFn) to determine its (LFnCdtB) anti-malignant character. LFnCdtB showed its ability to inhibit proliferation of several human cancerous cells by arresting cell cycle in the G2/M phase, followed by apoptosis. In LLC model, the cytotoxicity was moderate to low. Further study on A549 adenocarcinoma was reported by Yaghoobi et al. (2016). A combined formulation of pcDNA3.1, an expression vector and cdtB derived from Aggregatibacter actinomycetem-comitans was applied in A549 cells. Apoptosis was noted through caspase-9 activation. Cells evaluated with pcDNA3.1 alone demonstrated 16.5% cell death whereas the cells evaluated with pcDNA3.1/cdtB demonstrated about four-fold higher (63.4%) rate of cell death through apoptosis. Morphological changes were also observed by chromatin condensation, which changed the shape of cancer cells. Growth and proliferation were notably inhibited in a time-dependent manner when pcDNA3.1/cdtB combination was used.

Exotoxin A (ETA) is an exotoxin produced by *Pseudomonas aeruginosa*, a common gram-negative bacterium that is present in soil and water world-wide (WebMD, 2018). ETA works by catalyzing the ADP-ribosyltransferase of eukaryotic elongation factor-2 (eEF2) protein. This prevents protein synthesis on the targeted host (Yates et al., 2004). Because of this, recent studies focused to validate the anticancer potential of ETA on cancerous cells. Ellazeik et al. (2013) successfully enhanced ETA's toxicity through a mutation of the three wild-type isolates. *In vitro* breast carcinoma cells MCF-7 was treated with ETA and growth inhibition was noted at a relatively higher concentration. ETA prevented protein synthesis in MCF-7 cells. To further validate the efficacy of ETA on other malignant cells lines, Goldufsky et al. (2015) treated B16, HeLa, Calu-3, and EMT6 among others in diverse concentrations of ETA. The study revealed a higher concentration of lactate dehydrogenase (LDH), indicating a high level of cytotoxicity. The same research group conducted *in vivo* experiment using mice transplanted with B16 melanoma cells. On treatment with ETA, induction of cytotoxicity and tumor growth delay were observed. This determined the high efficacy of ETA as an anticancer agent.

5.4.2 ENZYMES

There are 20 essential amino acids that considered important and necessary for a healthy life. The human body can fully synthesize in enough quantities the following four essential amino acids: asparagine, glutamic acid, alanine, and aspartic acid, regardless if consumed in diet. The body cannot synthesize phenylalanine, threonine, valine, tryptophan, methionine, leucine, isoleucine, histidine, and lysine, which means they will need to be consumed through diet. The conditional essential amino acids are those that are synthesized under certain conditions like illness, stress, etc. These amino acids include arginine, cysteine, glutamine, tyrosine, glycine, ornithine, proline, and serine (Amino Acids, 2018). These amino acids serve as building blocks for several essential proteins and enzymes. Bacterial enzymes are important biocatalysts used by living organisms to catalyze their chemical reactions. Enzymes are made up of amino acids, which contain an amine (-NH2) and carboxyl (-COOH) group, along with a functional and R group. The human body contains thousands of different enzymes that are specifically tailored for specific reactions. Subsequently, many scientists focused on enzymes to discover a possible cure for cancers. Enzymes can be served as a therapeutic agent against cancer because enzymes are highly specific biocatalysts that may transform various substrates into high yield products (Vellard, 2003). Here, three important enzymes that showed anticancer activity *viz.* L-asparaginase, L-glutaminase, and L-arginase are highlighted (Table 5.2).

L-asparaginase, an enzyme found in the bacterium *Escherichia coli* or *Erwinia carotovora*. Primarily it is used in the treatment of ALL and other blood cancers. The enzyme works through selective hydrolyzation of L-asparagine, an extracellular amino acid, into L-aspartate and ammonia. This kills the lymphoblastic cells by depriving them of nutrients and inhibiting protein synthesis. Because of this reason, L-asparaginase is primarily studied in lymphoma (Jaccard et al., 2010), ovarian, and human microvascular endothelial cell lines (HMVEC) (Yu et al., 2012), and brain cancers (Panosyan et al., 2014). In a clinical phase II study (Jaccard et al., 2010), eleven patients out of the 18 patients suffering from extranodal NK/T-cell lymphoma, showed improvement when treated with L-asparaginase. Although very promising, common symptoms included cytopenia, hepatitis, and various allergens were noticed. The invasiveness of endothelial cells and cancerous cells are originated by the interactions between the extracellular membrane through integrins, mostly by the integrin Beta-1. In ovarian cancer and HMVEC, L-asparaginase selectively modified the extracellular

Beta1-integrin, which inhibits the integrin to initiate. By doing so, cancerous cells lose the ability to adhere to other cells, preventing invasion to occur. The overall conclusion is that L-aspartate reduces the growth of ovarian tumors through significant modifications of its microenvironment. In brain cancer, L-asparaginase (ASNase), inhibited cell growth in the brain-malignant cell lines *viz.* DAOY, p. 52, U87, and GBM-ES.

L-glutaminase is another vital enzyme that is primarily produced by *bacillus* and *pseudomonas* sp., while also be found in some fungi (Bind et al., 2017). It is extensively used in food industries, but recent research has determined that L-glutaminase has the potential to be a very powerful anti-cancer enzyme. The amino acid L-glutamine is essential for cellular function, especially for protein biosynthesis. The enzyme works by starving cancer cells through a conversion of L-glutamine to glutamic acid and ammonia. This catalytic reaction induced apoptosis because cancerous cells lack the glutamine biosynthesis enzyme, L-glutamine synthase (Unissa et al., 2015). A study conducted in 1966 by El-Asmar et al. (1966) revealed that normal cells were not affected by L-glutaminase because of the transformation of L-glutamine into glutamic acid. A study (Wise et al., 2010) determined that cancerous cells have an addiction to glutamine, as an important mitochondrial substrate, it takes part in NADPH production. For instance, pancreatic carcinoma MIA PaCa-2 and PANC-1 cells, showed dependence on L-glutamine in a study conducted in 1978 (Wu et al., 1978).

Arginase is a manganese-containing enzyme that is part of the ureohydrolase family. It carries out the role of a catalyst in the final step of the urea cycle, which transforms L-arginine into L-ornithine and urea. This might be the reason of the anticancer potential of L-arginase. Normal cells remain unaffected when Arginase converts L-arginine to ornithine because normal cells synthesize citrulline from ornithine but Cancerous cells are unable to do that. Studies showed that pegylated arginine deiminase (ADI-PEG2), a known anticancer enzyme that reduces arginine, can initiate cancerous cell death. Takaku et al. (1992, 1995) demonstrated that ADI could cause cell damage and ultimately death in HCC (hepatocellular carcinoma) and in deadly skin cancer, melanoma. *In vitro* MH134 liver cancer and Meth A fibrosarcoma cells lines were inhibited by 20 ng/ml of ADI. The reticence of cellular growth was due to lessening in L-arginine. ADI-PEG20 was tested in human trials on a single patient diagnosed with HCC (Curley et al., 2003). The study revealed that ADI-PEG20 could promote cellular death without directly effecting normal healthy cells. The optimal dose, which was the one that reduced tumor size was 160 IU/

m². This concluded that ADI-PEG20 might have antitumor activity with low-toxicity with higher potency. Subsequently, new investigations were conducted to achieve a bigger view on the mechanistic action ofADI-PEG20. In a recent phase II trial (Tsai et al., 2013), on AML patients, treated with ADI-PEG20 was reported. The study involved 43 patients. Out of 43 patients, 22 passed away due to disease. Two patients showed complete response (CR) to ADI-PEG20. The studies conducted on both of those patients suggested that arginosuccinate synthetase played crucial part in developing malignancy. Additionally, some AML makers and c-MYC regulated genes were imperative factors to determine the response to ADI-PEG20.

TABLE 5.2 Enzymes with Anticancer Potential

Enzyme	Species	Cancer type	Type of anticancer evaluation	Mechanistic route	Reference
L-asparaginase	*Escherichia coli* or *Erwinia carotovora*	Leukemia Lymphoma Ovarian Brain	*In vivo human models In vitro* human cell lines	Apoptosis through inhibition of protein synthesis through hydrolyzation of L-asparagine	(Jaccard et al., 2010; Panosyan et al., 2014; Yu et al., 2012)
L-glutaminase	*Bacillus* and *Pseudomonas sp.*	pancreatic	*In vivo* murine models *In vitro* human cell lines	Nutrient starvation through the conversion of L-gluatmine to glutamic acid and ammonia Apoptosis	(El-Asmar et al., 1966; Unissa et al., 2015; Wise et al., 2010; Wu et al., 1978)
L-arginase	Non-hepatic tissues	Liver melanoma	*In vivo* human models *In vitro* human cell lines	Cell death through the reduction of L-arginine	(Curley et al., 2003; Takaku et al., 1992; Takaku et al., 1995; Tsai et al., 2013)

5.5 BACTERIOCINS

Bacteriocins are antimicrobial peptides, produced by various bacteria (Cotter et al., 2013), first discovered by Andre Gratia in 1925. Around this time, antibiotics related to bacteriocins were also discovered (Gartia et al., 2000). Overall, their discovery revolutionized the way we see nature and opened a new chapter in drug discovery research. Classification of bacteriocins vary widely, usually depended on which bacteria they were derived from, either gram-negative or gram-positive. Gram-negative bacteria-derived bacteriocins are classified through their molecular weights: high or low (Djamel et al., 2011). Gram-positive bacteria-originated bacteriocins are classified into three classes: I, II, and III, based on molecular weights (Cotter et al., 2006). Class I bacteriocins are small lantibiotics (<5 kDA) that contain polycyclic thioether amino acids, specifically lanthionine (Hyungjae et al., 2010). Class II bacteriocins are small (<10 kDA) peptides that cannot be modified. Class III bacteriocins are large (>30 kDA) proteins. Although most classifications are through size and class type, some bacteriocins are classified through their mechanism of action, which varies widely but usually targets cell membrane (Cotter et al., 2013). Because of their wide range and action potential, many bacteriocins are being studied to evaluate their effects on cancers cells.

Bovicin HC5 is a lantibiotic (type I) that is specifically produced from the *Streptococcus bovis* bacterium (Klein et al., 1977). Structure- and function-wise, bovicin HC5 is like nisin, an antibacterial peptide that consists of 34 amino acids. Both bacteriocins possess anticancer potential. An *in vitro* study (Paiva et al., 2012)to evaluate the cytotoxicity of bovicin HC5 and nisin, on three different cell lines: Vero cells (Monkey kidney epithelial cell), MCF-7 (human breast adenocarcinoma), and HepG2 (human liver hepatocellular carcinoma) was reported. When treated with bovicin HC5, the IC_{50} reported was 279.39 μM for MCF-7 cells and 289.30 μM for HepG2 cells whereas for nisin the IC_{50}'s were better, i.e., 105.46μM for MCF-7 and 112.25 μM for HepG2 cells. Regardless the toxicity, cell shrinkage and a reduction of cell number was observed. Additionally, it was reported that bovicin HC5 could only exhibit its expected cytotoxic effect at high concentration to execute its biological activity.

Colicins are class III bacteriocins, originated from specific strains of the bacterium *Escherichia coli* or other similar *Enterobacteriaceae* (Feldgarden et al., 1999). The mechanism of action in bacterial cells has been studied, revealing that colicins have the potential to cause cell death through membrane translocation and cell binding. If colicins could kill bacterial cells, do they have the potential to destroy cancerous cells? A study conducted by Chumchalová (2003)to validate the effectiveness of four types of colicins *viz.* A, E1, E3, and U, on eleven types of malignant (human) cells and also in fibroblast cells that carried P53 gene mutations. The cell lines of interest were MRC5 (standard fibroblast, wild-type p53), MCF7, and ZR75 (breast carcinoma, wild-type p53). The following cell types had a detection of p53: breast carcinoma: BT549, BT474, MDA-MD-231, SKBR3, and T47D, colon carcinoma: HT29, osteosarcoma: HOS, leiomyosarcoma: SKUT-1, and fibrosarcoma: HS913T. The most promising colicins were colicin A and E1, as both showed inhibition of cellular growth to most of the tested cell lines. Colicine A showed 16–56% inhibition, while colicin E1 showed an inhibition of 17–40%. One problem with colicine A was that it also inhibited 36% of the normal MRC5 fibroblast cells. Colicine E3 and U showed no significant changes in cell cycles. Cell cycle was altered notably in cell lines HS913T and MRC5 produced by Colicin A.

5.6 ANIMAL

Animal-derived products are crucial components in complementary, alternative, and traditional medicine which dates to the earliest existence of human civilization. Despite all the numerous innovations and advancements of modern medicine, animal-derived natural products continue to be an essential component to discover new drugs, not just for cancer, but for various other diseases. Natural products from animals may also have diverse therapeutic effects in treating numerous ailments and diseases. Of the 252 essential and WHO recommended chemical drugs about 8.7% are from animals and 27 out of 150 prescription drugs frequently used in the US have animal origin (Costa-Neto, 2005). Current studies are evaluating the venom of bees and snakes as a form of anticancer treatment. Study results have determined that bee venom can be effective in treating various diseases starting from skin diseases to malignant tumors (Son et al., 2007). Two components of bee venom are melittin and apamin. Melittin is the major constituent of bee venom is actually a protein that contains 26 amino acid residues (Ma et al., 2017). Melittin is observed to inhibit neoplastic cellular growth through apoptosis. Apamin, one of the minutest neurotoxins in bee venom, consists of only 10 amino acids with two disulfide bonds. Apamin at a minimum concentration inhibits Ca2+- activated K+ channels (Lazdunski et al., 1988). In one study, MCF7 cells were treated with bee venom and inhibition of cell division was observed. This study tried to reveal the mechanistic route, although difficult, some insight was achieved. Bee venom initiated the generation of reactive oxygen species (ROS) and subsequent dysfunction of the mitochondrial membrane potential (Azm). Liberation of cytochrome c was observed, which increased the overall levels of caspase-9 and Poly(ADP-ribose) polymerase (PARP), resulted in apoptosis. In addition, bee venom could induce S-phase arrest in MCF7 cells that might be the possible reason for up-regulation of p53, p. 21, p. 27, and the exposition of Cdk2. This observation concluded that bee venom had the ability of damaging DNA in MCF7 cancerous cell lines. Bee venom was also evaluated against human glioblastoma cells. Initially, Gajski et al. (2015) studied the integrated activity of *cis*-platin with bee venom in human glioblastoma A1235 cell lines. It revealed that bee venom alone could generate a notable cytotoxicity, specifically by causing necrosis. Since cisplatin is one of the commonest anticancer agents for treating many different cancer types, it was incorporated in the study to evaluate how it would behave with bee venom in glioblastoma cancer cells. The results disclosed that bee venom could synergize the action of cisplatin in cancerous cells. This meant a less amount of cisplatin could generate the expected results, which would reduce drug resistance and side effects. A recent study was focused (Sisakht et al., 2017), to validate bee venom worked at a molecular-level. The result concluded that the expression of matrix metalloproteinase-2 in A172 glioblastoma cells was linked to bee venom. Matrix metalloproteinase-2 (MMP-2), a member of the metalloproteinase enzyme family, is expressed in glioblastoma and some other cancer types. The A172 cells were sequentially treated with elevated concentration of bee venom. The IC_{50} was detected as 28.5 µg/ml, having a decrease in the matrix metalloproteinase-2 level, the mode of cell death was apoptosis.

Another very important class of animal-derived venoms is very familiar to the people *viz.* snake venom. Snake venom is known to contain many different biomolecules, some proteins are common but some are species-specific. In fact, the proteins are authoritative for bioactivity of a venom (Leon et al., 2011; Gomes et al., 2007). Depending on the snake venom, they can provide a neurotoxic, cardiotoxic, hemotoxic, or cytotoxic effect in human body. A neurotoxic effect could block the communication between neurons and prevent the binding of acetylcholine (Vonk et al., 2011). If a cardiotoxic effect is triggered, it could prevent heart muscle contraction by binding the cells located in the heart together (Yanget et al., 2005). A hemotoxic effect caused by snake venom could have potential to destroy healthy red blood cells, effecting the circulation system and the muscle tissues greatly (Yamazaki et al., 2007). The final and most studied form of snake venom action is its cytotoxic effect. This is because cytotoxicity targets specific cellular sites, which effects cell membranes greatly. Antineoplastic validation of the venom extracted from viper, *Macrovipera lebetina turanica*, with ovarian tumor cells PA-1 and SK-OV3 was reported (Song et al., 2012). The IC_{50}s were around 4.5 µg/mL and cell death took place in a dose-dependent way (0~10 µg/mL) through apoptosis. The apoptosis was likely caused by overexpression of caspase-3 and Bax proteins. For apoptosis to occur, up-regulation of apoptotic-inducing protein and down-regulation of anti-apoptotic protein are essential. Here, the up-regulating proteins were Bax and capase-3, while the down-regulating proteins was Bcl-2. The venom inhibited the DNA-binding activity of signal transducer and activator of transcription 3 (STAT3), which is an important transcription factor. In untreated PA-1 and SK-OV3cells, there was a high binding activity of (NF-κB), a protein complex, which provides a major contribution in controlling the DNA transcription. Snake venom inhibited NF-κB*via* inhibition of p. 50 and p65, NF-κB subunits. Additionally, a venom-treated composition of NF-κB, salicylic acid, and STAT3 showed extended inhibition cell proliferation. This concluded that apoptosis could be started in PA-1 and SK-OV3 cells by inhibition of NF-κB and STAT3 signal.

Most scorpions are venomous arthropods and are found around the globe, except Antarctica. There are over 1,500 different scorpions' species have been known so far. Early studies regarding scorpion venom have indicated anti-inflammatory and antimicrobial properties (Biswas et al., 2012). Subsequent research has reported that scorpion toxin might possess antineoplastic potential but controversy arose. In Cuba, there is a commercial anticancer drug namely *Labiofam*, generated with the venom extracted from blue scorpion, *Rhopalurus juneus*. The respective pharma company claimed that their product was tested on more than 10,000 cancer patients, showed positive results (Vidatox 30-CH, 2013). They registered the product under the name of Vidatox 30-CH, which is only found in Cuba. To validate their claim, Giovannini et al. (2017)evaluated Vidatox on HCC cells that include BRl-3, HepG2, and Snu449. *In vivo*, remarkable tumor growth was reported in Vidatox-treated mice which clearly invalidated the claim. Also, Vidatox-treated mice had a larger rate of liver degerneation in comparison to control. Furthermore, the company that sells Vidatox claims that their product prevents angiogenesis. The same study proved that Vidatox increased VEGFR2 expression, the gene responsible for initiation and propagation of angiogenesis. The final remarks indicate that Vidatrox is not effective to treat HCC type cancer patients rather it may cause negative effect. Currently, the venoms of the scorpion species that are under active investigation are the *Androctonus crassicauda* (AC), *Androctonus bicolor* (BC), and the *Leiurus quinquestriatus* (LQ). Antineoplastic efficacy of their venoms is believed to be enhanced using liposomes, small vesicles made up of phospholipids and lipids (Poy et al., 2016). Liposomal delivery is becoming an essential part of new medical drug discovery because it can successfully delivery drug in the target. In a study (Al-Asmari et al., 2017) scorpion venom was used to treat colorectal cancer with a nano-liposomal approach. The scorpion venom of choice was AC, BC, and LQ, while the colorectal cell line of choice used was HCT-8. Liposomes were prepared with the previously mentioned scorpion venom types. They report that venom-loaded liposomes demonstrated promising anticancer properties. The most notable points were apoptosis, and lower cancer cell survivability. HCT-8 was also treated with scorpion venom alone but lower efficacy was noted than nanoliposomal venom.

5.7   ANTICANCER PHENOLIC COMPOUNDS

Phenols are organic compounds that consist of one (or more) hydroxyl groups are directly connected to aromatic caobon (s) by sigma bond. Additionally, many phenolic compounds contain various functional groups e. g. esters and glycosides. There are over 8000 phenolic compounds have been found in nature so far. In plants, phenolics range from simple phenols to very complex structures. This makes them one of the most broadly distributed secondary metabolites which are plant-derived compounds but plants do not use for their regular requirement. Instead, plants use these compounds for protection (defense mechanism) and even for competition for a better survival. Secondary metabolites are classified based on their biosyntheticroutes. In plants, four majors secondary metabolites' classes are phenolics, terpenoids, alkaloids, and steroids. Phenolic compounds can differ significantly in size and complexicity. To achieve better stability, phenolics are very often adjoined with esters and glycosides. There are generally two categories of phenolics: simple phenolics and complex phenolic compounds. Simple phenolic compounds consist of benzoic and cinnamic acid derivatives. Complex phenolic compounds consist of stilbenes, tannins, lignans, lignins, xanthones, and flavonoids. Because of high abundance and diverse biological activities, flavonoids are discussed as a separate entity (Figure 5.8).

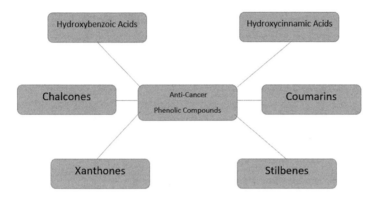

FIGURE 5.8 Anticancer phenolic compounds and their subdivisions.

5.7.1 HYDROXYCINNAMIC ACIDS

Most hydroxycinnamic acids are derived from caffeic, ferulic, *p*-coumaric, and caffeic acid phenethyl ester (CAPE) (Figure 5.9). This class is derived mainly from the phenylpropanoids. L-phenylalanine is converted to cinnamic acid by the enzyme phenylalanine ammonia-lyase (PAL) in plants. Through complex enzymatic hydroxylation and methylation, caffeic, ferulic, *p*-coumaric, and sinapic acids are biosynthesized. These newly formed acids are biosynthesized to defend the plants. Caffeic, ferulic, and *p*-coumaric acids are found plentifully in nature than sinapic acids. Overall, hydroxycinnamic acid scaffold is seen in many anticancer agents. For example, CAPE (2-phenylethyl (2*E*)-3-(3,4-dihydroxyphenyl) acrylate), a bioactive compound generated in the propolis in honeybee hives and in several plants, is reported to have anti-cancer activity. Research conducted by Wadhwa et al. (2016) showed that CAPE is quite unstable in bio-environments. To stabilize and enhance the antineoplastic efficiency of CAPE, it is necessary to adjoin it with gamma-cyclodextrin (γ-CD), a cyclic molecule formed by eight amylose-sugar scaffolds. Since cyclo-dextrins are hydrophobic on the inside with a hydrophilic outside, they can be formulated with other compounds, increasing their bioavailability

(Morrison et al., 2013). A report (Ishida et al., 2018)comparing the cyto-toxic effect of CAPE and CAPE-γ-CD on a variety of carcinoma cell lines was published. It was concluded that CAPE alone is capable of activating DNA damage through the up-regulation of tumor suppressors GADD45α and p53. CAPE-yCD provided similar results when compared to CAPE, but the effects were much more cytotoxic.

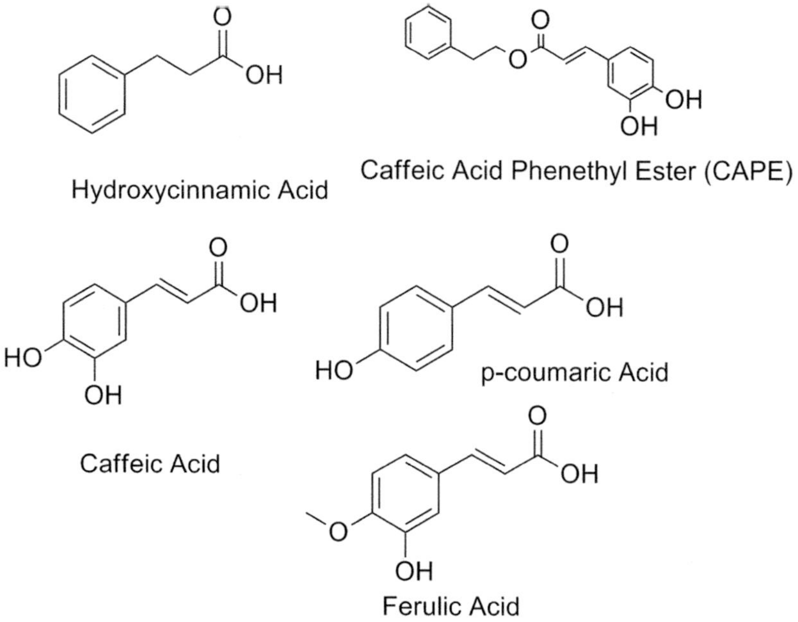

Hydroxycinnamic Acid   Caffeic Acid Phenethyl Ester (CAPE)

Caffeic Acid   p-coumaric Acid

Ferulic Acid

FIGURE 5.9   Anticancer hydroxycinnamic acids.

According to the WHO, an estimated 530,000 cases of cervical cancer will be diagnosed throughout the world in 2019. The number of women is expected to die from cervical cancer would be about 270,000 (Cervical Cancer, 2018). Cervical cancer can be treated quickly when diagnosed early and effective treatment is given. Early diagnosis is the key issue, but the majority of cervical cancer occurrences happen in low income counties, where early screening is unavailable (Cervical Cancer, 2018). Cervical cancer is mostly treated with combined therapy and surgery (Treating Cervical Cancer, 2018). The proficiency of caffeic acid as an anticancer agent (HeLa cells) was investigated. Caffeic acid, a simple phenolic compound biosynthesized and stored in the bark of the *Eucalyptus globulus* tree (Santos et al., 2011), was administered on HeLa cells. The research concluded that caffeic acid successfully inhibited proliferation of HeLa cells when treated in a control manner. Further observations confirmed caffeic acid's ability to reduce Bcl-2, increase p53 protein, and generate cytochrome c in cells. Being an anti-apoptotic protein Bcl-2 prevents apoptosis whereas p53 promotes apoptosis. Caffeic acid demonstrated its ability to reduce Bcl-2 and to up-regulate p53. The generation of cytochrome c could effectively promote caffeic acid-prompted apoptosis in cancer (HeLa) cells. Further research is required to validate the antitumor ability of caffeic acid in other cancer types (Chang et al., 2010).

The WHO estimated 1.5 million women would be detected with breast cancer (Breast Cancer, 2018). Treatment of this cancer is planned depending on the degree of advancement (stages). For local breast cancers, surgery followed by radiation are the method of choice. For advanced breast cancers, systemic treatments such as chemo, hormone, and/or targeted therapies are appropriate choices (Treating Breast Cancer, 2018). The WHO estimates that 627,000 women would die from this cancer worldwide (Breast Cancer, 2018). Recent research has put 4-hydroxy-3-methoxycinnamic acid (often called as ferulic acid) on the spotlight. Ferulic acid was first isolated from the plant *Ferula asafoetida*, a native in eastern and central parts of Asia including Afghanistan, India (Kashmir), and Iran (Yardley, 2010). MDA-MB-231 (breast cancer)cells were treated in different concentrations of ferulic acid that inhibited cell formation and persuaded apoptosis. Ferulic acid could significantly increase caspase-3 activity to foster apoptosis. Other studies conducted on MDA-MB-231 revealed that ferulic acid could inhibit tumor metastasis (Zhang et al., 2016).

There are roughly 1.4 million new cases of colon cancer are detected every year that resulted in about 700,000 deaths (Gandomani et al., 2017). The risk of colon cancer is increasing, primarily due to poor dieting and lack of exercise. The mortality and prognosis depend on the phase it is diagnosed. Statistically, individuals diagnosed at early stages are far likely to receive a better prognosis than those diagnosed at a later stage (Colorectal Cancer, 2018). New and novel medicines are needed to treat this disease. Recent studies to assess the antineoplastic efficacy of *p*-coumaric acid on HCT-15 and HT 29 (colorectal cancer) cells revealed that *p*-coumaric acid could success-fully prevent cell multiplication and growth. Under the electron microscope, *p*-coumaric acid-treated HCT-15 cells showed a shrinkage in size compared to untreated HCT-15assumed possible apoptosis. It (*p*-coumaric acid) did not affect *in vitro* colony formation either in HCT-15 or intestinal epithelial cells (IEC) (Jaganathan et al., 2013).

5.7.2 HYDROXYBENZOIC ACIDS

Hydroxybenzoic acids are produced by the shikimate (ester of shikimic acid) pathway in various plants, bacteria, and fungi. The pathway consists of seven enzymes: DAHP synthase, 3-dehydroquinate synthase, 3-dehydroquinate dehydratase, shikimate dehydrogenase, shikimate kinase, EPSP synthase, and chorismate synthase. Hydroxybenzoic acids' core structure contains a C6-C1 skeleton with seven carbon molecules. Some hydroxybenzoic acids are gallic, ellagic, and syringic acids (Figure 5.10). Tannins are a subgroup of hydroxybenzoic acids and two kinds of tannins are found: hydrolyzable and non-hydrolyzable. Hydrolysable tannins usually contain a central carbohydrate moiety. Hydrolyzable tannins are complex and usually biosynthesized by the condensation of various flavans. Hydroxybenzoic acids that contain anticancer activity are Gallic, ellagic, and syringic acid. Two tannins *viz.* cuphiin D1 and Oenothein Bhave been reported to contain limited anticancer activities. Their mechanism of inhibition is unknown.

Gallic acid (3,4,5-trihydroxybenzoic acid, GA) occurs naturally in various plant types and showed antineoplastic efficacy in HCC cells. The majority of liver cancers are HCC type, ranking recently as the third leading cancer in the world. In 2018, over 782,000 people died with this disease worldwide (Cicalese, 2018). In China, it is the second leading cause of cancer death. People with cirrhosis, obesity, diabetes, heavy drinkers, or infected with hepatitis B (HBV) or hepatitis C (HCV) have higher chance to develop HCC. Since Gallic acid caused apoptosis in HL-60RG, dRLh-84, and KB cell lines, it was assumed to cause apoptosis in SMMC-7721 and HepG2 cells, which are also HCC cell lines. Cell lines are a bunch of cells developed to form a single cell causing an overall identical genetic makeup (Kaur et al., 2012). *In vivo* studies on nude mice, models GU145 and 22Rv1, have confirmed GA's ability to prevent cell division and growth in prostate cancer cells, following apoptosis (Kaur et al., 2009). Apoptosis is programmed cell death which is a part of organism's natural development. Two types of apoptotic pathways are possible: extrinsic and intrinsic. Extrinsic pathway commonly refers to receptor death that imitates by transmembrane death receptors. Intrinsic apoptosis is mitochondrial-mediated cell death which occurs when mitochondria liberates cytochrome c to the cytoplasm. This causes cytochrome c to generate a protein complex with the protein APAF1, which causes APAF1 to link to ATP/dATP to form apoptosomes. Apoptosomes activate caspase-9, forming an elevation of caspase level (Pathway maps, 2018). The HepG2 and SMMC-7721 cells are prevented through GA by mitochondrial-mediated death (Sun et al., 2016). To determine if GA's cytotoxic effects only cancer-selective, normal hepatocyte cells (HL-7702), were treated with GA. Possibly GA may not affect normal hepatocyte cells. GA inhibits the cellular functions of HepG2 and SMMC-7721, accordingly cell growth is prevented. It also causes apoptosis in SMMC-7721.

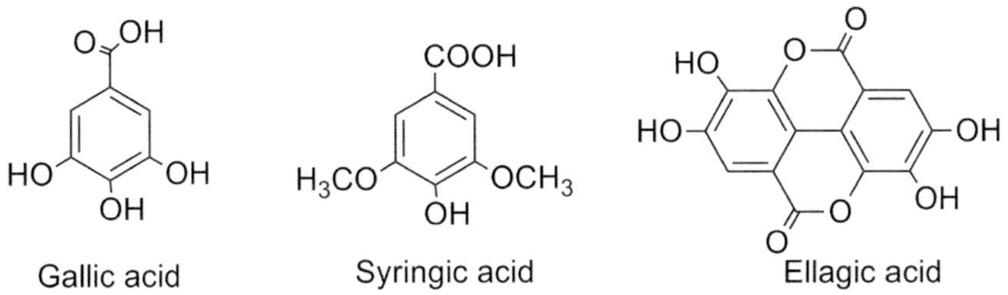

| Gallic acid | Syringic acid | Ellagic acid |

FIGURE 5.10   Anticancer hydroxybenzoic acids.

Ellagic acid (EA) has demonstrated enough evidences that support its neoplastic efficiency. EA showed (Cheng et al., 2017) its ability of preventing cell proliferation in drug-resistant PANC-1 (pancreatic carci-noma) with an IC_{50} about 5 µg/ml. PANC-1 cells were carefully chosen due to there are highly drug-resistant characteristics. Additionally, EA inhibited cellular migration of PANC-1 cells, preventing invasion and metastasis. *In vivo* study in nude mice with transplanted xenograft of PANC-1was reported to reveal EA's potential to inhibit growth in a relatively more complex system. No levels of EA toxicity were found. EA successfully enhanced mice survival rate as dosage increased. EA altered or modified the G_1 phase, promoting inhibition of PANC-1 cells. Furthermore, EA reduced inflammation factors that lead to epithelial-mesenchymal transi-tion (EMT) which is associated with the tumor's widening and metastasis. The outcome of this research helped to judge EA's antineoplastic ability for various other cancers.

5.7.3   STILBENES

Stilbenes indicate approximately 400 polyphenols that contain a 1,2-diphenyl-ethylene moiety and containa C6-C2-C6 skeletal in their structures. They are presentin lesser number of heterogenous plant species because the enzyme, stilbene synthase, is not widespread in nature (Riviere et al., 2012). The most famous anticancer stilbene is resveratrol (*trans*-3,4',5-trihydroxystilbene) (Figure 5.11). This compound is stored in grapes, and even in peanuts. Resveratrol is a chemo-preventive and chemotherapeutic agent against some cancer types. In nature, resveratrol is found in two existing isomers, *trans*-resveratrol, and *cis*-resveratrol. The most studied (*in vitro*)form of resveratrol is *trans*-resveratrol. Resveratrol metabolizes quickly in the intestine and liver by specific enzymes. Since resveratrol has low bioavailability, its use as anticancer agent can be somewhat limited. To compensate the limitation, two chemically modified resveratrol derivatives namely acetylresveratrol and polydatin have been synthesized that have much higher bioavailability than resveratrol. The chemical names of acetyl resveratrol and polydatin are 3,5,4'-tri-O-acetylresveratrol 3,5,4'-trihydroxy-stilbene-3-β-mono-D-glucoside respectively. These compounds showed preventive potential against ovarian cancers. To understand how resveratrol works it is important to understand its pharmacodynamics.

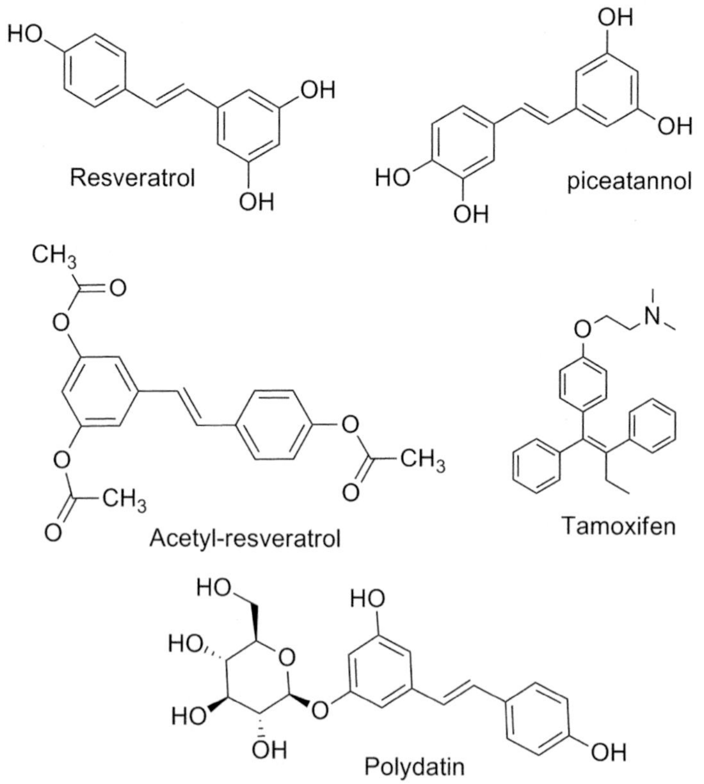

FIGURE 5.11   Anticancer stilbenes.

In human, resveratrol behaves as a suppressive agent on selective signaling pathways found in malignant tumor cells. For example, resveratrol successfully targets epidermal growth factors (EGF) and its related receptors (EGF-R), and tumor growth factor-beta (TGF-β) in ovarian malignancy. EGF-R is a transmembrane tyrosine kinase and it promotes regular cell growth. This results in 3D cell aggregation, which later from small tumors. Although EGF-R in living system is required but high-level causes threat of malignancy. When resveratrol acts in the EGF-R pathway, it demonstrates an antigrowth effect against EGF-R/Her-2-(+) and (–) ovarian malignant cells. Resveratrol alone presents itself as a gamble; depending on the malignant cell type it may prevent or accelerate the cellular growth. Consequently, using resveratrol in humans is still a controversial topic. Instead, acetylresveratrol, and polydatin have been under investigation to identify their anticancer properties. Polydatin has shown good anti-inflammatory and moderate antineoplastic efficacy (*in vitro*) (Riviere et al., 2012; Zhang et al., 2008). Acetylresveratrol and polydatin may be needed in high concentrations to execute inhibition in certain pathways. The mechanistic investigation with acetyl-resveratrol are very limited. Regardless, a few studies demonstrated acetyl-resveratrol could inhibit cell division of cancerous cells by working in conjugation (synergistic effect) with other drugs (Marel et al., 2008; Hogg et al., 2015). In addition, acetyl-resveratrol works as a cchemopreventivein lower concentrations by increasing the levels of mRNA antioxidant proteins (Tan et al., 2012). Intensive research is required to confirm the efficiencies of resveratrol, acetyl-resveratrol, polydatin, and other anticancer stilbenes.

5.7.4 COUMARINS

Coumarins are a derivative from phenolic acids that contain a benzene ring fused with pyrone (a six-membered oxygen heterocycle). Coumarins are also considered as aromatic delta-lactone derivatives. They are originated in plants belong to *Rutaceae* and *Umbelliferae* and in essential oils like lavender and cassia. Some microorganisms, namely *Streptomyces* and *Aspergillus* (aflatoxins) contain novobiocin &coumermycin, a form of less common natural coumarins. Other coumarins such as coumarin itself, natural analogs of warfarin, psoralidin, 4-hydroxycoumarin, and 7-hydroxycoumarin (umbelliferone) are common. *In vivo*, coumarin could inhibit stomach cancer in mice. Warfarin (synthetic coumarin) has been using as an anticoagulant, but patients that are treated with warfarin and vatalanib, a protein kinase angiogenesis inhibitor, can reduce advanced solid cancer significantly (Mody, 2018). Psoralidin instigates apoptosis in TNF-alpha-related-apoptosis-inducing-ligand (TRAIL)-mediated cervical cancer. 4-hydroxycoumarin, and umbelliferone are both effective in lung, ovarian, colon, brain, and skin cancers (Bronikowska et al., 2012). A few other Coumarin derivatives including a previously known anticancer agent RKS262 derivative, an analog of Nifurtimox demonstrated anticancer potential. RKS262 has shown efficacy against neuroblastoma in *in vitro/ in vivo* testing. In addition, RKS262 is promising in ovarian malignancy, specifically OVCAR-3 cells, human ovarian epithelial adenocarcinoma (Singh et al., 2011). It works by reducing the potential of Bcl-xl and Mcl-1 through the reduction of mitochondria-transmembrane-depolarization (Benci et al., 2012).

As mentioned before, while NSCLC accounts for about 85% of all lung cancers but there are limited drugs/effective treatments against NSCLC due to its stagnancy in advance lung cancer stages. Severalcoumarin derivatives are under preclinical exploration to validate their antineoplastic effica-cyin NSCLC. For instance, Osthole (7-methoxy-8-(3- methyl-2-butenyl) coumarin), daphnetin (7,8-dihydroxycoumarin), and umbelliprenin are notable antineoplastic coumarin derivatives, capable of treating NSCLC. Osthole an anticancer agent from *Cnidium monnieri* plant is capable to stop lung cancer growth. It helps to arrest G_2/M and speeds upapoptosis (Xu et al., 2011). Daphnetin (7,8-dihydroxycoumarin), suppresses Akt/NF-κB signaling pathways causing apoptosis in lung cancer. Umbelliprenin is a natural compound isolated from the plant genus *Ferula* (Wang et al., 2013). It adopts apoptosis in QU-DB (large lung tumor cells) and A549 adenocarci-noma (Khanghanzadeh et al., 2012) cells.

Prostate cancers mostly affect older men. Almost all prostate cancers are adenocarcinomas, that develop from the gland cells (American Cancer Society, 2018). Although most prostate cancers are treatable, new research is being conducted on scopoletin (6-methoxy-7-hydroxycoumarin). Scopoletin, a natural coumarinoid, originated in several plantspecies, especially the *Aster tataricus* family. It helps to arrest G_2/M and adopts apoptosis in prostate malignant (LNCaP) cells by inhibiting the expression of cycline D_1. Another coumarinoid, psoralidin enhances the potential of TRAIL in LNCaP cells. An uracil derivative, 5- fluorouracil (5-FU), is used for first line treatment of colon cancer but resistance is developed in many caes because of a high level of TS expression (Van et al., 1999) and possibly the over expression of Bcl-Xl and Bcl-2 proteins (Violette et al., 2002; Liu et al., 1999). Because of this resistance, research on developing new drugs to treat cancer is in great need.

Two newly synthesized anticancer compounds derived from natural coumarins are DMAC (5,7-dihydroxy-4-methyl-6-(3-methylbutanoyl)-coumarin) and di-coumarin polysulfide SV25. In recent studies, DMAC promoted apoptosis in HCT-116 and LoVo colon cancer cell lines. DMAC can be used with 5-FU to increase the potency of 5-FU (synergistic effect) toward various types of colon cancers (Klenkar et al., 2015). Dicoumarin polysulfide SV25 has can primarily arrest the G_2/M phase in HCT 116 cell line (Saidu et al., 2012) and eventually stops propagation. Figure 5.12 presents the structures of various anticancer coumarins and their derivatives.

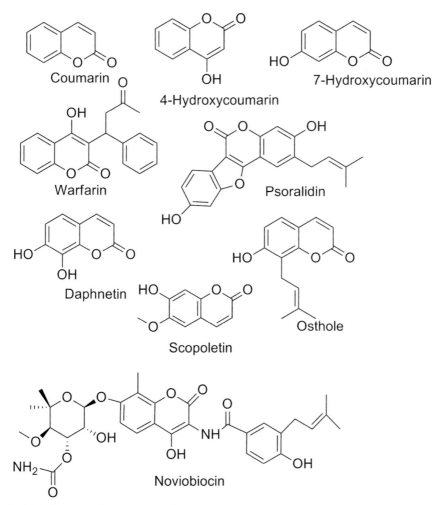

FIGURE 5.12 Anticancer coumarins.

5.7.5 XANTHONES

Xanthones are a group of naturally occurring compounds. Currently, over 200 xanthones are knownworldwide (see Figure 5.13). About 50 xanthone derivatives are originated in the pericarp of the purple mangosteen plant (*Garcinia mangostana*). The mangosteen plant uses xanthones and tannins to promote astringe effect on invading insects and fungi (Demirkiran et al., 2007). Xanthones are oxyheterocyclic ketones that can be classified into five subgroups: simple oxygenated xanthones, prenylated, xanthonolignoids, xanthone glycosides, and miscellaneous (Matsumoto et al., 2004). Recent research (*in vitro*) in leukemia, glioblastoma, and even melanoma cell lines has shown great promise of α-mangostin (axanthones derivative) for further study. The active compound α-mangostin is an important xanthone derivative that can be found within the skin of the mangosteen fruit. It was thought to contain limited anticancer activity, until a group of researchers reported an α-mangostin-induced apoptosis in human leukemia (HL60) cell lines. Matsumoto et al., (2004) conducted research on α-mangostin. It was concluded that α-mangostin could increase caspase-3 level in HL60 cells to adopt apoptosis by targeting the mitochondria in early cancer development. This effect, in turn, decreasesthe membrane potential (deltapsim). In HCC cells, it was (Hsieh et al., 2013) reported that α-mangostin could promote nuclear chromatin condensation and arrest in the sub-G1 phase to stop cellular division.

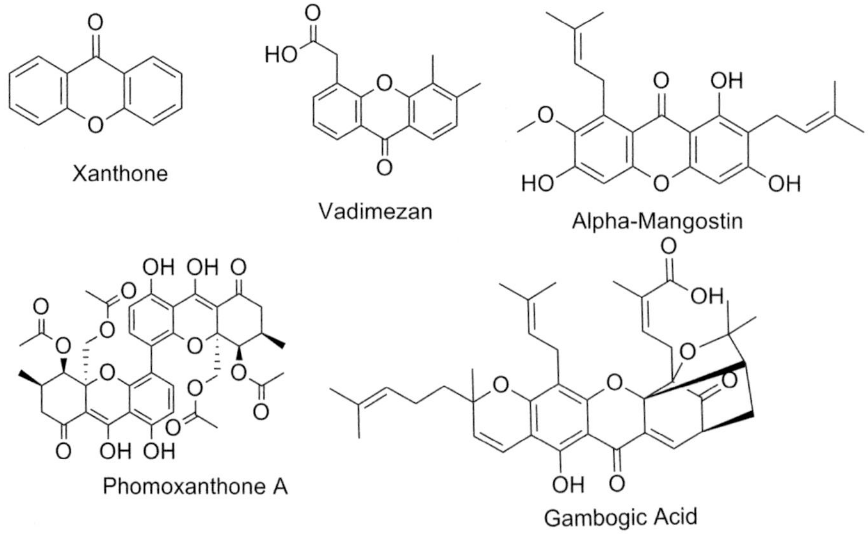

FIGURE 5.13   Anticancer xanthones.

In vivo studies uncovered that alpha-mangostin caused apoptosis in mice transplanted with human hepatoma SK-Hep-1 cell lines through p38 MARk signaling pathway that is induced by alpha-mangostin. Matsumoto et al. (2005) reported alpha-mangostin-assisted inhibition of cell growth in CLC-1 (colon malignancy) cell lines at 20 μM concentration. Furthermore, Cessation of cell-cycle following the expression of cyclins, dcd2, and p. 27 and intrinsic pathway were observed. Synergistic effect involving alpha-mangostin with 5_FU was studied (Akao et al., 2008) at variable concentrations of alpha-mangostin with 5-FU. This combination could successfully reduce the levels of two important proteins cyclin D1 and c-Myc (essential during the cell-cycle phase) at 2-5 μM concentrations. Apoptosis took place at about 15 μM concentration.

5.7.6 CHALCONES

Chalcones are polyphenols having diverse bioactivities. In nature, chalcones consist of two aromatic rings joined together by a three-carbon α, β-unsaturated carbonyl systems (Orlikova et al., 2011) (Figure 5.14). Chalcones area previl-iged class of molecules in *fabaceae* (pea) family. Many chalcones demonstrated various cytotoxic activities through multiple mechanisms which include angiogenesis, apoptosis, and cessasion of cell cycle. Their antineoplastic efficiencies are documented in some malignant cells, e.g., leukemia (HL60), stomach (AZ521), and lung (A549) cancers (Karthikeyan et al., 2014). One of the prominent medicinally active dihydrochalcones is phloretin. This impor-tant chalcone is isolated from the apple tree leaves and is found to be useful in skin rejuvenation. Recent studies conducted on phloretin with cisplatin have revealed amazing results in lung malignant like A549, Calu-1, H838, and H520 cell lines. Phloretin (a dihydrochalcone) significantly decreased the expression of the essential protein Bcl-2 in H520 cell lines. Studies further demonstrated that phloretin activated caspase-3 and -9 to adopt apoptosis in H520 neoplastic cells. In H828 malignant cells, phloretin inhibited MMP-2 and-9. This obser-vation was also documented in A549. A similar study conducted in Calu-1 neoplastic cells demonstrated that cisplatin with the aid of phloretin led to higher degree of apoptosis than cisplatin alone (Ma et al., 2015).

5.8 FLAVONOIDS

Flavonoids are secondary metabolites, biosynthesized in plants and also in fungi. Flavonoids are the most plentiful number of phenolic compounds in nature. Flavonoids consist of a C6-C3-C6 skeleton, a 15-carbon skeleton that contains two phenyl rings (A & B) and a heterocyclic ring (C). Flavonoids are biosynthesized mainly through acetate and shikimate pathways. They are rich in the leaves and skin of many fruits and plants. It is considered that flavonoids are also useful to protect skin from detrimental UV radiation. Flavonoids are categorized into many subclasses. These subclasses include flavonols, flavones, flavanols, flavanones, and isoflavones (Figure 5.15). Many studies recorded to validate biological potential of flavonoids that demonstrated anti-inflammatory (Yamamoto et al., 2001), anti-microbial (Cushnie et al., 2011), and anti-cancer activities (Ruela de Sousa et al., 2007). Early studies conducted on flavonoids determined their anticancer activities through a series of inhibitions, cell proliferation, and regulation of lymphocyte activations (Hendriks et al., 2003; Middleton et al., 2000; Verbeek et al., 2004). In this section, a concise description of anticancer activities of various subclasses of flavonoids will be presented.

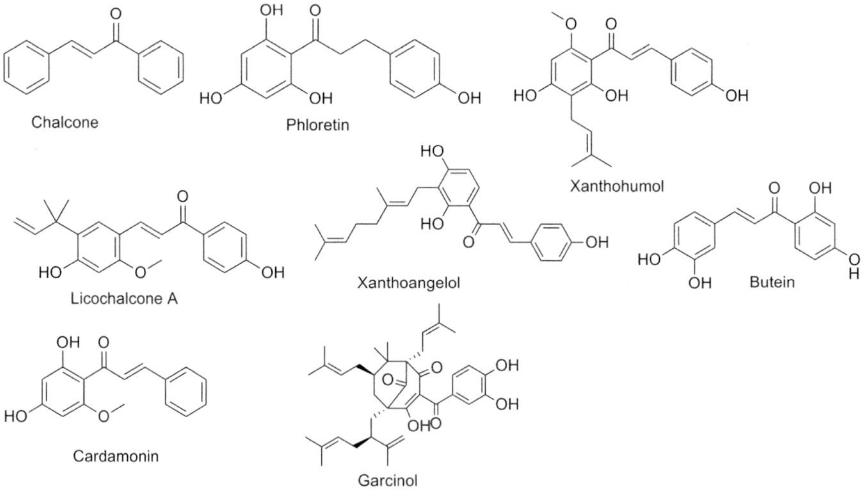

FIGURE 5.14   Anticancer chalcones.

5.8.1 FLAVONOLS

Flavonols contain 3-hydroxy-2-phenylchromen-4-one pharmacophore having 3-hydroxyflavone backbone. They are commonly found in certain teas, berries, and wines. Their exact quantity (amount) greatly depended on particular species and regional climate. Perhaps the most common

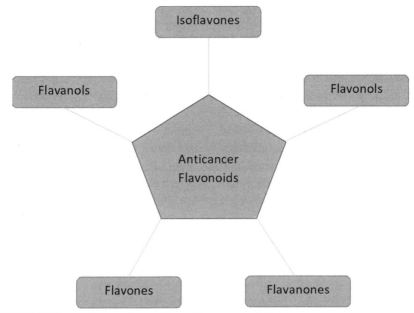

FIGURE 5.15   Anticancer flavonoids and its related subdivisions.

flavonol compounds are quercetin and kaempferol (Figure 5.16). Quercetin (3,3′,4′,5,7-pentahydroxyflavone) is found in abundance in apples, honey, and citric fruits. Recent studies conducted on quercetin indicated its ability to combat cancers through multi-targeted mechanisms (Erlund et al., 2004; Hendriks et al., 2003; Ramos, 2007). In various controlled concentrations, quercetin was able to overturn growth of the malignant tumor in ovarian, lung, colorectal, and breast cancers. Furthermore, daily intake of quercetin might be cancer-preventive. In one study, quercetin was validated, *in vitro* and animal model, on nine different malignant cells that involved CT-26 (colon carcinoma), LNCap (prostate carcinoma), PC3 (prostate adenocarcinoma), PC12 (adrenal medulla from rat), MCF-7 (breast carcinoma). U266B1 (B lymphocyte), Raji (B lymphocyte), CHO (epithelial ovary cells from Chinese hamster), and MOLT-4 (leukemia T lymphoblast). Studies concluded that quercetin-initiated apoptosis in CT-26, LNCaP, MOLT-4, and Raji in various concentrations. Animal models were subjected for CT-26 and MCF-7 malignancies since they demonstrated a higher sensitivity to quercetin. In animal model CT-26 and MCF-7 tumor growths were principally reduced although it could not prevent PC3 and CHO cell growth even at strong concentrations (Jakubowicz-Gil et al., 2005). Kaempferol, a similar flavonol compound, works in a similar manner. Kaempferol (3,4′,5,7-tetrahydroxyflavone), originated in the coniferophyte family, could prevent malignant cell growth as reported recently. In these experiments, kaempferol successfully regulated cyclin-dependent kinase 1 (CDK1), cyclin B, and p53 in MCF-7 and HeLa malignancies (Hashemzaei et al., 2017; Xu et al., 2008). Cyclin B binds to CDK1 to form a maturation promoting factor (MPF). This MPF promotes the transition of G_2 to M phase in cell cycle. By regulating CDK1 and cyclin B, kaempferol can initiate apoptosis to occur in cancerous cells. In a different study, conducted on U-2 OS osteosarcoma cells (bone cancer cells), kaempferol suppressed transcription factors AP-1 and ERK p38 which inhibited cancer metastasis (Chen et al., 2001).

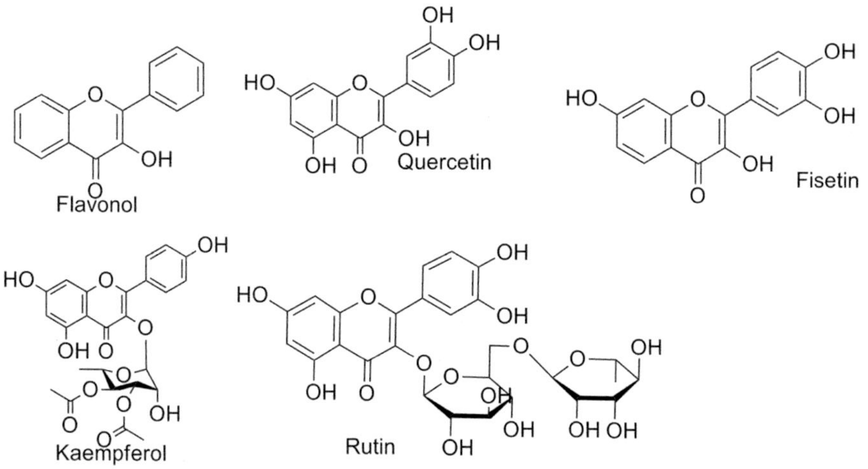

FIGURE 5.16   Anticancer flavonols.

5.8.2   FLAVONES

Flavones contain 2-phenylchromen-4-one backbone with an oxyhetero-cyclic enone system in their structure. These are sometimes derived from flavanones and are commonly found in spices, purple/red color vegetables and fruits, occasionally in 7-O-glycosidic form. Flavones protect the plants from harmful UV radiation and some flavones possess anti-inflammatory, antiviral, and antineoplastic activities. A common example of an anticancer flavone is chrysin. Chrysin (5,7-dihydroxyflavone) is isolated from the blue crown passionflower (Figure 5.17). According to Walle et al., chrysin contains a low bioavailability (Walle et al., 2001) and is quickly excreted (Nabavi et al., 2015). Regardless of its low bioavailability, research has been conducted to determine chrysin's effect on cancerous cell lines. Bahadori et al. (2016) conducted a study to evaluate the antineoplastic efficiency of chrysin in CT-26 (colon cancer) cells, both *in vitro* and animal models. The CT-26 cells were subjected to diverse concentrations of chrysin and up to 50% cellular growth inhibition was observed, compared to untreated cells. Furthermore, chrysin's cytotoxic properties adopted apoptosis *in vivo*. Khoo et al. (2010) carried out *in vitro* antineoplastic activity of chrysin in human cervical cancer, breast carcinoma, prostate cancer, esophageal squamous carcinoma, malignant glioma, and leukemia cells. Zhang et al. (2004) reported chrysin-induced prevention of cellular growth through adaptation of apoptosis in HeLa cells. The study also revealed subsequent drepression of the proliferating cell nuclear antigen (PCNA) caused by chrysin. Further research conducted on HeLa cells by Brandenstein et al., (2008) reportedthat chrysin could induce p38 and NFkappaB/65 activation. In malignant glioma (U87-MG), prostate, and breast malignant cells chrysin could inhibit proliferation and induced apoptosis (Nabavi et al., 2015). In addition, chrysin was capable to synergize the competence of wogonin to treat patients suffering from overe xpression in AKR1C1/1C2 induced by IL-6 in NSCLC (Wang et al., 2007). The same research also reported that chrysin disrupted G_2/M phase in SW480 cells.

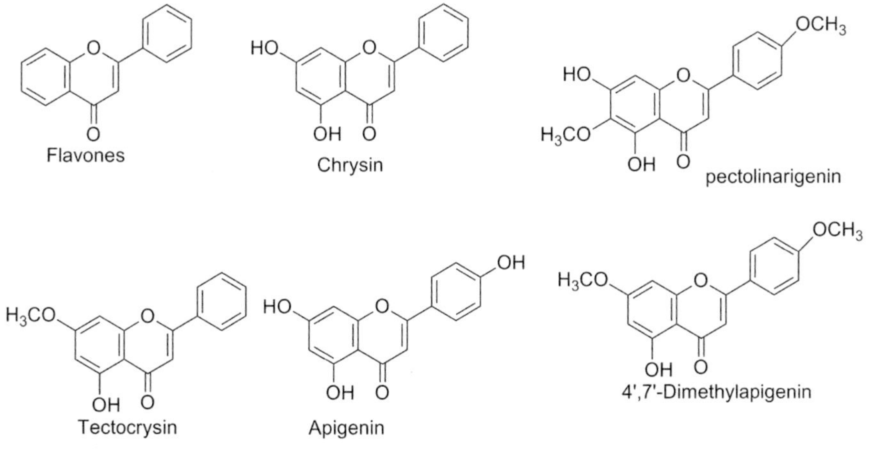

FIGURE 5.17   Anticancer flavones.

5.8.3   FLAVANOLS

Flavanols (sometimes calledFlavan-3-ols) are derivatives of flavans and are known for being the most complex class of flavonoid. They contain a 2-phenyl-3,4-dihydro-2*H*-chromen-3-ol structural core and are found abundantly in tea (*Camellia sinensis*), and cocoa (*Theobroma cacao*)(Song et al., 2008). Most flavanols contain a catechin backbone, which consist of two benzene rings (A and B) and a dihydropyran heterocycle (C). A hydroxyl group is also present and is attached on carbon3 (Figure 5. 18). Catechin itselfis known to contain anticancer properties and can prevent the metastasis in melanoma. This fact encouraged the scientists worldwide to facilitate chemical modification of catechin to synthesize semi-synthetic anticancer compounds with higher potency and less side-effects. Acommon catechin under investigation is epigallocatechin-3-gallate (EGCG), abundantly found in green tea. Green tea is aboundantly present in *Camellia sinensis* leaves, which is native to Southeastern and Eastern Asia. Notable research describing the anticancer properties of EGCG was reported (Min et al., 2014). EGCG inhibited VEGF, an imperative angiogenic factorin HT29 colon cancer*in vivo* (Jung et al., 2001). Similar effects of EGCG were seen in breast and pancreatic cancer cells. In skin cancer, ECGC increased the cytotoxic T-lymphocyte infiltration, that reduced (shrank) tumor size (Mantena et al.,

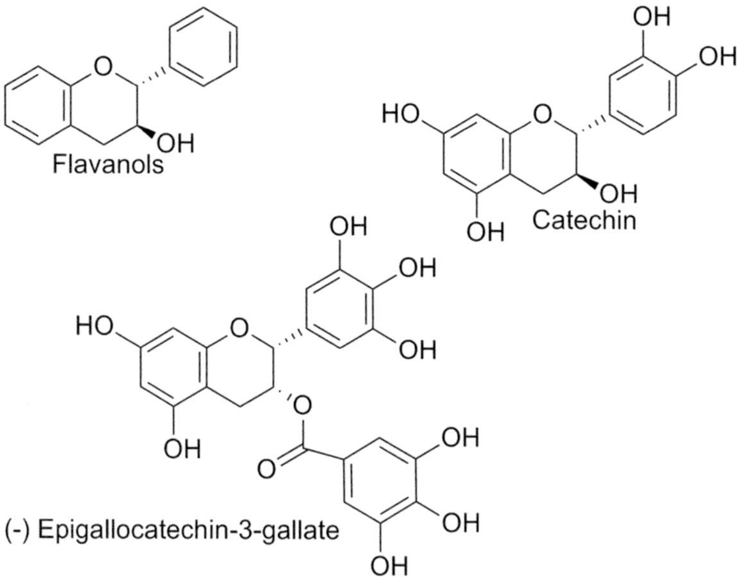

FIGURE 5.18   Structure and stereochemistry of anticancer flavanols.

2005). Since pre-clinical studies of EGCG were successful in treating breast, pancreatic, and skin cancers, further studies were conducted to observe its effects on metastatic cancers. *In vivo* studies conducted on UV-B-induced skin tumors revealed EGCG could inhibit protein expression and functional activity of matrix metalloproteinase (MMP)-2 (Bishayee et al., 2009).

5.8.4 FLAVANONES

The presence of flavanones in nature is quite limited. Flavanones contain two phenyl rings, an oxyheterocyclic ketone and are optically active (Figure 5.19). Many flavanones were isolated in glycosidic form. Flavanones are found mainly in citrus fruits and contain diverse health benefits together with anticancer activity. Various flavanones such as hesperidin (30,5,9-dihydroxy-40-methoxy-7-orutinosyl flavanone) demonstrated notable activity against cancers. Hesperidin, first isolated in 1827, containsnon-toxic effect on normal cells but high toxicity on cancer cells. Banjerdpongchai et al. (2016)reported that IC_{10}, IC_{20}, and IC_{50} concentrations of hesperidin could trigger apoptosis in HepG2 cells. Hesperidin was capable to activate caspase -9, -8, and -3 pathways (was monitored through colorimetric studies) to promote apoptosis in HepG-2cells. Furthermore, hesperidin was capable to regulate the expression of the apoptotic proteins in MSTO-211H mesothelioma cells. A down-regulation of Bcl-xl and Bax, and an increase in PARP and Bid, and the cleavage of caspase-3 are possible reason of apoptotic activity introduced by hesperidinin MSTO-211H cells. Finally, the study showed that hesperidin could successfully reduce mesothelioma cell growth through by inhibiting SP1 transcription factor.

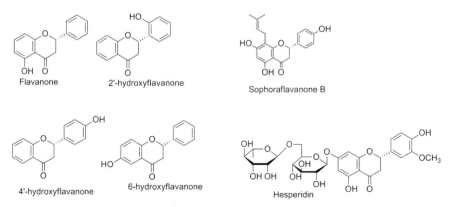

FIGURE 5.19 Structure and stereochemistry of anticancer flavanones.

5.8.5 ISOFLAVONES

Isoflavones are a subcategory of flavonoids, also called phytoestrogens, as many of these compounds have estrogenic activity in human. Isoflavonoids have a B-ring attached to a C3 carbon (C6-C3-C6 backbone) (Figure 5.20). Isoflavones can be found in foods which include dairy products, oils, and legumes. Upto now, many isoflavones have been tested for their anticancer potential. Their anticancer potential arises from their functional similarity to estrogens, which bind to estrogen receptor (ER)-beta rather than ER-alpha. Two commonly known isoflavones *viz.* genistein and daidzein are the major isoflavones in soybean. Ardito et al. (2017)assayed the efficacy of genistein in tongue (oral) cancer, which is an aggressive SCC, although having a comparatively reasonable 5-year survival rate which is 84% for non-spread cancer, 64% for that has spread to nearby lymph nodes, and 39% for cancer that has metastasized to other regions of the body. Early stage indications are jaw pain, lump inside the mouth, white or red gums, and difficulty chewing or moving the tongue, before they are fully diagnosed (Tongue Cancer, 2018). As limited chemotherapeutic agents are available for this cancer, new drugs are essential to ensure complete eradication (Oral Cancer: Your chances for recovery (prognosis), 2018). Genistein at 20-50 μM concentration inhibited cell adhesion by preventing cells to stink with one another. Cell life was reduced significantly at a moderate concentration of genistein. Metastasis was also inhibited due to down-regulation of an a few important proteins, like OCT4.

Daidzein (7-hydroxy-3-(4-hydroxyphenyl)-4*H*-chromen-4-one) is another significant isoflavone from legumes. Itscytotoxic potential was assayed inMG-63, HeLa, A549, BEL-7402, and HepG2 cells (Han et al., 2015). Cell cycle was arrested (G_2/M phase) in hepatoma BEL-7402 cell lines. Additionally, there was a down-regulation of Bcl-2, Bcl-x, and Baid proteins and an over expression of Bim protein tempted apoptosis. Daidzain also significantly elevated the ROS level, while promoting an overall inhibition of the mitochondrial membrane potential. In MG-63, HeLa, A549, BEL-7402, and HepG2 cells, the IC_{50} values of genistein were>100, >100, 97.9±11.3, 59.7±8.1, and >100 respectively. Accordingly, genisteindoes not have effective cytotoxicity in MG-63, HeLa, and A549 cells. This study indicated that daidzein could be chemically modified targeting more active semi-synthetic and subsequent validation of their effects on hepatoma patients.

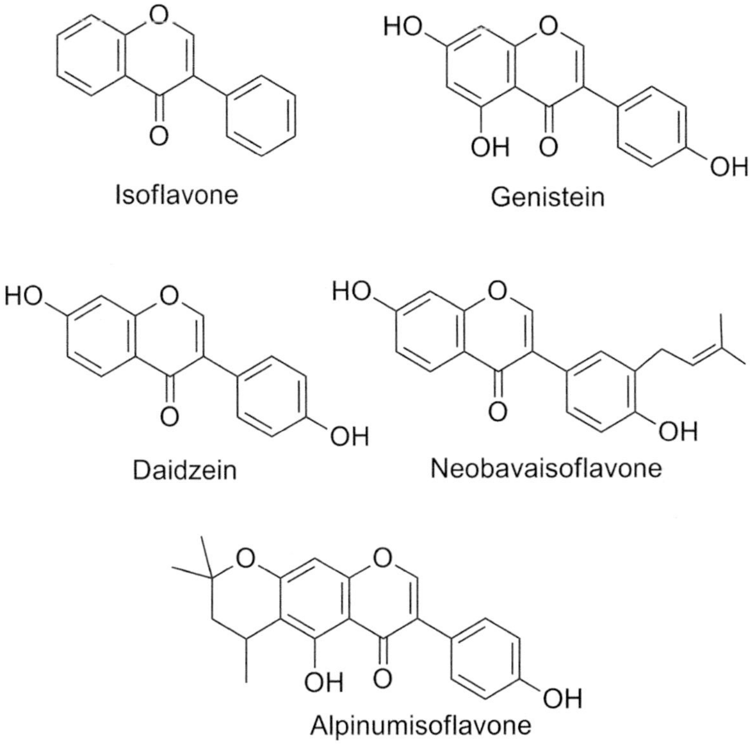

FIGURE 5.20 Anticancer isoflavones.

5.9 TERPENOIDS

Terpenoids are a large class of natural compounds composed of two or more isoprene (2-methyl-1,3-butadiene, C_5H_8)units. The isoprene units built the carbon skeletal of terpenes. Terpenoids contain a general formula $(C_5H_8)n$ and termed as monoterpenoids (C10), sesquiterpenoids (C15), diterpenoids (C20), sesterterpenoids (C25), triterpenoids (C30), sesquiterpenoids (C35), tetraterpenoids(C40) and polyterpenoids (>C40) (Figure 5.21). Terpenes are secondary metabolites, which are basically synthesized from isopentyl pyrophosphate and dimethylallyl pyrophosphate catalized by the enzyme terpene synthase. Terpenoids alone consist of over 50,000 compounds, making them a rich reservoir for new pharmacologically favored molecules.

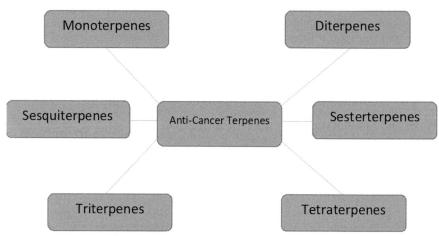

FIGURE 5.21   Anticancer terpenes and its subclasses.

5.9.1   *MONOTERPENOIDS*

Monoterpenoids consist of two isoprene units and have a skeletal formula of $C_{10}H_{16}$. In nature, they are found as acyclic, monocyclic, or bicyclic forms. Some monoterpenoids have medicinal efficiency as NK-kB signaling inhibitors. Many monoterpenoids have been isolated and investigated to determine their anticancer abilities (Figure 5.22). Limonene, a common monoterpenoid, is found in various citrus fruits and exists in nature as L-limonene and D-limonene, two enantiomeric forms. In human prostate carcinoma DU-145 cell lines D-limonene synergized the efficacy of docetaxel which resulted in substantial inhibition of cell growth (Bishayee et al., 2009). Intracellular ROS studies were also conducted with Limonene. ROS are natural byproducts of oxygen that are essential in cell signaling (Valluru et al., 2014). In DU-145 cells, ROS levels were notably elevated on treatment with D-limonene & docetaxel combination. ROS was inhibited by rotenone, a mitochondrial ROS blocker, indicating a non-mitochondrial and mitochondria-mediated change. This might explain limonene's ability to trigger docetaxel-induced apoptosis (Bishayee et al., 2009). Although promising antineoplastic efficacy was documented but more in-depth research is essential to conclude the safety (side-effects) of limonene in human. Further observation on anticancer monoterpenoids bring cantharidin into the picture. Cantharidin is extracted from many species of blister beetles, making it one of a few anticancer agents derived from animal origin. It demonstrated impressive anticancer activity (*in vitro*) against leukemia, hepatoma, and bladder carcinoma (Chen et al., 2002; Huan et al., 2012; Huh, et al., 2004). Cantharidin can trigger extracellular signal-regulated kinase ERK-1/2 (classical MAP kinases), p38, and JNK in U937 human leukemic cells (Huh et al., 2012). In human bladder carcinoma T24 cells, it has been proved that cantharidin could induce acute cytotoxicity through caspase 3-induced apoptosis (Huan et al., 2012). Although cantharidin has demonstrated promising results in various studies but more investigations are required to determine its safety in health cells.

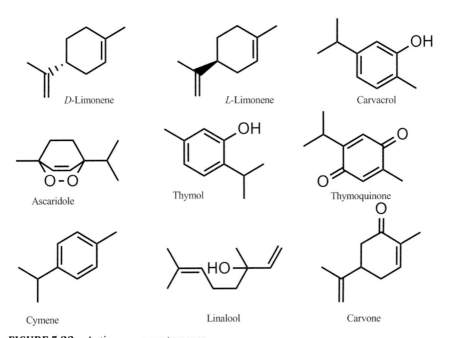

FIGURE 5.22   Anticancer monoterpenes.

5.9.2   *SESQUITERPENOIDS*

Sesquiterpenoids are biosynthesized from farnesyl pyrophosphate and are naturally abundant, primarily present in many plant and fungal species (Figure 5.23). Artemisinin, a prominent sesquiterpenoid lactone, isolated from *Artemisia annua* L. Artemisinin and its chemically-modified derivatives (artemisinin-combination therapies or ACT) are being considered as standard and broadly administered remedy for malignant malaria (*Plasmodium falciparum*) nowadays. Malaria is a mosquito-borne ailment triggered by the *plasmodium* parasites. An estimated 216 million cases and 445,000 deaths caused by malaria every year (Malaria, 2018). Artemisinin is highly effectual in treating malaria, but studies have also indicated the antineoplatic potential of artemisinin. Artemisinin has many semi-synthetic derivatives, e.g., artesunate, artemether, and dihydroartemisinin (DHA), etc. The first study on artemisinin to validate its antitumor potential was conducted in 1993. Since then, many research studies were recorded. The US NCI found that an artemisinin derivative namely artesunate could inhibit tumor growth in leukemia, colorectal cancer, melanoma, breast cancer, etc. (Efferth, 2006; O'Neill et al., 2010). In hepatoma cells, DHA, and artemisinin successfully inhibited proliferation while minimally affecting normal cells (Hou et al., 2008). When carboplatin, a well-known anticancer agent, was along with artemisinin and DHA, an enhanced cytotoxic effect was observed in ovarian

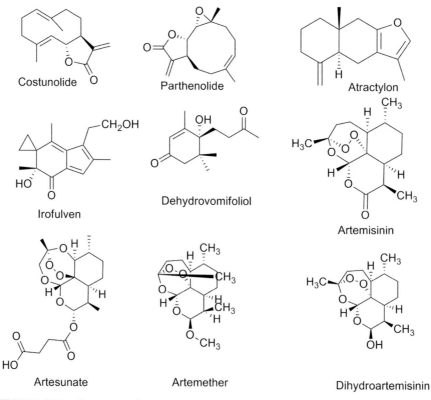

FIGURE 5.23   Structure and stereochemistry of anticancer sesquiterpenes.

cancer (Chen et al., 2009). Another notable derivative of artemisinin is artemether. Artemether showed promising results in treating U87 gastric malignant cells at 400 μM concentration (Wang et al., 2013). In metastatic gastric cancer, VCAM-1 is overexpressed significantly although the exact role of VCAM-1 in gastric cancer metastasis is unknown. In Wang's study, VCAM-1 was silenced with shRNA in U87 cancer cells. When the arte-mether was administered to the silenced VCAM-1 containingU87 cancer cells, a higher degree of metastasis prevention was observed. Breast and glioma cancer cells showed similar results (Haynes et al., 2013; Mercer et al., 2007). Overall, artemisinin, and its derivatives work by targeting multi-cancer functions, synergize other anticancer agents, and promote regulation of important biomolecules.

5.9.3　DITERPENOIDS

Diterpenoids consist of four isoprene subunits and contain a C20 skeleton in general. Diterpenoids are can be found plentiful in many terrestrial plants, fungi, and few animals. *Euphorbia fischeriana* Steud, a spurge family known to contain many anticancer diterpenoids (Barrero et al., 2011). In *E. fisch-eriana*, diterpenoid subtype *ent*-abietane is found. Jolkinolide B, a member of the *ent*-abietane subtype, could trigger apoptosis in hepatic, breast, gastric, and cervical tumor cells (Wang et al., 2013). Jolkinolide B inhibited the JAK2/STAT3 pathway in HL-60 and THP-1 cells by downregulation of JAK2/STAT3 and bcl-2, two important regulating proteins. Furthermore, jolkinolide B caused an upregulation of Bax and cytosolic cytochrome c, which triggered certain caspase to instigate apoptosis in leukemic cells (Lin et al., 2012; Wang et al., 2011). Jolkinolide B also inhibited metastasis in human breast cancer MDA-MB-23 cell lines through the detachment of fibronectin. Fibronectin is a glycoprotein having ability to alter cell growth, migration, and metastasis; it can also bind to integrins. Inhibition of fibro-nectin can greatly reduce cell migration and related metastasis (Pankov et al., 2002). Future research is being conducted on diterpenoid's ability to detach fibronectin from cancerous cells. Many other anticancer diterpenoids, e.g., prostratin and 17-Acetoxyjolkinolide B have demonstrated antineoplastic ability. Prostratin has the potential to reduce cellular proliferation selectively in breast tumors. While 17-acetoxyjolkinolide B inhibits NF-κB and subse-quently promotes apoptosis (Figure 5.24).

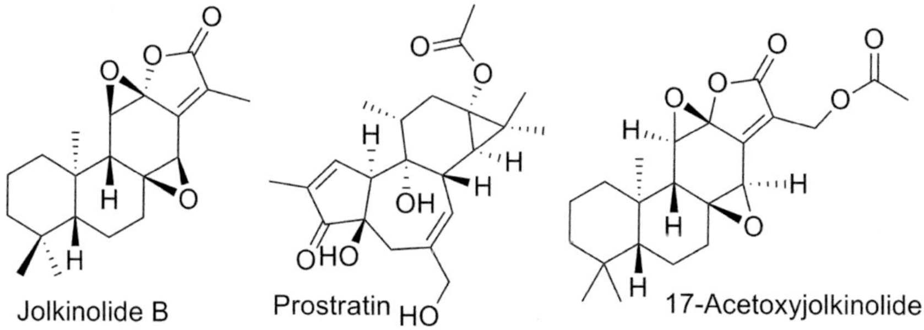

FIGURE 5.24   Structure and stereochemistry of anticancer diterpenes.

5.9.4   SESTERTERPENOIDS

Sesterterpenoids are a large class of complex botanicals, primarily present in plants, marine lives, and fungi and occasionally in bacteria. Sesterterpenoids are popularly classified with the number of carbocycles they contain in their structure; these classifications are linear, monocarbocyclic, bicarbocyclic, tricarbocyclic, and tertracarbocyclic (Hog et al., 2012; Liu et al., 2007). Linear sesterpenoids have relatively simple structures and few of them have antineoplastic potential. A well-studied subclass of linear sesterterpenoids is furano sesterterpenoids, found in *Ircinia* sponges. A recent study conducted by Aria et al. (2010), evaluated the anticancer activity of furospinosulin-1, a type of furanosesterter penoid, in DU145 cells. Furospinosulin-1 demonstrated selective antiproliferative activity under hypoxic condition at various concentrations. In animal (mice) model of sarcoma S180 cell lines, furospinosulin-1 demonstrated promising anticancer activity by downregulating insulin-like growth factor-2-*gene* (*IGF-2*) which was triggered by the inhibition under hypoxic condition due to docking of nuclear proteins to the transcription factor SP1 sequence.

Several sesterterpenoids are reported from sponges like the genus *Luffariella*, where many monocarbocyclic sesterterpenoids are present in the form of manoalides. Manoalides are metabolites, were first reported from *Luffariella variabilis* in 1980. Initially manoalides were evaluated as calcium channel blocker and bactericidal against *Streptomyces pyogenes* and *Staphylococcus aureus*. Later, 24-*n*-propyl-*O*-manoalide, a derivative of manoalide, demonstrated significant cytotoxicity in HCT-116 cells (Zhou et al., 2006) although the mechanism of cell death is unknown. Another sesterterpeniod, aikupikoxide C was isolated from the marine sponge *Negombata corticate*. This compound along with nuapapuin B, and nuapapuin B methyl ester and six norterpenes (one carbon lesser than terpene): negombatoperoxides A & B, negombatoperoxides C and D, negombatodiol, and negombatolactone, were evaluated (Chao et al., 2010) to determine their antineoplastic activity. Aikupikoxide C demonstrated good to excellent (in vitro) activity against MCF-7, HepG2, Hep3B, A-549, and MDA-MB-231 cell lines showing IC_{50} values 5.9, 0.9, 41.3, 0.6, and 0.3 µM respectively.

The genus *Salvia* contains vast variety of plants having several species. In the scientific world, this genus is highly appreciated because it houses many medicinally privileged compounds. Up to now, 24 tricyclic sesterterpenoids are reported only from one species *Salvia dominica* (Piaz et al., 2009). Out of the 24 isolated sesterterpenoids, 18 demonstrated significant interaction with tubulin-trysine ligase (TTL), an enzyme that dominates the tyrosi nation cycle of tubulin (specifically the C-terminal) and this interaction is responsible for antineoplastic efficacy. Another example of anticancer tricarbocyclic sesterterpenoids containing plant is *Leusceptrum canum*. Two significant anticancer tricarbocyclic sesterterpenoids, leucosesterterpenone, and leucosterlactone, were segregated from *Leuseptrum canum*. Leucosesterterpenone showed its potential to inhibit fibroblast growth factor-2 (FGF2), which is indispensable for mitogenic and angiogenic activities (Hussain et al., 2008). Both the compounds developed a complex with the fibroblast growth factor-2-receptor-1 and provided a down-regulation of ERK1/2 phosphorylation although leucosterlactone itself did not exhibit good anticancer activity.

5.9.5 TRITERPENOIDS

Triterpenoids consist of six isoprene units that contain 30 carbons'skeleton in general. In nature, they can be found as a single molecule or as glycoside or in their ester form. Many triterpenoids are known as NF-κβ inhibitors and their anticancer effects have well been documented. There are over 20,000 naturally occurring triterpenoids are known so far. They are widespread in terrestrial plants, and in marine lives such as sea cucumbers, marine algae, sponges, and even marine-derived fungi. In this section, the discussion will be confined to recent developments of antineoplastic triterpenoids.

Sea cucumbers contain an elongated body and are found worldwide. They are important, not just to preserve the ecosystem rolling, but because they contain various useful triterpenoids beneficial to mankind. Sea cucumbers are still being used as traditional Asian folk remedies to treat several ailments (Sea Cucumber, 2018). Most triterpenoids isolated from sea cucumbers are saponins and triterpenoid glycosides, that contain various anticancer properties. For instance, frondoside A, a triterpenoid isolated from *Cumaria frondosa*, an Atlantic Sea cucumber. Frondoside A has been tested on certain pancreatic and breast cancers. It is well-known that pancreatic cancinomais the deadliest cancer in this planet and the only cancer-type having 5-year survival in single digit (7.7%). Usually, the common drug is gemcitabine although gemcitabine is resistant to at least five different subtypes of pancreatic cancer. Frondoside Asynergizes gemcitabine's efficacy (Shemaili ct al., 2014) in pancreatic cancer AsPC-1 and S2013cell lines. Investigation carried out through *in vivo* mouse models concluded that gemcitabine and frondoside A in combination were highly effective in treating AsPC-1 and S2013 subtype of pancreatic cancer than individual drug alone. The antineoplastic potential of frondoside A in various breast cancer cell lines (Attoub et al., 2013; Marzouqi et al., 2011)such as MDA-MB-231 and MCF-10A (benign tumor) and MCF-7 have been validated. In MDA-MB-231cell lines, frondoside A increased the sub-G1 cell fraction that induces apoptosis through the activation of p53. Additional observations concluded that frondoside A had the potential to reduce tumor growth, cell invasion, and metastasis. In MCF-7 cell lines, frondoside A-induced cell invasion, metastasis, and angiogenesis. Moreover, Frondoside A could potentially enhance the potency of paclitaxel through synergism. These studies provide evidence to believe that apparently frondoside A has potential to treat breast cancers. After skin cancer, prostate cancer is the most common cancer in men. According to the American Cancer Society (Key Statistics for Prostate Cancer, 2018), about 1 in 9 men will be diagnosed with at least one type of prostate cancer in their lifetime. Studies conducted by Dyshlovoy et al. (2015), demonstrated that frondoside A might be effective in treating prostate cancer types PC-3, DUI145, LNCaP, and 22Rv1. Frondoside A up-regulated the proteins Bax, Bad, and PTENwhich stimulated apoptosis. Suppression of survival and Bcl-1 proteins, which prevent apoptosis was also noted. These mechanistic pathways indicate frondoside A as a double-edged sword against prostate cancer. *In vivo* administration of frondoside A showed notable depletion of tumor size in PC-3 and DU145 type of prostate cancers. This auspicious triterpene might become an essential anticancer agent in the future.

5.10 CHEMOSENSITIZERS

A chemosensitizer is an agent used to make tumor cells more sensitive to chemotherapeutic agents. Every day synthetic chemo sensitizers are being used in cancer chemotherapy and it has been becoming a serious concern that cancer is undergoing evolution and is slowly developing resistance to chemotherapy. Because certain types of cancers are developing resistance to synthetic chemo sensitizers, some natural products are under investigation for possible future use as chemosensitizers such as resveratrol, berberine, and bleomycin. Resveratrol continues to be studied as a chemo sensitizer for cancer chemotherapeutics. A study (Fulda et al., 2004) reported possible chemosensitization prospective of resveratrol. Resveratrol induces chemosensitization through modification and subsequent cessation of the cell cycle. Cessation of cell cycle lowers down the concentration of survivin (recombinant) protein, which at high concentration prevents apoptosis. Furthermore, resveratrol-triggered apoptosis is independent of p53 gene mutations. The study concluded that resveratrol could sensitize neuroblastoma, prostate carcinoma, breast carcinoma, pancreatic carcinoma, leukemia, and glioblastoma increasing the apoptotic possibility of certain drugs like taxol and actinomycin D (dactinomycin). Resveratrol could cause modification of p53 gene as well (Gatouillat et al., 2010). Resveratrol down-regulated cyclin D1/cdk4 and overexpressed p53 genes in *vivo* murine model of B16 melanoma which resulted in relatively higher survival rate. More research in this field was conducted to validate the chemosensitizing ability of resveratrol by modulation of cell survival proteins. Resveratrol could sensitize neoplastic cells by caspase activation and regulating *NF-κβ*genes (Buhrmann et al., 2015).

Studies also carried out to optimize the dose of resveratrol as chemosensitizer. Kumazaki et al. (2013) reported relatively higher dosage of resveratrol synergistically promoted 5-FU-mediated apoptosis. In contrast, a lower dosage of resveratrol inhibited 5-FU-triggered apoptosis. In a similar fashion resveratrol affected ovarian and uterine cancer cells by enhancing the growth inhibitory actions of cisplatin and doxorubicin (adriamycin). Asides from ovarian and uterine cancer cells, resveratrol acted as chemosensitizer in leukemic cells to facilitate apoptosis. In addition, resveratrol targeted SIRT1 specifically in breast malignant cells by functioning as a SIRT1-activating compound.

Berberine, a pharmacologically relevant isoquinoline alkaloid, is present in several plants such as *Berberis aristata* (also known as Indian barberry), European barberry, goldenseal, goldthread, Oregon grape, phellodendron, and tree turmeric. It is an AMPK activator that reflects its neoplastic potential. Studies revealed berberine as an effective alternative in anti-metastasis therapy of certain cancers which include gastric, oral, bladder, liver, colon, prostate, and breast(McCubrey et al., 2017). Recently, berberine has shown its ability of preventing metastatic growth in lung cancer by down regulation of the transforming growth factor. In addition, berberine could act as radio-sensitizer *in vitro* by inhibiting VEGF and HIF-1α in prostate cancer (Pan et al., 2017). Berberine itself showed anti-metastatic potential by targeting EMT in metastatatic prostate cancer. Berberine demonstrated inhibitory effect by repressing a panel of mesenchymal genes which are EMT growth-regulators. Furthermore, berberine, could regulate bone morphogenetic protein 7, NODAL, and Snail (*SNAI1 gene*) proteins (survival biomarkers) inprostate cancer (Liu et al., 2015). To determine the anti-metastasis activity of berberine, the expression levels of EMT mediators in PC-3 and DU145 (treated with berberine) cell lines were examined (as cancer metastasis happens due to cellular motility) by using an immunoblot analysis (Liu et al., 2015). This study identified berberine as anti-metastasis agent. To understand the role of berberine on suppression of *EMT genes* multiple experiments were carried out. Berberine down-regulated five *EMT*-related genes, relevant in migration and invasion of prostate cancer.

Furthermore, research conducted by Pan et al. (2017) identified the required dosages of berberine in breast malignancy. Long-term chemo-therapy causes adverse (toxic) effects in breast cancer. In addition, long-term chemotherapy resulted in multi-drug resistance toward cancer. It was shown that berberine could overcome DOX (doxorubicin) resistance. A low dosage of berberine potentially enhanced DOX sensitivity in drug-resistance breast neoplastic cells especially through the AMPK-HIF-1α-P-gp pathway and a high dosage of berberine showed that it could directly induce apoptosis following AMPK-p53 pathway with *HIF-1α* expression. To study the effectiveness of berberine in breast neoplasm, multiple pre-clinical experiments were pursued starting from SRB assay, apoptosis detection, Western blot (*in vitro*) to subcutaneously transplanted nude mice tumor model (*in vivo*). These experiments confirmed berberine's extensive influence on cell growth inhibition if a low dosage is administered whereas high dosage of berberine showed instigation of apoptosis in multi-drug resistant MCF-7/MDR cell lines. An additional study was conducted to confirm if low dose of berberine could sensitize DOX chemotherapy in MCF-7/MDR cell lines. To perform this experiment, MCF-7 and MCF-7/MDR cells were exposed to diverse concentrations of DOX for a stipulated time. It resulted that a low dosage of berberine enhanced the sensitivity of DOX-induced chemotherapy in MCF-7/MDR cell lines. Berberine at a low-dose activated AMPK and suppressedHIF-1α and P-gp proteins. Overall, this study concluded that berberine could sensitize drug-resistant breast cancer to DOX chemotherapy at a lower dose (as a chemosensitizer) and could induce apoptosis directly at a comparatively higher dose (as an anticancer agent).

5.11 CONCLUSION

Cancer is a serious issue that is being confronted worldwide, regardless of scientific/social advancements and affecting socio-economic conditions abruptly. According to Mayo Clinic, cancer can be significantly reduced by following healthy food habits, be physically active, protecting skin from the sun (UV), get immunized, not sharing needles, safe sex, regular medical exams, and quitting tobacco (Cancer Prevention, 2018). Even after in-depth knowledge of all these preventive-measures, cancer is still the second leading cause of disease-related death around the world. The current era demands potentially novel and selective anticancer agents from natural sources. Mother Nature provides us with the richest and most complex biodiversity. Compounds that are arduous or almost impossible to synthesize, can be isolated either from native or exotic plants, microbes, marine organisms, or animals. Phytoceuticals are an intimate part of human civilization and it is impossible to say exactly when menfolk started to take plant portions to cure various diseases. In fact, natural products are evolved for self-defense but amazingly that are being used for alleviating diseases from the beginning of human civilization. Pharmacological investigation of natural resources is the most reliable and traditional routes to discover new and novel natural drugs. Two major drawbacks of natural drug discovery include high cost and extreme time-consumption (Littleton et al., 2005). As the majority of natural sources are still unexplored, effective studies should be continued to isolate and evaluate (pre-clinical and clinical) possible anticancer agents from the vast natural origin to combat our war against one of the most dreadful enemies in the world, cancer.

ACKNOWLEDGMENTS

We are thankful to the Department of Chemistry, University of Texas Rio Grande Valley for providing library and related facilities.

KEYWORDS

- ATP binding cassette
- bioactive botanicals
- cancer
- chemotherapy
- heterocycles
- natural drugs

REFERENCES

Abidi, A., (2013). Cabazitaxel: A novel taxane for metastatic castration-resistant prostate cancer-current implications and future prospects. *J. Pharmacol. Pharmacother., 4*, 230.

Aicher, T. D., Buszek, K. R., Fang, F. G., Forsyth, C. J., Jung, S. H., Kishi, Y., Matelich, M. C., et al., (1992). Total synthesis of halichondrin B and norhalichondrin B. *J. Am. Chem. Soc., 114*, 3162–3164.

Akao, Y., Nakagawa, Y., Iinuma, M., & Nozawa, Y., (2008). Anti-cancer effects of xanthones from pericarps of mangosteen. *International Journal of Molecular Sciences, 9*, 355–370.

Akl, H., Vervloessem, T., Kiviluoto, S., Bittremieux, M., Parys, J. B., Smedt, H. D., & Bultynck, G., (2014). A dual role for the anti-apoptotic Bcl-2 protein in cancer: Mitochondria versus endoplasmic reticulum. *Biochim. Biophys. Acta. Mol. Cell Res., 1843*, 2240–2252.

Al-Asmari, A. K., Ullah, Z., Al Balowi, A., & Islam, M., (2017). *In vitro* determination of the efficacy of scorpion venoms as anti-cancer agents against colorectal cancer cells: A nano-liposomal delivery approach. *Int. J. Nanomed., 12*, 559–574.

American Cancer Society, (2016). *What is Non-Small Cell Lung Cancer?* https://www.cancer.org/cancer/non-small-cell-lung-cancer/about/what-is-non-small-cell-lung-cancer.html (accessed on 20 May 2020).

American Cancer Society. *What is Prostate Cancer?* https://www.cancer.org/cancer/prostate-cancer/about/what-is-prostate-cancer.html (accessed on 20 May 2020).

American Lung Association, (2018). *Health Effects of Secondhand Smoke.* http://www.lung.org/stop-smoking/smoking-facts/health-effects-of-secondhand-smoke.html (accessed on 20 May 2020).

Amino Acids, (2018). *U.S. National Library of Medicine.* https://medlineplus.gov/ency/article/002222.htm (accessed on 20 May 2020).

An Efficacy Study of Milataxel (TL139) Administered Orally for Malignant Mesothelioma (TL139), (2008). *U.S. National Library of Medicine.* https://www.clinicaltrials.gov/ct2/show/NCT00685204 (accessed on 20 May 2020).

An Efficacy Study of Ortataxel in Recurrent Glioblastoma (Ortataxel), (2016). *U.S. National Library of Medicine.* https://clinicaltrials.gov/ct2/show/NCT01989884 (accessed on 20 May 2020).

Arai, M., Kawachi, T., Setiawan, A., & Kobayashi, M., (2010). Hypoxia-selective growth inhibition of cancer cells by furospinosulin-1, a furanosesterterpene isolated from an Indonesian marine sponge. *Chem. Med. Chem., 5*, 1919–1926.

Ardito, F., Pellegrino, M. R., Perrone, D., Troiano, G., Cocco, A., & Lo Muzio, L., (2017). *In vitro* study on anti-cancer properties of genistein in tongue cancer. *Onco. Targets Ther., 10*, 5405–5415.

Ardizzoni, A., Manegold, C., Debruyne, C., Gaafar, R., Buchholz, E., Smit, E. F., Lianes, P., Ten, V. G., Bosquee, L., Legrand, C., Neumaier, C., & King, K., (2003). European organization for research and treatment of cancer (EORTC) 08957 phase II Study of topotecan in combination with cisplatin as second-line treatment of refractory and sensitive small cell lung cancer. *Clin. Cancer Res., 9*, 143–150.

Armstrong, D. K., Spriggs, D., Levin, J., Poulin, R., & Lane, S., (2005). Hematologic safety and tolerability of topotecan in recurrent ovarian cancer and small cell lung cancer: An integrated analysis. *Oncologist, 10*, 686–694.

Ashkenazi, A., (2015). Targeting the extrinsic apoptotic pathway in cancer: Lessons learned and future directions. *J. Clin. Investig., 125*, 487–489.

Attoub, A., Arafart, K., Gelaunde, A., Sultan, M. A. A., Bracke, M., Collin, P., Takahashi, T., Adrian, T. E., Wever, O. D., & Frondoside, A., (2013). Suppressive effects on lung cancer survival, tumor growth, angiogenesis, invasion, and metastasis. *PLoS One., 8*, e53087.

Avato, P., Migoni, D., Argentieri, M., Fanizii, F. P., & Tava, A., (2017). Activity of saponins from medicago species against HeLa and MCF-7 Cell Lines and their capacity to potentiate cisplatin effect. *Anticancer Agents Med. Chem., 17*(11), 1508–1518.

Bachran, C., Hasikova, R., Leysath, C. E., Sastalla, I., Zhang, Y., Fattah, R. J., Liu, S., & Leppla, S. H., (2014). Cytolethal distending toxin B as a cell-killing component of tumor-targeted anthrax toxin fusion proteins. *Cell Death and Dis., 5*, e1003.

Bacterial Toxins, (2018). *General Bacteriology*. http://generalbacteriology.weebly.com/bacterial-toxins.html (accessed on 20 May 2020).

Bahadori, M., Baharara, J., & Amini, E., (2016). Anticancer properties of chrysin on colon cancer cells, *in vitro* and *in vivo* with modulation of caspase-3,-9, Bax and Sall4. *Iran. J. Biotechnol., 14*, 177–184.

Banjerdpongchai, R., Wudtiwai, B., Khaw-on, P., Rachakhom, W., Duangnil, N., & Kongtawelert, P., (2016). Hesperidin from *citrus* seed induces human hepatocellular carcinoma HepG2 cell apoptosis via both mitochondrial and death receptor pathways. *Tumor Biol., 37*, 227–237.

Barrero, R. A., Chapman, B., Yang, Y., Moolhuijzen, P., Keeble-Gagnère, G., Zhang, N., Tang, Q., Bellgard, M. I., & Qiu, D., (2011). De novo assembly of Euphorbia fischeriana root transcriptome identifies prostratin pathway related genes. *BMC Genomics, 12*, 600.

Basili, S., & Moro, S., (2009). Novel camptothecin derivatives as topoisomerase I inhibitors. *Expert Opin. Ther. Pat., 19*, 555–574.

Beeram, M., Papadopoulous, K., Patnaik, A., Qureshi, A., & Tolcher, A. W., (2010). Phase I dose-ranging, pharmacokinetic (PK) study of tesetaxel, a novel orally active tubulin-binding agent. *J. Clin. Oncol., 28*, e13075–e13075.

Bellmunt, J., Théodore, C., Demkov, T., Komyakov, B., Sengelov, L., Daugaard, G., Caty, A., et al., (2009). Phase III trial of vinflunine plus best supportive care compared with best supportive care alone after a platinum-containing regimen in patients with advanced transitional cell carcinoma of the urothelial tract. *J. Clin. Oncol., 27*, 4454–4461.

Benci, K., Mandic, L., Suhina, T., Sedic, M., Klobucar, M., Pavelic, S. K., Pavelic, K., et al., (2012). Novel coumarin derivatives containing 1,2,4-Triazole, 4,5-dicyanoimidazole and purine moieties: Synthesis and evaluation of their cytostatic activity. *Molecules, 17,* 11010–11025.

Bennouna, J., Delord, J. P., Campone, M., & Nguyen, L., (2008). Vinflunine: A new microtubule inhibitor agent. *Clin. Cancer Res., 14,* 1625–1632.

Bennouna, J., Fumoleau, P., Armand, J. P., Raymond, E., Compone, M., Delgado, F. M., Puozzo, C., & Marty, M., (2003). Phase I and pharmacokinetic study of the new vinca alkaloid vinflunine administered as a 10-min infusion every 3 weeks in patients with advanced solid tumors. *Ann. Oncol., 14,* 630–637.

Bind, P., Singhu, R., Madhavan, A., Abraham, A., Mathew, A. K., Beevi, U. A., Sukumaran, R. K., et al., (2017). Recent developments in l-glutaminase production and applications – An overview. *Bioresour. Technol., 245,* 1766–1774.

Bishayee, A., & Rabi, T., (2009). D-Limonene sensitizes docetaxel-induced cytotoxicity in human prostate cancer cells: Generation of reactive oxygen species and induction of apoptosis. *J. Carcinog., 8,* 9.

Bissery, M. C. R., Bouchard, H., Riou, J. F., Vrignaud, P., Combeau, C., Bourzat, J. D., Commercon, A., & Lavelle, F., (2000). Preclinical evaluation of TXD258, a new taxoid. *Cancer Res., 41,* 214.

Biswas, A., Gomes, A., Sengupta, J., Datta, P., Singha, S., Dasgupta, A. K., & Gomes, A., (2012). Nanoparticle-conjugated animal venom-toxins and their possible therapeutic potential. *J. Venom Res., 3,* 15–21.

Brandenstein, M. G. V., Abety, A. N., Depping, R., Roth, T., Koehler, M., Dienes, H. P., & Fries, J. W., (2008). A p38–p65 transcription complex induced by endothelin-1 mediates signal transduction in cancer cells. *Biochim. Biophys. Acta Mol. Cell Res., 1783,* 1613–1622.

Breast Cancer, (2018). World Health Organization. http://www.who.int/cancer/prevention/diagnosis-screening/breast-cancer/en/ (accessed on 20 May 2020).

Bronikowska, J., Szliszka, E., Jaworska, D., Czuba, Z. P., & Krol, W., (2012). The *Coumarin psoralidin* enhances anticancer effect of tumor necrosis factor-related apoptosis-inducing ligand (TRAIL). *Molecules, 17,* 6449–6464.

Buhrmann, C., Shayan, P., Kraehe, P., Popper, B., Goel, A., & Shakibaei, M., (2015). Resveratrol induces chemosensitization to 5-fluorouracil through up-regulation of intercellular junctions, epithelial-to-mesenchymal transition and apoptosis in colorectal cancer. *Biochem. Pharmacol., 98,* 51–68.

Camptothecin. *Drugbank.* https://www.drugbank.ca/drugs/DB04690 (accessed on 20 May 2020).

Centers for Disease Control and Prevention, (2019). *What Are the Risk Factors for Lung Cancer?* https://www.cdc.gov/cancer/lung/basic_info/risk_factors.html (accessed on 20 May 2020).

Cervical Cancer, (2018). World Health Organization. http://www.who.int/cancer/prevention/diagnosis-screening/cervical-cancer/en/ (accessed on 20 May 2020).

Cha, T. L., Qiu, L., Chen, C. T., Wen, Y., & Hung, M. C., (2005). Emodin down-regulates androgen receptor and inhibits prostate cancer cell growth. *Cancer Res., 65,* 2287–2295.

Chan, T., Chang, C., Koonchanok, N., & Geahlen, R., (1993). Selective inhibition of the growth of ras-transformed human bronchial epithelial cells by emodin, a protein-tyrosine kinase inhibitor. *Biochem. Biophys. Res. Commun., 193,* 1152–1158.

Chang, C. J., Ashendel, C. L., Chan, T. C. K., Geahlen, R. L., Mclaughlin, J., & Waters, D. J., (1999). Oncogene signal transduction inhibitors from medicinal plants. *Pure Appl. Chem.*, *71*, 1101–1104.

Chang, W. C., Hsieh, C. H., Hsiao, M. W., Lin, W. C., Hung, Y. C., & Ye, J. C., (2010). Caffeic acid induces apoptosis in human cervical cancer cells through the mitochondrial pathway. *Taiwan. J. Obstet. Gynecol., 49*, 419–424.

Chao, C. H., Chou, K. J., Wang, G. H., Wu, Y. C., Wang, L. H., Chen, J. P., Sheu, J. H., & Sung, P. J., (2010). Norterpenoids and related peroxides from the Formosan marine sponge *Negombata corticata*. *J. Nat. Prod.*, *73*, 1538–1543.

Charles, B., (2018). The Canadian Medical Hall of Fame. http://www.cdnmedhall.org/inductees/dr-charles-beer (accessed on 20 May 2020).

Chaudhary, A., & Singh, N., (2011). Contribution of world health organization in the global acceptance of Ayurveda. *J. Ayurveda Integr. Med., 2*, 179–186.

Chen, H. J., Lin, C. M., Lee, C. Y., Shih, N. C., Peng, S. F., Tsuzuki, M., Amagaya, S., et al., (2013). Kaempferol suppresses cell metastasis via inhibition of the ERK-p38-JNK and AP-1 signaling pathways in U-2 OS human osteosarcoma cells. *Oncol. Rep., 30*, 925–932.

Chen, H., Hsieh, W., Chang, W., & Chung, J., (2004). Aloe-emodin induced *in vitro* G2/M arrest of cell cycle in human promyelocytic leukemia HL-60 cells. *Food Chem. Toxicol., 42*, 1251–1257.

Chen, T., Li, M., Zhang, R., & Wang, H., (2009). Dihydroartemisinin induces apoptosis and sensitizes human ovarian cancer cells to carboplatin therapy. *J. Cell. Mol. Med., 13*, 1358–1370.

Chen, Y. N., Chen, J. C., Yin, S. C., Wang, G. S., Tsauer, W., Hsu, S. F., & Hsu, S. L., (2002). Effector mechanisms of norcantharidin-induced mitotic arrest and apoptosis in human hepatoma cells. *Int. J. Cancer., 100*, 158–165.

Cheng, H., Lu, C., Tang, R., Pan, Y., Bao, S., Qiu, Y., & Xie, M., (2017). Ellagic acid inhibits the proliferation of human pancreatic carcinoma PANC-1 cells *in vitro* and *in vivo*. *Oncotarget., 8*, 12301–12310.

Cheng, P., Clive, D. L., Fernandopulle, S., & Chen, Z., (2013). Racemic marinopyrrole B by total synthesis. *Chem. Commun., 49*, 558–560.

Choi, C. H., Lee, Y. Y., Song, T. J., Park, H. S., Kim, M. K., Kim, T. J., Lee, J. W., et al., (2010). Phase II study of belotecan, a camptothecin analog, in combination with carboplatin for the treatment of recurrent ovarian cancer. *Cancer, 117*, 2104–2111.

Christenhusz, M. J., & Byng, J. W., (2016). The number of known plant species in the world and its annual increase. *Phytotaxa., 261*, 201–217.

Chumchalová, J., (2003). Human tumor cells are selectively inhibited by colicins. *Folia Microbiol., 48*, 111–115.

Cicalese, L., (2018) *Hepatocellular Carcinoma. Medscape.* https://emedicine.medscape.com/article/197319-overview (accessed on 20 May 2020).

Colorectal Cancer, (2018). American Cancer Society. https://www.cancer.org/cancer/colon-rectal-cancer/detection-diagnosis-staging/survival-rates.html (accessed on 20 May 2020).

Correia, J., & Lobert, S., (2001). Physiochemical aspects of tubulin-interacting antimitotic drugs. *Curr. Pharm. Des., 7*, 1213–1228.

Costa-Neto, E. M., (2005). Animal-based medicines: Biological prospection and the sustainable use of zootherapeutic resources. *An. Acad. Bras. Cienc., 77*, 33–43.

Cotter, P. D., Hill, C., & Ross, R. P., (2006). What's in a name? Class distinction for bacteriocins. *Nat. Rev. Microbiol., 4*, 160.

Cotter, P. D., Ross, R. P., & Hill, C., (2013). Bacteriocins—a viable alternative to antibiotics? *Nat. Rev. Microbiol., 11*, 95–105.

Curley, S. A., Bomalaski, J. S., Ensor, C. M., Holtsberg, F. W., & Clark, M. A., (2003). Regression of hepatocellular cancer in a patient treated with *Arginine deiminase. Hepatogastroenterology, 50*, 1214–1216.

Cushnie, T. P. T., & Lamb, A. J., (2011). Recent advances in understanding the antibacterial properties of flavonoids. *Int. J. Antimicrob. Agents., 38*, 99–107.

Desai, A. G., Qazi, G. N., Ganju, R. K., El-Tamer, M., Singh, J., Saxena, A. K., Bedi, Y. S., et al., (2008). Medicinal plants and cancer chemoprevention. *Curr. Drug Metab., 9*, 581–591.

Desjardins, J. P., Abbott, E. A., Emerson, D. L., Tomkinson, B. E., Leray, J. D., Brown, E. N., Hamilton, M., et al., (2001). Biodistribution of NX211, liposomal lurtotecan, in tumor-bearing mice. *Anti-Cancer Drugs, 12*, 235–245.

Djamel, D., & Rebuffat, S., (2011). Bacteriocins from gram-negative bacteria: A classification? *Prokaryotic Antimicrobial Peptides* (pp. 55–72). Springer: New York, NY.

Dyshlovoy, S. A., Menchinskaya, E. S. L., Venz, S., Rast, S., Amann, K., Hauschild, J., Otte, K., Kalinin, V. I., et al., (2015). The marine triterpene glycoside frondoside a exhibits activity *in vitro* and *in vivo* in prostate cancer. *Int. J. Cancer., 238*(10), 2450–2465.

Efferth, T., (2003). Molecular modes of action of artesunate in tumor cell lines. *Mol. Pharmacol., 64*, 382–394.

Eichhorn, J. M., Alford, S. E., Hughes, C. C., Fenical, W., & Chambers, T. C., (2013). Purported Mcl-1 inhibitor marinopyrrole A fails to show selective cytotoxicity for Mcl-1-dependent cell lines. *Cell Death Dis., 4*, e880.

Ekman, S., Harmenberg, J., Frödin, J. E., Bergström, S., Wassberg, C., Eksborg, S., Larsson, O., Axelson, M., Jerling, M., Abrahmsen, L., Hedlund, Å., Alvfors, C., Ståhl, B., & Bergqvist, M., (2015). A novel oral insulin-like growth factor-1 receptor pathway modulator and its implications for patients with non-small cell lung carcinoma: A phase I clinical trial. *Acta Oncol., 55*, 140–148.

El-Asmar, F. A., Greenberg, D. M., & St. Amand, G., (1966). Studies on the mechanism of inhibition of tumor growth by the enzyme glutaminase. *Cancer Res., 26*, 116–122.

Ellazeik, Y. A. O., Abdelgawad, S. S., Elsawy, E. M. A., & El-Wassef, A. M., (2013). Augmenting anticancer potential of exotoxin by mutating *Pseudomonas aeruginosa. Am. J. Sci., 9*, 312–321.

Elumalai, P., Gunadharini, D. N., Sentikumar, K., Banudevi, S., Arunkumar, R., Benson, C. S., Sharmila, G., & Arunakaran, J., (2012). Induction of apoptosis in human breast cancer cells by nimbolide through extrinsic and intrinsic pathway. *Toxicol. Lett., 215*, 131–142.

Erlund, I., (2004). Review of the flavonoids quercetin, hesperetin, and naringenin. Dietary sources, bioactivities, bioavailability, and epidemiology. *Nutr. Res., 24*, 851–874.

Ertel, F., Nguyen, M., Roulston, A., & Shore, G. C., (2013). Programming cancer cells for high expression levels of Mcl1. *EMBO Rep., 14*, 328–336.

Etiévant, C., Kruczynski, A., Barret, J. M., Tait, A., Kavallaris, M., & Hill, B., (2001). Markedly diminished drug resistance-inducing properties of vinflunine (20,'20'-difluoro-3,'4'-dihydrovinorelbine) relative to vinorelbine, identified in murine and human tumor cells *in vivo* and *in vitro. Cancer Chemother. Pharmacol., 48*, 62–70.

Facompre, M., Tardy, C., Bal-Mahieu, C., Colson, P., Perez, C., Manzanares, I., Cuevas, C., & Bailly, C., (2003). Lamellarin B: A novel potent inhibitor of topoisomerase I. *Cancer Res., 63*, 7392–7399.

FDA Approves Trabectedin to Treat Two Types of Soft Tissue Sarcoma, (2015). National Cancer Institute. https://www.cancer.gov/news-events/cancer-currents-blog/2015/fda-trabectedin-sarcoma (accessed on 20 May 2020).

Feldgarden, M., & Riley, M. A., (1999). The phenotypic and fitness effects of colicin resistance in *Escherichia coli* K-12. *Evolution, 53*, 1019.

Fulda, S., & Debatin, K. M., (2004). Sensitization for anticancer drug-induced apoptosis by the chemopreventive agent resveratrol. *Oncogene., 23*, 6702–6711.

Gajski, G., Čimbora-Zovko, T., Rak, S., Osmak, M., & Garaj-Vrhovac, V., (2015). Antitumour action on human glioblastoma A1235 cells through cooperation of bee venom and cisplatin. *Cytotechnology., 68,* 1197–1205.

Gandomani, H. S., Yousefi, S. M., Aghajani, M., Hafshejani, A. M., Tarazoj, A. A., Pouyesh, V., & Salehiniya, H., (2017). Colorectal cancer in the world: Incidence, mortality, and risk factors. *Biomed. Res. Ther., 4*(10), 1656–1675.

Gatouillat, G., Balasse, E., Joseph-Pietras, D., Morjani, H., & Madoulet, C., (2010). Resveratrol induces cell-cycle disruption and apoptosis in chemoresistant B16 melanoma. *J. Cell. Biochem., 110*, 893–902.

Gidding, C. E. M., Kellie, S. J., Kamps, W. A., & De Graff, N. S. S., (1999). Vincristine revisited. *Crit. Rev. Oncol. Hematol., 29*, 267–287.

Giddings, L. A., & Newman, D. J., (2013). Microbial natural products: Molecular blueprints for antitumor drugs. *J. Ind. Microbiol. Biotechnol., 40*, 1181–1210.

Gilbert, M. R., Supko, J. G., Batcherlor, T., Lesser, G., Fisher, J. D., Piantadosi, S., & Grossman, S., (2003). Phase I clinical and pharmacokinetic study of irinotecan in adults with recurrent malignant glioma. *Clin. Cancer Res., 9*, 2940–2949.

Giovannini, C., Baglioni, M., Baron, T. M., Cescon, M., Bolondi, L., & Gramantieri, L., (2017). Venom from Cuban blue scorpion has tumor activating effect in hepatocellular carcinoma. *Sci. Rep., 7*, 44685.

Goldstein, L. J., (1996). MDR1 gene expression in solid tumors. *Eur. J. Cancer., 32*, 1039–1050.

Goldufsky, J., Wood, S., Hajihossainlou, B., Rehman, T., Majdobeh, O., Kaufman, H. L., Ruby, C. E., & Shafikhani, S. H., (2015). *Pseudomonas aeruginosa* exotoxin T induces potent cytotoxicity against a variety of murine and human cancer cell lines. *Indian J. Med. Microbiol., 64*, 164–173.

Gomes, A., Choudhury, S. R., Saha, A., Mishra, R., Giri, B., Biswas, A., Debnath, A., & Gomes, A., (2007). A heat stable protein toxin (drCT-I) from the Indian Viper (*Daboia russelli*) venom having antiproliferative, cytotoxic and apoptotic activities. *Toxicon., 49*, 46–56.

Gratia, J. P., (2000). *André gratia*: A Forerunner in microbial and viral genetics. *Genetics, 156*, 471–476.

Grohar, P. J., Griffin, L. B., Yeung, C., Chen, Q. R., Pommier, Y., Khanna, C., Khan, J., & Helman, L. J., (2011). Ecteinascidin 743 interferes with the activity of EWS-FLI1 in Ewing sarcoma cells 1, 2. *Neoplasia., 13*, 145–153.

Guerra, L., Cortes-Bratti, X., Guidi, R., & Frisan, T., (2011). The Biology of the cytolethal distending toxins. *Toxins, 3*, 172–190

Hagiwara, H., & Sunada, Y., (2004). Mechanism of texane neurotoxicity. *Breast Cancer, 11*(1), 82–85.

Han, B. J., Li, W., Jiang, G. B., Lai, S. H., Zhang, C., Zeng, C. C., & Liu, Y. J., (2015). Effects of daidzein in regards to cytotoxicity *in vitro*, apoptosis, reactive oxygen species level, cell cycle arrest and the expression of caspase and Bcl-2 family proteins. *Oncol. Rep., 34,* 1115–1120.

Harvey, A. L., (2008). Natural products in drug discovery. *Drug Discov. Today, 13,* 894–901.

Hashemzaei, M., Far, A. D., Yari, A., Heravi, R. E., Tabrizian, K., Taghdisi, S. M., & Rezaee, R., (2017). Anticancer and apoptosis-inducing effects of quercetin *in vitro* and *in vivo. Oncol. Rep., 38,* 819–828.

Haste, N. M., Hughes, C. C., Tran, D. N., Fenical, W., Jensen, P. R., Nizet, V., & Hensler, M. E., (2011). Pharmacological properties of the marine natural product marinopyrrole a against methicillin-resistant *Staphylococcus aureus. Antimicrob. Agents Chemother., 55,* 3305–3312.

Hata, A. N., Engelman, J. A., & Faber, A. C., (2015). The BCL-2 family: Key mediators of the apoptotic response to targeted anti-cancer therapeutics. *Cancer Discovery, 5,* 475–487.

Haynes, R. K., Cheu, K. W., N'DA, D., Coghi, P., & Monti, D., (2013). Considerations on the mechanism of action of artemisinin antimalarials: Part 1-The 'carbon radical' and 'heme' hypotheses. *Infect. Disord. Drug Targets, 13,* 217–277.

Hendriks, J. J. A., De Vries, H. E., Van, D. P. S. M. A., Van, D. B. T. K., Van, T. E. A. F., & Dijkstra, C. D., (2003). Flavonoids inhibit myelin phagocytosis by macrophages: A structure-activity relationship study. *Biochem. Pharmacol., 65,* 877–885.

Henkel, J. S., Baldwin, M. R., & Barbieri, J. T., (2010). Toxins from bacteria. *EXS, 100,* 1–29.

Heterocycles, III, & Mahmud, T. H. K., (2007). *Activities in Topics in Heterocyclic Chemistry Bioactive* (Vol. 9, pp. 139–178). Springer, Berlin, Heidelberg: NY.

Hog, D. T., Webster, R., & Trauner, D., (2012). Synthetic approaches toward sesterterpenoids. *Nat. Prod. Rep., 29,* 752–779.

Hogg, S. J., Chitcholtan, K., Hassan, W., Sykes, P. H., & Garrill, A., (2015). Resveratrol, acetyl-resveratrol, and polydatin exhibit antigrowth Activity against 3D cell aggregates of the SKOV-3 and OVCAR-8 ovarian cancer cell lines. *Obstet. Gynecol. Int.,* 279591.

Hou, J., Wang, D., Zhang, R., & Wang, H., (2008). Experimental therapy of hepatoma with artemisinin and its derivatives: *In vitro* and *In vivo* activity, chemosensitization, and mechanisms of action. *Clin. Cancer Res., 14,* 5519–5530.

Hsieh, S. C., Huang, M. H., Cheng, C. W., Hung, J. H., Yang, S. F., & Hsieh, Y. H., (2013). α-Mangostin induces mitochondrial dependent apoptosis in human hepatoma SK-Hep-1 cells through inhibition of p38 MAPK pathway. *Apoptosis, 18,* 1548–1560.

Hu, S., Zhou, Q., Wu, W. R., Duan, Y. X., Gao, Z. Y., Li, Y. W., & Lu, Q., (2016). Anticancer effect of deoxypodophyllotoxin induces apoptosis of human prostate cancer cells. *Oncol. Lett., 12,* 2918–2923.

Huan, S. K. H., Lee, H. H., Liu, D. Z., Wu, C. C., & Wang, C. C., (2012). Cantharidin-induced cytotoxicity and cyclooxygenase 2 expression in human bladder carcinoma cell line. *Toxicology, 298,* 65.

Huang, X. C., Xiao, X., Zhang, Y. K., Talele, T. T., Salim, A. A., Chen, Z. S., & Capon, R. J., (2014). Lamellarin O, a pyrrole alkaloid from an Australian marine sponge, *Ianthella* sp., reverses BCRP mediated drug resistance in cancer cells. *Mar. Drugs., 12,* 3818–3837.

Hughes, C. C., Kauffman, P. R., Jensen, P. R., & Fenical, W., (2010). Structures, reactivities, and antibiotic properties of the marinopyrroles A-F. *J. Org. Chem., 75,* 3240–3250.

Hughes, C. C., Prierto-Davo, A., Jensen, P. R., & Fenical, W., (2008). The marinopyrroles, antibiotics of an unprecedented structure class from a marine streptomyces sp. *Org. Lett., 10,* 629–631.

Huh, J. E., Kang, K. S., Chae, C., Kim, H. M., Ahn, K. S., & Kim, S. H., (2004). Roles of p38 and JNK mitogen-activated protein kinase pathways during cantharidin-induced apoptosis in U937 cells. *Biochem. Pharmacol., 67*, 1811–1818.

Hussain, S., Slevin, M., Matou, S., Ahmed, N., Choudhary, I., Ranjit, R., West, D., & Gaffney, J., (2008). Anti-angiogenic activity of sesterterpenes; natural product inhibitors of FGF-2-induced angiogenesis. *Angiogenesis, 11*, 245–256.

Huyck, T. K., Gradishar, W., Manuguid, F., & Kirkpatrick, P., (2001). *Eribulin mesylate. Nat. Rev. Drug Discovery., 10*, 173–174.

Hyungjae, L., & Kim, H. Y., (2010). Lantibiotics, class I bacteriocins from the genus *Bacillus. J. Microbiol. Biotechnol., 21*, 229–235.

Ishida, Y., Gao, R., Shah, N., Bhargava, P., Furune, T., Kaul, S. C., Terao, K., & Wadhwa, R., (2018). Anticancer activity in honeybee propolis: Functional insights to the role of caffeic acid phenethyl ester and its complex with γ-cyclodextrin. *Integr. Cancer Ther., 17*, 867–873.

Jaccard, A., Gachard, N., Marin, B., Rogez, S., Audrain, M., Suarez, F., Tilly, H., et al., (2010). Efficacy of L-asparaginase with methotrexate and dexamethasone (AspaMetDex regimen) in patients with refractory or relapsing extranodal NK/T-cell lymphoma, a phase 2 study. *Blood, 117*, 1834–1839.

Jaganathan, S. K., Supriyanto, E., & Mandal, M., (2013). Events associated with apoptotic effect of p-Coumaric acid in HCT-15 colon cancer cells. *World J. Gastroenterol., 19*, 7726–7734.

Jakubowicz-Gil, J., Paduch, R., Piersiak, T., Glowniak, K., Gawron, A., & Kandefer-Szerszen, M., (2005). The effect of quercetin on pro-apoptotic activity of cisplatin in HeLa cells. *Biochem. Pharmacol., 69*, 1343–1350.

Jeon, W., Jeon, Y. K., & Nam, M. J., (2012). Apoptosis by aloe-emodin is mediated through down-regulation of calpain-2 and ubiquitin-protein ligase E3A in human hepatoma Huh-7 cells. *Cell Biol. Int., 36*, 163–167.

Jinadasa, R. N., Bloom, S. E., Weiss, R. S., & Duhamel, G. E., (2011). Cytolethal distending toxin: A conserved bacterial genotoxin that blocks cell cycle progression, leading to apoptosis of a broad range of mammalian cell lineages. *Microbiology, 157*, 1851–1875.

Joshi, P., Vishwakarma, R. A., & Bharate, S. B., (2017). Natural alkaloids as P-gp inhibitors for multidrug resistance reversal in cancer. *Eur. J. Med. Chem., 138*, 273–292.

Jung, Y. D., Kim, M. S., Shin, B. A., Chay, K. O., Ahn, B. W., Liu, W., Bucana, C. D., et al., (2001). EGCG, a major component of green tea, inhibits tumor growth by inhibiting VEGF induction in human colon carcinoma cells. *Br. J. Cancer., 84*, 844–850.

Karthikeyan, C., Moorthy, N. S. H. N., Ramasamy, S., Vanam, U., Elangovan, M., Karunagaren, D., & Trivedi, P., (2014). Advances in chalcones with anticancer activities. *Recent Pat. Anti-Cancer Drug Discov., 10*, 97–115.

Kartner, N., Riordan, & Ling, V., (1983). Cell surface P-glycoprotein associated with multidrug resistance in mammalian cell lines. *Science, 221*, 1285–1288.

Kaur, G., & Dufour, J. M., (2012). Cell lines: Valuable tools or useless artifacts. *Spermatogenesis, 2*, 1–5.

Kaur, M., Velmurugan, B., Rajamanickam, S., Agarwal, R., & Agarwal, C., (2009). Gallic Acid, an active constituent of grape seed extract, exhibits anti-proliferative, pro-apoptotic and anti-tumorigenic effects against prostate carcinoma xenograft growth in nude mice. *Pharm. Res., 26*, 2133–2140.

Key Statistics for Prostate Cancer, (2018). American Cancer Society. https://www.cancer.org/cancer/prostate-cancer/about/key-statistics.html (accessed Sep 23, 2018)

Key Statistics for Brain and Spinal Cord Tumors, (2018). American Cancer Society. https://www.cancer.org/cancer/brain-spinal-cord-tumors-adults/about/key-statistics.html (accessed on 20 May 2020).

Khanghanzadeh, N., Mojtahedi, Z., Ramezani, M., Erfani, N., & Ghaderi, A., (2012). Umbelliprenin is cytotoxic against QU-DB large cell lung cancer cell line but anti-proliferative against A549 adenocarcinoma cells. *DARU J. Pharm. Sci., 20,* 69.

Khoo, B. Y., Chua, S. L., & Balaram, P., (2010). Apoptotic effects of chrysin in human cancer cell lines. *Int. J. Mol. Sci., 11,* 2188–2199.

Klein, R. S., Recco, R. S., Catalano, M. T., Edberg, S. C., Casey, J. I., & Steigbigel, N. H., (1977). Association of *Streptococcus bovis* with carcinoma of the colon. *N. Engl. J. Med., 91,* 560.

Klenkar, J., & Molnar, M., (2015). Natural and synthetic coumarins as potential anticancer agents. *J. Chem. Pharm. Res., 7,* 1223–1238.

Kruczynski, A., Etiévant, C., Perrin, D., Chansard, N., Duflos, A., & Hill, B. T., (2002). Characterization of cell death induced by vinflunine, the most recent *Vinca* alkaloid in clinical development. *Br. J. Cancer, 86,* 143–150.

Kumazaki, M., Noguchi, S., Yasui, Y., Iwasaki, J., Shinohara, H., Yamada, N., & Akao, Y., (2013). Anti-cancer effects of naturally occurring compounds through modulation of signal transduction and miRNA expression in human colon cancer cells. *J. Nutr. Biochem., 24,* 1849–1858.

Larotaxel Compared to Continuous Administration of 5-FU in Advanced Pancreatic Cancer Patients Previously Treated with A Gemcitabine-Containing Regimen (PAPRIKA), (2016). *U.S. National Library of Medicine.* https://clinicaltrials.gov/ct2/show/NCT00417209 (accessed on 20 May 2020).

Lazdunski, M., Fosset, M., Hugues, M., Mourre, C., Renaud, J. F., Romey, G., & Schmid-Antomarchi, H., (1988). The apamin-sensitive Ca^{2+}-dependent K^+ channel molecular properties, differentiation, and endogenous ligands in mammalian brain. *Calcium in Drug Action* (Vol. 83, pp. 164–171). Springer, Berlin, Heidelberg.

Lee, H. Z., Hsu, S. L., Liu, M. C., & Wu, C. H., (2001). Effects and mechanisms of aloe-emodin on cell death in human lung squamous cell carcinoma. *Br. J. Pharmacol., 134,* 11–20.

Leon, G., Sanchez, L., Hernandez, A., Villalta, M., Herrera, M., Segura, A., Estrada, R., & Gutierrez, J. M., (2011). Immune response towards snake venoms. *Inflamm. Allergy Drug Targets., 10,* 381–398.

Leukaemia, (2017). *Cancer Research UK.* https://www.cancerresearchuk.org/about-cancer/leukaemia (accessed on 20 May 2020).

Leukemia & Lymphoma Society, (2018). *Facts and Statistics.* https://www.lls.org/facts-and-statistics/facts-and-statistics-overview (accessed on 20 May 2020).

Lin, Y., Cui, H., Xu, H., Yue, L., Xu, H., Jiang, L., & Liu, J., (2012). Jolkinolide B induces apoptosis in MDA-MB-231 cells through inhibition of the PI3K/Akt signaling pathway. *Oncol. Rep., 27,* 1976–1980.

Littleton, J., Rogers, T., & Falcone, D., (2005). Novel approaches to plant drug discovery based on high throughput pharmacological screening and genetic manipulation. *Life Sci., 78,* 467–475.

Liu, C. H., Tang, W. C., Sia, P., Huang, C. C., Yang, P. M., Wu, M. H., Lai, I. L., & Lee, K. H., (2015). Berberine inhibits the metastatic ability of prostate cancer cells by suppressing epithelial-to-mesenchymal transition (EMT)-associated genes with predictive and prognostic relevance. *Int. J. Med. Sci., 12*, 63–71.

Liu, R., Page, C., Beidler, D. R., Wicha, M. S., & Núñez, G., (1999). Overexpression of Bcl-x(L) promotes chemotherapy resistance of mammary tumors in a syngeneic mouse model. *Am. J. Pathol., 155*, 1861–1867.

Liu, Y. Q., Li, W. Q., Morris-Natschke, S. L., Qian, K., Yang, L., Zhu, G. X., & Lee, K. H., (2015). Perspectives on biologically active camptothecin derivatives. *Med. Res. Rev., 35*, 753–789.

Liu, Y., Wang, L., Jung, J. H., & Zhang, S., (2007). Sesterterpenoids. *Nat. Prod. Rep., 24*, 1401–1429.

Lovitt, C. J., Shelper, T. B., & Avery, V. M., (2018). Doxorubicin resistance in breast cancer cells is mediated by extracellular matrix proteins. *BMC Cancer, 18*, 41.

Lubran, M. M., (1988). Bacterial toxins. *Ann. Clin. Lab. Sci., 18*, 58–71.

Ma, L., Wang, R., Nan, Y., Li, W., Wang, Q., & Jin, F., (2015). Phloretin exhibits an anticancer effect and enhances the anticancer ability of cisplatin on non-small cell lung cancer cell lines by regulating expression of apoptotic pathways and matrix metalloproteinases. *Int. J. Oncol., 48*, 843–853.

Ma, R., Mahadevappa, R., & Kwok, H. F., (2017). Venom-based peptide therapy: Insights into anti-cancer mechanism. *Oncotarget., 8*, 100908–100930.

Macmillian Cancer Support, (2018). *Cancer and Cell Types.* https://www.macmillan.org.uk/information-and-support/understanding-cancer/cancer-and-cell-types.html (accessed on 20 May 2020).

Majumder, D., Das, A., & Saha, C., (2017). Catalase inhibition an anti cancer property of flavonoids: A kinetic and structural evaluation. *Int. J. Biol. Macromol., 104*, 929–935.

Malaria, (2018). *Centers for Disease Control and Prevention.* https://www.cdc.gov/parasites/malaria/index.html (accessed on 20 May 2020).

Mamtani, R., & Vaughn, D. J., (2014). Vinflunine in the treatment of advanced bladder cancer. *Expert Rev. Anticancer Ther., 11*, 13–20.

Mantena, S. K., Roy, A. M., & Katiyar, S. K., (2005). Epigallocatechin-3-gallate inhibits photocarcinogenesis through inhibition of angiogenic factors and activation of CD8+ T cells in tumors. *Photochem. Photobiol., 81*, 1174.

Marel, A. K., Lizard, G., Lizard, J. C., Latruffe, N., & Delmas, D., (2008). Inhibitory effects of *trans*-resveratrol analogs molecules on the proliferation and the cell progression of human colon tumoral cells. *Mol. Nutr. Food Res., 52*, 538–48

Martarelli, D., Pompei, P., & Mazzoni, G., (2009). Inhibition of adrenocortical carcinoma by diphtheria toxin mutant CRM197. *Chemotherapy, 55*, 425–432.

Marzouqi, N. A., Iratri, R., Nemmar, A., Arafat, K., Ahmed, M., Sultan, A., Yasin, J., et al., (2011). Frondoside A inhibits human breast cancer cell survival, migration, invasion, and the growth of breast tumor xenografts. *Eur. J. Pharmacol., 668*, 25–34.

Matsumoto, K., Akao, Y., Ohguchi, K., Ito, T., Tanaka, T., Iinuma, M., & Nozawa, Y., (2005). Xanthones induce cell-cycle arrest and apoptosis in human colon cancer DLD-1 cells. *Bioorg. Med. Chem., 13*, 6064–6069.

Matsumoto, K., Akao, Y., Yi, H., Ohguchi, K., Ito, T., Tanaka, T., Kobayashi, E., et al., (2004). Preferential target is mitochondria in alpha-mangostin-induced apoptosis in human leukemia HL60 cells. *Bioorg. Med. Chem., 12*, 5799–5806.

Mayo Clinic. (2018). *Cancer Prevention: Seven Tips to Reduce Your Risk.* https://www.mayoclinic.org/healthy-lifestyle/adult-health/in-depth/cancer-prevention/art-20044816 (accessed on 20 May 2020).

McCubrey, J. A., Lertpiriyapong, K., Steelman, L. S., Abrams, S. L., Yang, L. V., Murata, R. M., Rosalen, P. L., et al., (2017). Effects of resveratrol, curcumin, berberine and other nutraceuticals on aging, cancer development, cancer stem cells and microRNAs. *Aging., 9,* 1477–1536.

Mercer, A. E., Maggs, J. L., Sun, X. M., Cohen, G. M., Chadwick, J., Oneill, P. M., & Park, B. K., (2007). Evidence for the involvement of carbon-centered radicals in the induction of apoptotic cell death by artemisinin compounds. *J. Biol. Chem., 282,* 9372–9382.

Mérino, D., Khaw, S. L., Glaser, S. P., Anderson, D. J., Belmont, L. D., Wong, C., Yue, P., et al., (2012). Bcl-2, Bcl-x_L, and Bcl-w are not equivalent targets of ABT-737 and navitoclax (ABT-263) in lymphoid and leukemic cells. *Blood, 119,* 5807–16.

Middleton, A., Kandaswami, C., & Theoharides, T. C., (2000). The effects of plant flavonoids on mammalian cells: Implications for inflammation, heart disease, and cancer. *Pharmacol. Rev., 52,* 673–751.

Min, K. J., & Kwon, T. K., (2014). Anticancer effects and molecular mechanisms of epigallocatechin-3-gallate *Integr. Med. Res., 3,* 16–24.

Mody, R. (2018). *Palbociclib in Treating Patients with Relapsed or Refractory Rb Positive Advanced Solid Tumors, Non-Hodgkin Lymphoma, or Histiocytic Disorders with Activating Alterations in Cell Cycle Genes (A Pediatric MATCH Treatment Trial).* U.S. National Library of Medicine. https://clinicaltrials.gov/ct2/show/NCT03526250 (accessed on 20 May 2020).

Montaser, R., & Luesch, H., (2011). Marine natural products: A new wave of drugs? *Future Med. Chem., 3*(12), 1475–1489.

Moore, M. R., Jones, C., Harker, G., Lee, F., Ardalan, B., Saif, M. W., Hoff, P., et al., (2006). Phase II trial of DJ-927, an oral tubulin depolymerization inhibitor, in the treatment of metastatic colorectal cancer. *J. Clin. Oncol., 24,* 3591–3591.

Morrison, P. W. J., Connon, C. J., & Khutoryanskiy, V. V., (2013). Cyclodextrin-mediated enhancement of riboflavin solubility and corneal permeability. *Mol. Pharmaceutics, 10,* 756–762.

Munro, M. H. G., Blunt, J. W., Dumdei, E. J., Hickford, S. J. H., Lill, R. E., Li, S., Battershill, C. N., & Duckworth, A. R., (1999). The discovery and development of marine compounds with pharmaceutical potential. *J. Biotechnol., 70,* 15–25.

Nabavi, S. F., Braidy, N., Habtemariam, S., Orhan, I. E., Daglia, M., Gortzi, O., & Nabavi, S. M., (2015). Neuroprotective effects of chrysin: From chemistry to medicine. *Neurochem. Int., 90,* 224–231.

National Cancer Institute. (2005). *Halichondrin B (NSC 609395) E7389 (NSC 707389).* https://dtp.cancer.gov/timeline/flash/success_stories/s4_halichondrinb.htm (accessed on 20 May 2020).

NCI Dictionary of Cancer Terms, (2018). National Cancer Institute. https://www.cancer.gov/publications/dictionaries/cancer-terms/def/topoisomerase-inhibitor?redirect=true (accessed on 20 May 2020).

Newman, D. J., & Cragg, G. M., (2016). Natural products as sources of new drugs from 1981 to 2014. *J Nat Prod., 79,* 629–661.

Nishida, N., Yano, H., Nishida, T., Kamura, T., & Mojiro, M., (2006). Angiogenesis in cancer. *Vasc. Health Risk Manage, 2,* 213–219.

Nogales, E., & Wang, H. W., (2006). Structural intermediates in microtubule assembly and disassembly: How and why? *Curr. Opin. Cell Biol., 18*, 179–184.

O'Neill, P. M., Barton, V. E., & Ward, S. A., (2010). The molecular mechanism of action of Artemisinin-the debate continues. *Molecules, 15*, 1705–1721.

Oral Cancer: Your Chances for Recovery (Prognosis), (2018). The University of Rochester Medical Center. https://www.urmc.rochester.edu/encyclopedia/content.aspx?ContentTypeI D=34&ContentID=BoraD5 (accessed on 20 May 2020).

Orlikova, B., Tasdemir, D., Golais, F., Dicato, M., & Diederich, M., (2011). Dietary chalcones with chemopreventive and chemotherapeutic potential. *Genes Nutr., 6*, 125–147.

Ortataxel, (2018). *Drugbank*. https://www.drugbank.ca/drugs/DB11669 (accessed on 20 May 2020).

Paiva, A. D., Oliveira, M. D. D., Paula, S. O. D., Baracat-Pereira, M. C., Breukink, E., & Mantovani, H. C. F., (2012). Toxicity of bovicin HC5 against mammalian cell lines and the role of cholesterol in bacteriocin activity. *Microbiology, 158*, 2851–2858.

Pan, Y., Zhang, F., Zhao, Y., Shao, D., Zheng, X., Chen, Y., He, K., et al., (2017). Berberine enhances chemosensitivity and induces apoptosis through dose-orchestrated AMPK signaling in breast cancer. *J. Cancer., 8*, 1679–1689.

Pankov, R., & Yamada, K. M., (2002). Fibronectin at a glance. *J. Cell Sci., 115*, 3861–3863.

Panosyan, E. H., Wang, Y., Xia, P., Lee, W. N. P., Pak, Y., Laks, D. R., Morre, T. B., et al., (2014). Asparagine depletion potentiates the cytotoxic effect of chemotherapy against brain tumors. *Mol. Cancer Res., 12*, 694–702.

Pathway Maps, (2018). *Life Sciences Research*. http://pathwaymaps.com/maps/373 (accessed on 20 May 2020).

Pecere, T., Gazzola, M. V., Mucignat, C., Parolion, C., Vecchia, F. D., Cavaggioni, A., Basso, G., Diaspro, A., Salvato, B., Carli, M., & Palu, G., (2000). Aloe-emodin is a new type of anticancer agent with selective activity against neuroectodermal tumors. *Cancer Res., 60*, 2800–2804.

Piaz, F. D., Vassallo, A., Lepore, L., Tosco, A., Bader, A., & Tommasi, N. D., (2009). Sesterterpenes as tubulin tyrosine ligase inhibitors. First insight of structure-activity relationships and discovery of new lead. *J. Med. Chem., 52*, 3814–3828.

Pichersky, E., & Gang, D. R., (2000). Genetics and biochemistry of secondary metabolites in plants: An evolutionary perspective. *Trends Plant Sci., 5*, 439–445.

Pommier, Y., Leo, E., Zhang, H., & Christophe, M., (2010). DNA topoisomerases and their poisoning by anticancer and antibacterial drugs. *Cell Chem. Biol., 17*, 421–433.

Pourroy, B., Carre, M., Honore, S., Bournarel-Rey, V., Kruczynski, A., Braind, C., & Bradguer, D., (2004). Low concentrations of vinflunine induce apoptosis in human sk-n-sh neuroblastoma cells through a postmitotic g1 arrest and a mitochondrial pathway. *Mol. Pharmacol., 66*, 580–591.

Poy, D., Akbarzadeh, A., Shahmabadi, H. E., Ebrahimifar, M., Farhangi, A., Zarabi, M. F., Akbari, A., et al., (2016). Preparation, characterization, and cytotoxic effects of liposomal nanoparticles containing cisplatin: An *in vitro* study. *Chem. Biol. Drug Des., 88*, 568–573.

Rajabi, M., & Mousa, S. A., (2017). The role of angiogenesis in cancer treatment. *Biomedicines, 5*, 24.

Ramanathan, R. K., Picus, J., Raftopoulos, H., Bernard, S., Lockhart, A. C., Frenette, G., Macdonald, J., Melin, S., Berg, D., Brescia, F., Hochster, H., & Cohn, A., (2007). A phase II study of milataxel: A novel taxane analog in previously treated patients with advanced colorectal cancer. *Cancer Chemother. Pharmacol., 61*, 453–458.

Ramos, A., (2007). Effects of dietary flavonoids on apoptotic pathways related to cancer chemoprevention. *J. Nutr. Biochem., 18*, 427–442.

Ren, L., Li, Z., Dai, C., Zhao, D., Wang, Y., Ma, C., & Liu, C., (2018). Chrysophanol inhibits proliferation and induces apoptosis through NF-κB/cyclin D1 and NF-κB/Bcl-2 signaling cascade in breast cancer cell lines. *Mol. Med. Rep., 17*, 4376–4382.

Riviere, C., Pawlus, A. D., & Merillon, J. M., (2012). Natural stilbenoids: Distribution in the plant kingdom and chemotaxonomic interest in Vitaceae. *Nat. Prod. Rep., 29*, 1317–1333.

Roche, M., Kyriakou, H., & Seiden, M., (2006). Drug evaluation: Tesetaxel: An oral semisynthetic taxane derivative. *Curr. Opin. Invest. Drugs, 7*, 1092–1099.

Rowinsky, E. K., (1997). The development and clinical utility of the taxane class of antimicrotubule chemotherapy agents. *Annu. Rev. Med., 48*, 353–374.

Ruela, D. S. R. R., Queiroz, K. C. S., Souza, A. C. S., Gurgueira, S. A., Augusto, A. C., & Miranda, M. A., (2007). Phosphoprotein levels, MAPK activities, and NFκB expression are affected by fisetin. *J. Enzyme Inhib. Med. Chem., 22*, 439–444.

Saidu, N. E. B., Valiente, S., Bana, E., Kirsch, G., Bagrel, D., & Montenarh, M., (2012). *Coumarin polysulfides* inhibit cell growth and induce apoptosis in HCT116 colon cancer cells. *Bioorg. Med. Chem., 20*, 1584–159.

Santos, S. A. O., Freire, C. S. R., Domingues, M. R. M., Silvestre, A. J. D., & Neto, C. P., (2011). Characterization of phenolic components in polar extracts of eucalyptus globulus labill. Bark by high-performance liquid chromatography-mass spectrometry. *J. Agric. Food Chem., 59*, 9386–9393.

Schutz, F. A., Bellmunt, J., Rosenberg, J. E., & Choueiri, T. K., (2011). Vinflunine: Drug safety evaluation of this novel synthetic vinca alkaloid. *Expert Opin. Drug Saf., 10*, 645–653.

Science Daily, (2018). *Blood Flow is a Major Influence on Tumor Cell Metastasis.* https://www.sciencedaily.com/releases/2018/04/180409120442.htm (accessed on 20 May 2020).

Sea Cucumber, (2018). American Cancer Society. https://web.archive.org/web/20100714040906/, http:/www.cancer.org/Treatment/TreatmentsandSideEffects/ComplementaryandAlternativeMedicine/PharmacologicalandBiologicalTreatment/sea-cucumber (accessed on 20 May 2020).

Sebastian, D. B. J., Oudard, S., Ozguroglu, M., Hansen, S., Macheils, J. P., Kocak, I., Bodrogi, I., et al., (2010). Prednisone plus cabazitaxel or mitoxantrone for metastatic castration-resistant prostate cancer progressing after docetaxel treatment: A randomised open-label trial. *Lancet, 376*, 1147–1154.

Shamas-Din, A., Kale, J., Leber, B., & Andrews, D. W., (2013). Mechanisms of Action of Bcl-2 Family proteins. *Cold Spring Harbor Perspect. Biol., 5*, a008714.

Shemaili, J. A., Mensah-Brown, E., Parekh, K., Thomas, S. A., Attoub, S., Hellman, B., Nyberg, F., et al., (2014). Frondoside A enhances the antiproliferative effects of gemcitabine in pancreatic cancer. *Eur. J. Cancer., 50*, 1391–1398.

Singh, R. K., Lange, T. S., Kim, K. K., & Brard, L., (2011). A coumarin derivative (RKS262) inhibits cell-cycle progression, causes pro-apoptotic signaling and cytotoxicity in ovarian cancer cells. *Invest. New Drugs, 29*, 63–72.

Sisakht, M., Mashkani, B., Bazi, A., Ostadi, H., Zare, M., Avval, F. Z., Sadeghnia, H. R., Mojarad, M., Nadri, M., Ghorbani, A., & Soukhtanloo, M., (2017). Bee venom induces apoptosis and suppresses matrix metaloprotease-2 expression in human glioblastoma cells. *Rev. Bras. Farmacogn., 27*, 324–328.

Sisodiya, P. S., (2013). Plant derived anticancer agents: A review. *Int. J. Res. Dev. Pharm. L. Sci., 2*, 293–308.

Son, D. J., Lee, J. W., Lee, Y. H., Song, H. S., Lee, C. K., & Hong, J. T., (2007). Therapeutic application of anti-arthritis, pain-releasing, and anti-cancer effects of bee venom and its constituent compounds. *Pharmacol. Ther.*, *115*, 246–270.

Song, J. K., Jo, M. R., Park, M. H., Song, H. S., An, B. J., Song, M. J., Han, S. B., & Hong, J. T., (2012). Cell growth inhibition and induction of apoptosis by snake venom toxin in ovarian cancer cell via inactivation of nuclear factor κB and signal transducer and activator of transcription 3. *Arch. Pharmacal Res.*, *35*, 867–876.

Song, W. O., & Chun, O. K., (2008). Tea is the major source of flavan-3-ol and flavonol in the, U. S., diet. *J. Nutr.*, *138*, 1543S–1547S.

Stathopoulos, G. P., Christodoulou, C., Stathopoulos, J., Skarlos, D., Rigatos, S. K., Giannakakis, T., Armenaki, O., et al., (2005). Second-line chemotherapy in small cell lung cancer in a modified administration of topotecan combined with paclitaxel: A phase II study. *Cancer Chemother. Pharmacol.*, *57*, 796–800.

Sternberg, C. N., Skoneczna, I. A., Castellano, D., Theodore, C., Blais, N., Voog, E., Bellmunt, J., et al., (2013). Larotaxel with Cisplatin in the first-line treatment of locally advanced/ metastatic urothelial tract or bladder cancer: A randomized, active-controlled, phase III trial (CILAB). *Oncology, 85*, 208–215.

Sun, G., Zhang, S., Xie, Y., Zhang, Z., & Zhao, W., (2016). Gallic acid as a selective anticancer agent that induces apoptosis in SMMC-7721 human hepatocellular carcinoma cells. *Oncol. Lett.*, *11*, 150–158.

Takaku, H., Matsumoto, M., Misawa, S., & Miyazaki, K., (1995). Anti-tumor activity of *Arginine deiminase* from *Mycoplasma arginini* and its growth-inhibitory mechanism. *Jpn. J. Cancer Res.*, *86*, 840–846.

Takaku, H., Takase, M., Abe, S. I., Hayashi, H., & Miyazaki, K., (1992). *In vivo* anti-tumor activity of *Arginine deiminase* purified from *Mycoplasma arginini*. *Int. J. Cancer.*, *51*, 244–249.

Tan, X. L., Marquardt, G., Massimi, A. B., Shi, M., Han, W., & Spivack, S. D., (2012). High-throughput library screening identifies two novel NQO1 Inducers in human lung cells. *Am. J. Respir. Cell Mol. Biol.*, *46*, 365–371.

Tesetaxel as First-line Therapy for Metastatic Breast Cancer, (2012). *U.S. National Library of Medicine.* https://clinicaltrials.gov/ct2/show/NCT01221870 (accessed on 20 May 2020).

Tongue Cancer, (2018). *Cancer Treatment Cancer of America.* https://www.cancercenter. com/oral-cancer/types/tab/tongue-cancer/ (accessed on 20 May 2020).

Torisel (temsirolimus) Injection, (2018). *How Torisel is Given.* https://www.torisel.com/ given-intravenously (accessed on 20 May 2020).

Treating Breast Cancer, (2018). American Cancer Society. https://www.cancer.org/cancer/ breast-cancer/treatment.html (accessed on 20 May 2020).

Tsai, H. J., Jiang, S. S., Hung, W. C., Borthakur, G., Lin, S. F., Pemmaraju, N., Jabbour, E., et al., (2017). A phase ii study of *Arginine deiminase* (ADI-PEG20) in relapsed/refractory or poor-risk acute myeloid leukemia patients. *Sci. Rep.*, *7*, 11253.

Types of Cancer, (2017). Cancer Research UK. https://www.cancerresearchuk.org/what-is-cancer/how-cancer-starts/types-of-cancer (accessed on 20 May 2020).

Unissa, R., Sudakar, M., & Reddy, A. S. K., (2015). Selective isolation and molecular identification of l-arginase producing bacteria from marine sediments. *World J. Pharm. Pharm. Sci.*, *4*, 998–1006.

Valluru, L., Dasari, S., & Wudayagiri, R., (2014). Free radicals and antioxidants in human health: Current status and future prospects. *Oxid. Antioxid. Med. Sci.*, *3*, 15.

Van, T. B., Pinedo, H. M., Van, H. Y., Smid, K., Telleman, F., Schoenmakers, P. S., Van, D. W. C. L., et al., (1999). Thymidylate synthase level as the main predictive parameter for sensitivity to 5-fluorouracil, but not for Folate-based thymidylate synthase inhibitors, in 13 nonselected colon cancer cell lines. *Clin. Cancer Res., 5,* 643–654.

Vellard, M., (2003). The enzyme as drug: Application of enzymes as pharmaceuticals. *Curr. Opin. Biotechnol., 14,* 444–450.

Verbeek, R., Plomp, A. C., Van, T. E. A. F., & Van, N. J. M., (2004). The flavones luteolin and apigenin inhibit *in vitro* antigen-specific proliferation and interferon-gamma production by murine and human autoimmune T cells. *Biochem. Pharmacol., 68,* 621–629.

Verschraegen, C. F., Kudelka, A. P., Hu, W., Vincent, M., Kavanagh, J. J., Loyer, E., Bastien, L., Duggal, A., & Jager, R. D., (2004). A phase II study of intravenous exatecan mesylate (DX-8951f) administered daily for 5 days every 3 weeks to patients with advanced ovarian, tubal or peritoneal cancer resistant to platinum, taxane and topotecan. *Cancer Chemother. Pharmacol., 53,* 1–7.

Vidatox 30CH, (2013). *What is Vidatox?* http://vidatoxromania.ro/en/what-is-vidatox/ (accessed on 20 May 2020).

Violette, S., Poulain, L., Dussaulx, E., Pepin, D., Faussat, A. M., Chambaz, J., Lacorte, J. M., Staedel, C., & Lesuffleur, T. C., (2002). Resistance of colon cancer cells to long-term 5-fluorouracil exposure is correlated to the relative level of Bcl-2 and Bcl-X(L) in addition to Bax and p53 status. *Int. J. Cancer., 98,* 498–504.

Vonk, F. J., Jackson, K., Doley, R., Madaras, F., Mirtschin, P. J., & Vidal, N., (2011). Snake venom: From fieldwork to the clinic: Recent insights into snake biology, together with new technology allowing high-throughput screening of venom, bring new hope for drug discovery. *Bio Essays, 33,* 269–279.

Wadhwa, R., Nigam, N., Bhargava, P., Dhanjal, J. K., Goyal, S., Grover, A., Sundar, D., Ishida, Y., Terao, K., & Kaul, S. C., (2016). Molecular characterization and enhancement of anticancer activity of caffeic acid phenethyl ester by γ cyclodextrin. *J. Cancer., 7,* 1755–1771.

Walle, T., Otake, Y., Brubaker, J. A., Walle, U. K., & Halushka, P. V., (2001). Disposition and metabolism of the flavonoid chrysin in normal volunteers. *Br. J. Clin. Pharmacol., 51,* 143–146

Wan, X., Shen, N., Mendoza, A., Khanna, C., & Helman, L. J., (2006). CCI-779 Inhibits Rhabdomyosarcoma xenograft growth by an antiangiogenic mechanism linked to the targeting of mTOR/Hif-1a/VEGF signaling. *Neoplasia., 8,* 394–401.

Wang, H. W., Lin, C. P., Chiu, J. H., Chow, K. C., Kuo, K. T., Lin, C. S., & Wang, L. S., (2007). Reversal of inflammation-associated dihydrodiol dehydrogenases (AKR1C1 and AKR1C2) overexpression and drug resistance in nonsmall cell lung cancer cells by wogonin and chrysin. *Int. J. Cancer., 120,* 2019–2027.

Wang, J. H., Zhang, K., Niu, H. Y., Shu, L. H., Yue, D. M., Li, D. Y., & He, P., (2013). Jolkinolide B from Euphorbia fischeriana Steud induces in human leukemic cells apoptosis via JAK2/STAT3 pathways *Int. J. Clin. Pharmacol. Ther., 51,* 170–178.

Wang, J. H., Zhou, Y. J., Bai, X., & He, P., (2011). Jolkinolide B from Euphorbia fischeriana Steud induces apoptosis in human leukemic U937 cells through PI3K/Akt and XIAP pathways. *Mol. Cells., 32,* 451–457.

Wang, Y. B., Hu, Y., Li, Z., Wang, P., Xue, Y. X., Yao, Y. L., Yu, B., & Liu, Y. H., (2013). Artemether combined with shRNA interference of vascular cell adhesion molecule-1 significantly inhibited the malignant biological behavior of human glioma cells. *PLoS One,* *8*, e60834.

Wang, Y., Li, C. F., Pan, L. M., & Gao, Z. L., (2013). 7,8-Dihydroxycoumarin inhibits A549 human lung adenocarcinoma cell proliferation by inducing apoptosis via suppression of Akt/NF-κB signaling. *Exp. Ther. Med., 5*, 1770–1774.

WebMD, (2018). *What Is a Pseudomonas Infection?* https://www.webmd.com/a-to-z-guides/ pseudomonas-infection#1 (accessed on 20 May 2020).

WebMD. *Renal Cell Carcinoma.* https://www.webmd.com/cancer/renal-cell-carcinoma#1 (accessed on 20 May 2020).

Wise, D. R., & Thompson, C. B., (2010). Glutamine addiction: A new therapeutic target in cancer. *Trends Biochem. Sci., 35*, 427–433.

World Health Organization, (2018). *Cancer.* http://www.who.int/cancer/en/ (accessed on 20 May 2020).

Wu, M. C., Arimura, G. K., & Yunis, A. A., (1978). Mechanism of sensitivity of cultured pancreatic carcinoma to asparaginase. *Int. J. Cancer., 22*, 728–733.

Xu, W., Liu, J., Li, C., Wu, H. Z., & Liu, Y. W., (2008). Kaempferol-7-O-β-d-glucoside (KG) isolated from *Smilax china* L. rhizome induces G2/M phase arrest and apoptosis on HeLa cells in a p53-independent manner. *Cancer Lett., 264*, 229–240.

Xu, X., Zhang, Y., Qu, D., Jiang, T., & Li, S., (2011). Osthole induces G2/M arrest and apoptosis in lung cancer A549 cells by modulating PI3K/Akt pathway. *J. Exp. Clin. Cancer Res., 30*, 30.

Yaghoobi, H., Bandehpour, M., & Kazemi, B., (2016). Apoptotic effects of the B subunit of bacterial cytolethal distending toxin on the A549 lung cancer cell line. *Asian Pac. J. Cancer Prev., 17*, 299–304.

Yamamoto, Y., & Gaynor, R. B., (2001). Therapeutic potential of inhibition of the NF-κB pathway in the treatment of inflammation and cancer. *Journal of Clinical Investigation, 107*, 135–142.

Yamanaka, K., Ryan, K. S., Gulder, T. A., Hughes, C. C., & Moore, B. S., (2012). Flavoenzyme-catalyzed atropo-selective n,c-bipyrrole homocoupling in marinopyrrole biosynthesis. *J. Am. Chem. Soc., 134*, 12434–12437.

Yamazaki, Y., & Morita, T., (2007). Snake venom components affecting blood coagulation and the vascular system: Structural similarities and marked diversity. *Curr. Pharm. Des., 13*, 2872–2886.

Yang, S. H., Chien, C. M., Lu, M. C., Lu, Y. J., Wu, Z. Z., & Lin, S. R., (2005). Cardiotoxin III induces apoptosis in K562 cells through a mitochondrial-mediated pathway. *Clin. Exp. Pharmacol. Physiol., 32*, 515–520.

Yardley, D. A., (2010). Visceral disease in patients with metastatic breast cancer: Efficacy and safety of treatment with ixabepilone and other chemotherapeutic agents. *Clin. Breast Cancer, 10*, 64–73.

Yates, S. P., & Merrill, A. R., (2004). Elucidation of eukaryotic elongation factor-2 contact sites within the catalytic domain of pseudomonas aeruginosa exotoxin A. *Biochem. J., 379*, 563–572.

Yeh, F. T., Wu, C. H., & Lee, H. Z., (2003). Signaling pathway for aloe-emodin-induced apoptosis in human H460 lung nonsmall carcinoma cell. *Int. J. Cancer., 106*, 26–33.

Yu, M., Henning, R., Walker, A., Kim, G., Perroy, A., Alessandro, R., Virador, V., & Kohn, E. C., (2012). L-asparaginase inhibits invasive and angiogenic activity and induces autophagy in ovarian cancer. *J. Cell. Mol. Med., 16*, 2369–2378.

Zhang, T., Chen, X., Qu, L., Wu, J., Cui, R., & Zhao, Y., (2004). Chrysin and its phosphate ester inhibit cell proliferation and induce apoptosis in Hela cells. *Bioorg. Med. Chem., 12,* 6097–6105.

Zhang, W., Li, Q., Zhu, M., Huang, Q., Jia, Y., & Bi, K., (2008). Direct determination of polydatin and its metabolite in rate excrement samples by high-performance liquid chromatography. *Chem. Pharm. Bull., 56,* 1592–1595.

Zhang, X., Lin, D., Jiang, R., Li, H., Wan, J., & Li, H., (2016). Ferulic acid exerts antitumor activity and inhibits metastasis in breast cancer cells by regulating epithelial to mesenchymal transition. *Oncol Rep, 36,* 271–278.

Zhang, Y., Schulte, W., Pink, D., Phipps, K., Zijlstra, A., Lewis, J. D., & Waisman, D. M., (2010). Sensitivity of cancer cells to truncated diphtheria toxin. *PLoS One, 5,* e10498.

Zhou, G. X., & Molinski, T. F., (2006). Manoalide derivatives from a sponge, *Luffariella* sp. *J. Asian Nat. Prod. Res., 8,* 15–20.

Zilla, M. K., Nayak, D., Amin, H., Nalli, Y., Rah, B., Chakraborty, S., Kitchlu, S., Goswami, A., & Ali, A., (2014). 4′-Demethyl-deoxypodophyllotoxin glucoside isolated from *Podophyllum hexandrum* exhibits potential anticancer activities by altering Chk-2 signaling pathway in MCF-7 breast cancer cells. *Chem.-Biol. Interact, 224,* 100–107.

CHAPTER 6

Current and Future Perspectives of Marine Drugs for Cancer Disorders: A Critical Review

BHASKARAN MAHENDRAN,[1] THIRUMALARAJU VAISHNAVI,[2] VISHAKANTE GOWDA,[1] JOHURUL ISLAM,[3] NARAHARI RISHITHA,[3] ARUNACHALAM MUTHURAMAN,[3,4] and RAJAVEL VARATHARAJAN[4]

[1]*Department of Pharmaceutics, JSS Academy of Higher Education and Research, Mysuru – 570015, Karnataka, India*

[2]*Department of Pharmacy Practice, JSS Academy of Higher Education and Research, Mysuru – 570015, Karnataka, India*

[3]*Department of Pharmacology, JSS Academy of Higher Education and Research, Mysuru – 570015, Karnataka, India*

[4]*Pharmacology Unit, Faculty of Pharmacy, AIMST University, Semeling, Bedong – 08100, Kedah, Darul Aman, Malaysia, E-mail: arunachalammu@gmail.com (A. Muthuraman)*

ABSTRACT

Cancer is an unchecked growth of normal cells via accelerating cell division; inhibition of cell cycles (arrest); and activation of programmed cell death. Current cancer chemotherapeutic agents are mainly obtained natural sources from plants like vincristine, vinblastine, etoposide, and paclitaxel; from animals, i.e., antibodies; from marine sources, i.e., cytarabine, aplidine, and dolastatin 10; and from micro-organisms such as dactinomycin, bleomycin, and doxorubicin. Conventional medicines of anti-cancer agents are expensive and produce numerous adverse effects and organ failure. The natural anti-cancer medicines are better than synthetic anticancer agents due to efficacy and reduction of side effects. In addition, some of the newer research reports revealed that some of the nutraceuticals agents possess a better therapeutic action against the cancer cell. However, there is no strong evidence of the anti-cancer actions of marine drugs. Further, the clear molecular mechanisms of marine compounds are not explored yet. This book chapter makes understanding marine drugs and possible molecular action on cancer cells. Therefore, it can be open the Pandora's Box to discover the newer marine drugs can treatment for cancer disorders.

6.1 INTRODUCTION

Cancer leads to major life-threatening genetic disorders and is caused by the changes in varieties of genes. The changes of abnormal genetic materials alter the cell function, cell division, and programmed cell death process (apoptosis). The changes of genes may be hereditary from parents or lifestyle modification of persons. Exposures by certain environmental factors are also responsible for the damage of deoxyribonucleic acid (DNA) and mutation of genes (Jackson and Bartek, 2009). Such exposures are chemicals: tobacco smoke, nitrosamines, asbestos fibers and polycyclic aromatic hydrocarbons and benzene; diet: high-salt diet, aflatoxin B1, gluten-free diet and betel nut; infection: By on co-viruses (i.e., human papillomavirus, Epstein–Barr virus, hepatitis B&C viruses, Kaposi's sarcoma herpes virus, and human T-cell leukemia virus-1), by bacterial infection (i.e., *Helicobacter pylori)*, parasitic infections (i.e., liver flukes, *Schistosoma haematobium*, *Clonorchis sinensis* and *Opisthorchis viverrini*; radiation: ionizing ultraviolet rays and non-ionizing radiofrequency radiation, and electric power transmission; heavy metal, i.e., cobalt, and nickel and crystalline silica exposure; hormones: high level of estrogen & progesterone in women and higher and lowest levels of testosterone in men are causing the varieties of cancer in human being (Xiang et al., 2017). In addition, autoimmune diseases like celiac disease, Crohn's disease, and ulcerative colitis also causing intestinal cancers. The inherited and lifestyle factors make the three main types of gene alterations, i.e., proto-oncogenes, tumor suppressor genes, and DNA repair genes (Chiurchiu and Maccarrone, 2011). In a normal cell, proto-oncogenes, and tumor suppressor genes are responsible for the regulation of normal cell growth and cell division; DNA repair genes are responsible for fix the damaged DNA (Okazaki et al., 2003). In a cancer cell, these genes are overactive than normal leads to enhance the rapid growth of cells activates the uncontrolled cell division and increase the cell life span (Reya and Clevers, 2005). Further, it also produces the paracrine function to develop additional mutations in other genes. The prevalence rate of cancer is higher in various countries especially in developing countries (zur Hausen, 2002). As per the report of WHO, cancer is the second leading cause of death in the world after cardiovascular disease (Nagai and Kim, 2017). According to the current WHO report, in 2018: 9.6 million deaths are accounted with cancer disorders. The most common form of cancer is lung cancers. The death rate with lung cancer is 2.09 million in 2018 (https://www.who.int/news-room/fact-sheets/detail/cancer). Therefore, the rising of global burden with cancer disorders needs to manage with more potential and effective anti-cancer agents without compromising the adverse drug reactions. Generally, cancer disorders are treated with various natural and synthetic molecules (Dela Cruz et al., 2011). However, their management is still complicated and cause life-threatening side effects. Therefore it makes the higher challenging and emerging situation in the health sector.

Furthermore, the types of cancer can be classified based on location, cell type, and microscopical features. The epithelial cells cancers also called carcinoma. This type is very common in all kind of tissue cancer like breast, prostate, lung, pancreas, and colon. The cancer formation in connective tissue such as fat, cartilage, bone, and nerve called sarcoma (Righi et al., 2007). It is developed from mesenchymal cells of bone marrow (outside). The development of cancer growth in hematopoietic cells and lymph nodes are called lymphoma and leukemia respectively. Even, embryonic tissue cancer growth called blastoma (Balkwill, 2004). In addition, germ cell cancer growth also called germ cell tumor. Microscopical features based cancer cells are also called spindle cell carcinoma, small-cell carcinoma and giant cell carcinoma (Martinelli et al., 2017). The cancer disorders are treated with various categories of medicines like synthetic molecules like alkylating agents (mechlorethamine and methotrexate), antimetabolites (fluorouracil, gemcitabine, and 6-thioguanine), and hormones (raloxifene, toremifene, and fulvestrant); natural medicines from plant, i.e., vinblastine, vincristine, paclitaxel, and podophyllotoxins; and dietary source, i.e., allicin, curcumin, lycopene, genistein, capsaicin, resveratrol, diosgenin, gingerol, catechins, eugenol, Vitamin C, Vitamin E, D-limonene, and beta carotene (Soo et al., 2017). In addition, immunotherapy, radiation, surgery, and laser therapies also have a significant role in the management of cancer disorders. However, still, all categories of medicines are causing life-threatening toxic effects (Janicek and Averette, 2001). Comparatively, plant-based medicines are shown potential action on cancer disorders with fewer side effects. Therefore, newer medicines from additional natural sources are preferred for the finding of lead molecules for cancer disorders. This book chapter is focused to explore the possible preventive action of marine drugs in cancer cell proliferation.

6.2 IMPORTANCE OF MARINE RESOURCES FOR PREVENTION OF CANCER DISORDERS

Various marine drugs are identified to treat multiple life-threatening disorders. Currently, the FDA approved the various marine-derived products to treats cancer disorders. Such agents are eribulin mesylate (used for metastatic breast cancer), cytarabine (treatment of leukemia), brentuximab vedotin (used for anaplastic T-cell malignant lymphoma and Hodgkin's lymphoma), and trabectidine (treatment of ovarian cancer and soft tissue sarcoma) (Grignani et al., 2018; Fleeman et al., 2019; Sandoval-Sus et al., 2019). Furthermore, marine-derived products are documented to produce the multi-targeted action on cancer cells by more potent and efficacy. In addition, around 1500 natural molecules are isolated from marine origin, and tested for their potency with various *in-vivo* biological screening methods. Moreover, these marine resources are also showing promising role in the management of cancer cell proliferation (Fleeman et al., 2019). Marine drugs have potent anticancer property via inhibition of inflammatory signaling pathway, i.e., NF-κB, phosphorylation of p65 and inhibitory kappa B (IκB) in breast cancer cells and leukemia. Some of the marine products regulate the topoisomerase-I activity, pro-angiogenic actions, and epidermal growth factor receptor (EGFR) protein expression. These are the major target for the development of human cancer cell proliferation (Dyshlovoy and Honecker, 2018; Malve 2016). Hence, the marine drugs based discovery of newer molecules is important to treat multiple cancer cell proliferation.

6.3 MARINE DRUGS: AN OVERVIEW

Various medicines are obtained from the marine sources and it produces the potential therapeutic actions for the prevention of progression of cardiovascular and neurological disorders. In addition, some of the molecules are also documented to produce anti-cancer effects. Because, marine molecules possess the potential antioxidant, anti-inflammatory, immunomodulatory, anti-apoptotic, and cytotoxic effects (Clark, 1996). The following sections are expressed the possible action of marine drugs to prevents cancer cell proliferation.

6.3.1 ANTICANCER AGENTS FROM MARINE FLORA

Marine floras like seaweeds, bacteria, cyanobacteria, actinobacteria, micro-algae, fungi, mangroves plants and other halophytes are widely spread over the oceanic regions. It has a large quantity of bioactive and structural similarity of conventional anti-cancer drugs. The major chemical constituents, i.e., polyphenols, *and* sulfated polysaccharides are present in marine floras. These compounds possess the antioxidant, immune-stimulatory, and anti-tumor actions (Sithranga Boopathy and Kathiresan, 2010). In addition, it also activates the macrophages, enhances the apoptosis of cancer cells and prevents the oxidative stress associated damage of DNA. This action known to prevents cancer cells proliferation and control the cancer cell division. However, the evidence of marine floras on cancer cell is limited (Valko et al., 2006). Therefore, marine floras area major marine source for the isolation of lead compound used for cancer disorders. The microbial floras are present in various invertebrates like marine fish and chordate phylum (Sable et al., 2017). These invertebrates have medicinal properties for various disorders, especially for cancer disorder. Such invertebrates are sponges, soft corals, sea hares, sea fans, bryozoans, nudibranchs, and tunicates. Some of the compounds derived from the marine organisms have antioxidant property and anticancer activities (Lindequist, 2016). Therefore, it is expected to prevent cancer cell proliferation.

6.3.2 ANTICANCER AGENTS FROM MARINE ALGAE (SEAWEED)

Seaweeds are the main sources of vitamins, iodine, protein, and minerals. The metabolites of seaweeds have a promising role in treating cancer cell proliferation. In addition, *Halimeda* sp. (Chlorophyceae) has a large number of polyphenols like catechin, epicatechin, epigallocatechin gallate, and gallic acid. These compounds are documented to the shown antioxidant, anti-tumor, and immunomodulating actions (Bocanegra et al., 2009). Further, edible seaweed, i.e., *Palmaria palmate* produces the anticancer activity with reduction of oxidative stress. The alcoholic extract of *Acanthophora spicifera* (red algae) inhibits the growth of Ehrlich's ascites carcinoma cells. This report is similar to that of 5-fluorouracil treatment. Therefore, it considers as anti-cancer agents. Similarly, various marine algae like *Ulva reticulata, Gracilariaf oliifera*, and *Padina boergesenii* are reported to produce cyto-toxic action (Patra and Muthuraman, 2013). Moreover, it also produces the anticancer action on adenocarcinoma cell via immunomodulatory action.

The low-molecular-weight fucoidan of *Ascophyllum nodosum* also reported to produces the anti-proliferative effect on malignant and normal cells like fibroblasts, sigmoid colon adenocarcinoma cells, and smooth muscle cells via antitumor, anticancer, antimetastatic, and fibrinolytic action (Huang and Pardee, 1999). Furthermore, a variously isolated compound of marine algae, i.e., stylopoldione from *Stypodium* sp.; and condriamide-A from the *Chondria* sp. have produced the cytotoxic effect in colorectal cancer cells and human nasopharyngeal (Sung et al., 2005). In addition, caulerpenyne from *Caulerpa* sp.; meroterpenes and usneoidone from *Cystophora* sp.; eckol, phlorofucofuroeckol A, dieckol, and 8,8 –bieckol of brown alga (*Eisenia bicyclis*) also produced the anticancer, antitumor, and anti-proliferating effects (Popovich and Kitts, 2004). Further, these compounds possess the various molecular mechanisms for an anticancer activity like modulation of apoptotic activity via caspase-7, caspase-8, Bax, Bcl-xL, and poly (ADP-ribose) polymerase (PARP). PARP is inducers of apoptosis of cancer cells via DNA cleavage (Elmore, 2007). However, the large exposure and clinical evidence remain to be explored. Further, seaweeds reduce the higher level of potassium in plasma leads to prevent the abnormalities of blood pressure (Srinath Reddy and Katan, 2004). Clinically, the seaweed-derived alginic acid is producing the glucose-lowering effects in diabetic patients. The dietary fibers of seaweed reported on antioxidant, anti-mutagenic, anticoagulant, and antitumor activities. In addition, it also plays an important role in the modification of lipid metabolism and protein modification in the cancer cell.

6.3.3   ANTICANCER AGENTS FROM MICRO-ALGAE

Marine blue-green algae also are known as cyanobacteria are possessed the novel bioactive compounds and are used for the various pharmaceutical applications (Abed et al., 2009). The major marine cyanobacteria compounds, i.e., scytonemin has potential anticancer effects due to its potential inhibitory action of serine/threonine kinase in cancer cells (Strebhardt and Ullrich, 2006). This molecule mainly isolated from the cyanobacterium *Stigonema* sp. In addition, scytonemin regulates the formation of cytoskeletal proteins, i.e., a mitotic spindle which is required for cell division process which leads to reducing the proliferation of fibroblasts and endothelial cancer cells (Matutes, 2018). In addition, indole, and phenanthridine alkaloids of microalgae, i.e., calothrixin A and B have potential anticancer effects at nanomolar concentrations (Sithranga Boopathy and Kathiresan, 2010). The extracts of *Calothrix* also inhibit the growth of human HeLa cancer cells in a dose-dependent manner (Gonzalez-Resendiz et al., 2018). Further, newer molecules, i.e., curacin-A (isolated from *Lyngbya majuscula* shown potent antiproliferative action on the breast, colon, and renal cancer cells via inhibits of tubulin polymerization process (Giordano et al., 2018). The isolation of largazole is also reported to produce the antiproliferative actions and it is obtained from *Symploca* Sp (Taori et al., 2008). The apratoxins also acts on adenocarcinoma cells and this compound is isolated from *Lyngbya boulloni* (Swain et al., 2015). Similarly, coibamide A is also shown cytotoxic effects on lung and neuro-2a cells. This compound is isolated from *Leptolyngbya lagerheimii* (Calderon et al., 1999). Currently, cyanovirin, cryptophycin 1 & 8, and borophycin, have anticancer potential in a variety of cancer cells (Sithranga Boopathy and Kathiresan, 2010). Borophycin (boron-containing metabolite) is isolated from cyanobacterial strains, i.e., *Nostoclinckia, Streptomyces antibioticus,* and *N. spongiaeforme var.* tenue (Dembitsky et al., 2011). Cryptophycin 1 also possesses the anticancer potential and it is isolated from *Nostoc*Sp. It mainly acts in solid tumors (Luesch et al., 2001). Therefore, the source of microalgae can play a key role in the innovation of marine drugs for cancer disorders.

6.3.4 ANTICANCER AGENTS FROM MARINE BACTERIA

Marine microorganisms have multiple varieties of newer genes. Marine bacteria are known to develop the secondary metabolites and it has potential anti-inflammatory action, e.g., pseudopterosins, topsentins, scytonemin, and manoalide; anticancer action, e.g., bryostatins, discodermolide, eleutherobin, and sarcodictyin; and antibiotic actions, e.g., marinone. In addition, probiotic bacteria like bifido bacteria and lactobacilli are documented to control the pathogenic microbes. And, it releases the antibacterial protein called bacteriocin (Malve, 2016) and anticancer substances (Lordan et al., 2019). Experimentally, the dietary supplements of lactobacilli have reduced the progress of colon cancer via variation of the immune system and reduction of proinflammatory cytokines (Tuohy et al., 2003). This cell regulatory function is occurred due to inhibition of NFκB pathways, and rising of anti-inflammatory cytokines, i.e., IL-10, IgA & host defense peptides, i.e., β-defensin 2 (Bian et al., 2017). Some reports are revealed that marine microbial toxins are useful for the regulation of neurophysiological and neuropharmacological action. The major metabolite marine bacteria, i.e., macrolactin-A has inhibited the proliferation of melanoma cancer cells, herpes simplex virus (HSV) infection and T-lymphocytes associated human immunodeficiency virus (HIV) replication (Valyi-Nagy et al., 2018). Another peptide, i.e., kahalalide F (KF) of *Elysia rubefescens* induces the cytotoxicity and inhibits the cell cycle (G1 phase) pattern of the cancer cells. Further, KF has potential anti-proliferative effects on breast and colon cancer cells (Choi et al., 2009). However, more extensive studies are essential to isolate and explore the anti-cancer potential of marine bacteria compounds.

6.3.5   ANTICANCER AGENTS FROM ACTINOMYCETES

Actinomycetes are mainly developed from the soil source. The sediment of marine lands especially in Laguna de Terminos, Gulf of Mexico has potent anti-cancer molecules, i.e., gutingimycin. The high polar compound of gutingimycin is trioxacarcin derivatives and it isolated from *Streptomyces* species. Further, this species also has large yields of trioxacarcins D &F molecules. These compounds are also documented to shown potential antibiotic and anticancer actions via inhibition of cancer cell proteasome function (Tan et al., 2015). In addition to that, the newer marine compound, i.e., thiocoraline (depsipeptide) has anticancer action via inhibits RNA synthesis process of cancer cells. From *Micromonospora marina* it was isolated and it is located in Mozambique Strait. It selectively acts on colon and lung cancer cell lines and moreover in melanoma (Newman and Cragg, 2004). Moreover, it exerts the anti-proliferative action on defective p53 associated colon cancer cell lines (Efferth et al., 2001). Hence, thiocoraline claimed as an anticancer agent for colon cancer. However, the more extensive investigation is required to use the marine actinomycetes derived molecules in a human being.

6.3.6   ANTICANCER AGENTS FROM MARINE FUNGI

Marine fungi are one of the species live in estuarine or marine environments. These marine fungi are sporadically or wholly submerged in seawater. Normally, marine fungi are occupied the terrestrial or freshwater and it also capable to sporulate in a marine environment. Different marine habitats support to very different fungal communities (Shavandi et al., 2019). Marine fungi can exist parasitic or saprobic on algae, parasitic or saprobic on animals, saprobic on dead wood or saprobic on plants. The filamentous fungi of marine origin have the highest bioactive metabolites. Three marine fungi general, i.e., penicillin, *Aspergillus,* and *Fusarium* are possessed multiple pharmacological actions including anticancer effects (Dedeyan et al., 2000). There are limited pharmacological studies are reported about marine fungi based medicines. The lignicolous fungus, i.e., *Leptosphaeria oraemaris* has bioactive molecules, i.e., leptosphaerin, polyketides, leptosphaerolide, *o*-dihydroquinone derivative and leptosphaerodione (Deshmukh et al., 2017). The secondary metabolites of marine fungi, i.e., acremonin A (isolated from *Acremonium* sp.) and xanthone (isolated from *Wardomyces anomalus*) are also producing the antioxidant and anticancer effects (Konig et al., 2006). Therefore, marine fungus is one of the sources to discover newer molecules to treats the cancer disorders.

6.3.7 ANTICANCER AGENTS FROM MANGROVE PLANTS

Mangrove plants are specialized properties to grow in different environmental setup. The roots of this plant are developed from mud gathered and trees grow upward. This plant plays a large and important role in the maintaining of ecosystems. It's grown in the saltwater zones, i.e., between water and land (Hardoim et al., 2015). Further, it has properties of physiological adaptations to overcome the various environmental problems like anoxia, high salinity, and frequent tidal inundation. Therefore, it has variable secondary metabolites depends on season and environmental stimuli. Thus, it has a variety of anti-cancer molecules. In folk medicine, mangroves parts play an important role in treats the multiple ailments (Kerry et al., 2018). The mangrove plants list and their compounds for the anticancer activity are summarized in Table 6.1. Sulfur-containing an alkaloid, 1,2-dithiolane (brugine, isolated form *Bruguiera sexangula*) is used for the arresting of tumor cell growth. In addition, tannin of this plant also showed potential anti-cancer actions against the lung carcinoma. Further, the ribose derivative of 2-benzoxazoline (isolated from *Acanthus ilicifolius*) is also shown anticancer as well as antiviral activity (Sithranga Boopathy and Kathiresan, 2010). Moreover, the tea extraction of *Ceriops decandra* is produced by the dimethyl benz[a]anthracine-induced oral cancer (Das et al., 2015, Sithranga Boopathy and Kathiresan, 2010). In spite of enriched resources of secondary metabolites, the major mangrove floras are also had additional anticancer compounds, whereas the screening of these compounds is still unexplored. Therefore, the detailed study of mangrove plants and flora based discovery of newer anti-cancer molecules are urgent requirements to bring the lead compounds for cancer therapy.

TABLE 6.1   List of Mangrove Plants and Their Compounds for Anticancer Actions

Sl. No.	Mangrove Plants	Anti-Cancer Compounds
1.	*Ceriops decandra*	Quinine
2.	*Pongamia pinnata*	Lanceolatin B, chalcone, flavonoid, and polyhydroxylates
3.	*Avicennia officinalis*	Triterpene, and betulinic acid
4.	*Avicennia marina*	Napththoquinones, avicequinone, stenocarpoquinone, and iridoidglycosides
5.	*Bauhinia variegate*	Steroids, triterpenoids, and flavonoids
6.	*Heritieria fomes*	Phenolic compounds
7.	*Excoecaria agallocha*	Diterpenes, Tannin,, and Excoecarin
8.	*Xylocarpus mekongenesis*	Xylomolin, and xyloeccensin
9.	*Bruguiera gymnorrhiza*	Brugine
10.	*Calophylum inophylum*	Bioflavonoids, neoflavonoids, xanthone, and benzophenones
11.	*Bruguiera sexangula*	Brugine, tropine, isobutyric, and benzoic acid
12.	*Avicienna alba*	Naphthoquinone, and avicequinone
13.	*Caesalpinia bonduc*	Caesalpines
14.	*Agiceras corniculatusm*	Hydroxyquinone
15.	*Acanthus ilicifolius*	–
16.	*Acanthus ebracteatus*	–
17.	*Conocarpus erectus*	–
18.	*Phoenix paludosa*	–
19.	*Thespesia populnea*	–

6.4   SUMMARY OF MARINE DRUGS FOR ANTICANCER ACTIVITY

Marine drugs are shown the potential action on the cancer cell. The details of marine drugs source and their active constituents are explored in the previous section. The summarized pattern of marine drugs and their sources has been mentioned in Table 6.2.

6.5   MECHANISMS OF MARINE DRUGS FOR ANTICANCER ACTIVITY

The marine drugs mainly act on genetic materials of cancel cells. These activities are similar to that of the conventional anticancer agent. The

TABLE 6.2　Source of Marine drugs for Anticancer Actions

Sl. No.	Marine Drugs	Sources	References
1.	Microviridin and siatoxin	*Microcystis aeruginosa*	Genuário et al., 2019; Rao et al., 2002
2.	Daunorubicin	*Streptomyces peucetius*	Qiao et al., 2019
3.	Apratoxins	*Cyanobacteria* sp.	Bajpai et al., 2018; Li et al., 2019
	Apratoxin A	*Lyngbya boulloni*	
4.	Cyptophycin 1	*Nostoc linckia*	Kitagaki et al., 2015
5.	Cryptophycin 8	*Nostocspongiaeforme*	Cazzamalli et al., 2018
6.	Stypoldione	*Stylopodium* sp.	Jerković et al., 2019
7.	Condriamide A	*Chondria* sp.	Sithranga Boopathy and Kathiresan, 2010
8.	Caulerpenyne	*Caulerpa* sp.	Alves et al., 2018
9.	Meroterpenes and Usneoidone	*Cystophora* sp.	Kerry et al., 2018; Zong et al., 2018
10.	Largazole	*Symploca* sp.	Song et al., 2018
11.	Coibamide A	*Leptolyngbya* sp.	Kumar et al., 2012
12.	Scytonemin	*Stigonema* sp.	Fan et al., 2018
13.	Fucoidan	*Ascophyllum nodosum*	Silchenko et al., 2018
14.	Phloroglucinol (eckol), phlorofucofuroeckol A, dieckol, 8,8-bieckol	*Palmaria palmata* *Eisenia bicyclis*	Abbas et al., 2018; Pangestuti et al., 2018; Park et al., 2018
15.	Borophycin	*Nostoc linckia, Nostoc spongiaeformevar*	Arai et al., 2004; Bajpai et al., 2018
16.	Crude extracts	*Acanthophora spicifera, Sargassum thunbergii*	Sali et al., 2018; Jin et al., 2019

induction of cancer cell DNA damage is the most important steps to achieve anticancer actions (Edmondson et al., 2014). Free radicals have double edge sward functions. One side protects the cells by stimulating the immune system. In another way, it increases the oxidative stress to cellular environment leads to induce the DNA mutation, alteration of the free radical defensive system and inactivates the normal cellular apoptotic signals (Sosa et al., 2013). These lead to enhance cancer cell proliferation. Natural plant-based medicines are ameliorated the cancer cell proliferation via interference of free radicals. Antioxidants are defending against the free radicals via superoxide dismutase (SOD), glutathione peroxidase (GPx), and catalase regulation. The natural vitamins E & C, and carotenoids are had therapeutic potential for cancer cell proliferation (Valko et al., 2006). In addition, certain minerals like selenium, copper, manganese, and zinc are played catalytic actions on cancer cells via alteration of oxidative defense enzymes (Fang et al., 2002). Further, it protects the normal cell from free radical associated abnormalities of membrane lipids and phospholipid integrity. In cancer cells, these actions are reversible direction leads to alter the cell membrane integrity (Rice-Evans and Burdon, 1993). These reversible actions of natural antioxidants are due to alteration of multiple cell signal pathways via its own free radical action. Example, vitamin C donates the more stable hydrogen atom via L-ascorbate anion and it is effective against the hydrogen peroxide, superoxide radical anion, hydroxyl radical, and singlet oxygen (Kumar et al., 2012). Another metabolite of vitamin C is dehydroascorbate and it is unstable but can be recycled like vitamin E via reduced glutathione and nicotinamide adenine dinucleotide phosphate (NADPH) action with hexose monophosphate shunt pathway (Chaudiere and Ferrari-Iliou, 1999). Therefore, chronic radical scavenging action of antioxidants in cancer cells can reduce the DNA mutation and activates the DNA repairing process (Valko et al., 2004).

Moreover, free radical also has interactive properties with normal and cancer cell immune system and an apoptotic cell process. The programmed cell death in normal cells regulates the cell cycle pattern via maturation, autophagy, and cell division (Lotze et al., 2007). These actions are also initiated with immunological agents. It may be the endogenous and/or exogenous mechanism. Further, apoptotic machinery is also triggered by the expression of various post-translational proteins by the balanced action of anti- and pro-apoptotic factors (Valko et al., 2007). The up-regulation of anti-apoptotic proteins, down-regulation of pro-apoptotic proteins, and reduction of caspases protein expression alters the speed of apoptosis action in cancer as well as normal cells (Lukiw et al., 2009). In cancer cells, avoidance of apoptosis process is identified and it facilitates the cancer cell proliferation via inhibition of cell differentiation, enhancing of angiogenesis, and raise the cancer cell motility, invasion, and metastasis process (Egeblad and Werb, 2002). Therefore, the induction of the apoptosis process in a cancer cell is one of the treatment strategies to control cancer cell proliferation. The radiation therapy and chemotherapy like tamoxifen are shown induction of mitigation and apoptosis of cancer cell growth (Dunn et al., 1997). Many plant extracts and phytoconstituents like curcumin and resveratrol are inducing the apoptosis of cancer cells via up-regulation of immune surveillance and activation of macrophage

(Sithranga Boopathy and Kathiresan, 2010). Marine drugs like fucoidans are also known to activate the macrophages and induction of apoptosis via immunological reactions. Further, fucoidan also induces the cytokine production like interleukin-1 (IL-1) and interferon-γ (IFN-γ) leads to enhances the functions of T-lymphocyte, B-lymphocyte, natural killer cell (NK cell) and macrophage via primary antibody response (Sithranga Boopathy and Kathiresan, 2010). Similarly, the cancer cells are regulated by multiple cell signaling processes. The details of marine drugs and their molecular mechanisms are explored in Table 6.3.

TABLE 6.3 List of Marine-Derived Compounds That Have Exhibited Potential as Cancer Therapies

Sl. No.	Marine Source	Compounds	Molecular Mechanisms	References
1.	Bryozoan	Bryostatin-1	PKC activator	Khan and Nelson, 2018
2.	*Dolabella auricularia*	Aplysiatoxin	PKC activator	Irie and Yanagita, 2014
3.	Marine sponge	(3R)-icos-(4E)-en-1-yn-3-ol and (3R)-14-methyldocos-(4E)-en-1-yn-3-ol	IGF-1Rβ	Song et al., 2018
		Hymenialdisine and Debromohymenialdisine	CDK1, CDK2, CDK5,	Phutdhawong et al., 2012
		Fascaplysin	CDK4	Lyakhova et al., 2018
		Palinurin	GSK-3β	Gao et al., 2018
		Manzamine A		Gao et al., 2018
		PMH-1 and PMH-2		Song et al., 2018
		Halenaquinone	HDACs	Shih et al., 2015
		Spongiacidin C	USP-7	Zhou et al., 2018
		Sulawesins A–C		Afifi et al., 2017
		12β-(3'β-hydroxybutanoyloxy)-20,24-dimethyl-24-oxo-scalara-16-en-25-al	Hsp90	Lai et al., 2016
		Sipholane triterpenoids	P-gp	Hamed et al., 2019

TABLE 6.3   *(Continued)*

Sl. No.	Marine Source	Compounds	Molecular Mechanisms	References
		Panicein A hydroquinone	Patched (Multidrug Transporter Ptch1)	Hasanovic and Mus-Veteau, 2018
		Swinhosterol B	PXR agonist	Li et al., 2018
		Mycalamide A	Protein synthesis inhibitor	Venturi et al., 2012
		Pyrroloiminoquinone alkaloids	HIF-1α/p300	Harris et al., 2018
		Stellettin A and Stellettin B	p50/p65	Song et al., 2018
		Psammaplin A	HDAC1	Cincinelli et al., 2018
4.	Sponges and Echinoderms	PXR antagonists 20 and 24	PXR agonist	Cerra et al., 2018
5.	Marine tunicate	Meridianin A-G	GSK-3β	Nisha et al., 2016
6.	Marine fungi	Pannorin, Alternariol, and Alternariol-9-methylether	GSK-3β	Wiese et al., 2016
7.	Red Sea soft coral	Pachycladin A	EGFR and PKC	Song et al., 2018
		Waixenicin A	TRPM7	Liu et al., 2018
8.	Marine echinoderm	1'-deoxyrhodoptilometrin (SE11), and (S)-(−)-rhodoptilometrin	IGF-1R, FAK, EGFR, ErbB2, and ErbB4	Watjen et al., 2017
9.	Marine red alga	BDDPM	FGFR2,3, VEGFR2, PDGFRα, PKB/Akt, eNOS	Bajpai et al., 2018
		Allolaurinterol	eIF4A ATPase activity	Peters et al., 2018
10.	Marine red alga bacterium	Diacetoxyscirpenol	HIF-1α	Choi et al., 2016
11.	Marine cyanobacteria	Hoiamide D	p53/MDM2	Malloy et al., 2012
		Largazole	HDAC1	Chen et al., 2018
		Carmaphycin A and CarmaphycinB	Proteasome inhibitor	Trivella et al., 2014
		Apratoxin A	Hsp90	Kang et al., 2018

TABLE 6.3 *(Continued)*

Sl. No.	Marine Source	Compounds	Molecular Mechanisms	References
		Biselyngbyaside	Calcium channel blocker	Morita et al., 2015
12.	Marine bacterium	Chromopeptide A	HDAC1,2,3,8	Sun et al., 2017
13.	Marine actinomycete bacteria	Salinosporamide A	20S proteasome	Groll et al., 2018
14.	Marine organisms	Metal-based 2, 3-indolinedione	26S proteasome	Zhang et al., 2014
15.	Antarctic fungus	HDN-1	Hsp90	Rong and Yang, 2018
16.	Indian gorgonian octocoral	Elisabatin A	eIF4A ATPase activity	Tillotson et al., 2017

Abbreviation: Akt, serine/threonine-specific protein kinase; CDK, cyclin-dependent kinase; EGFR; eIF4A, eukaryotic initiation factor-4A; eNOS, endothelial nitric oxide synthase; ErbB2, receptor tyrosine-protein kinase; FAK, focal adhesion kinase; FGFRs, fibroblast growth factor receptors; GSK, glycogen synthase kinase; HDAC, histone deacetylases; HIF, hypoxia-inducible factor; Hsp90, heat shock protein 90; IGF-1R, insulin-like growth factor 1 receptor; MDM, mouse double minute; p53, tumor antigen p53; PDGFR, platelet-derived growth factor receptor; PK-B, protein kinase B; PKC, protein kinase C; PXR, pregnane X receptor; S (in 26S proteasome), Svedberg sedimentation coefficient; TRPM7, Transient receptor potential cation channel Subfamily M Member 7; USP, ubiquitin specific protease; and VEGFR, vascular endothelial growth factor receptor.

The summarized Table 6.3 explored the marine-derived compounds and their molecular mechanism for the prevention of cancer cell proliferation.

6.6   CLINICAL TRIAL STATUS OF MARINE PRODUCTS

Various marine-derived products are identified as potential therapeutic efficacy *in vitro* and *in vivo*. Some of the marine-derived products are under clinical trials use to treat cancer disorders. The ongoing marine drugs in clinical trials are plinabulin, plitidepsin, glembatumumabvedotin, and lurbinectedin at phase III levels; mafodotin, depatuxizumab, polatuzumab vedotin, tisotumab AGS-16C3F, PM184, vedotin, enfortumabvedotin, and monomethyl auristatin F at phase II levels; GSK2857916, ABBV-085, ABBV-399, ABBV-221, ASG-67E, ASG-15ME, bryostatin, marizomib, and SGN-LIV1A at phase I levels (Dyshlovoy and Honecker, 2018; Mayer,

2017; Doronina et al., 2018). Furthermore, the additional marine-derived products are pipelines for the clinical trials. Some of newer molecules are identified in different marine sources. These molecules proved their potency and efficacy in various ailments in laboratory animals. Therefore, marine-derived natural molecules are expected to manage multiple cancer disorders.

6.7 FUTURE DIRECTION

The present data make an understanding of marine drugs is to possess anti-cancer actions. Some of the active constituents are identified and isolated from the marine source and it explored the anticancer actions. Hence, it has ample scope to treats cancer disorders. However, the detailed mechanisms (specificity and selectivity of targets and cancer cells) are limited. Further, the toxicological and pharmacological profile and limitations of marine drugs usage in clinical trials need to be evaluated with preclinical (in the animal) as well as in human beings. This book chapter can open Pandora's Box to bring out the newer molecules from marine source to treat cancer disorders.

6.8 CONCLUSION

The present literature revealed that marine molecules possess effective anticancer properties. Some of the molecules are investigated in different cancer cell lines as well as in experimental animals. However, the effective utilization of these molecules can be possible after completion of further investigation about toxicity, pharmacokinetic, and safety studies. Therefore, further studies are required to improve the anticancer activity of marine drugs against different cancer cell proliferation.

ACKNOWLEDGMENTS

The authors are thankful to JSS College of Pharmacy, JSS Academy of Higher Education and Research, Mysuru, India; and Faculty of Pharmacy, AIMST University, Semeling, Malaysia for support to make this book chapter.

KEYWORDS

- **anti-cancer agents**
- **deoxyribonucleic acid**
- **immunotherapy**
- **mangrove plants**
- **marine drugs**
- **nutraceuticals**

REFERENCES

Abbas, M., Saeed, F., & Suleria, H. A. R., (2018). Marine bioactive compounds: Innovative trends in food and medicine. In: *Plant-and Marine-Based Phytochemicals for Human Health: Attributes, Potential, and Use* (p. 61). Taylor & Francis Group, United Kingdom.

Abed, R. M., Dobretsov, S., & Sudesh, K., (2009). Applications of cyanobacteria in biotechnology. *J Appl. Microbiol., 106*(1), 1–12.

Afifi, A. H., Kagiyama, I., El-Desoky, A. H., Kato, H., Mangindaan, R. E. P., De Voogd, N. J., Ammar, N. M., Hifnawy, M. S., & Tsukamoto, S., (2017). Sulawesins A-C, furano sesterterpene tetronic acids that inhibit usp7, from a *Psammocinia* sp. marine sponge. *J. Nat. Prod., 80*(7), 2045–2050.

Alves, C., Silva, J., Pinteus, S., Gaspar, H., Alpoim, M. C., Botana, L. M., & Pedrosa, R., (2018). From marine origin to therapeutics: The antitumor potential of marine algae-derived compounds. *Front Pharmacol.,* p. 9.

Arai, M., Koizumi, Y., Sato, H., Kawabe, T., Suganuma, M., Kobayashi, H., Tomoda, H., & Omura, S., (2004). Boromycin abrogates bleomycin-induced G2 checkpoint. *J. Antibiot., 57*(10), 662–668.

Bajpai, V., Shukla, S., Kang, S. M., Hwang, S., Song, X., Huh, Y., & Han, Y. K., (2018). Developments of cyanobacteria for nano-marine drugs: Relevance of nanoformulations in cancer therapies. *Mar. Drugs., 16*(6), 179.

Balkwill, F., (2004). Cancer and the chemokine network. *Nat. Rev. Cancer., 4*(7), 540–550.

Bian, T., Li, H., Zhou, Q., Ni, C., Zhang, Y., & Yan, F., (2017). Human beta-defensin 3 reduces TNF-alpha-induced inflammation and monocyte adhesion in human umbilical vein endothelial cells. *Mediators Inflamm., 2017,* 8529542.

Bocanegra, A., Bastida, S., Benedi, J., Rodenas, S., & Sanchez-Muniz, F. J., (2009). Characteristics and nutritional and cardiovascular-health properties of seaweeds. *J. Med. Food., 12*(2), 236–258.

Calderon, F. H., Bonnefont, A., Munoz, F. J., Fernandez, V., Videla, L. A., & Inestrosa, N. C., (1999). PC12 and neuro 2a cells have different susceptibilities to acetylcholinesterase-amyloid complexes, amyloid 25–35 fragment, glutamate, and hydrogen peroxide. *J. Neurosci. Res., 56*(6), 620–631.

Cazzamalli, S., Figueras, E., Petho, L., Borbely, A., Steinkuhler, C., Neri, D., & Sewald, N., (2018). *In vivo* antitumor activity of a novel acetazolamide-cryptophycin conjugate for the treatment of renal cell carcinomas. *ACS Omega., 3*(11), 14726–14731.

Cerra, B., Carotti, A., Passeri, D., Sardella, R., Moroni, G., Di Michele, A., Macchiarulo, A., et al., (2018). Exploiting chemical toolboxes for the expedited generation of tetracyclic quinolines as a novel class of PXR agonists. *ACS Med. Chem. Lett.*

Chaudiere, J., & Ferrari-Iliou, R., (1999). Intracellular antioxidants: From chemical to biochemical mechanisms. *Food Chem. Toxicol., 37*(9–10), 949–962.

Chen, Q. Y., Chaturvedi, P, R., & Luesch, H., (2018). Process development and scale-up total synthesis of largazole, a potent class I histone deacetylase inhibitor. *Org. Process Res. Dev., 22*(2), 190–199.

Chiurchiu, V., & Maccarrone, M., (2011). Chronic inflammatory disorders and their redox control: From molecular mechanisms to therapeutic opportunities. *Antioxid. Redox. Signal., 15*(9), 2605–2641.

Choi, H. J., Lim, D. Y., & Park, J. H., (2009). Induction of G1 and G2/M cell cycle arrests by the dietary compound 3,3'-diindolylmethane in HT-29 human colon cancer cells. *BMC Gastroenterol., 9*, 39.

Choi, Y. J., Shin, H. W., Chun, Y. S., Leutou, A. S., Son, B. W., & Park, J. W., (2016). Diacetoxyscirpenol as a new anticancer agent to target hypoxia-inducible factor 1. *Oncotarget., 7*(38), 62107–62122.

Cincinelli, R., Musso, L., Artali, R., Guglielmi, M., Bianchino, E., Cardile, F., Colelli, F., Pisano, C., & Dallavalle, S., (2018). Camptothecin-psammaplin A hybrids as topoisomerase I and HDAC dual-action inhibitors. *Eur. J. Med. Chem., 143*, 2005–2014.

Clark, A. M., (1996). Natural products as a resource for new drugs. *Pharm Res., 13*(8), 1133–1144.

Das, G., Gouda, S., Mohanta, Y. K., & Patra, J. K., (2015). Mangrove plants: A potential source for anticancer drugs. *Indian J. Geomarine Sci., 44*(05), 666–672.

Dedeyan, B., Klonowska, A., Tagger, S., Tron, T., Iacazio, G., Gil, G., & Le Petit, J., (2000). Biochemical and molecular characterization of a laccase from *Marasmius quercophilus*. *Appl. Environ. Microbiol., 66*(3), 925–929.

Dela, C. C. S., Tanoue, L. T., & Matthay, R. A., (2011). Lung cancer: Epidemiology, etiology, and prevention. *Clin. Chest. Med., 32*(4), 605–644.

Dembitsky, V. M., Al Quntar, A. A., & Srebnik, M., (2011). Natural and synthetic small boron-containing molecules as potential inhibitors of bacterial and fungal quorum sensing. *Chem. Rev., 111*(1), 209–237.

Deshmukh, S. K., Prakash, V., & Ranjan, N., (2017). Marine fungi: A source of potential anticancer compounds. *Front Microbiol., 8*, 2536.

Doronina, S. O., Bovee, T. D., Meyer, D. W., Miyamoto, J. B., Anderson, M. E., Morris-Tilden, C. A., & Senter, P. D., (2008). Novel peptide linkers for highly potent antibody-auristatin conjugate. *Bioconjug. Chem., 19*(10), 1960–1963.

Dunn, S. E., Hardman, R. A., Kari, F. W., & Barrett, J. C., (1997). Insulin-like growth factor 1(IGF-1) alters drug sensitivity of HBL100 human breast cancer cells by inhibition of apoptosis induced by diverse anticancer drugs. *Cancer Res., 57*(13), 2687–2693.

Dyshlovoy, S. A., & Honecker, F., (2018). Marine compounds and cancer: 2017 updates. *Mar. Drugs., 16*(2), 41.

Edmondson, R., Broglie, J. J., Adcock, A. F., & Yang, L., (2014). Three-dimensional cell culture systems and their applications in drug discovery and cell-based biosensors. *Assay Drug Dev. Techn., 12*(4), 207–218.

Efferth, T., Dunstan, H., Sauerbrey, A., Miyachi, H., & Chitambar, C. R., (2001). The antimalarial artesunate is also active against cancer. *Int. J. Oncol., 18*(4), 767–773.

Egeblad, M., & Werb, Z., (2002). New functions for the matrix metalloproteinases in cancer progression. *Na. Rev. Cancer., 2*(3), 161–74.

Elmore, S., (2007). Apoptosis: A review of programmed cell death. *Toxicol. Pathol., 35*(4), 495–516.

Fan, M., Nath, A., Tang, Y., Choi, Y. J., Debnath, T., Choi, E. J., & Kim, E. K., (2018). Investigation of the anti-prostate cancer properties of marine-derived compounds. *Mar. Drugs., 16*(5), 160.

Fang, Y. Z., Yang, S., & Wu, G., (2002). Free radicals, antioxidants, and nutrition. *Nutrition, 18*(10), 872–879.

Fleeman, N., Bagust, A., Duarte, R., Richardson, M., Nevitt, S., Boland, A., Kotas, E., et al., (2019). Eribulin for treating locally advanced or metastatic breast cancer after one chemotherapy regimen: An evidence review group perspective of a nice single technology appraisal. *Pharmacoecon Open.* (In press)

Gao, Y., Zhang, P., Cui, A., Ye, D. Y., Xiang, M., & Chu, Y., (2018). Discovery and anti-inflammatory evaluation of benzothiazepinones (BTZs) as novel non-ATP competitive inhibitors of glycogen synthase kinase-3β (GSK-3β). *Bioorg. Med. Chem., 26*(20), 5479–5493.

Genuário, D. B., Vaz, M. G., Santos, S. N., Kavamura, V. N., & Melo, I. S., (2019). Cyanobacteria from Brazilian extreme environments: Toward functional exploitation. In: *Microbial Diversity in the Genomic Era* (pp. 265–284). Elsevier, Netherlands.

Giordano, D., Costantini, M., Coppola, D., Lauritano, C., Nunez, P. L., Ruocco, N., Di Prisco, G., Ianora, A., & Verde, C., (2018). Biotechnological applications of bioactive peptides from marine sources. *Adv. Microb. Physiol., 73*, 171–220.

Gonzalez-Resendiz, L., Johansen, J. R., Escobar-Sanchez, V., Segal-Kischinevzky, C., Jimenez-Garcia, L. F., & Leon-Tejera, H., (2018). Two new species of Phyllonema (Rivulariaceae, Cyanobacteria) with an emendation of the genus. *J. Phycol., 54*(5), 638–652.

Grignani, G., D'Ambrosio, L., Pignochino, Y., Palmerini, E., Zucchetti, M., Boccone, P., Aliberti, S., et al., (2018). Trabectedin and olaparib in patients with advanced and non-resectable bone and soft-tissue sarcomas (TOMAS): An open-label, phase 1b study from the Italian sarcoma group. *Lancet Oncol., 19*(10), 1360–1371.

Groll, M., Nguyen, H., Vellalath, S., & Romo, D., (2018). (−)-Homosalinosporamide A and its mode of proteasome inhibition: An X-ray crystallographic study. *Mar, Drugs, 16*(7), 240.

Hamed, A. R., Abdel-Azim, N. S., Shams, K. A., & Hammouda, F. M., (2019). Targeting multidrug resistance in cancer by natural chemosensitizers. *Bull. Natl. Res. Cent., 43*(1), 8.

Hardoim, P. R., van Overbeek, L. S., Berg, G., Pirttila, A. M., Compant, S., Campisano, A., Doring, M., & Sessitsch, A., (2015). The hidden world within plants: Ecological and evolutionary considerations for defining functioning of microbial endophytes. *Microbiol. Mol. Biol. Rev., 79*(3), 293–320.

Harris, E., Strope, J., Beedie, S., Huang, P., Goey, A., Cook, K., Schofield, C., Chau, C., Cadelis, M., & Copp, B., (2018). Preclinical evaluation of discorhabdins in antiangiogenic and antitumor models. *Mar. Drugs., 16*(7), 241.

Hasanovic, A., & Mus-Veteau, I., (2018). Targeting the multidrug transporter Ptch1 potentiates chemotherapy efficiency. *Cells, 7*(8), 107.

https://www.who.int/news-room/fact-sheets/detail/cancer (accessed on 20 May 2020).

Huang, L., & Pardee, A. B., (1999). Beta-lapachone induces cell cycle arrest and apoptosis in human colon cancer cells. *Mol. Med., 5*(11), 711–720.

Irie, K., & Yanagita, R. C., (2014). Synthesis and biological activities of simplified analogs of the natural PKC ligands, bryostatin-1 and aplysiatoxin. *Chem. Rec., 14*(2), 251–267.

Jackson, S. P., & Bartek, J., (2009). The DNA-damage response in human biology and disease. *Nature, 461*(7267), 1071–1078.

Janicek, M. F., & Averette, H. E., (2001). Cervical cancer: Prevention, diagnosis, and therapeutics. *CA Cancer J. Clin., 51*(2), 92–114.

Jerković, I., Kranjac, M., Marijanović, Z., Roje, M., & Jokić, S., (2019). Chemical diversity of headspace and volatile oil composition of two brown algae (*Taonia atomaria* and *Padina pavonica*) from the Adriatic sea. *Molecules, 24*(3), 495.

Jin, W., Wu, W., Tang, H., Wei, B., Wang, H., Sun, J., Zhang, W., & Zhong, W., (2019). Structure analysis and anti-tumor and anti-angiogenic activities of sulfated galactofucan extracted from *Sargassum thunbergii. Mar. Drugs., 17*(1), 52.

Kang, H., Choi, M. C., Seo, C., & Park, Y., (2018). Therapeutic properties and biological benefits of marine-derived anticancer peptides. *Int. J. Mol. Sci., 19*(3), 919.

Kerry, R. G., Pradhan, P., Das, G., Gouda, S., Swamy, M. K., & Patra, J. K., (2018). Anticancer potential of mangrove plants: Neglected plant species of the marine ecosystem. In: *Anticancer Plants: Properties and Application* (pp. 303–325). Springer, Singapore.

Khan, T. K., & Nelson, T. J., (2018). Protein kinase C activator bryostatin-1 modulates proteasome function. *J. Cell. Biochem.*

Kitagaki, J., Shi, G., Miyauchi, S., Murakami, S., & Yang, Y., (2015). Cyclic depsipeptides as potential cancer therapeutics. *Anticancer Drugs, 26*(3), 259–271.

Konig, G. M., Kehraus, S., Seibert, S. F., Abdel-Lateff, A., & Muller, D., (2006). Natural products from marine organisms and their associated microbes. *Chem. Bio Chem., 7*(2), 229–238.

Kumar, H., Lim, H. W., More, S. V., Kim, B. W., Koppula, S., Kim, I. S., & Choi, D. K., (2012). The role of free radicals in the aging brain and Parkinson's disease: Convergence and parallelism. *Int. J. Mol. Sci., 13*(8), 10478–10504.

Lai, K. H., Liu, Y. C., Su, J. H., El-Shazly, M., Wu, C. F., Du, Y. C., Hsu, Y. M., Yang, J. C., Weng, M. K., Chou, C. H., Chen, G. Y., Chen, Y. C., & Lu, M. C., (2016). Antileukemic scalarane sesterterpenoids and meroditerpenoid from carteriospongia (*Phyllospongia*) sp., induce apoptosis via dual inhibitory effects on topoisomerase II and Hsp90. *Sci. Rep., 6*, 36170.

Li, J., Tang, H., Kurtán, T., Mándi, A., Zhuang, C. L., Su, L., Zheng, G. L., & Zhang, W., (2018). Swinhoeisterols from the South China Sea sponge *Theonellas winhoei*. *J. Nat. Prod., 81*(7), 1645–1650.

Li, T., Ding, T., & Li, J., (2019). Medicinal purposes: Bioactive metabolites from marine-derived organisms. *Mini-Rev. Med. Chem., 19*(2), 138–164.

Lindequist, U., (2016). Marine-derived pharmaceuticals-challenges and opportunities. *Biomol. Ther., 24*(6), 561–571.

Liu, K., Xu, S. H., Chen, Z., Zeng, Q. X., Li, Z. J., & Chen, Z. M., (2018). TRPM7 over expression enhances the cancer stem cell-like and metastatic phenotypes of lung cancer through modulation of the Hsp90α/uPA/MMP2 signaling pathway. *BMC Cancer, 18*(1), 1167.

Lordan, R., Tsoupras, A., & Zabetakis, I., (2019). The potential role of dietary platelet-activating factor inhibitors in cancer prevention and treatment. *Adv. Nutr.*

Lotze, M. T., Zeh, H. J., Rubartelli, A., Sparvero, L. J., Amoscato, A. A., Washburn, N. R., Devera, M. E., Liang, X., Tor, M., & Billiar, T., (2007). The grateful dead: Damage-associated molecular pattern molecules and reduction/oxidation regulate immunity. *Immunol. Rev., 220*, 60–81.

Luesch, H., Yoshida, W. Y., Moore, R. E., Paul, V. J., & Corbett, T. H., (2001). Total structure determination of apratoxin A, a potent novel cytotoxin from the marine cyanobacterium *Lyngbya majuscula*. *J. Am. Chem. Soc., 123*(23), 5418–5423.

Lukiw, W. J., Cui, J. G., Li, Y. Y., & Culicchia, F., (2009). Up-regulation of micro-RNA-221 (miRNA-221; chr Xp11.3) and caspase-3 accompanies down-regulation of the survivin-1 homolog BIRC1 (NAIP) in glioblastoma multiforme (GBM). *J. Neuro-Oncol., 91*(1), 27–32.

Lyakhova, I., Bryukhovetsky, I., Kudryavtsev, I., Khotimchenko, Y. S., Zhidkov, M., & Kantemirov, A., (2018). Antitumor activity of fascaplysin derivatives on glioblastoma model *in-vitro*. *Bull. Exp. Biol. Med.*, 1–7.

Malloy, K. L., Choi, H., Fiorilla, C., Valeriote, F. A., Matainaho, T., Gerwick, W. H., & Hoiamide, D., (2012). A marine cyanobacteria-derived inhibitor of p53/MDM2 interaction. *Bioorg. Med. Chem. Lett., 22*(1), 683–688.

Malve, H., (2016). Exploring the ocean for new drug developments: Marine pharmacology. *J. Pharm. Bioallied. Sci., 8*(2), 83–91.

Martinelli, C., Lengert, A. V. H., Carcano, F. M., Silva, E. C. A., Brait, M., Lopes, L. F., & Vidal, D. O., (2017). MGMT and CALCA promoter methylation are associated with poor prognosis in testicular germ cell tumor patients. *Oncotarget., 8*(31), 50608–50617.

Matutes, E., (2018). The 2017 WHO update on mature T- and natural killer (NK) cell neoplasms. *Int. J. Lab Hematol., 40*(1), 97–103.

Mayer, A. M. S., Rodríguez, A. D., Taglialatela-Scafati, O., & Fusetani, N., (2017). Marine pharmacology in 2012–2013: Marine compounds with antibacterial, antidiabetic, antifungal, anti-inflammatory, antiprotozoal, antituberculosis, and antiviral activities; affecting the immune and nervous systems, and other miscellaneous mechanisms of action. *Mar. Drugs., 15*(9).

Morita, M., Ogawa, H., Ohno, O., Yamori, T., Suenaga, K., & Toyoshima, C., (2015). Biselyngbyasides, cytotoxic marine macrolides, are novel and potent inhibitors of the Ca (2+) pumps with a unique mode of binding. *FEBS Lett., 589*(13), 1406–1411.

Nagai, H., & Kim, Y. H., (2017). Cancer prevention from the perspective of global cancer burden patterns. *J. Thorac. Dis., 9*(3), 448–451.

Newman, D. J., & Cragg, G. M., (2004). Marine natural products and related compounds in clinical and advanced preclinical trials. *J. Nat. Prod., 67*(8), 1216–1238.

Nisha, C. M., Kumar, A., Vimal, A., Bai, B. M., & Pal, D., (2016). Docking and ADMET prediction of few GSK-3 inhibitors divulges 6-bromoindirubin-3-oxime as a potential inhibitor. *J. Mol. Graph Model, 6*5, 100–107.

Okazaki, I. M., Hiai, H., Kakazu, N., Yamada, S., Muramatsu, M., Kinoshita, K., & Honjo, T., (2003). Constitutive expression of AID leads to tumorigenesis. *J. Exp. Med., 197*(9), 1173–1181.

Pangestuti, R., Siahaan, E., & Kim, S. K., (2018). Photoprotective substances derived from marine algae. *Mar Drugs, 16*(11), 399.

Park, S. R., Kim, J. H., Jang, H. D., Yang, S. Y., & Kim, Y. H., (2018). Inhibitory activity of minor phlorotannins from *Ecklonia cava* on α-glucosidase. *Food Chem., 257*, 128–134.

Patra, S., & Muthuraman, M. S., (2013). *Gracilaria edulis* extract induces apoptosis and inhibits tumor in Ehrlich ascites tumor cells *in vivo*. *BMC Complement Altern. Med., 13*, 331.

Peters, T. L., Tillotson, J., Yeomans, A. M., Wilmore, S., Lemm, E., Jiménez-Romero, C., Amador, L. A., Li, L., Amin, A. D., & Pongtornpipat, P., (2018). Target-based screening against EIF4A1 reveals the marine natural compound elatol as a novel inhibitor of translation initiation with *in vivo* anti-tumor activity. *Clin. Can Res., 3645*.

Phutdhawong, W., Eksinitkun, G., Ruensumran, W., Taechowisan, T., & Phutdhawong, W. S., (2012). Synthesis and anticancer activity of 5,6,8,13-tetrahydro-7H-naphtho[2,3-a][3]-benzazepine-8,13-diones. *Arch Pharm. Res., 35*(5), 769–777.

Popovich, D. G., & Kitts, D. D., (2004). Ginsenosides 20(S)-protopanaxadiol and Rh2 reduce cell proliferation and increase sub-G1 cells in two cultured intestinal cell lines, Int-407 and Caco-2. *Can J. Physiol. Pharm., 82*(3), 183–190.

Qiao, X., Gan, M., Wang, C., Liu, B., Shang, Y., Li, Y., & Chen, S., (2019). Tetracenomycin X exerts antitumour activity in lung cancer cells through the down regulation of cyclin D1. *Mar. Drugs., 17*(1), 63.

Rao, P. V., Gupta, N., Bhaskar, A. S., & Jayaraj, R., (2002). Toxins and bioactive compounds from cyanobacteria and their implications on human health. *J. Environ. Biol., 23*(3), 215–224.

Reya, T., & Clevers, H., (2005). Wnt signalling in stem cells and cancer. *Nature, 434*(7035), 843–850.

Rice-Evans, C., & Burdon, R., (1993). Free radical-lipid interactions and their pathological consequences. *Prog. Lipid Res., 32*(1), 71–110.

Righi, L., Volante, M., Rapa, I., Scagliotti, G. V., & Papotti, M., (2007). Neuro-endocrine tumors of the lung: A review of relevant pathological and molecular data. *Virchows Arch., 451*(1), S51–S59.

Rong, B., & Yang, S., (2018). Molecular mechanism and targeted therapy of Hsp90 involved in lung cancer: New discoveries and developments. *Int. J. Oncol., 52*(2), 321–336.

Sable, R., Parajuli, P., & Jois, S., (2017). Peptides, peptidomimetics, and polypeptides from marine sources: A wealth of natural sources for pharmaceutical applications. *Mar. Drugs, 15*(4).

Sali, V. K., Malarvizhi, R., Manikandamathavan, V., & Vasanthi, H. R., (2018). Isolation and evaluation of phytoconstituents from red alga *Acanthophora spicifera* as potential apoptotic agents towards A549 and HeLa cancer cells lines. *Algal Res., 32*, 172–181.

Sandoval-Sus, J. D., Brahim, A., Khan, A., Raphael, B., Ansari-Lari, A., & Ruiz, M., (2019). Brentuximabvedotin as frontline treatment for HIV-related extra cavitary primary effusion lymphoma. *Int. J. Hematol.* (In press)

Shavandi, A., Hou, Y., Carne, A., McConnell, M., & Bekhit, A. E. A., (2019). Marine waste utilization as a source of functional and health compounds. *Adv. Food Nutr. Res., 87*, 187–254.

Shih, S. P., Lee, M. G., El-Shazly, M., Juan, Y. S., Wen, Z. H., Du, Y. C., Su, J. H., Sung, P. J., Chen, Y. C., Yang, J. C., Wu, Y. C., & Lu, M. C., (2015). Tackling the cytotoxic effect of a marine polycyclic quinone-type metabolite: Halenaquinone induces molt 4 cells apoptosis via oxidative stress combined with the inhibition of HDAC and topoisomerase activities. *Mar. Drugs., 13*(5), 3132–3153.

Silchenko, A. S., Rasin, A. B., Kusaykin, M. I., Malyarenko, O. S., Shevchenko, N. M., Zueva, A. O., Kalinovsky, A. I., Zvyagintseva, T. N., & Ermakova, S. P., (2018). Modification of native fucoidan from *Fucus evanescens* by recombinant fucoidanase from marine bacteria Formosa algae. *Carbohydr. Polym., 193*, 189–195.

SithrangaBoopathy, N., & Kathiresan, K., (2010). Anticancer drugs from marine flora: An overview. *J. Oncol.,* 214186.

Song, X., Xiong, Y., Qi, X., Tang, W., Dai, J., Gu, Q., & Li, J., (2018). Molecular targets of active anticancer compounds derived from marine sources. *Mar. Drugs., 16*(5).

Soo, V. W., Kwan, B. W., Quezada, H., Castillo-Juarez, I., Perez-Eretza, B., Garcia-Contreras, S. J., Martinez-Vazquez, M., Wood, T. K., & Garcia-Contreras, R., (2017). Repurposing of anticancer drugs for the treatment of bacterial infections. *Curr. Top Med. Chem., 17*(10), 1157–1176.

Sosa, V., Moline, T., Somoza, R., Paciucci, R., Kondoh, H., & ME, L. L., (2013). Oxidative stress and cancer: An overview. *Ageing Res. Rev., 12*(1), 376–390.

Srinath, R. K., & Katan, M. B., (2004). Diet, nutrition and the prevention of hypertension and cardiovascular diseases. *Public Health Nutr., 7*(1a), 167–186.

Strebhardt, K., & Ullrich, A., (2006). Targeting polo-like kinase 1 for cancer therapy. *Nat. Rev. Cancer, 6*(4), 321–330.

Sun, J. Y., Wang, J. D., Wang, X., Liu, H. C., Zhang, M. M., Liu, Y. C., Zhang, C. H., Su, Y., Shen, Y. Y., Guo, Y. W., Shen, A. J., & Geng, M. Y., (2017). Marine-derived chromopeptide A, a novel class I HDAC inhibitor, suppresses human prostate cancer cell proliferation and migration. *Acta Pharmacol. Sin., 38*(4), 551–560.

Sung, F. L., Poon, T. C., Hui, E. P., Ma, B. B., Liong, E., To, K. F., Huang, D. P., & Chan, A. T., (2005). Antitumor effect and enhancement of cytotoxic drug activity by cetuximab in nasopharyngeal carcinoma cells. *In Vivo, 19*(1), 237–245.

Swain, S. S., Padhy, R. N., & Singh, P. K., (2015). Anticancer compounds from cyanobacterium *Lyngbya* species: A review. *Antonie Van Leeuwenhoek., 108*(2), 223–265.

Tan, L. T., Ser, H. L., Yin, W. F., Chan, K. G., Lee, L. H., & Goh, B. H., (2015). Investigation of antioxidative and anticancer potentials of streptomyces sp. MUM256 isolated from Malaysia mangrove soil. *Front Microbiol., 6*, 1316.

Taori, K., Paul, V. J., & Luesch, H., (2008). Structure and activity of largazole, a potent antiproliferative agent from the Floridian marine cyanobacterium *Symploca* sp. *J. Am. Chem. Soc., 130*(6), 1806–1807.

Tillotson, J., Kedzior, M., Guimaraes, L., Ross, A. B., Peters, T. L., Ambrose, A. J., Schmidlin, C. J., Zhang, D. D., Costa-Lotufo, L. V., Rodriguez, A. D., Schatz, J. H., & Chapman, E., (2017). ATP-competitive, marine derived natural products that target the DEAD box helicase, eIF4A. *Bioorg. Med. Chem. Lett., 27*(17), 4082–4085.

Trivella, D. B., Pereira, A. R., Stein, M. L., Kasai, Y., Byrum, T., Valeriote, F. A., Tantillo, D. J., Groll, M., Gerwick, W. H., & Moore, B. S., (2014). Enzyme inhibition by hydroamination: Design and mechanism of a hybrid carmaphycin-syringolinenone proteasome inhibitor. *Chem. Biol., 21*(6), 782–791.

Tuohy, K. M., Probert, H. M., Smejkal, C. W., & Gibson, G. R., (2003). Using probiotics and prebiotics to improve gut health. *Drug Discov. Today, 8*(15), 692–700.

Valko, M., Izakovic, M., Mazur, M., Rhodes, C. J., & Telser, J., (2004). Role of oxygen radicals in DNA damage and cancer incidence. *Mol. Cell. Biochem., 266*(1–2), 37–56.

Valko, M., Leibfritz, D., Moncol, J., Cronin, M. T., Mazur, M., & Telser, J., (2007). Free radicals and antioxidants in normal physiological functions and human disease. *Int. J. Biochem. Cell Biol., 39*(1), 44–84.

Valko, M., Rhodes, C. J., Moncol, J., Izakovic, M., & Mazur, M., (2006). Free radicals, metals and antioxidants in oxidative stress-induced cancer. *Chem. Biol Interact., 160*(1), 1–40.

Valyi-Nagy, T., Fredericks, B., Ravindra, A., Hopkins, J., Shukla, D., & Valyi-Nagy, K., (2018). Herpes simplex virus type 1 infection promotes the growth of a subpopulation of tumor cells in 3D uveal melanoma cultures. *J. Virol.*

Venturi, V., Davies, C., Singh, A. I., Matthews, J. H., Bellows, D. S., Northcote, P. T., Keyzers, R. A., & Teesdale-Spittle, P. H., (2012). The protein synthesis inhibitors mycalamides A and E have limited susceptibility toward the drug efflux network. *J. Biochem. Mol. Toxicol., 26*(3), 94–100.

Watjen, W., Ebada, S. S., Bergermann, A., Chovolou, Y., Totzke, F., Kubbutat, M. H. G., Lin, W., & Proksch, P., (2017). Cytotoxic effects of the anthraquinone derivatives 1'-deoxyrhodoptilometrin and(S)-(-)-rhodoptilometrin isolated from the marine echinoderm *Comanthus* sp. *Arch Toxicol., 91*(3), 1485–1495.

Wiese, J., Imhoff, J. F., Gulder, T. A., Labes, A., & Schmaljohann, R., (2016). Marine fungi as producers of benzocoumarins, a new class of inhibitors of glycogen-synthase-kinase 3beta. *Mar. Drugs., 14*(11).

Xiang, X., Qin, H. G., You, X. M., Wang, Y. Y., Qi, L. N., Ma, L., Xiang, B. D., Zhong, J. H., & Li, L. Q., (2017). Expression of P62 in hepatocellular carcinoma involving hepatitis B virus infection and aflatoxin B1 exposure. *Cancer Med., 6*(10), 2357–2369.

Zhang, P., Bi, C., Schmitt, S. M., Li, X., Fan, Y., Zhang, N., & Dou, Q. P., (2014). Metal-based 2,3-indolinedione derivatives as proteasome inhibitors and inducers of apoptosis in human cancer cells. *Int. J. Mol. Med., 34*(3), 870–879.

Zhou, J., Wang, J., Chen, C., Yuan, H., Wen, X., & Sun, H., (2018). USP7: Target validation and drug discovery for cancer therapy. *Med. Chem., 14*(1), 3–18.

Zong, Y., Wang, W., & Xu, T., (2018). Total synthesis of bioactive marine meroterpenoids: The cases of liphagal and frondosin B. *Mar. Drugs., 16*(4), 115.

Zurhausen, H., (2002). Papilloma viruses and cancer: From basic studies to clinical application. *Nat Rev. Cancer., 2*(5), 342–350.

CHAPTER 7

Edible Pulses: Part of A Balanced Diet to Manage Cancer

VANDANA GARG, KRIPI VOHRA, and HARISH DUREJA

Department of Pharmaceutical Sciences,
Maharshi Dayanand University, Rohtak, Haryana – 124001, India,
E-mail: vandugarg@rediffmail.com (V. Garg)

ABSTRACT

Nowadays, the management of cancer using pulses is gaining popularity. Pulses are widely consumed due to their high nutritional benefits. They are rich in protein and have high carbohydrate and fiber levels but the fat content and glycemic index in low. In addition, they are rich in bioactive compounds (polyphenols, saponins, triterpenes, etc.) which contribute to improving various health issues. Literature revealed that the pulses (*Lens culinaris, Vigna unguiculata, Vicia faba, Vigna angularis, Phaseolus vulgaris, Pisum sativum, Vigna mungo, Vigna radiata, Cajanuscaja,* and *Cicer arietinum*) possess the ability to treat and/or prevent various types of cancer, mainly breast and colon cancer. Despite being such an easily available, cost-effective, and reliable contributor in the human diet, good health as well as management of cancer, pulses are still underutilized and warrant the attention of oncologists. Therefore, the present review focuses on the nutritional features of edible pulses as well as their potential contribution to management or decreasing the risk of developing cancer. The role of pulses in treating other diseases such as hyperlipidemia, cardiovascular diseases, etc. has also been discussed. The review highlights that the consumption of pulses should be increased in human's diet worldwide to reduce the risk of developing cancer and to manage other diseases including cancer.

7.1 INTRODUCTION

Cancer, being associated with abnormal proliferation of cells, has been considered the second most leading cause of mortality worldwide. About 8.8 million deaths occurred in 2015 due to cancer. World Health Organization estimates one death, out of six deaths, due to cancer at a global level (WHO, 2018). Despite the use of numerous conventional anticancer therapies, the disadvantages associated with them as well as tumor resistance towards anticancer drugs has been the major loophole (Ke and Shen, 2017).

Pulses are edible seeds of the Leguminosae family and include beans, peas, and lentils. These are an essential part of a balanced diet from several years (FAO, 2010; FAO, 1994; Leterme and Muñoz, 2002; Mudryj et al., 2014). Pulses are high in carbohydrates (around 50-65%) and protein content while they have a low glycemic index (GI) (McCrory et al., 2010; Ofuya and Akhidue, 2005). They contain both insoluble and soluble fiber, mono-, and polyunsaturated fat and sterols (Iqbal et al., 2006; Tosh and Yada, 2010). Other vital nutrients present in pulses include macronutrients and micronutrients including niacin, iron, folate, zinc, riboflavin, selenium, pyridoxine, thiamin, and vitamins including A, B, C, and E (Patterson et al., 2009; Winham et al., 2008).

In addition to the high nutritional value of pulses, they possess diverse therapeutic value due to the presence of bioactive phytochemicals (Mudryj et al., 2014). High consumption of pulses is even recommended for prophylaxis of dreadful diseases such as cancer (Venter and Van Eyssen, 2001). Many studies also support the positive correlation between the consumption of pulses and reduced risk of developing cancer (Bennink, 2002; El-Aassar et al., 2014; Hafidh et al., 2012; Mo'ez Al-Islam et al., 2009; Patel, 2014; Shomaf and Takruri, 2012; Stanisavljevic et al., 2016; Vohra et al., 2015; Xu and Chang, 2012; Zhao et al., 2012). Pulses also possess the antioxidant property and improve serum lipid profiles. These also reduce risk of developing cardiovascular diseases (CVD) and assist persons with diabetes in maintaining a healthy blood glucose and optimum insulin levels (Mudryj et al., 2014). Therefore, the present study reviews the potential of various pulses in treating cancer. In addition, the nutritional value of pulses and other health benefits associated with the consumption of pulses have also been discussed.

7.2 EDIBLE PULSES

The pulses recognized by Food and Agriculture Organization (FAO) exclude the ones used for extraction of oil. Only 11 pulses are recognized by FAO, i.e., beans (dry and dry broad), peas (dry, chickpeas, black-eyed, pigeon), lentils, bambara groundnut, vetch, and lupins (FAO, 1994; FAO, 2010). However, human diet does not include all pulses. The ones included in human diet are chickpeas (*Cicer arietinum)*, pigeon pea (*Cajanuscajan*), lima bean (*Vigna lunatus*), lentil (*Lens culinaris*), dry broad bean (*Vicia faba*), mung bean (*Vigna radiata)*, urad bean (*Vigna mungo*), field peas (*Pisum sativum*), adzuki bean (*Vigna angularis*), rice bean (*Vigna umbellata*), kidney bean (*Phaseolus vulgaris*), moth bean (*Vigna acontifolia*), and dry cowpea (*Vigna unguiculata*) (Singh and Basu, 2012).

7.3 NUTRITIONAL VALUE OF PULSES

The present review has discussed a brief about nutritional value of pulses in general. Pulses are considered as an essential part of a balanced diet by various health agencies (HC, 2013; USDHHS, 2010). As per the United States Department of Agriculture, consumption of pulses (2.5 to 3.5 cups) per week is essential (USDHHS, 2010). Moreover, pulses, in combination with another complimenting protein source, are considered comparable to meat, hence, form an appropriate food diet for vegetarians (Winham et al., 2008). Figure 7.1 represents the major nutritional constituents present in pulses.

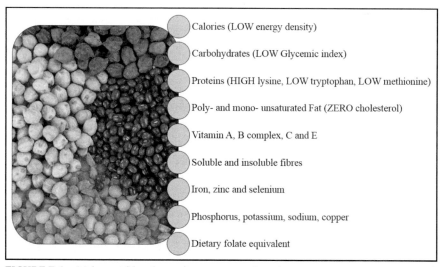

Calories (LOW energy density)

Carbohydrates (LOW Glycemic index)

Proteins (HIGH lysine, LOW tryptophan, LOW methionine)

Poly- and mono- unsaturated Fat (ZERO cholesterol)

Vitamin A, B complex, C and E

Soluble and insoluble fibres

Iron, zinc and selenium

Phosphorus, potassium, sodium, copper

Dietary folate equivalent

FIGURE 7.1 Major nutritional constituents present in pulses.

According to Australian Export Grains Innovation Centre, starch, protein, crude fibers, ash, and fat constitute as 44%, 29%, 9%, 4% and 2% of dry matter of pulses, while 12% of dry matter of pulses contain soluble fiber and other constituents (Siah, 2016). Pulses consist of complex carbohydrates and high fibers; therefore, they are low in GI. The GI of pulses varies based on their processing; however, their GI is still low as compared to food rich in carbohydrates (McCrory et al., 2010; Mudryj et al., 2014; Ofuya and Akhidue, 2005). Cooked pulses have lower GI as compared to that of canned pulses (Atkinson et al., 2008; Mudryj et al., 2014; Wolever et al., 1987). The fiber content of pulses is high, while they are a poor source of fat as compared to oil seeds (Siah, 2016). It is interesting to note that pulses such as chickpeas, lentils, and beans do not contain cholesterol (Mudryj et al., 2014). Pulses are a rich source of proteins and constitute $1/10^{th}$ of the total global dietary protein requirements. They are rich in an amino acid named lysine. In comparison, cereals contain low lysine content. Inversely, some amino acids, namely methionine and tryptophan are low in pulses but are abundantly found in cereals (Boye et al., 2010; Iqbal et al., 2006; Mudryj et al., 2014). Pulses are also micronutrient-rich food source (Winham, Webb, and Barr, 2008). Iron content is high; however, it bounds to phytates present in pulses, which results in reduced absorption of iron, leading to iron deficiency (Petry et al., 2010; Sandberg, 2002). Pulses also contain a high amount of magnesium, potassium, zinc, phosphorus, sodium, copper, and selenium. Vitamin levels (thiamin, niacin, folate, riboflavin, pyridoxine, vitamin A, vitamin E) are also high in pulses (CFIA, 2011; USDA, 2012). Dried pulses are a poor source of vitamin C; however, the sprouted pulses contain more vitamin C as compared to dried ones (Riddoch et al., 1998).

Pulses are harvested in million pounds and even exported internationally to meet the needs. Pulses are processed to prepare food products such as bakery items, breads, pasta, snacks, soups, tortillas, meat, etc. They, additionally, are an essential constituent in frozen dough foods. Keeping in view the nutritional value of pulses, they are also used in baby foods and sports food products (Asif et al., 2013). Pulses are used in food preparations as back as the 16th century. People prepared roasted pulses' flour, pancakes, fried pulses dishes, etc. Japanese consume pulses in the form of chocolates and cakes, while Americans prepare various platters such as pulao and biryani. Chinese and people of Mediterranean countries consume pulses in the form of soups and stew. Indians prepare curries and chats from pulses (Siah, 2016). As pulses may not be available in all the regions, other sources can be consumed to gain similar nutrition (Table 7.1) (Lupien, 1997).

TABLE 7.1 Other Sources for Nutritional Constituents Present in Pulses

Nutritional Constituents	Role in Human Body	Other Sources
Starch	Fuel for energy	Cereals, starchy fruits and roots such as potatoes
Fiber	Bulking of diet, habitat for bacterial flora and easy elimination of waste	Cereals, starchy fruits and roots, citrus fruits
Protein	Growth and repair	Egg, meat, seafood, milk, cheese, yogurt, liver
Carbohydrates	Fuel for energy	Cereals, milk, potatoes, starchy food
Vitamins	Development, metabolism, and protection	Fruits and vegetables
Minerals	Development, metabolism, and protection	Fruits, vegetables, whole grains, milk, and milk products, meat, nuts, whole grains, liver
Fat	Fuel for energy	Oil seeds

7.4 ANTICANCER POTENTIAL OF PULSES

A strong association between pulses and cancer prevention as well as treatment has been reviewed by Vohra et al., 2015 (Vohra et al., 2015). The phyto-composition of pulses, including nutrients and bioactive components, is positively connected with decreased risk of cancer development (Figure 7.2) (Dahl et al., 2012; Vohra et al., 2015). Various food agencies such as United States Food and Drug Administration and cancer institutes (Canadian Cancer Society, World Cancer Research Fund) support the use of pulses to reduce the risk of cancer (Guenther et al., 2006; Venter and Van Eyssen, 2001).

Most of the edible pulses are pharmacologically active against cancer due to the presence of bioactive anticancer phytoconstituents present in them. Moreover, the studies have also been reported which have evaluated their anticancer potential. The mechanism of action of phytoconstituents present in pulses is presented in Figure7.3 (Ganesan and Xu, 2017; Hayat et al., 2014; Luna-Vital et al., 2015).

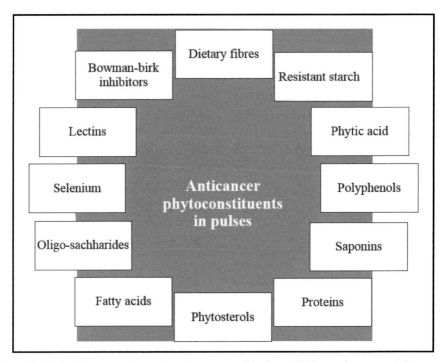

FIGURE 7.2   Phytoconstituents in pulses responsible for exhibiting anticancer activity.

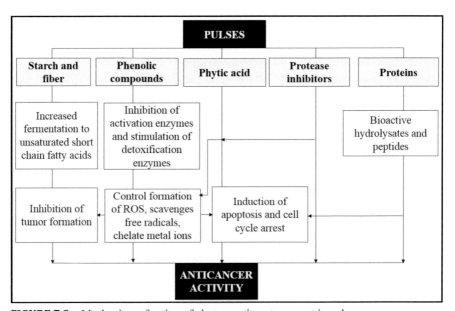

FIGURE 7.3   Mechanism of action of phytoconstituents present in pulses.

The anticancer activity of various edible pulses has been discussed in subsections.

7.4.1   *CICER ARIETINUM*

C. arietinum, also known as chickpea, is a nutritional pulse with medicinal properties. It was processed to isolate an antifungal protein, C-25 which also possessed cytotoxic property. It was able to reduce the proliferation of human oral carcinoma cells (half maximal inhibitory concentration; IC_{50} value: 37.5 µg/mL). Moreover, the protein did not affect the normal human peripheral blood mononuclear cells at a concentration as high as 600 µg/mL (Kumar et al., 2014). Maggi and his colleagues (2012) isolated seven protease inhibitor concentrates (PICs) from chickpeas and other various leguminous seeds to evaluate their effect on the breast (MDA-MB-231) and prostate (PC-3 and LNCaP) cancer cells. The PIC obtained from *C. arietinum* inhibited the growth of breast cancer and prostate cancer cells at all concentrations tested during the study (Magee et al., 2012).

7.4.2   *CAJANUS CAJAN*

C. cajan is also is known as pigeon pea. An isoflavone, cajanol (5-hydroxy-3-(4-hydroxy-2-methoxyphenyl)-7-methoxychroman-4-one) isolated from *C. cajan* roots, is reported to exhibit anticancer activity against human breast cancer cells (MCF-7) in a time and dose-dependent manner. Cajanol showed cytotoxic effects at an IC_{50} value of 83.42, 58.32, and 54.05 micro M after 24, 48 and72 hours of treatment, respectively (Luo et al., 2010).

7.4.3   *VIGNA RADIATA*

V. radiate is also known as green gram and mung bean. The proteins isolated from *V. radiata* (48 kDa) exhibit anticancer properties (Chen et al., 2017). A study evaluated cytotoxic effects of methanol extracts obtained from sprouts against HeLa and hepatocellular carcinoma (HepG-2) cell lines. (Hafidh et al., 2012). *In vitro* study of *V. radiata* exhibited cytotoxicity in a dose-dependent manner against various cancer cell lines including digestive system related CAL27, AGS, HepG-2, SW480, and Caco-2, ovary related SKOV-3and MCF-7 cancer cell lines (Xu and

Chang, 2012). Trypsin inhibitors obtained from *V. radiata* also possessed anticancer activity and showed cytotoxic effects against human colon cancer cells (SW480) (Zhao et al., 2012). Vitexin and isovitexin isolated from beans were evaluated on melanoma cell line (B16F1) of mouse and they reported to exert inhibitory effects against melanogenesis (Jeong et al., 2016). Cytotoxic effect of fermented as well as non-fermented beans (mung bean and soybean) were compared where the fermented bean was found to exhibit cytotoxic activity against MCF-7 cells. They were found to arrest the G0/G1 phase followed by apoptosis. They also induced proliferation of splenocyte and improved the serum interleukin-2 and interferon-γ levels (Ali et al., 2016). The use of fermented bean as a potential chemopreventive agent against breast cancer cells (Yac-1, 4T1) has also been reported. Beans have reported chemopreventive effects in female Balb/C mice (Yeap et al., 2013).

7.4.4   VIGNA MUNGO

V. mungo, also known as urad bean and black gram, possess anticancer activity. The methanol extract obtained from the stem and leaves of Urad bean possesses cytotoxic activity. This was confirmed by brine shrimp lethality bioassay, where the lethal concentration 50 (LC_{50}) value of leaf extract (4.52 µg/ml) and stem extract (3.25µg/ml) was compared with standard vincristine sulfate (0.67µg/ml) against *Artemia salina* (Nasrin et al., 2015).

7.4.5   PISUM SATIVUM

P. sativum, also known as field pea, is a well-known anticancer agent. Two studies have been conducted (Clemente et al., 2005) and 2012 (Clemente et al., 2012) that compared the anticancer activity of protease inhibitors, rTl1B and rTl2B, obtained from *P. sativum* seed with Bowman-Birk trypsin-chymotrypsin inhibitor. Both studies confirmed the antiproliferative properties of protease inhibitor against human colorectal adenocarcinoma (HT-29). Lectins isolated from pea seeds, leaves, and buds also exhibit antiproliferative property to MCF-7, HepG-2, larynx (HEP-2), and (HCT-116) colon cancer cells (El-Aassar, Hafez, El-Deeb, and Fouda, 2014; Patel, 2014). The phenolic compounds extracted from pea seeds also possess anticancer activity against colon (LS174), breast (MDA-MB-453), lung (A594), and blood (K562) cancer cells (Stanisavljevic et al., 2016).

7.4.6   *PHASEOLUS VULGARIS*

Kidney beans, pinto beans, and jamapa beans are the other names of *P. vulgaris* seeds. Aparicio-Fernández et al. have been continuously working on beans to evaluate their anticancer potential. In 2005, they reported that phenolic compounds extracted from recently harvested or stored *P. vulgaris* seeds possess antimutagenic activity against aflatoxin B1 (AFB1). They also reported that the harvested beans were superior to stored beans (Aparicio-Fernández et al., 2005). In the subsequent years, they also reported the anticancer activity of 100% methanol crude extract against HeLa cells and of Toyopearl and silica gel fractions against human premalignant keratinocytes (HaCaT) (Aparicio-Fernández et al., 2006; Aparicio-Fernández et al., 2008). Anticancer property of polyphenols present in *P. vulgaris* was also confirmed by Cardador-Martinez et al. in 2002 and 2006 (Cardador-Martinez et al., 2006; Cardador-Martinez et al., 2002). A positive association between the antioxidant and anticancer activity of *P. vulgaris* has also been established (Rocha-Guzmán et al., 2007). An *in vivo* study conducted to evaluate the anticancer potential of seeds in rats with colon carcinogenesis reported a 44–75% reduction in colon carcinogenesis in rats fed with beans (Bennink, 2002).

7.4.7   *VIGNA ANGULARIS*

V. angularis is also known as adzuki bean, possesses anticancer activity. The hot aqueous extract of *V. angularis* has been found potent against various cancers (Itoh et al., 2002). An ethanol (40%) fraction of hot-water extract also possessed chemopreventive effects against benzo(a)pyrene-induced tumorigenesis in the forestomach of mice (Itoh et al., 2004). The same fraction also showed cytotoxic effects against melanoma cells (B16-BL6) (Itoh et al., 2005). The heat-stable extract obtained from adzuki beans is also active against human leukemia cells (U937) (Nakaya et al., 2012).

7.4.8.   *VICIA FABA*

V. faba, also known as faba beans or dry broad bean, contains phytoconstituents which are responsible for the anticancer activity. Phenolic compounds containing extract inhibited the growth of many human cancer cell lines including bladder cancer cell line (BL-13), AGS, HepG-2, and HT-29. However, the negligible effect was observed on the non-transformed human cells (CCD-18Co). The extract showed apoptotic effects on acute promyelo-cytic leukemia (HL-60) cells (Siah et al., 2012).

7.4.9 *VIGNA UNGUICULATA*

Dry cowpea/*V. unguiculata* possess antiproliferative properties. A Bowman-Birk protease inhibitor isolated from the seeds showed cytotoxic and cyto-static effects on the MCF-7 breast cancer cells (Joanitti et al., 2010).

7.4.10 *LENS CULINARIS*

L. culinaris, also known as lentils possess anticancer/chemopreventive prop-erties against colon cancer (Mo'ez Al-Islam et al., 2009; Shomaf and Takruri, 2012). An *in vivo* study was conducted to evaluate the chemo-preventive ability of raw and cooked lentils in whole and split form on 60 Fisher 344 male rats with chemically-induced colonic cancer. The study reported a reduction of 43–57%in total colonic lesions and neoplasms in lentil groups as compared to control group. Lentils in the raw split, cooked whole, and cooked split forms reduced severe dysplasia significantly ($p = 0.0022$, reduction: 0–10%), while 20% reduction in dysplasia was observed using raw whole lentils as compared to the control group (Shomaf and Takruri, 2012). Protein fraction isolated from lentil seeds are also active against liver and cervical cancer (IC_{50} value= 15 µg/ml and 24.8 µg/ml, respectively) (Badria, 2014). The lentil's lectins can also prevent cancer as they were found active against nasopha-ryngeal carcinoma CNE1 and CNE2 cell lines at a dose of 0.125 mg/ml ($p<0.001$) and 1.00 mg/ml ($p<0.004$), respectively (Chan et al., 2015).

Overall, pulses are reported to be active against breast and colon cancer. This may also be supported by the phytoconstituents present in pulses. In addition, many other types of cancer can also be treated by consuming pulses as presented in Table 7.2.

7.5 OTHER HEALTH BENEFITS OF PULSES

Besides anticancer activity, pulses are found to be linked with other protec-tive effects. They reduce the risk of developing type 2 diabetes, heart disease, hypertension, Parkinson's disease, and Alzheimer's disease (Alcalay et al., 2012; Esposito et al., 2010; Fung et al., 2009; Scarmeas et al., 2009; Willett

TABLE 7.2 Types of Cancer Reported to be Treated by Pulses

Type of Cancer	Pulse	References
Oral	*Cicer arietinum*	Kumar et al., 2014
	Vigna radiata	Xu and Chang, 2012
Breast	*Cicer arietinum*	Magee et al., 2012
	Cajanus cajan	Luo et al., 2010
	Vigna radiata	Ali et al., 2016; Xu and Chang, 2012; Yeap et al., 2013
	Pisum sativum	El-Aassar, Hafez, El-Deeb, and Fouda, 2014; Patel, 2014; Stanisavljevic, Ilic, Matic, Jovanovic, Cupic, Dabic, Natic, and Tesic, 2016
	Vigna unguiculata	Joanitti, Azevedo, and Freitas, 2010
Prostate	*Cicer arietinum*	Magee et al., 2012
Cervical	*Cicer arietinum*	Kim et al., 2013
	Vigna radiata	Hafidh et al., 2012
	Phaseolus vulgaris	Aparicio-Fernández et al., 2008
Hepatocellular	*Vigna radiata*	Hafidh et al., 2012
	Pisum sativum	El-Aassar et al., 2014
	Vicia faba	Siah et al., 2012
	Lens culinaris	Badria, 2014
Gastric	*Vigna radiata*	Xu and Chang, 2012
Colon	*Vigna radiata*	Xu and Chang, 2012; Zhao et al., 2012
	Pisum sativum	El-Aassar et al., 2014; Patel, 2014; Stanisavljevic et al., 2016; Bennink, 2002
		Mo'ez Al-Islam et al., 2009; Shomaf and Takruri, 2012
	Phaseolus vulgaris	
	Lens culinaris	
Colorectal	*Vigna radiata*	Xu and Chang, 2012
	Pisum sativum	Clemente et al., 2012; Clemente et al., 2005
Ovary	*Vigna radiata*	Xu and Chang, 2012
Melanoma	*Vigna radiata*	Jeong et al., 2016
	Vigna angularis	Itoh et al., 2005
Larynx	*Pisum sativum*	El-Aassar et al., 2014; Patel, 2014
Blood	*Pisum sativum*	Stanisavljevic et al., 2016
Lung	*Pisum sativum*	Stanisavljevic et al., 2016
	Vigna angularis	Itoh et al., 2005

TABLE 7.2 *(Continued)*

Type of Cancer	Pulse	References
Adenocarcinoma	*Phaseolus vulgaris*	Aparicio-Fernández et al., 2006
Premalignant keratinocytes	*Phaseolus vulgaris*	Aparicio-Fernández et al., 2006
Stomach	*Vigna angularis*	Itoh et al., 2004
	Vicia faba	Siah et al., 2012
Histiocytic lymphoma	*Vigna angularis*	Nakaya et al., 2012
Acute promyelocytic leukemia	*Vigna faba*	Siah et al., 2012
Nasopharyngeal	*Lens culinaris*	Chan et al., 2015
Bladder	*Vicia faba*	Siah et al., 2012

et al., 1995). The epidemiological and animal studies also report a negative correlation between pulses and the development of various diseases (Iqbal et al., 2006; Winham, Webb, and Barr, 2008). The pulses responsible for imparting each activity have been mentioned in Table 7.3 and brief has been discussed in the following subsections.

TABLE 7.3 Health Benefits of Pulses

Sl. No.	Indication	Pulses	References
1.	Hypertension	*Lens culinaris*	Hanson, 2013
		Pisum sativum	Li et al., 2011
		Cajanuscajan	Nawaz et al., 2017
		Cicer arietinum	Yust et al., 2003
		Vicia faba	Siah et al., 2012
2.	Inflammation	*Cajanuscajan*	Hassan et al., 2016
		Cicer arietinum	Shafeen, 2012
		Pisum sativum	Hermsdorff et al., 2011
		Phaseolus vulgaris	Oomah et al., 2010
3.	Aphrodisiac	*Cicer arietinum*	Sajja et al., 2014
4.	Estrogenic	*Cicer arietinum*	Wikhe et al., 2013

TABLE 7.3 *(Continued)*

Sl. No.	Indication	Pulses	References
5.	Diabetes	*Cicer arietinum*	Yadav et al., 2009
		Pisum sativum	Marinangeli and Jones, 2011;
		Vicia faba	Marinangeli et al., 2009
		Lens culinaris	Siah et al., 2012
			Al-Tibi et al., 2010
6.	Hyperlipidemia	*Cicer arietinum*	Yust et al., 2012
		Pisum sativum	Sandstrom et al., 1994
		Phaseolus vulgaris	Anderson et al., 1990;
		Lens culinaris	Anderson et al., 1984; Shutler et al., 1989; Al-Tibi et al., 2010; Vohra et al., 2016
7.	Gastrointestinal disorders	*Cicer arietinum*	Dalal et al., 2011
		Pisum sativum	Dahl et al., 2003; Veenstra et al., 2010
8.	Convulsion	*Cicer arietinum*	Sardari et al., 2015
9.	Hepatoprotective	*Cicer arietinum*	Santhoshi et al., 2013
10.	Antimicrobial	*Cicer arietinum*	Kan et al., 2010
		Pisum sativum	Dominika et al., 2011; Świątecka et al., 2010
		Lens culinaris	Shenkarev et al., 2014; Wang and Ng, 2007; Zadernowski et al., 1992
11.	Diuretic	*Cicer arietinum*	Divya and Banda, 2014
12.	Obesity	*Vigna angularis*	Kim et al., 2015; Liu et al., 2017
		Viviafaba	
			Siah et al., 2012
13.	Rheumatoid arthritis	*Vigna angularis*	Oh et al., 2013
14.	Parkinson's disease	*Vicia faba*	Rabey et al., 1993
15.	Nephrotoxicity	*Lens culinaris*	Adikay, 2016
16.	Human immune-deficiency virus	*Phaseolus vulgaris*	Ye et al., 2001

7.5.1   *CORONARY HEART DISEASE (CHD)*

According to the first National Health and Nutrition Examination Survey Epidemiologic Follow-up Study, pulses can reduce the risk of developing

CHD. A 22% lower risk of CHD is also reported in adults consuming pulse four times a week as compared to the ones consuming once per week (Bazzano et al., 2001). The nutritional and bioactive phyto-chemicals present in the pulses including protein, fiber, starch, vitamins, minerals, oligosaccharides, isoflavones, phospholipids, antioxidants, etc. are responsible to exhibit cardio-protective property (Anderson and Major, 2002).

7.5.2 HYPERTENSION

The dietary approaches to stop hypertension (DASH diet) (Winham, Webb, and Barr, 2008), as well as the Gluten-Free Diet, for patients with Celiac disease (Kupper, 2005), recommend the use of pulses in their diet. Phytochemicals such as fiber present in pulses normalize blood pressure (BP) levels and reduce the risk of developing CVD (Anderson et al., 1994; Papanikolaou and Fulgoni, 2008). Studies also report an inverse association between pulse consumption and systolic BP in American adults (Papanikolaou and Fulgoni, 2008). A recent meta-analysis including eight human trials concluded reduction in BP in population consuming pulses including lentils, chickpeas, peas, beans (Jayalath et al., 2014). Another study reported that increased consumption of beans is inversely proportional to diastolic BP among Costa Rican adults (Mattei et al., 2011). A relationship of lentils (in comparison to dried beans, peas, and chickpeas) with decreased BP in hypertensive rats has also been reported (Hanson et al., 2014).

7.5.3 HYPERLIPIDEMIA

Consumption of pulses contributes in normalizing lipid profile as they enhance high-density lipoprotein levels and reduce triglyceride, low-density lipoprotein (LDL) as well as total cholesterol (TC) levels (Anderson and Major, 2002; Bazzano, et al., 2001; Iqbal et al., 2006; Patterson et al., 2009; Winham and Hutchins, 2007). Daily consumption of half a bowl of cooked beans has reported a reduction in TC by 8% (Finley et al., 2007). Meta-analyses also reported a reduction in LDL by consuming dietary pulses (Bazzano et al., 2001; Ha et al., 2014). American Heart Association also recommends the use of high-fiber foods such as pulses to reduce the risk of heart disease (Lichtenstein et al., 2006).

7.5.4   INFLAMMATION

Pulses contribute to lowering inflammatory biomarkers associated with the risk of CHD (Esmaillzadeh and Azadbakht, 2012). Consumption of pulses leads to stimulation of adipocytes which in turn secrete a cardioprotective hormone, adiponectin. Adiponectin shows anti-inflammatory properties in blood vessel cells and it decreases the risk of a heart attack in both, men (Pischon et al., 2004) as well as women (Teede et al., 2003).

7.5.5   DIABETES

Pulses, due to low GI and high-fiber content are the best food option for people with diabetes (Jenkins et al., 1981; Rizkalla et al., 2002). In addition, the resistant starch present in pulses also contributes to the improvement of glucose tolerance as well as insulin sensitivity (Jenkins et al., 2002). A meta-analysis of randomized controlled trials supported the positive role of pulses in lowering fasting blood glucose and insulin levels (Sievenpiper et al., 2009). Currently, the Canadian Diabetes Association also supports the consumption of high fiber foods and recommends consuming lentils, dried beans, and peas (CDA, 2012).

7.5.6   OBESITY

Pulses help in weight management by showing satiating effects due to the presence of fiber (Heaton, 1973; McCrory, Hamaker, Lovejoy, and Eichels-doerfer, 2010). A study confirmed the same by comparing the regular bread with pea fiber-containing bread (Lunde et al., 2011). Pulses also play an essential role in weight management by comparing the food intake capacity of humans consuming pasta and sauce alone and with the ones consuming the same along with lentils (Mollard et al., 2012).

7.5.7   HUMAN IMMUNO-DEFICIENCY VIRUS (HIV)

Lectins isolated from pulses may have the potential to inhibit HIV-1reverse transcriptase. Antifungal peptides present in *P. vulgaris* can inhibit HIV-1 reverse transcriptase *in vitro* (Ye et al., 2001). Moreover, the pulses also can enhance immune function (Hou et al., 2010).

7.5.8 OTHERS

Other health benefits associated with the consumption of pulses include aphrodisiac activity, estrogenic effect, etc.

7.6 CONCLUSION

Legumes/ pulses play an essential role in human nutrition as well as human health. They are a rich source of bioactive compounds, therefore, possess high therapeutic value. Among the different pharmacological activities imparted by pulses, management of cancer is gaining the attention of oncologists as pulses are readily available, cost-effective, and high on nutritional value. Other associated benefits include reduced risk of CVS related diseases, obesity, Parkinson's disease, etc. However, still, many of the pulses are underutilized. Therefore, there is a need to conduct more human studies to evaluate the anticancer potential of pulses while clinical studies should be conducted to confirm the anticancer activity of pulses with *in vitro* anticancer potential. Moreover, pulse-based food products should be developed to increase its consumption in regions where pulses are not much used as a part of a balanced diet. Formulation of nutraceuticals using these pulses may also contribute in the management of cancer.

ACKNOWLEDGMENT

The authors would like to acknowledge the financial assistance provided by Haryana State Council for Science and Technology, Panchkula under research & development scheme (No. HSCST/R & D/2017/66).

KEYWORDS

- **anticancer**
- **coronary heart disease**
- **glycemic index**
- **legumes**
- **phytoconstituents**
- **pulses**

REFERENCES

Adikay, S., (2016). Phytoremedial effect of *Lens culinaris* against doxorubicin-induced nephrotoxicity in male Wistar rats. *Int. J. Green Pharm., 10*(03), 172–177.

Alcalay, R. N., Gu, Y., Mejia-Santana, H., Cote, L., Marder, K. S., & Scarmeas, N., (2012). The association between Mediterranean diet adherence and Parkinson's disease. *Mov. Disord., 27*(6), 771–774.

Ali, N. M., Yeap, S. K., Yusof, H. M., Beh, B. K., Ho, W. Y., Koh, S. P., Abdullah, M. P., et al., (2016). Comparison of free amino acids, antioxidants, soluble phenolic acids, cytotoxicity and immunomodulation of fermented mung bean and soybean. *J. Sci. Food Agric., 96*(5), 1648–1658.

Al-Tibi, A. M., Takruri, H. R., & Ahmad, M. N., (2010). Effect of dehulling and cooking of lentils (*Lens culinaris* L.) on serum glucose and lipoprotein levels in streptozotocin-induced diabetic rats. *Malays. J. Nutr., 16*(3), 409–18.

Anderson, J. W., & Major, A. W., (2002). Pulses and lipaemia, short- and long-term effect: Potential in the prevention of cardiovascular disease. *Br. J. Nutr., 88*(3), S263–S271.

Anderson, J. W., Gustafson, N. J., Spencer, D. B., Tietyen, J., & Bryant, C., (1990). Serum lipid response of hypercholesterolemic men to single and divided doses of canned beans. *Am. J. Clin. Nutr., 51*(6), 1013–1019.

Anderson, J. W., Smith, B. M., & Gustafson, N. J., (1994). Health benefits and practical aspects of high-fiber diets. *Am. J. Clin. Nutr., 59*(5), 1242S–1247S.

Anderson, J. W., Story, L., Sieling, B., Chen, W., Petro, M. S., & Story, J., (1984). Hypocholesterolemic effects of oat-bran or bean intake for hypercholesterolemic men. *Am. J. Clin. Nutr., 40*(6), 1146–1155.

Aparicio-Fernández, X., García-Gasca, T., Yousef, G. G., Lila, M. A., González, D. M. E., & Loarca-Pina, G., (2006). Chemopreventive activity of polyphenolics from black Jamapa bean (*Phaseolus vulgaris* L.) on HeLa and HaCaT cells. *J. Agric. Food Chem., 54*(6), 2116–2122.

Aparicio-Fernández, X., Manzo-Bonilla, L., & Loarca-Piña, G. F., (2005). Comparison of antimutagenic activity of phenolic compounds in newly harvested and stored common beans *Phaseolus vulgaris* against aflatoxin B1. *J. Food. Sci., 70*(1), S73-S78.

Aparicio-Fernández, X., Reynoso-Camacho, R., Castaño-Tostado, E., García-Gasca, T., De Mejía, E. G., Guzmán-Maldonado, S. H., Elizondo, G., et al., (2008). Antiradical capacity and induction of apoptosis on HeLa cells by a *Phaseolus vulgaris* extract. *Plant Foods Hum. Nutr., 63*(1), 35–40.

Asif, M., Rooney, L. W., Ali, R., & Riaz, M., (2013). Application and opportunities of pulses in food system: A review. *Crit. Rev. Food Sci. Nutr., 53*(11), 1168–1179.

Atkinson, F. S., Foster-Powell, K., & Brand-Miller, J. C., (2008). International tables of glycemic index and glycemic load values. *Diab. Care, 31*(12), 2281–2283.

Badria, F. A., (2014). Anticancer activity of plant-derived proteins against human tumor cell lines. *Drug Discov. Ther., 2*(13), 60–69.

Bazzano, L. A., He, J., Ogden, L. G., Loria, C., Vupputuri, S., Myers, L., & Whelton, P. K., (2001). Legume consumption and risk of coronary heart disease in US men and women: NHANES I epidemiologic follow-up study. *Arch. Intern. Med., 161*(21), 2573–8.

Bennink, M., (2002). Consumption of black beans and navy beans (*Phaseolus vulgaris*) reduced azoxymethane-induced colon cancer in rats. *Nutr. Cancer, 44* (1), 60–65.

Boye, J., Zare, F., & Pletch, A., (2010). Pulse proteins: Processing, characterization, functional properties and applications in food and feed. *Food Res. Intl., 43*(2), 414–431.

Cardador-Martinez, A., Albores, A., Bah, M., Calderon-Salinas, V., Castano-Tostado, E., Guevara-Gonzalez, R., Shimada-Miyasaka, A., & Loarca-Pina, G., (2006). Relationship among antimutagenic, antioxidant and enzymatic activities of methanolic extract from common beans (*Phaseolus vulgaris* L.). *Plant Foods Hum. Nutr., 61*(4), 161–168.

Cardador-Martinez, A., Castano-Tostado, E., & Loarca-Pina, G., (2002). Antimutagenic activity of natural phenolic compounds present in the common bean (*Phaseolus vulgaris*) against aflatoxin B 1. *Food Addit. Contam., 19*(1), 62–69.

CFIA Canadian Food Inspection Agency. (2011). *Nutrient Content Claims.* http://www.inspection.gc.ca/english/fssa/labeti/guide/ch7be.shtml (accessed on 20 May 2020).

Chan, Y. S., Yu, H., Xia, L., & Ng, T. B., (2015). Lectin from green speckled lentil seeds (*Lens culinaris*) triggered apoptosis in nasopharyngeal carcinoma cell lines. *Chin. Med., 10*(1), 25.

Chen, Z., Wang, J., Liu, W., & Chen, H., (2017). Physicochemical characterization, antioxidant, and anticancer activities of proteins from four legume species. *J. Food Sci. Tech., 54*(4), 964–972.

Clemente, A., Carmen Marin-Manzano, M., Jimenez, E., Carmen, A. M., & Domoney, C., (2012). The anti-proliferative effect of TI1B, a major bowman-birk isoinhibitor from pea (*Pisum sativum* L.), on HT29 colon cancer cells is mediated through protease inhibition. *Br. J. Nutr., 108*(1), S135–S144.

Clemente, A., Gee, J. M., Johnson, I. T., Mackenzie, D. A., & Domoney, C., (2005). Pea (*Pisum sativum* L.) protease inhibitors from the bowman-birk class influence the growth of human colorectal adenocarcinoma HT29 cells *in vitro*. *J. Agric. Food Chem., 53*(23), 8979–8986.

Dahl, W. J., Foster, L. M., & Tyler, R. T., (2012). Review of the health benefits of peas (*Pisum sativum* L.). *Br. J. Nutr., 108*(1), S3–10.

Dahl, W. J., Whiting, S. J., Healey, A., Zello, G. A., & Hildebrandt, S. L., (2003). Increased stool frequency occurs when finely processed pea hull fiber is added to usual foods consumed by elderly residents in long-term care. *J. Am. Diet Assoc., 103*(9), 1199–1202.

Dalal, K., Singhroha, S., Ahlawat, S., & Patra, A., (2011). Anti-diarrheal activity of roots of *Cicer arietinum* Linn. *Int. J. Res. Pharm. Bio. Sci., 2*(1), 268–270.

Divya, S., & Banda, T., (2014). Evaluation of anti-diuretic and anti-nephrolithiatic activities of ethanolic seeds extract of *Cicer arietinum* in experimental rats. *IJPRD, 5*(12), 9–12.

Dominika, Ś., Arjan, N., Karyn, R. P., & Henryk, K., (2011). The study on the impact of glycated pea proteins on human intestinal bacteria. *Int. J. Food Microbiol., 145*(1), 267–272.

El-Aassar, M. R., Hafez, E. E., El-Deeb, N. M., & Fouda, M. M., (2014). Microencapsulation of lectin anti-cancer agent and controlled release by alginate beads, biosafety approach. *Int. J. Biol. Macromol., 69*, 88–94.

Esmaillzadeh, A., & Azadbakht, L., (2012). Legume consumption is inversely associated with serum concentrations of adhesion molecules and inflammatory biomarkers among Iranian women. *J. Nutr., 142*(2), 334–9.

Esposito, K., Maiorino, M. I., Ceriello, A., & Giugliano, D., (2010). Prevention and control of type 2 diabetes by Mediterranean diet: A systematic review. *Diabetes Res. Clin. Pract., 89*(2), 97–102.

Finley, J. W., Burrell, J. B., & Reeves, P. G., (2007). Pinto bean consumption changes SCFA profiles in fecal fermentations, bacterial populations of the lower bowel, and lipid profiles in blood of humans. *J. Nutr., 137*(11), 2391–8.

Food and Agriculture Organization (FAO). (2020). *Crops Statistics: Concepts, Definitions and Classifications.* http://www.fao.org/economic/the-statistics-division-ess/methodology/

methodology-systems/crops-statistics-concepts-definitions-and-classifications/en/ (accessed on 20 May 2020).

Food and Agriculture Organization (FAO) (1994). *Definition and Classification of Commodities: Pulses and Derived Products*. http://www.fao.org/es/faodef/fdef04e.htm (accessed on 20 May 2020).

Fung, T. T., Rexrode, K. M., Mantzoros, C. S., Manson, J. E., Willett, W. C., & Hu, F. B., (2009). Mediterranean diet and incidence of and mortality from coronary heart disease and stroke in women. *Circulation, 119*(8), 1093–100.

Ganesan, K., & Xu, B., (2017). A critical review on phytochemical profile and health promoting effects of mung bean (*Vigna radiata*). *Food Sci. Hum. Welln., 7*(1), 11–33.

Guenther, P. M., Dodd, K. W., Reedy, J., & Krebs-Smith, S. M., (2006). Most Americans eat much less than recommended amounts of fruits and vegetables. *J. Am. Diet Assoc., 106*(9), 1371–9.

Ha, V., Sievenpiper, J. L., De Souza, R. J., Jayalath, V. H., Mirrahimi, A., Agarwal, A., Chiavaroli, L., Mejia, S. B., Sacks, F. M., & Di Buono, M., (2014). Effect of dietary pulse intake on established therapeutic lipid targets for cardiovascular risk reduction: A systematic review and meta-analysis of randomized controlled trials. *Canad. Med. Assoc. J., 186*(8), E252–E262.

Hafidh, R. R., Abdulamir, A. S., Bakar, F. A., Jalilian, F. A., Abas, F., & Sekawi, Z., (2012). Novel molecular, cytoxical, and immunological study on promising and selective anticancer activity of Mung bean sprouts. *BMC Complement Altern. Med., 12*(1), 208.

Hanson, M. G., (2013). *The Effect of Lentils (Lens culinaris) on Hypertension and Hypertension–Associated Vascular Remodeling in the Spontaneously Hypertensive Rat*. https://mspace.lib.umanitoba.ca/xmlui/bitstream/handle/1993/22140/Hanson_Matthew. pdf?sequence=1&isAllowed=y (accessed on 20 May 2020).

Hanson, M. G., Zahradka, P., & Taylor, C. G., (2014). Lentil-based diets attenuate hypertension and large-artery remodeling in spontaneously hypertensive rats. *Br. J. Nutr., 111*(4), 690–698.

Hassan, E. M., Matloub, A. A., Aboutabl, M. E., Ibrahim, N. A., & Mohamed, S. M., (2016). Assessment of anti-inflammatory, antinociceptive, immunomodulatory, and antioxidant activities of *Cajanus cajan* L. seeds cultivated in Egypt and its phytochemical composition. *Pharm. Biol., 54*(8), 1380–1391.

Hayat, I., Ahmad, A., Masud, T., Ahmed, A., & Bashir, S., (2014). Nutritional and health perspectives of beans (*Phaseolus vulgaris* L.): An overview. *Crit. Rev. Food Sci. Nutr., 54*(5), 580–592.

HC Health Canada (2020). *Canada's Food Guide to Healthy Eating: Tips for Meat and Alternatives*. https://www.unlockfood.ca/en/Articles/Canada-s-Food-Guide/How-Many-Meat-and-Alternatives-Do-You-Need.aspx#:~:text=Examples of one food guide serving of meat and alternatives:&text=75 g (½ cup) cooked poultry like chicken, duck,canned tuna* or canned salmon (accessed on 20 May 2020).

Heaton, K., (1973). Food fiber as an obstacle to energy intake. *Lancet, 302*(7843), 1418–1421.

Hermsdorff, H. H., Zulet, M. A., Abete, I., & Martinez, J. A., (2011). A legume-based hypocaloric diet reduces proinflammatory status and improves metabolic features in overweight/obese subjects. *Eur. J. Nutr., 50*(1), 61–69.

Hou, Y., Hou, Y., Yanyan, L., Qin, G., & Li, J., (2010). Extraction and purification of a lectin from red kidney bean and preliminary immune function studies of the lectin and four Chinese herbal polysaccharides. *Bio. Med. Res. Int.,* 1–9.

Iqbal, A., Khalil, I. A., Ateeq, N., & Khan, M. S., (2006). Nutritional quality of important food legumes. *Food Chem., 97*(2), 331–335.

Itoh, T., Itoh, Y., Mizutani, M., Fujishiro, K., Furuichi, Y., Komiya, T., & Hibasami, H., (2002). A hot water extract of adzuki (*Vigna angularis*) induces apoptosis in cultured human stomach cancer cells. *J. Jpn. Soc. Food Sci.*

Itoh, T., Itoh, Y., Mizutani, M., Fujishiro, K., Furuichi, Y., Komiya, T., & Hibasami, H., (2004). Hot-water extracts from adzuki beans (*Vigna angularis*) suppress not only the proliferation of KATO III cells in culture but also benzo(a)pyrene-induced tumorigenesis in mouse forestomatch. *J. Nutr. Sci. Vitaminol. (Tokyo), 50*(4), 295–299.

Itoh, T., Umekawa, H., & Furuichi, Y., (2005). Potential ability of hot water adzuki (*Vigna angularis*) extracts to inhibit the adhesion, invasion, and metastasis of murine B16 melanoma cells. *Biosci. Biotechnol. Biochem., 69*(3), 448–454.

Jayalath, V. H., De Souza, R. J., Sievenpiper, J. L., Ha, V., Chiavaroli, L., Mirrahimi, A., Di Buono, M., et al., (2014). Effect of dietary pulses on blood pressure: A systematic review and meta-analysis of controlled feeding trials. *Am. J. Hypertens., 27*(1), 56–64.

Jenkins, D. J., Kendall, C. W., Augustin, L. S., & Vuksan, V., (2002). High-complex carbohydrate or lente carbohydrate foods? *Am. J. Med., 113*(9), 30–37.

Jenkins, D. J., Wolever, T. M., Taylor, R. H., Barker, H., Fielden, H., Baldwin, J. M., Bowling, A. C., Newman, H. C., Jenkins, A. L., & Goff, D. V., (1981). Glycemic index of foods: A physiological basis for carbohydrate exchange. *Am. J. Clin. Nutr., 34*(3), 362–366.

Jeong, Y. M., Ha, J. H., Noh, G. Y., & Park, S. N., (2016). Inhibitory effects of mung bean (*Vigna radiata*L.) seed and sprout extracts on melanogenesis. *Food Sci. Biotech., 25*(2), 567–573.

Joanitti, G. A., Azevedo, R. B., & Freitas, S. M., (2010). Apoptosis and lysosome membrane permeabilization induction on breast cancer cells by an anticarcinogenic Bowman-Birk protease inhibitor from *Vigna unguiculata* seeds. *Cancer Letters, 293*(1), 73–81.

Kan, A., Özçelik, B., Kartal, M., Özdemir, Z., & Özgen, S., (2010). *In vitro* antimicrobial activities of *Cicer arietinum*L (Chickpea). *Trop. J. Pharm. Res., 9*(5), 475–481.

Ke, X., & Shen, L., (2017). Molecular targeted therapy of cancer: The progress and future prospect. *Front. Lab. Med., 1*(2), 69–75.

Kim, E. K., Kim, Y. S., Hwang, J. W., Lee, J. S., Moon, S. H., Jeon, B. T., & Park, P. J., (2013). Purification and characterization of a novel anticancer peptide derived from Ruditapes philippinarum. *Process Biochem., 48*(7), 1086–1090.

Kim, M., Park, J. E., Song, S. B., & Cha, Y. S., (2015). Effects of black adzuki bean (*Vigna angularis*) extract on proliferation and differentiation of 3T3-L1 preadipocytes into mature adipocytes. *Nutrients, 7*(1), 277–292.

Kumar, S., Kapoor, V., Gill, K.,Singh, K., Xess, I., Das, S. N., & Dey, S., (2014). Antifungal and antiproliferative protein from *Cicer arietinum:* A bioactive compound against emerging pathogens. *Biomed. Res. Int.*, 387203.

Kupper, C., (2005). Dietary guidelines and implementation for celiac disease. *Gastroenterol., 128*(4), S121–S127.

Leterme, P., & Muñoz, L. C., (2002). Factors influencing pulse consumption in Latin America. *Br. J. Nutr., 88*(S3), 251–254.

Li, C., & Uppal, M. (2010). Canadian diabetes association national nutrition committee clinical update on dietary fibre in diabetes: food sources to physiological effects. *Canadian Journal of Diabetes, 34*(4), 355–361.

Li, H., Prairie, N., Udenigwe, C. C., Adebiyi, A. P., Tappia, P. S., Aukema, H. M., Jones, P. J., & Aluko, R. E., (2011). Blood pressure lowering effect of a pea protein hydrolysate in hypertensive rats and humans. *J. Agric. Food Chem., 59*(18), 9854–9860.

Lichtenstein, A. H., Appel, L. J., Brands, M., Carnethon, M., Daniels, S., Franch, H. A., Franklin, B., Kris-Etherton, P., et al., (2006). Diet and lifestyle recommendations revision 2006: A scientific statement from the American heart association nutrition committee. *Circulation, 114*(1), 82–96.

Liu, R., Zheng, Y., Cai, Z., & Xu, B., (2017). Saponins and Flavonoids from Adzuki Bean (*Vigna angularis* L.) ameliorate high-fat diet-induced obesity in ICR mice. *Front. Pharmacol., 8*, 687.

Luna-Vital, D. A., Mojica, L., De Mejía, E. G., Mendoza, S., & Loarca-Piña, G., (2015). Biological potential of protein hydrolysates and peptides from common bean (*Phaseolus vulgaris* L.): A review. *Food. Res. Int., 76*, 39–50.

Lunde, M. S., Hjellset, V. T., Holmboe-Ottesen, G., & Høstmark, A. T., (2011). Variations in postprandial blood glucose responses and satiety after intake of three types of bread. *J. Nutr. Metabol.*, 1–7.

Luo, M., Liu, X., Zu, Y., Fu, Y., Zhang, S., Yao, L., & Efferth, T., (2010). Cajanol, a novel anticancer agent from *Pigeonpea* [*Cajanus cajan* (L.) Millsp.] roots, induces apoptosis in human breast cancer cells through a ROS-mediated mitochondrial pathway. *Chem. Biol. Interact., 188*(1), 151–60.

Lupien, J., (1997). *Agriculture Food and Nutrition for Africa: A Resource Book for Teachers of Agriculture*. Rome: FAO. Available at: http://www.fao.org/docrep/w0078e/w0078e00. HTM (accessed on 20 May 2020).

Magee, P. J., Owusu-Apenten, R., McCann, M. J., Gill, C. I., & Rowland, I. R., (2012). Chickpea (*Cicer arietinum*) and other plant-derived protease inhibitor concentrates inhibit breast and prostate cancer cell proliferation *in vitro. Nutr. Cancer, 64*(5), 741–748.

Marinangeli, C. P., & Jones, P. J., (2011). Whole and fractionated yellow pea flours reduce fasting insulin and insulin resistance in hypercholesterolaemic and overweight human subjects. *Br. J. Nutr., 105*(1), 110–117.

Marinangeli, C. P., Kassis, A. N., & Jones, P. J., (2009). Glycemic responses and sensory characteristics of whole yellow pea flour added to novel functional foods. *J. Food Sci., 74*(9), S385-S389.

Mattei, J., Hu, F. B., & Campos, H., (2011). A higher ratio of beans to white rice is associated with lower cardiometabolic risk factors in Costa Rican adults. *Am. J. Clin. Nutr., 94*(3), 869–876.

McCrory, M. A., Hamaker, B. R., Lovejoy, J. C., & Eichelsdoerfer, P. E., (2010). Pulse consumption, satiety, and weight management. *Adv. Nutr., 1*(1), 17–30.

Mo'ez Al-Islam, E. F., Takruri, H. R., Shomaf, M. S., & Bustanji, Y. K., (2009). Chemopreventive effect of raw and cooked lentils (*Lens culinaris* L) and soybeans (Glycine max) against azoxymethane-induced aberrant crypt foci. *Nutr. Res., 29*(5), 355–362.

Mollard, R. C., Zykus, A., Luhovyy, B. L., Nunez, M. F., Wong, C. L., & Anderson, G. H., (2012). The acute effects of a pulse-containing meal on glycaemic responses and measures of satiety and satiation within and at a later meal. *Br. J. Nutr., 108*(3), 509–517.

Mudryj, A. N., Yu, N., & Aukema, H. M., (2014). Nutritional and health benefits of pulses. *Appl Physiol. Nutr. Metab., 39*(11), 1197–1204.

Nakaya, K., Nabata, Y., Ichiyanagi, T., & An, W. W., (2012). Stimulation of dendritic cell maturation and induction of apoptosis in leukemia cells by a heat-stable extract from azuki bean (*Vigna angularis*), a promising immunopotentiating food and dietary supplement for cancer prevention. *Asian Pac. J. Cancer Prev., 13*(2), 607–611.

Nasrin, F., Paul, S., Zaman, S., & Koly, S., (2015). Study of antimicrobial and cytotoxic activities of *Vigna mungo* Linn. Hepper (Family-Leguminosae). *Pharma. Tutor, 3*(4), 40–46.

Nawaz, K. A. A., David, S. M., Murugesh, E., Thandeeswaran, M., Kiran, K. G., Mahendran, R., Palaniswamy, M., & Angayarkanni, J., (2017). Identification and *in silico* characterization of a novel peptide inhibitor of angiotensin converting enzyme from pigeon pea (*Cajanus cajan*). *Phytomed., 36*, 1–7.

Ofuya, Z., & Akhidue, V., (2005). The role of pulses in human nutrition: A review. *J. Appl. Sci. Environ. Mgt., 9*(3), 99–104.

Oh, H. M., Lee, S. W., Yun, B. R., Hwang, B. S., Kim, S. N., Park, C. S., Jeoung, S. H., Kim, H. K., Lee, W. S., & Rho, M. C., (2013). *Vigna angularis* inhibits IL-6-induced cellular signaling and ameliorates collagen-induced arthritis. *Rheumatol., 53*(1), 56–64.

Oomah, B. D., Corbé, A., & Balasubramanian, P., (2010). Antioxidant and anti-inflammatory activities of bean (*Phaseolus vulgaris* L.) hulls. *J. Agric. Food Chem., 58*(14), 8225–8230.

Papanikolaou, Y., & Fulgoni, V. L., (2008). 3rd, Bean consumption is associated with greater nutrient intake, reduced systolic blood pressure, lower body weight, and a smaller waist circumference in adults: Results from the National Health and Nutrition Examination Survey 1999–2002. *J. Am. Coll. Nutr., 27*(5), 569–576.

Patel, A., (2014). Isolation, characterization, and production of a new recombinant lectin protein from leguminous plants. *Biochem. Compds., 2*(1), 2.

Patterson, C., Maskus, H., & Dupasquier, C., (2009). Pulse crops for health. *Cer. Foods World, 54*(3), 108.

Petry, N., Egli, I., Zeder, C., Walczyk, T., & Hurrell, R., (2010). Polyphenols and phytic acid contribute to the low iron bioavailability from common beans in young women, 2. *J. Nutr., 140*(11), 1977–1982.

Pischon, T., Girman, C. J., Hotamisligil, G. S., Rifai, N., Hu, F. B., & Rimm, E. B., (2004). Plasma adiponectin levels and risk of myocardial infarction in men. *JAMA, 291*(14), 1730,1737.

Rabey, J. M., Vered, Y., Shabtai, H., Graff, E., Harsat, A., & Korczyn, A. D., (1993). Broad bean (*Vicia faba*) consumption and Parkinson's disease. *Adv. Neurol., 60*, 681–684.

Riddoch, C., Mills, C., & Duthie, G., (1998). An evaluation of germinating beans as a source of vitamin C in refugee foods. *Eur. J. Clin. Nutr., 52*(2), 115.

Rizkalla, S. W., Bellisle, F., & Slama, G., (2002). Health benefits of low glycaemic index foods, such as pulses, in diabetic patients and healthy individuals. *Br. J. Nutr., 88*(S3), 255–262.

Rocha-Guzmán, N. E., Herzog, A., González-Laredo, R. F., Ibarra-Pérez, F. J., Zambrano-Galván, G., & Gallegos-Infante, J. A., (2007). Antioxidant and antimutagenic activity of phenolic compounds in three different colour groups of common bean cultivars (*Phaseolus vulgaris*). *Food Chem., 103*(2), 521–527.

Sajja, R., Venkatesh, V., Suneetha, B., & Srinivas, N., (2014). Evaluation of aphrodisiac activity of methanolic extract of *Cicer arietinum* seeds in sexually sluggish male albino rats. *Int. J. Pharm., 4*(4), 309–313.

Sandberg, A. S., (2002). Bioavailability of minerals in legumes. *Br. J. Nutr., 88*(S3), 281–285.

Sandstrom, B., Hansen, L. T., & Sorensen, A., (1994). Pea fiber lowers fasting and postprandial blood triglyceride concentrations in humans. *J. Nutr., 124*(12), 2386–2396.

Santhoshi, K., Divya, S., Banda, T., & Ravi, K. V., (2013). Potential hepatoprotective effect of ethanolic seeds extract of *Cicer arietinum* against paracetamol induced hepatotoxicity. *J. Pharm. Res., 6*(9), 924.

Sardari, S., Amiri, M., Rahimi, H., Kamalinejad, M., Narenjkar, J., & Sayyah, M., (2015). Anticonvulsant effect of *Cicer arietinum* seed in animal models of epilepsy: Introduction of an active molecule with novel chemical structure. *Iran. Biomed. J., 19*(1), 45.

Scarmeas, N., Stern, Y., Mayeux, R., Manly, J. J., Schupf, N., & Luchsinger, J. A., (2009). Mediterranean diet and mild cognitive impairment. *Arch Neurol., 66*(2), 216–225.

Shafeen, S., (2012). Anti-inflammatory activity of *Cicer arietinum* seed extracts. *Asian J. Pharm. Clin. Res.*, 5, 64–68.

Shenkarev, Z. O., Gizatullina, A. K., Finkina, E. I., Alekseeva, E. A., Balandin, S. V., Mineev, K. S., Arseniev, A. S., & Ovchinnikova, T. V., (2014). Heterologous expression and solution structure of defensin from lentil *Lens culinaris*. *Biochem. Biophys. Res. Commun., 451*(2), 252–257.

Shomaf, M. S., & Takruri, H. R., (2012). Lentils (*Lens culinaris*, L.) attenuate colonic lesions and neoplasms in fischer 344 Rats. *Jordan Med. J., 45*(3), 231–239.

Shutler, S. M., Bircher, G. M., Tredger, J. A., Morgan, L. M., Walker, A. F., & Low, A., (1989). The effect of daily baked bean (*Phaseolus vulgaris*) consumption on the plasma lipid levels of young, normo-cholesterolaemic men. *Br. J. Nutr., 61*(2), 257–265.

Siah, S. D., Konczak, I., Agboola, S., Wood, J. A., & Blanchard, C. L., (2012). *In vitro* investigations of the potential health benefits of Australian-grown faba beans (*Vicia faba* L.): Chemopreventative capacity and inhibitory effects on the angiotensin-converting enzyme, α-glucosidase and lipase. *Br. J. Nutr., 108*(S1), S123–S134.

Siah, S., (2016). *Anti-Cancer Property of Pulses*. Australia: Australian Export Grains Innovation Centre.

Sievenpiper, J., Kendall, C., Esfahani, A., Wong, J., Carleton, A., Jiang, H., Bazinet, R., Vidgen, E., & Jenkins, D., (2009). *Effect of Non-Oil-Seed Pulses on Glycaemic Control: A Systematic Review and Meta-Analysis of Randomized Controlled Experimental Trials in People with and Without Diabetes*. Springer, Germany.

Singh, J., & Basu, P. S., (2012). Non-nutritive bioactive compounds in pulses and their impact on human health: An overview. *Food Nutr. Sci., 3*(12), 1664.

Stanisavljevic, N. S., Ilic, M. D., Matic, I. Z., Jovanovic, Z. S., Cupic, T., Dabic, D. C., Natic, M. M., & Tesic, Z., (2016). Identification of phenolic compounds from seed coats of differently colored European varieties of pea (*Pisum sativum* L.) and characterization of their antioxidant and *in vitro* anticancer activities. *Nutr. Cancer., 68*(6), 988–1000.

Świątecka, D., Kostyra, H., & Świątecki, A., (2010). Impact of glycated pea proteins on the activity of free-swimming and immobilized bacteria. *J. Sci. Food Agric., 90*(11), 1837–1845.

Teede, H. J., McGrath, B. P., DeSilva, L., Cehun, M., Fassoulakis, A., & Nestel, P. J., (2003). Isoflavones reduce arterial stiffness: A placebo-controlled study in men and postmenopausal women. *Arterioscler. Thromb. Vasc. Biol., 23*(6), 1066–1071.

Tosh, S. M., & Yada, S., (2010). Dietary fibers in pulse seeds and fractions: Characterization, functional attributes, and applications. *Food Res. Int., 43*(2), 450–460.

United States Department of Health and Human Services, (USDHHS). *Dietary Guidelines for Americans*. https://www.hhs.gov/fitness/eat-healthy/dietary-guidelines-for-americans/index.html (accessed on 20 May 2020).

USDA, A. R. S. USDA National Nutrient Database for Standard Reference, Release 25. *Nutrient Data Laboratory Home Page.* www.ars.usda.gov/ba/bhnrc/ndl (accessed on 20 May 2020).

Veenstra, J., Duncan, A., Cryne, C., Deschambault, B., Boye, J., Benali, M., Marcotte, M., Tosh, S., Farnworth, E., & Wright, A., (2010). Effect of pulse consumption on perceived flatulence and gastrointestinal function in healthy males. *Food Res. Int., 43*(2), 553–559.

Venter, C., & Van, E. E., (2001). More legumes for better overall health. *Magnesium, 172,* 280.

Vohra, K., Dureja, H., & Garg, V., (2015). An insight of pulses: From food to cancer treatment. *J. Pharmacogn. Nat. Prod., 1*(108), 2472–0992.1000108.

Vohra, K., Gupta, V. K., Dureja, H., & Garg, V., (2016). Antihyperlipidemic activity of *Lens culinaris* Medikus seeds in triton WR-1339 induced hyperlipidemic rats. *J. Pharmacogn. Nat. Prod., 2,* 117.

Wang, H., & Ng, T., (2007). An antifungal peptide from red lentil seeds. *Peptides, 28*(3), 547–552.

Wikhe, M., Zade, V., Dabadkar, D., & Patil, U., (2013). Evaluation of the abortifacient and estrogenic activity of *Cicer arietinum* leaves on female albino rat. *J. Bioinnov., 2*(3), 105–113.

Willett, W. C., Sacks, F., Trichopoulou, A., Drescher, G., Ferro-Luzzi, A., Helsing, E., & Trichopoulos, D., (1995). Mediterranean diet pyramid: A cultural model for healthy eating. *Am J. Clin. Nutr., 61*(6), 1402S–1406S.

Winham, D. M., & Hutchins, A. M., (2007). Baked bean consumption reduces serum cholesterol in hypercholesterolemic adults. *Nutr. Res., 27*(7), 380–386.

Winham, D., Webb, D., & Barr, A., (2008). Beans and good health. *Nutr. Today, 43*(5), 201–209.

Wolever, T. M., Jenkins, D. J., Thompson, L. U., Wong, G. S., & Josse, R. G., (1987). Effect of canning on the blood glucose response to beans in patients with type 2 diabetes. *Hum. Nutr. Clin. Nutr., 41*(2), 135–140.

World Health Organization (WHO) (2018). *Cancer Factsheet.* http://www.who.int/mediacentre/factsheets/fs297/en/ (accessed on 20 May 2020).

Xu, B., & Chang, S. K., (2012). Comparative study on antiproliferation properties and cellular antioxidant activities of commonly consumed food legumes against nine human cancer cell lines. *Food Chem., 134*(3), 1287–1296.

Yadav, B. V., Deshmukh, T. A., Badole, S. L., Kadam, H., Bodhankar, S. L., & Dhaneshwar, S. R., (2009). Antihyperglycaemic activity of *Cicer arietinum* seeds. *Pharmacology Online, 3,* 748–757.

Ye, X. Y., Ng, T. B., Tsang, P. W., & Wang, J., (2001). Isolation of a homodimeric lectin with antifungal and antiviral activities from red kidney bean (*Phaseolus vulgaris*) seeds. *J. Protein Chem., 20*(5), 367–75.

Yeap, S. K., Mohd, Y. H., Mohamad, N. E., Beh, B. K., Ho, W. Y., Ali, N. M., Alitheen, N. B., Koh, S. P., & Long, K., (2013). *In vivo* immunomodulation and lipid peroxidation activities contributed to chemoprevention effects of fermented mung bean against breast cancer. *Evid. Based Complement. Alternat. Med.,* 1–7.

Yust, M. A. M., Pedroche, J., Giron-Calle, J., Alaiz, M., Millán, F., & Vioque, J., (2003). Production of ace inhibitory peptides by digestion of chickpea legumin with alcalase. *Food Chem., 81*(3), 363–369.

Yust, M. D. M., Millán-Linares, M. D. C., Alcaide-Hidalgo, J. M., Millán, F., & Pedroche, J., (2012). Hypocholesterolaemic and antioxidant activities of chickpea (*Cicer arietinum* L.) protein hydrolysates. *J. Sci. Food Agric., 92*(9), 1994–2001.

Zadernowski, R., Pierzynowska-Korniak, G., Ciepielewska, D., & Fornal, L., (1992). Chemical characteristics and biological functions of phenolic acids of buckwheat and lentil seeds. *Fagopyrum., 12*, 27–35.

Zhao, Y., Li, Z., Zhao, C., Fu, R., Wang, X., & Li, Z., (2012). Effects of recombinant mung bean trypsin inhibitor fragments on migration of colon cancer cell SW480. *J. Shanxi Univ. (Nat. Sci. Ed.), 1*, 29.

Part III

Natural Drugs with Heteroatoms

CHAPTER 8

Heterocyclic Drugs from Plants

DEBASISH BANDYOPADHYAY,[1,2] VALERIA GARCIA,[1] and
FELIPE GONZALEZ[1]

¹Department of Chemistry

*²School of Earth Environment and Marine Sciences (SEEMS),
University of Texas Rio Grande Valley, 1201 West University Drive,
Edinburg, Texas – 78539, USA, Tel.: +1(956)6653824,
Fax: +1(956)6655006, E-mail: debasish.bandyopadhyay@utrgv.edu*

ABSTRACT

Heterocyclic chemistry is a branch of organic chemistry which pertains to diverse studies on heterocycles. Natural heterocyclic compounds have begun dominating the fields of drug development research, and other sciences including materials science. Plants have been the source of healing since before dates and authentication of such research could be taken. Studies conducted with heterocyclic compounds have had led the researchers to investigate more and more plants due to occurrences of various phytoceuticals, found in plants having the capacity of treating a wide range of illnesses. Diverse plant species could be explored to investigate heterocyclic phytoceuticals, for example, the stem bark from *Bauhinia variegate*, the leaves from *Camellia sinensis*, or the rootstock of *Scutellaria baicalensis* and so on. Phytochemicals might be heterocyclic in nature and eventually some of the phytochemicals become phytoceuticals. Cancer, diabetes, and HIV are some of the widely known health-related threats to the human population. Natural heterocyclic drugs can play an important role to combat against these dreadful diseases. Moreover, heterocyclic botanicals can also be beneficial to treat cardiovascular

This chapter is respectfully dedicated to Professor (Dr.) Hassan Ahmad to honor his lifelong devotion to Chemical Sciences.

diseases, neurodegenerative diseases, inflammation, malaria, asthma, depression, psychosis, obesity, and many other ailments. This chapter summarizes the abundance, acquisition, and identification of relatively recent natural heterocyclic drugs and/or molecules that have demonstrated valid potential against several diseases. Incidence, prevalence, and possible preventive measures along with drug actions on major diseases have been summarized. To maintain the continuity and clarity of discussion, a few semi-synthetic and synthetic drugs have also been included in the discussion, pharmacokinetics, and pharmacodynamics of many medicinally privileged natural heterocyclic compounds and drugs are the key features of this chapter.

8.1 INTRODUCTION

Heterocyclic chemistry, a division of organic chemistry, discusses the structures, properties, isolation/semi-synthesis/synthesis, and the implementation of heterocycles in the real world. Heterocycles can be found in many terrestrial plant-derived natural products (botanicals). Based on the structure, organic compounds can be acyclic and cyclic. Again, cyclic compounds can be carbocyclic and heterocyclic. Heterocycles must possess a ring/cyclic structure that contains at least one hetero (non-carbon atom-like nitrogen, oxygen, sulfur, phosphorus, bismuth, etc.) atom as a ring constituent. Furthermore, a subclass of heterocycles that contain aromaticity is called heteroaromatics or heteroaromatic compounds. If heteroaromatic compounds contain two or more fused rings they are termed as polyheteroaromatic compounds or polyheteroaromatics. Many bioprivileged heterocycles are well-known commercial drugs, these heterocycles may differ in ring size, type, number, and the positioning (place) of heteroatom(s) in the ring.

The smallest ring for a heterocyclic moiety can be of three-membered where at least one heteroatom must be a ring-constituent. Three-membered heterocycles should be more reactive due to angle strain and subsequent steric and torsional strains which taken together is known as ring strain. Angle strain is generated due to distortion of the bonds from a regular tetrahedral angle (109°28′) whereas electronic repulsion between the covalent bonds (bonding electrons) generates torsional strain. Alternatively, if the functional groups or atoms are too close to obscure one another, a strain is generated, called steric strain. The strains of a ligand are crucial to bind a protein/enzyme in the human body. The three strains are present in a three-membered ring. There is no specific limit of ring-size and it can be from three- to nine-membered or even larger rings and fused rings are also possible. The ring constituents in a heterocyclic scaffold can be of any non-carbon elements except the alkali metals.

The general structure of heterocycles closely resembles with the organic cyclic compounds that integrate only carbons in the rings. While classifying heterocycles one of the easiest differences which can be considered is the number of bond separations between the heteroatoms.

8.2 MULTIFUNCTIONAL HETEROCYCLIC BIOACTIVE SCAFFOLDS

Heterocyclic moieties are found in diverse natural drugs that are an indispensable part of life including nucleic acids, amino acids, carbohydrates, vitamins, and alkaloids. Currently, many marketed drugs contain heterocyclic pharmacophores (Gomtsyan, 2012). A large abundance of pharma- and agrochemicals contain at the least one heterocyclic ring structure. To treat a disease effectively, the correct medication must be taken through appropriate route of administration. Oral medications work by traveling through the stomach and then being absorbed into the bloodstream. Once in the bloodstream the medication flows throughout the body supplying to the therapeutic area it needs.

Natural ailments have been used by the human being since the early stage of humanization. In the ancient period, there were shamans, traditional healers, and divinators, known for being the person to treat if there was a sickness affecting someone in their tribe. Furthermore, there were also herbalists could be found in a royal court or town. All these ancient medical professionals used plants as their main source of medical care. They would crush, dry, soak, and boil different plants and then use the extract to treat ailments. This usage gives no room to doubt that for thousands of years these botanicals possess a wide spectrum of bioactivities (Panche et al., 2016). This is why nowadays many ethno pharmacologists, botanists, microbiologists, and natural product chemists around the world have been searching for new drugs and dietary supplements through phytoceutical/nutraceutical investigations (Cowan et al., 1999).

Plants produce several categories of secondary metabolites like tannins, terpenoids, alkaloids, and flavonoids (Figure 8.1) that humans (and some animals) use as remedies to cure many diseases (Cowan et al., 1999). These secondary metabolites including antibiotics are biosynthesized as a part of defense mechanism of the host (Demain et al., 2000) by serving as weapons against insects, microbes, and other hurtful organisms, as transporting agents, symbiotic agent, sex hormones, or as differentiation effectors. Using of plant materials as an aid to increase the "healing powers "of ancient healthcare professionals was an primitive idea, however, modern scientists began to observe and experiment with plants and their properties to heal and prevent sickness by creating drugs (Gomtsyan, 2012).

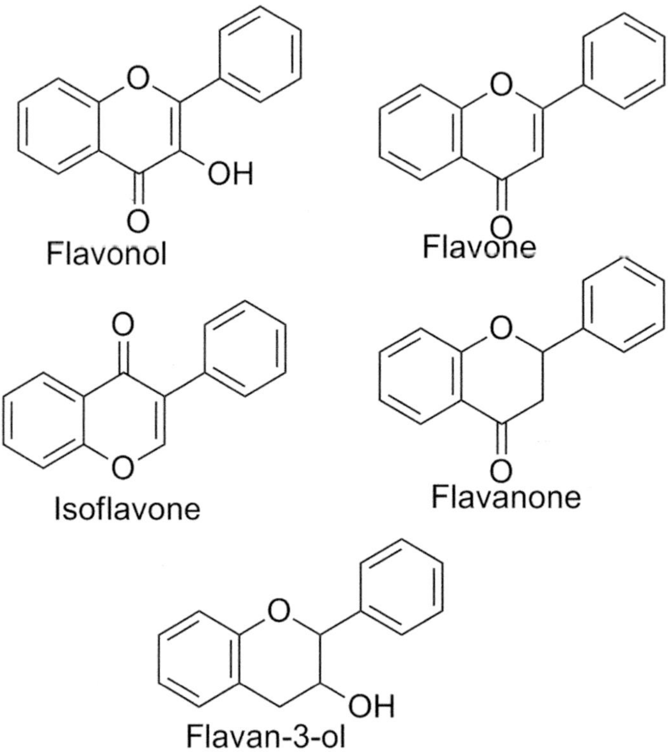

FIGURE 8.1   General structures of flavonoids.

8.2.1   *FLAVONOIDS*

Flavonoids, are a class of secondary metabolites that are largely found in the fruit or skin of various species of plants. Flavonoids primarily contain a C6-C3-C6 skeleton, made by two phenyl and anoxyhetero-cyclic rings, are abundant in various plants. Flavonoids can be divided into five majorsubclasses. These subclasses include flavones, flavonols, flavanones, flavanols, and isoflavones. (Figure 8.1). These subclasses contain structural and biological diversities. These characteristics are important for a pharmacological role in biogenesis and chemotaxonomic signification (Murkovic, 2003). A research conducted by Joray et al. (2015) reported five flavonoid compounds from the plant *Flourensiaoo-lepis* and determined the cytotoxic and antibacterial activities of these five flavonoids. Regionally known as "chilca" in the Argentinean region, the ethanol extract of *Flourensiaoolepis*, is recognized in containing both antibacterial and anticancer activities. Five out of 51 chemical compounds: 2,4-dihydroxychalcone, isoliquiritigenin, pinocembrin, 7-hydroxyflava-none, and 7,4-dihydroxy-3-methoxyflavanone have demonstrated *in vitro* anticancer activity in acute lymphoblastic leukemia (ALL) and chronic myeloid leukemia (CML) cell lines. Furthermore, 2,4-dihydroxychaclone, isoliquiritigenin, pinocembrin, 7-hydroxyflavanone demonstrated antibac-terial properties (Figure 8.2). The study reported that the bacteriostatic activity of 2,4-dihydroxychaclone against *P. mirabilis* bacteria resistant strain was better than a few commercial antibiotics.

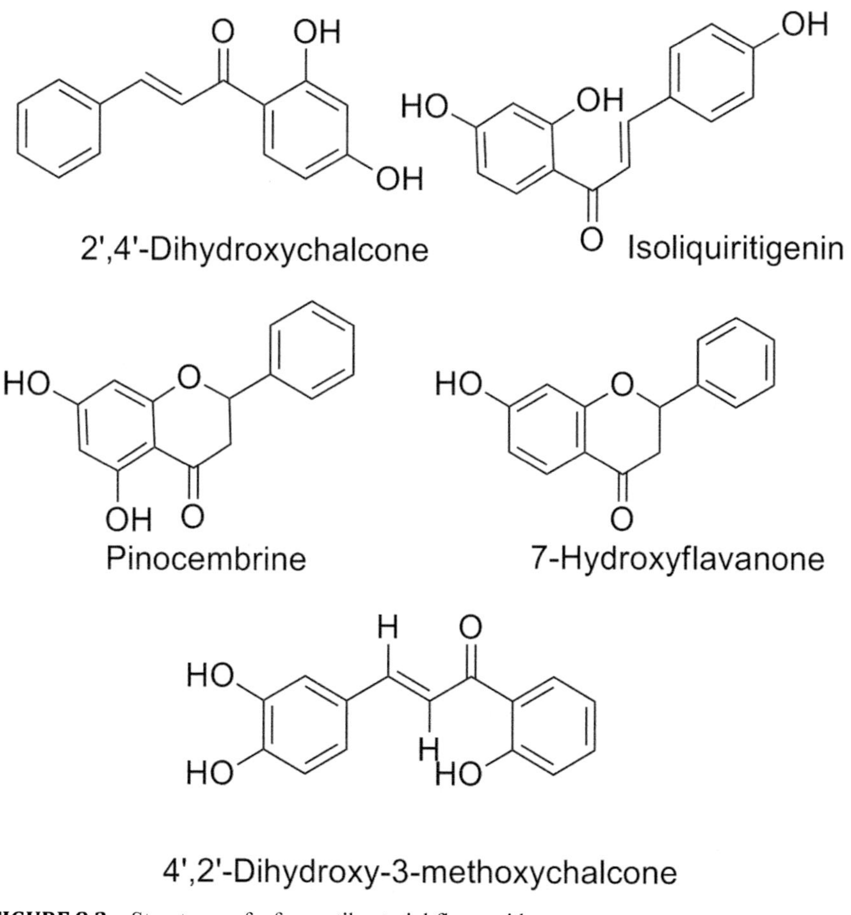

FIGURE 8.2   Structures of a few antibacterial flavonoids.

Flavones having a double bond in between carbon-2 and carbon-3 plays a pivotal role to execute biological and pharmacological activities. Flavones execute therapeutic efficacy in diseases such as cancer, neurodegenerative disorders, and metabolic diseases. Flavones are present in many food sources like fruits, herbs, and vegetation like parsley, thyme, herbs fruits, and even oranges. Many flavones have become an important part of the human diet due to multiple health-related benefits: both curative and preventive (Arredondo et al., 2015).

Flavones and related derivatives are very important to develop new and novel phytoceuticals. An important flavone is an apigenin, which has been tested and showed therapeutic promise of becoming a chemo-preventive agent (Shankar et al., 2017). Apigenin is known to have several medicinal properties that affect different molecular and cellular pathways on various diseases. This flavone also demonstrated anti-cancer, anti-inflammatory, and anti-oxidant properties. Apigenin lowered down the risk of several cancers and has synergized the chemotherapeutic outcome of some anticancer drugs (Shankar et al., 2017). Furthermore, apigenin has not only been used to prevent the risk of cancer but has also been studied for its activity to reduce neuroinflammation. Apigenin is capable of restoring cAMP which is the response element-binding protein, and this protein is essential for the up-regulation of brain-derived neurotrophic transcription factors (Bonetti et al., 2017). *In vivo* animal studies of this flavone showed an effect on inflammatory cytokines which are originated in certain immune cells and influence other cells.

Luteolin, another important flavone, was reported to demonstrate a negative biological effect by inducing a neuro abnormality commonly known as brain fog. Brain fog (or brain fatigue) is a mental health condition that reduces (sometimes severely) the ability of multi-tasking and can affect patients' short and long-term memories (Orhan, 2018). In contrast to luteolin, there are flavones that can be beneficial for such mental conditions. Certain plants contain flavone derivatives that aid with specific diseases. An example is the plant *Crataegusoxyacantha*. The species *Crataegusoxyacantha* is a well-known plant in Europe, is being used as a remedy for cardiovascular diseases. This species contains a higher abundance of polyphenolic compounds (Orhan, 2018) and is a valid source of oligomeric procyanidins and several flavonoids such as flavones and flavonols, which are primarily responsible for the medicinal activity. Other compound classes include fatty acids, sterols, phenolic acids, and triterpenes (Orhan, 2018). The plant aids various cardiovascular conditions including hypertension, heart failure, atherosclerosis, and mild alteration of cardiac rhythms.

Furthermore, it presents antimutagenic, antihyperglycemic, and immune-modulatory activities.

The root of *Scutellaria baicalensis* was studied and the root showed great potential as a neuroprotector. This plant did not execute any toxicity towards humans so far. The two major flavones that were isolated and characterized from the plant *Scutellaria baicalensis* are known as baicalein and oroxylin (Gasiorowski et al., 2011). These flavones have demonstrated cognitive and amnestic functions on the animals being studied for aging and degeneration of the brain. Baicalein is a flavone that has shown to aid the inhibition of neuronal amyloidogenic proteins and also helps out to induce the amyloid deposit. Additionally, baicalein is known to function as an anti-inflammatory, an anxiolytic, and takes away action of mild sedative agents. This plant also contains wogonin, a minor flavone constituent, having the potential to aid in brain tissue regeneration (Gasiorowski et al., 2011).

Flavones are also secondary metabolites like many other natural products and as such can undergo changes during their biosynthesis (Xu et al., 2016). In nature, synthesis of flavones is possible by a gene called *phenylalanine ammonia-lyase* (*PAL*) gene. Flavones produced by filamentous fungus *Inonotus baumii* has been studied with the enzyme *phenylalanine ammonia-lyase* to observe the function of the enzyme to produce *Inonotus baumii* (Lin et al., 2018). Overall the studies found similarities in their amino acid sequences and showed evidence of understanding enzyme phenylalanine ammonia-lyase in the flavone synthesis for future cloning (Lin et al., 2018).

8.2.2 PYRROLES

Pyrrole is a planar, aromatic, five-membered heterocyclic compound in which a nitrogen atom is present as a heteroatom. This compound contains four carbons and one nitrogen (Aromatic Heterocycles: Pyridine and Pyrrole, 2016). Pyrrole derivatives are used extensively as pharmaceu-ticals and build the structural subunit (scaffold) for many bio-privileged molecules. Plenty of natural compounds contain pyrrole moiety (Figure 8.3) as the central core, primarily in plants although this scaffold is present also in marine natural products. The pyrrole derivatives have demonstrated a wide variety of medicinal properties including antipsychotic, anticancer, antibacterial, antidiabetic, fungicidal, antiprotozoal, and others. There is always a surge in researching new pyrrole derivatives to discover new selective drugs. Efforts were also made to develop antibiotics to combat resistant strains of various bacteria. It is well-known that bacterial resistance is a growing problem because of over-use and also due to mal-prescription of current antibiotics. These incidents have eventually been creating drug-resistant bacteria. Natural pyrrole derivatives have the potential in developing effective drugs because they may act as intermediates in synthesizing chemically-modified pharmaceuticals of natural origin (Kaur et al., 2017). Numerous attempts in the isolation/synthesis/chemical modification of new bioactive pyrroles were taken. Pyrrole derivatives have a capable future in the medicinal and pharmaceutical industry and the patients. The heteroaromatic ring provides the pyrrole derivatives multifunctional abilities. The resonance of the heteroaromatic ring is also significant. Pyrrole derivatives are very important compounds in medicinal chemistry as they constitute the foundation of larger, more complicated compounds and drugs.

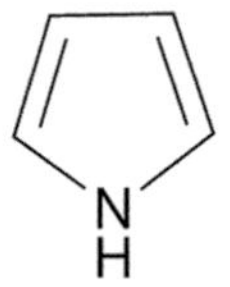

Pyrrole

FIGURE 8.3   Pyrrole: One of the most significant scaffolds found in drugs (both natural and synthetic).

Aromaticity is dependent upon the structure of a molecule and determines physical and chemical properties. As stated earlier, pyrroles possess a heteroaromatic ring having six π (pi) electrons and a heteroatom (N). Pyrrole is weakly basic because the nitrogen has a lone pair of electrons (also patriciate to fulfill Hückel's rule) that can accept protons. Thus, pyrroles are both aBrønsted and Lewis base. All the ring constituents are sp^2 hybridized having a delocalized π (pi) bonding system in. The sidewise overlap of un-hybridized orbital's results the delocalized π (pi)-electron cloud, which accounts for the aromatic character of pyrrole. For a molecule to be aromatic, it must be cyclic and planar, having delocalized π (pi)-electrons, and fulfilHuckel's rule of $(4n+2)\,\pi$ (pi)-electrons. Pyrrole obeys all these criteria as a heteroaromatic compound.

Pyrrole is the central core of tolmetin, and this scaffold is present in several drugs, including atorvastatin, ketorolac, and sunitinib. Notably, atorvastatin was the highest-selling drug in the last decade (Domgala et al., 2015). Severalpyrrole-centered molecules are anti-inflammatory, anti-microbial, anticancer, and antimalarial drugs (Wavhale et al., 2017). Some pyrrole derivatives also include enzyme inhibitors, cytotoxic agents, and antioxidants. There are also reports that describe pyrrole derivatives as antipsychotic and anti-diabetic.

While studying pyrrole derivatives, the structure-activity relationship abbreviated as SAR should be emphasized. SAR correlates the structure and medicinal activity of a molecule, which is a mandatory requisite in the pharmaceutical industry. SAR can easily identify the important functional groups and sub-units that play a major role to determine the medicinal effect of a drug. Efforts were taken to develop pyrrole-derived drugs to remove/reduce the harmful secondary responses of many current anti-diabetic drugs. Some anti-diabetic drugs work as dipeptidyl peptidase IV (DPP4) inhibitors. Attempts were taken to reduce the aftereffects through incorporation of pyrrole group to the inhibitor molecule (Mohamed et al., 2014). Replacement of a thienopyrimidine with a pyrrolopyrimidine in the dipeptidyl peptidase IV inhibitor increases activity as well as stability. The activity of these pyrrole derivatives was compared to a widely used medi-cation for diabetes, glimepiride. The pyrrole derivatives showed similar therapeutic activity.

Synthesized pyrrole based anti-inflammatory medication has effects similar to diclofenac (anti-inflammatory drug). The production of pyrrole derived anti-inflammatory compounds could eventually build a separate variety of anti-inflammatory drugs. This has many implications, mainly to eliminate the knock-on effect of NSAIDs. These pyrrole derivatives could be less harsh to the stomach and gastrointestinal tract (Fernandes et al., 2004) and might avoid hepatotoxicity (Gholap, 2016).

Pyrrole derivatives also showed antineoplastic activity. Many pyrrole derivatives reported as anti-microbial and anti-cancer drugs (Biava et al., 2003). A pyrrole derivative derived from *Staphylococcus*, pyrrole (1, 2*a*) pyrazine 1,4-dioneheahydro-3-(2-methyl propyl), is an antagonist to some bacteria that affect humans (Khazir et al., 2014). These derivatives also showed antitumor efficacy. These compounds also promote apoptosis. Induction of apoptosis is a safer and less harmful method as it produces less toxic and debilitating side effects. This certain strain (pyrrole derivative) could have some serious implications (Lalitha et al., 2016). These implica-tions could increase the life expectancy by reducing the harmful side effects, produced due to chemotherapy and radiation. These pyrrole-based new drugs could eventually replace harsh chemotherapy and invasive radiation therapy as a go-to treatment for cancer (Regina et al., 2014).

Current treatments for osteoporosis include testosterone and dihydrotestosterone; a steroidal approach to treat osteoporosis also known as hormone replacement therapy (HRT). Even though these treatments are beneficial to increasing bone density and muscle mass although some serious and potentially fatal side effects (hepatotoxicity) associated with these drugs (Sechi et al., 2004). To reduce such aftereffects of the steroidal approach, non-steroidal approach of treating osteoporosis known as non-steroidal selective androgen receptor modulator (SARM's) is necessary (Unwalla et al., 2017). Pyrrole derivatives might be fitted in this role (Micheli et al., 2003). A cyanopyrrole was studied and showed decent activity in promoting muscle growth with minimal prostate toxicity than other steroidal treatments.

A pyrrole derivative isolated from *Grifolafrondosa* showed activity both in diabetes and cancer. This derivative inhibits the alpha-glucosidase to demonstrate hypoglycemic effect (Chen et al., *2018)*. This derivative also works by inhibiting the protein, which promotes the cell growth and migration. *G. frondosa* is a mushroom and consumption of this mushroom increases preventative qualities (Ma et al., 2014). Acquiring this pyrrole derivative would be relatively cost effective, in contrasting the prices of current available treatments for both cancer and diabetes. Thus far, the pyrrole derivative found in *G. frondosa* showed capable antitumor properties and positive hypoglycemic effect. It improved the insulin resistance in various animal studies and it would be a much better and cost-effective method of treating diabetes in contrast to currently available expensive insulin pens and anti-diabetic medications. Such treatments cost patients up to plenty of dollars a month. A cost-effective method could mean less hospitalizations, amputations, and deaths of so many. The implications of developing a drug with concomitant preventative and curative properties are great, as it could potentially save both resources and lives.

Recently, some pyrrole derivatives have been characterized as having anti-tubercular effects (Kamal et al., 2013). The compound was reported back in 1998 and synthesized in 2008 and is currently in the clinical trial phase to finally become an anti-tuberculosis drug (Wavhale et al., 2017). All the currently prescribed anti-tubercular drugs are almost over 50 years outdated. The treatments are becoming less effective as time progresses. Some treatments may even take up to half a year to execute efficacy. Due to such a prolonged treatment period, it can develop serious toxicity which is reflected by several harmful effects (Denny et al., 2016). The development of this pyrrole-derived drug is incredibly important as it displayed an effective activity in antibacterial properties regarding tuberculosis. The newly constructed triazole linked pyrrole compounds did show some antimicrobial activity in tuberculosis, however, not as much as the already available commercial drugs (Devi et al., 2017). Notable development of other pyrrole derived anti-tubercular drugs are reported, specifically for resistant strains (Kasim et al., 2012). However, these efforts displayed a high amount of harmful side effects and poor absorption (consequently poor bioavailability) in the body. Still it is useful study because it may provide a foundation for the improvement and/or development of more potent anti-tuberculosis medications that are effective against the antibiotic-resistant strains of the disease (Raimondi et al., 2006). Efforts have been continuing to improve pharmacokinetics of the previously mentioned derivatives. These efforts have also included attempts to reduce the drug-generated toxicity (Wang et al., 2004).

Natural pyrrole derivatives may act as potassium-competitive acid blocker (P-CABs) as well by treating acid reflux and related issues effectively. A few could successfully compete with current heterocyclic medications like Nexium (esomeprazole), omeprazole, and pantoprazole. The derivatives demonstrated higher efficacy and lower toxicity. The P-CAB pyrrole derived hits could compete with other proton pump inhibitors (Mai et al., 2007). Also, the popular proton pump inhibitors are very effective during the day, however, at night the patient may experience recurrence of acid reflux. This is very common these drugs do not inhibit at night. The natural pyrrole derived P-CABs have studied and excellent therapeutic activities and low toxicity were noticed (Nishida et al., 2017). The newly identified lead could provide relief from acid reflux through the night. It may even cure the gastric reflux (Nishida et al., 2017). Up to now it showed success in animal models and the next step is to test its efficacy in human (clinical trials). Moreover, the lead showed prolong half-life *in vivo*. This might be alarming to researchers because it could possess dangerous implications in human trials. However, the problem was overcome by a semi-synthetic fluoro pyrrole derivative which showed similar therapeutic activity with better pharmacokinetics. The modified lead could generate a better therapeutic outcome in acid reflux and other gastric issues. The important findings showed the pyrrole derivative remains active for a longer period than other popular proton pump inhibitors. Even the lead may work better to reduce/terminate the effects of gastric acid reflux. Pyrrole derivatives are incredibly diverse and the scaffold induces therapeutic properties (Nishida et al., 2017).

8.3 HETEROCYCLIC ANTICANCER DRUGS FROM PLANTS

The word 'cancer' is still 'dreadful' to mankind. This starts with abnormal tissue growth due to uncontrolled cell division, that have irregular life spans, and eventually forms a malignant tumor(s). Normal cells conduct regular cellular functions and die, however, cancer stem cells keep growing and making new cells uncontrollably. Since the malignant cells continue to grow in abundance, they eventually suppress the normal cells to worsen the patients' health. The abnormal cellular growth can metastasize i.e. generating new malignant tumors to various parts of the body, fetching by blood and lymph (immune) systems.

Unfortunately, based on the nature of human cell/tissue there are more than 400 different types of neoplasm that fall under the categories of lung, liver, colon, pancreas, prostate, breast, cervical, esophageal, ovarian, skin cancer and others. If a specific kind of cancer for example breast cancer spreads to the lung, still considered as breast cancer because the invading cells are the same malignant cells that suppressed the normal breast cells (American Cancer Society, 2015). To be termed as lung cancer the abnormality has to originate in the lung, not in another organ.

The most common type of lung cancer is non-small cell lung cancer, abbreviated as NSCLC. Subtypes of NSCLC include SCC, adenocarcinoma, and large cell carcinoma (Non-Small Cell Lung Cancer, 2016). About 85% of lung cancer patients contain NSCLC (Lung cancer, 2018). On the other hand, about 10 to 15% of diagnosed lung cancers are SCLC. This leaves fewer than 5% to be lung carcinoid tumors, also called lung neuroendocrine tumors. Stomach cancer (gastric cancer) has the subtypes like adenocarcinoma, lymphoma, gastrointestinal stromal tumor (GIST), and carcinoid tumor (American Cancer Society, 2017). In 2018, the ACS estimated about 26,240 stomach cancer cases would be diagnosed only in one country, the United States (Key Statistics for Stomach Cancer, 2018). In Figure 8.4. different heterocyclic cancer drugs (natural & semi-synthetic)are presented.

Cancer has varying stages and each stage indicates an advancement of the disease to the next level. In stage I, the cancer growth is localized in its origin; stage 2 implies a larger growth and pre-invasion stage; stage 3 denotes that invasion to the neighboring tissues in ongoing and migration of the malignant cells have begun whereas stage 4 of the sickness indicates the terminal stage due to extensive invasion, angiogenesis, and metastasis. The creation of blood vessels in tumors is called angiogenesis whereas metastasis implies tumor formation in other organs.

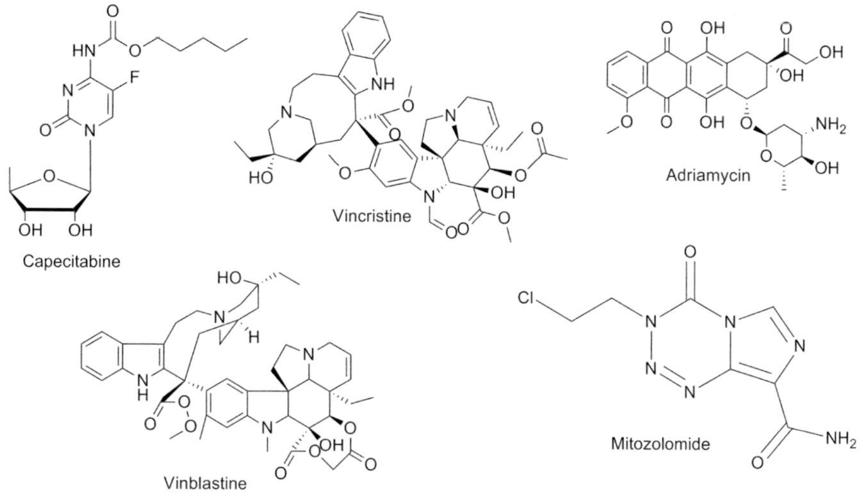

FIGURE 8.4 Frontline natural heterocyclic anticancer drugs.

8.3.1 CANNABINOIDS FROM CANNABIS SATIVA

Cannabis sativa, an annual herbaceous flowering plant native to eastern Asia, is greatly stigmatized for being one of the extreme hallucinogenic natural sources (Satterlund, 2015). The most hallucinating constituent is Δ-9-tetrahydrocannabinol (THC) (Figure 8.5). The drug is said to be "not currently accepted for medicinal use and has a high risk of potentially being abused" (Title 21 Code of Federal Regulations, 2018). In consideration of the foregoing, many states have approved medicinal marijuana. Unfortunately, prejudice runs deep and the patients, which truly need it for their cancer treatment are regarded unfairly by the people who disagree with its medicinal application. The stigma affects the patients which could be meaningful in their sickness. Many patients don't count marijuana as a legitimate treatment option, the patients are too ashamed to be judged (Satterlund, 2015). The drug, stigmatized or not, helps with various illnesses including cancer.

Lung cancer invades the primary organs and metastasizes to create tumors in other organs (How the Heart Works, 2018). Although cancer is malignant, it does have short and long-termed remedies. Adding on to this, cannabinoids, and their semi-synthetic compounds which can be found in cannabis. These have also medicinal uses in skin, lung, uterine, and prostate carcinomas (Fridlender et al., 2015). Cannabinoids might be useful for patients who have received radiation or chemotherapies and have been agonizing from the side effects like nausea, and loss of appetite. To reduce these symptoms and to provide temporary comfort, cannabinoids are used as support (Fridlender et al., 2015). When patients took chemotherapy their body undergoes great deals of stress and this stress on the body it can make the patient feel nauseous and other side effects (Marijuana and Cancer, 2017), the cannabinoids can help in treating those symptoms. It was reported that cannabinoids inhibit cellular division in neoplasm, initiate autophagy/apoptosis, and not only that but they also inhibit angiogenesis and metastasis (Śledziński, 2018). In another experiment cannabinoids demonstrated malignant growth inhibition in animal models by modulating key cell-signaling pathways that control cancer cell division and survival (Velasco et al., 2016). The mechanism of the antitumor effects can take place in different ways. The most widely known effect is apoptosis (Velasco et al., 2016).

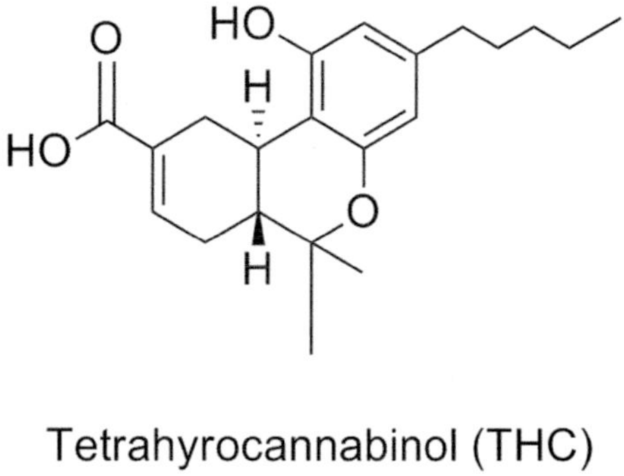

Tetrahyrocannabinol (THC)

FIGURE 8.5 Δ-9-tetrahydrocannabinol, also known as THC (molecular formula for THC is $C_{21}H_{30}O_2$).

Cannabinoids are also known for being able to increase blood flow and enhance cortical activity (Terpenes and the "Entourage Effect," 2018). Medical marijuana also helps in neuropathy (nerve damage), and anorexia (Medical Marijuana and Cancer, 2018). Cannabinoids work as appetite stimulants in debilitating health conditions such as cancer and AIDS (Amar, 2006). Medical marijuana can help cancer patients with their pain also. Notably, in contrast to current belief the clinical trials of the medicinal use of cannabinoids have actually been conducted with purified cannabinoids or also a single fraction of *Cannabis sativa* and not smoked as the stigma leads people to believe (Guzman, 2018). *Cannabis sativa* can have similar health outcomes as opioids as pain relievers, giving the patient relief but seldom demonstrates anti-inflammatory effect (Medical Marijuana and Cancer, 2018).

8.3.2 TAXANES FROM TAXUS BREVIFOLIA

Docetaxel better recognizes as Taxotere, is a heterocyclic plant-derived anticancer drug that is frequently used to treat cancers like stomach, lung (NSCLC) breast, brain, prostate, and others (General Cancer Information, 2016). Docetaxel is a semi-synthetic product of taxol (paclitaxel), separated from the bark of Pacific yew (*Taxus brevifolia*) giving insight as to its brand name taxotere. Taxol (paclitaxel) is commonly extracted from *Taxus brevifolia* and is required to treat ovarian, breast, and NSCLC. Paclitaxel can also treat AIDS-related Kaposi sarcoma (Paclitaxel, 2018). Both docetaxel and paclitaxel belong to 'taxanes' which is a diterpene class having a taxadiene core (Zhao et al., 2012). Taxanes are very efficient anticancer drugs. In contrast, diterpenes belong to a class called 'terpenoids.' Docetaxel is a frontline anticancer chemotherapeutic agent in this era (Choi et al., 2015).

Docetaxel inhibits cellular division in neoplastic tumors and ceases the spreading or migration of the malicious cells (General Cancer Information, 2016). The chemotherapy is administered for each time during an hour for three weeks intravenously (IV) as the pill doesn't work well because of poor bioavailability. The dose is patient-dependent that includes several factors (Taxotere: Drug Information, 2016) optimal dosage should be administered. Apart from the patient's height, weight, current health status and age, ethnicity also matters to optimize the effective dose. A study concluded that the Japanese population seems more susceptible to the toxicity of docetaxel which requires them to take a smaller dosage of the medication than the US/European patients. While the Japanese patients required 60 mg/m^2, the US/European patients had to administer a dosage of 75–100 mg/m^2 (Kenmotsu et al., 2015).

Taxotere (docetaxel) can be administered after surgery or during combined chemotherapy. It was regulated in a combined therapy to a patient having drug-resistant early-stage breast cancer after surgery (Taxotere: Drug Information, 2016). Experiments revealed that docetaxel along with DIM-CpPhC(6)H(5), a selective PPAR gamma modulator, showed potential activity in NSCLC(Ichite et al., 2009). Although paclitaxel and docetaxel share major parts of their structural and pharmacological aspects but differ in multiple functions, e.g., tubulin polymer generation, and docetaxel appear twice as active in depolymerization inhibition (Figure 8.6).

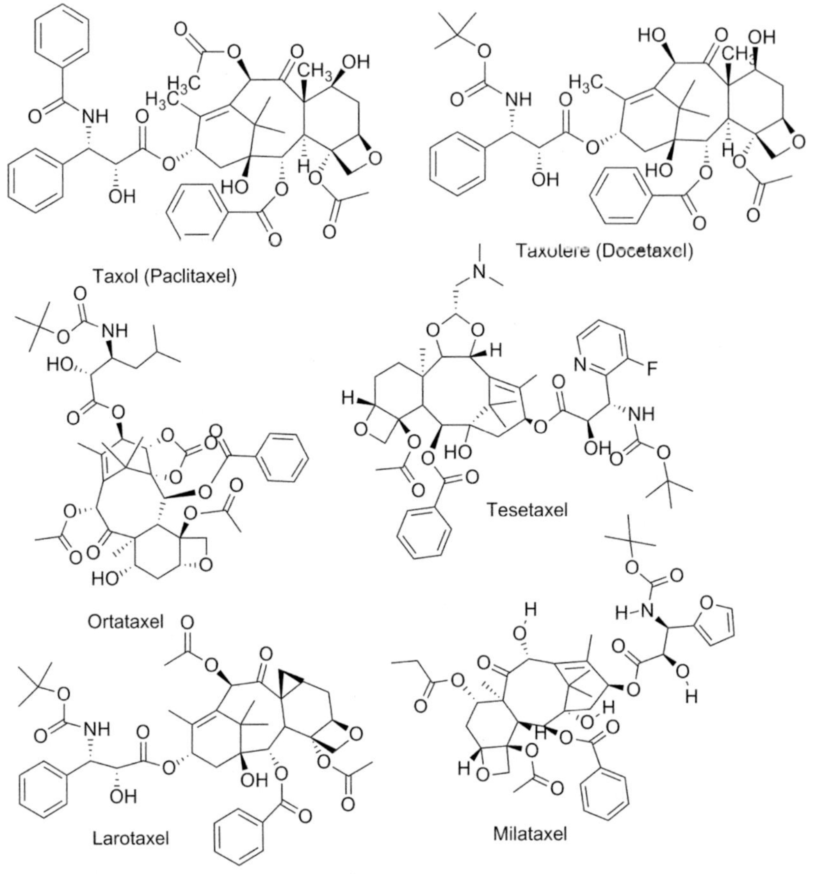

FIGURE 8.6 Taxane class of anticancer drugs: Natural or semi-synthetic drugs from the plant *Taxus*.

8.4 HETEROCYCLIC ANTIDIABETIC DRUGS FROM PLANTS

Diabetes mellitus is a devastating health situation where the body fails to *process* the glucose derived from the food intake. The consumed food is eventually converted into glucose to supply energy for biological processes. To generate energy from glucose, the pancreas produces a hormone called insulin which facilitates the bond breaking in glucose to produce energy (About Diabetes, 2017). For majority diabetic patients, body's capacity to generate or react to insulin is impaired, called insulin-dependent diabetic mellitus (IDDM). This disease is difficult to cure completely but highly preventable and treatable (Diabetes Basics, 2018). In 2015, about 30.3

million Americans (9.4% of the population) were identified as diabetic (Statistics About Diabetes, 2018) and 79,535 death certificates were issued listing diabetes as the 'cause of death.' This put diabetes at the seventh position in the list of 'leading causes of deaths in the United States.'

Broadly, diabetes can be classified into three categories: Type 1 (formerly known as IDDM), type 2 (formerly known as NIDDM), and gestational diabetes *viz* diabetes in the pregnant stage. Types 1 & 2 have different causes, however, for both types patients inherit a predisposition to diabetes which then triggers to ultimately become sick (Genetics of Diabetes, 2017). The fraction of people estimated to have type 1 is about 5% of the total diabetic population whereas 95% of the total diabetic population belongs to type 2. Patients diagnosed as type 1, must need to take insulin every day to survive (About Diabetes, 2017). When genetic considerations come into play, it is a common observation that type 1 diabetic patients inherit the risk components from, not one, but both of their parents (Genetics of Diabetes, 2017). In type 2, the body mislays its command on the production and/or usage of insulin properly and consequently the blood sugar becomes higher. Type 2 is more tightly linked with ancestry and lineage than type 1. With the foregoing in mind, lifestyle has a mighty influence in developing type 2, unlike type 1 diabetes. To prevent type 2 someone can get accustomed to healthy lifestyle changes which include maintaining body weight (if overweight), eating wholesome foods, and sustaining a steady schedule of exercise. Figure 8.7 shows a few frontline heterocyclic drugs, commonly used for treatment. Notably, gestational diabetes develops during pregnancy who has never previously diagnosed as diabetic. This diabetes might put the mother and the offspring at risk and the likelihood of developing type 2 diabetes is high for both (About Diabetes, 2017).

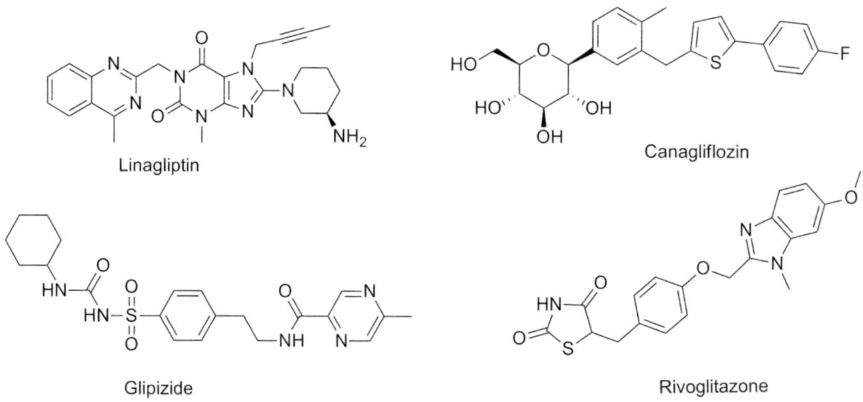

FIGURE 8.7   Frontline heterocyclic drugs for diabetes.

8.4.1 ROSEOSIDE CONSTITUENT OF BAUHINIA VARIEGATA

Bauhinia variegate, from the family *Febaceae,* has many names such as butterfly ash, mountain ebony, orchid tree, camels' foot, kachnar, and bauhinia (*Bauhinia variegate.* Weeds of Australia, 2016). It is a deciduous tree (giant flowering plant) that produces fragrant and colorful flower (SelecTree: Tree Detail, 2018). The tree has many regional names due to its abundance in many places around the world such as India, Bhutan, China, Nepal, Laos, Myanmar, Vietnam, and the north western part of Thailand (*Bauhinia variegate.* Weeds of Australia, 2016). People have been using many *Bauhinia* species traditionally as remedies in treating ailments like gastrointestinal tract (GIT) disorders, diabetes, infectious diseases, and inflammation (Ahmed et al., 2012). Interestingly, *B. variegate* is a (potential) environmental weed as narrated in invasive plant list (*Bauhinia variegate.* Weeds of Australia, 2016; FLEPPC List of Invasive Plant Species, 2017). The species being widely cultivated in many urban zones, e.g., in eastern Australia as a street tree. There are ten *Bauhinia* species having antidiabetic potential (A Review on Antidiabetic Potential of Genus *Bauhinia*, 2015).

Apart from the treatment of diabetes, the extract of *Bauhinia* species is also used in asthma, jaundice, tuberculosis, leprosy, and skin maladies (Garud et al., 2016). Adding on to this, aside from antidiabetic activity the leaf extract also demonstrated antieosinophilic activity (Mali et al., 2011). The leaves and stem bark extracts (made in boiling water) are normally administered in antidiabetic treatment wherein the active principle roseoside ($C_{19}H_{30}O_8$) is present (Mali et al., 2011). Besides roseoside (Figure 8.8), the extract contains alkaloids, carbohydrates, flavonoids, phenolic compounds, and very little volatile oils.

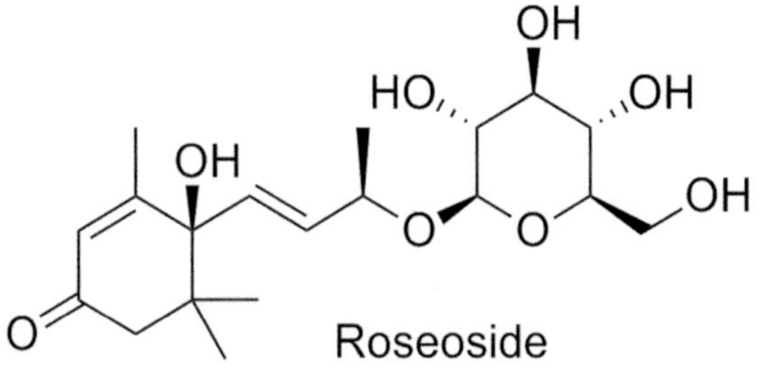

FIGURE 8.8 Structure of Roseoside: The major antidiabetic constituent of the plant *Bauhinia variegate* (also known as corchoionoside, $C_{19}H_{30}O_8$).

The lethal dose of the extract was reported to be at >5,000 mg/kg (Mali et al., 2011) which indicates that the use of this extract in a lower dose is relatively safer. A study was also conducted to evaluate the efficacy of roseoside on diabetes. The extract was effective against both type I and type II diabetes (Garud et al., 2016). *B. variegate* extract also decreased cholesterol, triglyceride, creatinine, and blood urea nitrogen level in type I and II diabetes (Garud et al., 2016). Furthermore, another experiment concluded that the crude extract and the major metabolite were showing an indication of increasing insulin secretion in a dose-dependent manner (Frankish et al., 2010).

8.4.2 EPIGALLOCATECHIN GALLATE OF CAMELLIA SINENSIS

Camellia sinensis (green tea), is a broadleaf evergreen plant from the family *Theaceae*, can be 15 feet high and 10 feet wide (*Camellia sinensis*, 2018). It is harvested commercially and ornamentally owing to its excellent foliage and small but attractive shaped fall flowers. Dark green and glossy leaves, white fragrant flowers, and small but smooth fruits are attractive. It contains many flavonoids and polyphenols along with few other composites such as saponins, caffeine, tannins, and even some vitamins (Rafieian-Kopaei et al., 2014).

Epigallocatechin gallate also known as EGCG or epigallocatechin-3-gallate, is an ester which is a derivative from epigallocatechin. EGCG is an ester of Gallic acid which is an organic acid also known as 3, 4, 5-trihydroxybenzoic acid ($C_7H_6O_5$). Through several reported studies green tea has drawn huge attention because of its medicinally-privileged polyphenols (Haidari et al., 2013). Green tea also contains catechins, a subcategory of polyphenols with antioxidant activity (Haidari et al., 2013). EGCG, EGC, ECG, and EC are the polyphenols found in green tea that possess a wider variety of health-promoting effects (Haidari et al., 2013). A catechin is a type of natural phenol and antioxidant. The chemical formula of EGCG is $C_{22}H_{18}O_{11}$. EGCG demonstrates its anticancer and anti-diabetic activities through antioxidant mechanisms (Lead Phytochemicals for Anticancer Drug Development, 2016). Green tea is a sought-after health drink, making it an exceedingly marketable, is taken daily by a multitude of people worldwide (Du et al., 2012). EGCG is more commonly known by its anticancer properties, however, it also helps to fight diabetes, some cardiovascular problems, and in some metabolic disorders (Du et al., 2012; Wolfram, 2007). Green tea could help to protect kidney in diabetes and to modulate the blood glucose level (Rafieian-Kopaei et al., 2014). A study conducted on rodents showed that diabetes was alleviated by using EGCG as a supplement. Rodents with type II diabetes were assessed after their ingestion of EGCG. The study showed that EGCG was able to modify glucose and the lipid metabolism in H4IIE cells in a positive manner. This beneficial modification markedly enhanced glucose tolerance in diabetic rodents (Wolfram et al., 2006).

In another animal study, the results supported falling down of blood glucose in the diabetes population (Haidari et al., 2013), owing to green tea consumption. Moreover, dietary supplementation alongside the catechins in green tea can improve TAC and decrease malondialdehyde (MDA) concentration (lipid peroxidation biomarker) in rat's liver, blood, and brain (Haidari et al., 2013). MDA being a result of lipid peroxidation which is yielded by the oxidative degradation of lipids. EGCG presents a nontoxic, effective, and inexpensive mode of supplementation for those who suffer from diabetes (Rafieian-Kopaei et al., 2014) (Figure 8.9).

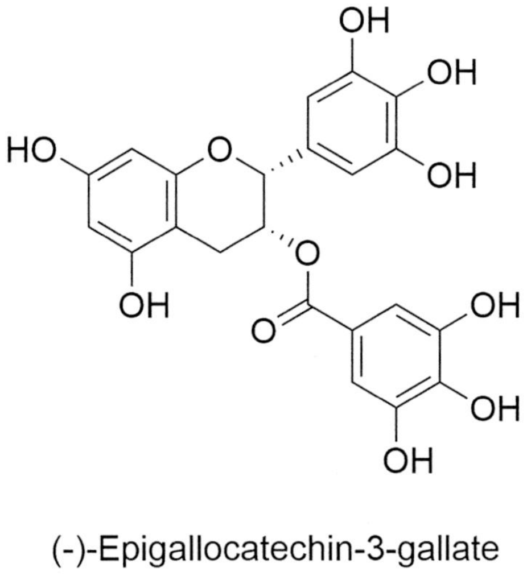

(-)-Epigallocatechin-3-gallate

FIGURE 8.9   Structure of (-)-epigallocatechin-3-gallate isolated from the plant *Camellia sinensis.*

8.5   HETEROCYCLIC ANTI-HIV DRUGS FROM PLANTS

Human immunodeficiency virus or HIV is a type of lentivirus. This virus debilitates a person's immune system. The immune system is crippled through destruction of essential cells which fight the diseases and infections that constantly attempt to strike human body (HIV basics, 2018). Lentivirus, a genus of retroviruses that generate long-standing and fatal diseases characterized by long incubation periods, in the mammalians including human. The basis of HIV was discovered when scientists identified a specific chimpanzee species in Central Africa (About HIV/AIDS, 2018). The chimpanzee version of the immunodeficiency virus is called SIV. Scientists concluded that SIV was transmitted onto humans when they hunted the simians and contamination of blood took place. Thus SIV was transformed to HIV. This virus was originated in Africa, became more prominent until it escalated worldwide. Body fluids are the carriers of HIV transmission. Subsequent attack is targeted to the immune system, or more specifically the body's CD4 cells (T cells). The CD4 cells are primarily responsible, alongside the immune system, to encounter the infections and other disease-containing entities (HIV.gov, 2017). When left untreated, HIV can destroy the CD4 cells, once again damaging body's propensity of inhibiting infections. Once the body is attacked by HIV, it loses the inherent defense mechanism and over time body becomes unable to defend against infection and subsequent illness. HIV makes the body weak and open to be attacked by other microbes/parasites that weaken the health even more.

In human, the lentivirus can cause chronic diseases which eventually causes AIDS during a time span of few years. This kind of virus integrates huge amount of viral RNA into the DNA of the host cell it has chosen and can infect non-dividing cells to ensure gene delivery (Cockrell et al., 2007). The virus attaches itself to its host through the glycoprotein and then it is randomly integrated into the cell's genome by the viral integrase or provirus. Consequently, transcription, and translation occur, and lastly the proteolytic processing of the precursor's polyproteins by viral protease and maturation of the virions take place (Lentivirus, 2018).

HIV has two principal types. There is HIV-1 which most commonly infects people, and HIV-2 which is relatively uncommon and is less infectious (HIV Strains and Types, 2018). Type 1 and 2 have several similarities in basic gene arrangement, modes of transmission, intracellular replication pathways, and clinical consequences which implies that both types are ended up with AIDS in humans (Nyamweya et al., 2013). With that being said, HIV-2 has a lower transmissibility rate and has a reduced likelihood of progression towards AIDS. Although HIV has two categories *viz* HIV-1 and 2 but the in-depth mechanism remains incompletely understood.

Based on intensity, HIV has three stages; the first stage being an acute HIV infection. In this stage, two to four weeks after infection, flu symptoms are seen. These flu-like symptoms last for a few weeks and the infected person become highly contagious owing to a high level of virus in blood although the patient might be unaware of the infection. Stage two is known as clinical latency, HIV inactivity, or dormancy. During this stage, known as asymptomatic HIV infection or chronic HIV infection, HIV is still active, but it however, reproduces itself in low levels. Like the first stage, the patient might or might not become sick during this period. Without proper medication the patient can survive up to a decade or longer although based on other parameters lifespan can be shorter. In this stage the body begins to suffer, and the viral load begins to increase whereas the CD4 cell count begins to decrease. The person will then move into stage three which is AIDS. The patient gets many severe illnesses which are called *opportunistic* illnesses because these illnesses take the opportunity of the body's poor immune system and at this stage patients can survive up to three years. At this stage (AIDS), CD4 counts must have dropped below 200 cells/mm or if the *opportunistic* illnesses are prevalent (About HIV/AIDS, 2018). At any time during the stages, the immune response to HIV-2 has been seen to be more protective against disease progression showing that pivotal immune factors limit viral pathology (Nyamweya et al., 2013). If those similar immune responses could be replicated in patients having HIV-1, it is possible that the patient's survival rate would increase and requirements for antiretroviral therapy might be reduced.

Furthermore, no effective cure has been discovered yet to provide complete recovery from this sickness (HIV Basics, 2018) but it can be controlled by proper medical care. The available medication is antiretroviral therapy, commonly called ART. Treatment with prescribed ART can reduce viral load to a smaller almost undetectable amount (About HIV/AIDS, 2018). When the viral load becomes undetectable, patient can survive longer with relatively healthy life (Figure 8.10).

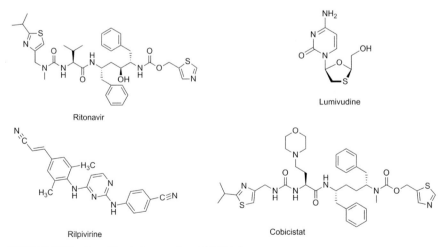

Ritonavir

Lumivudine

Rilpivirine

Cobicistat

FIGURE 8.10   Frontline heterocyclic anti-HIV drugs.

8.5.1 CALANOLIDE A (COUMARINOID) FROM CALOPHYL LUMLANIGERUM

Calanolides are coumarin derivatives (also known as coumarinoids) and are considered as bio-privileged compounds. An important calanolide is cala-nolide A which is (Figure 8.11) NNRT inhibitor (Hanna, 1999). Coumarin scaffold is wide-spread in nature and calanolides, belong to coumarinoid category, found in the *Calophyllum* genus (Kurapati et al., 2015). The compound (calanolide A) was studied with purified bacterial recombinant reverse transcriptase (RT) which identified calanolide A as a potent HIV-1 specific RT inhibitor (Currens et al., 1996; Kashman et al., 1992). This RT inhibitor is predominantly active against multiple concentrations of HIV-1 strains which encompasses nucleoside and non-nucleoside resistant variants. With that being said this particular compound (calanolide A) is essentially inactive against the rarer form of HIV which is HIV-2 (Hanna, 1999).

The natural calanolide A is isolated from the latex of *Calophyl lumla-nigerum* that is domicile to the tropical rainforest of Sarawak, Malaysia (Hanna, 1999). *C. lanigerum* is a flowering plant. Calanolide A has quite low availability in its natural form. However, the total synthesis of this couma-rinoid was developed to provide enough supply for research (Creagh et al., 2001). A similar compound of Calanolide A is (-)-Calanolide B (also known as Costatolide ($C_{22}H_{26}O_5$) was also isolated from the same plant species. The biological activity of Calanolide A is higher than (-)-Calanolide B (The Calophyllum Story, 2018).

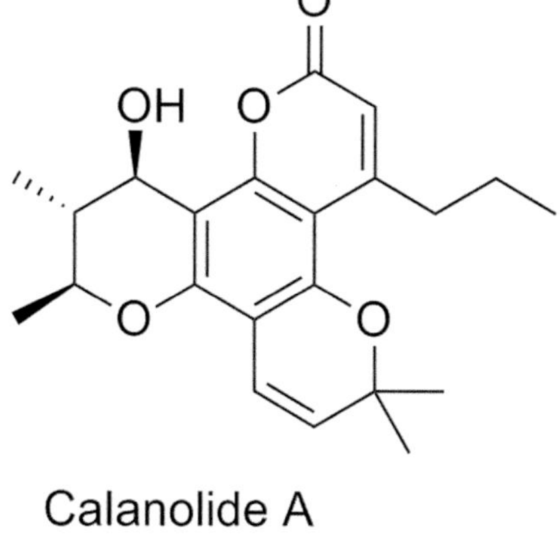

Calanolide A

FIGURE 8.11 Structure for (+)-Calanolide A; extracted from the plant *Calophyl lumlanigerum.*

Calanolide A is a distinctive NNRTI inhibitor as it can bind two different sites in RT those being an active site and a binding site of the enzyme (Currens et al., 1996). A further distinction of calanolide A with the binding of residual part of NNRTI's was noticed. It downregulated HIV-1 RT in a synergistic manner (Currens et al., 1996). An experiment was conducted to analyze the inhibition of HIV-1 RT caused by Calanolide A. In this experiment, the two templates/ primer systems that were examined were ribosomal RNA and homopolymeric rA-dT 12-18 (Currens et al., 1996). The data of the experiment demonstrated that Calanolide A bound itself near the enzyme binding site and affected dNTP binding. Calanolide A interacted with the binding domains of phosphonoformic acid (brand name: Foscavir, used to treat cytomegalovirus (CMV) eye infection in AIDS patients) and with 1-ethoxymethyl-5-ethyl-6-phenylthio-2-thiouracil expanding the evidence that it can bind two different sites. Calanolide A is take-norally (Hanna, 1999) with an oral dosage of 200 to 800 mg during the treatment (Creagh et al., 2001). Some viral life-cycle studies indicate that calanolide A acted early stage of infection (Currens et al., 1996). In addition, enzyme inhibition assays were conducted, and the results showed calanolide A effectively and selectively inhibiting recombinant type 1 HIV, but it did not inhibit cellular DNA polymerases or HIV-2 with the tested concentration range.

8.5.2 BAICALIN FROM SCUTELLARIA BAICALENSIS

Flavonoids, frequently found in fruits, vegetables, nuts, medicinal plants, and wine have varying phenolic structures. These are commonly known owing to their multifunctional pharmacological and nutraceutical benefits. Flavonoids have shown anti-inflammatory, anti-HIV, anti-oxidative, anti-carcinogenic, anti-mutagenic, and some other curative/preventive activities (Panche et al., 2016). Studies concerning flavonoids were done to support the efficacy of flavonoids in inhibiting both HIV-1& -2 (Mahmood et al., 1993) up to a certain level. The investigational flavonoids interacted irreversibly with envelope glycoprotein gp120 to deactivate viral infectivity and subsequently reduced infection (Mahmood et al., 1993). GP120 is a biomolecule found in HIV-1, a glycoprotein that participates in constructing the viral outer layer (Yoon et al., 2010). This molecule facilitates the entry of HIV into the host cell and helps to generate the infectious path of HIV. However, gp120 might also interact with CD4 cells. The glycoprotein showed an ease in endorsing the viral continuance by influencing the CD4 cells.

The flavonoid 'baicalin' is isolated from *Scutellaria baicalensis* (roots). *Scutellaria* (skullcap) are flowering plants belong to *mint* family. The flowering plant is domicile to various east-Asian countries and Russia, and it is harvested in several European countries (Zhao et al., 2016) as well. The plant has had been using by the people as a folklore medicine over 2,000 years (Zhao et al., 2016), both in Indian (Ayurveda) and Chinese medicinal systems and in modern herbal medications (Orzechowska et al., 2014) in treating asthma, diarrhea, hypertension, insomnia, respiratory infections, pain, and inflammation (Zhao et al., 2016; Herbs Used in Asthma Treatment – An Overview, 2018). In addition, *Scutellaria* has immunomodulatory effects by stimulating antibody formation or by inhibiting white blood cell activity either by promoting or suppressing their functions (Immunomodulation, 2018). The *S. baicalensis* root extract is enriched in baicalin could stimulate the non-specific antiviral immunity and assess the resistance of peripheral blood leukocytes and bone marrow cells to vesicular stomatitis virus infection.

Apart from baicalin, *Scutellaria baicalensis* is a source of other bio-privileged molecules, e.g., wogonin, wogonoside, and baicalein which inhibit the release of histamine from mast cells *in vitro* (Orzechowska et al., 2014; Herbs Used in Asthma Treatment – An Overview, 2018). These flavones have shown to have anti-carcinogenic, antibacterial, antiviral, antioxidant, neuroprotective, and anti-HIV effects (Zhao et al., 2016). Baicalin inhibits the copying of HIV-1 in a concentration-dependent manner (Kitamura et al., 1998). Baicalin also possesses effective anticancer activity in different hematological malignancies (Chen et al., 2014). Studies have recently found that baicalin is capable to form complexes with selected chemokines, attenuate their capacity to bind, and to activate receptors on the cell surface (Li et al., 2000). This study proposed that baicalin hinders the interaction of HIV-1 with chemokine coreceptors and that also blocks the entry of HIV-1 target cells. Accordingly, baicalin can be considered as a building block to developing other novel anti-HIV-1 agents (Li et al., 2000) (Figure 8.12).

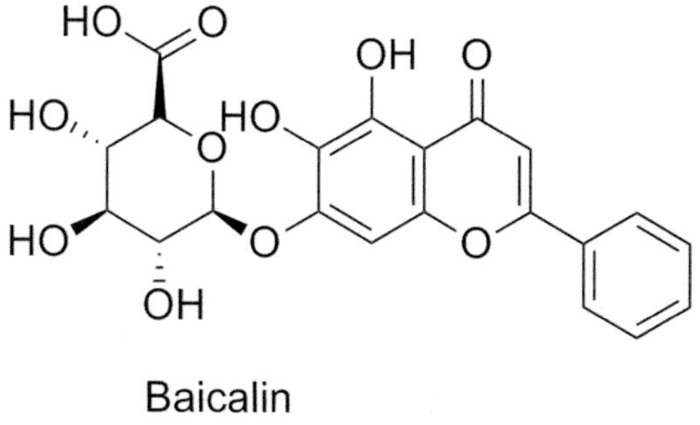

Baicalin

FIGURE 8.12   Baicalin, a flavonoid isolated from the root of *Scutellaria baicalensis*.

8.6   HETEROCYCLIC ANTIMICROBIAL DRUGS FROM PLANTS

Microbes or microorganisms (microscopic organisms) can be bacteria, fungi, algae, protozoa, or viruses which can be in a single-celled form or in a colony of cells (Informed Health Online, 2016). These microorganisms are incredibly small, consequently invisible in bare eyes, and a high-resolution microscope is required to properly examine them. These organisms can live in different habitats. Each microorganism is specifically adapted in a certain habitat where it chooses to be lived in. Many of them adapted at ambient temperature but some being adapted at high temperatures whereas some others at low temperatures. Furthermore, microbes are a large part of every day's life, found in water, soil, and air by encompassing all habitats (Panche et al., 2016). Human is practically accommodated trillions of microbes, with the foregoing being said including bacteria, protozoa, viruses, and some fungi. In each human cell the body hosts 10 microorganisms on average and they can cause the host sick or contribute positively by detoxifying harmful chemicals and helping to develop the immune system (Stark, 2010). However, microorganisms like protozoa, and other tiny living creatures cause ailments such as toxoplasmosis and malaria.

Microorganisms can cause diseases although all microbes are not dangerous. Some bacteria prefer a warm habitat while others prefer colder settings. Some bacteria that live on/in human body can even support us to stay healthy. In human stomach there are innumerable bacteria including lactic acid bacteria which help in digesting food and medicines, are popularly known as *good bacteria* that come with food such as yogurt. Only a small percentage of bacteria cause various diseases, e.g., diarrhea, colds, or tonsillitis. Viruses have no cells, just contain nucleic acid, either DNA or RNA (but not both), and a protein coat, which encases the nucleic acid. Some viruses are also enclosed by an envelope formed by fat and protein molecules. Once they find suitable host cells, the infection starts. In its infective form, outside the cell, a virus particle is called virion. Some viruses attack the healthy good cells and cause ailments from influenza to more serious diseases like AIDS. For reproduction, the viruses must attack healthy cells, they cannot reproduce by their own. Sometimes viruses do not respond to medication, instead vaccine helps the body to make it less susceptible to the virus by recognizing them through vaccine. Fungi are generally associated with yeast, mold, or the edible fungi like mushrooms. However, fungi can also cause diseases, e.g., mycoses, ringworm, histoplasmosis, talaromycosis, and others. Although fungi cause many diseases but without fungi the mankind would never be blessed by penicillin, discovered in 1928 (Discovery and Development of Penicillin, 2018). Antimicrobial medications are drugs that remove/reduce microbial infections. The medicines can be antibiotics, antifungals, antivirals, and others (Antimicrobial Drugs, 2018). Most antibiotics work by inhibiting the targeted bacterial cells (Figure 8.13).

Ciprofloxacin and fluconazole are antibiotics that are frequently used to treat bacterial and fungal infections respectively. Ciprofloxacin (Figure 8.13) is an antibiotic belong to the fluoroquinolone class of antibiotics (Ciprofloxacin, 2018). The primary mechanism is the inhibition of bacterial DNA gyrase (Campoli et al., 1988). The drug has demonstrated higher potency than other fluoroquinolones (Lebel, 1988). It is also prescribed in treating pneumonia, gonorrhea, typhoid, infectious diarrhea, skin infection, bone, joint, abdomen, prostate, bronchitis, urinary tract, and skin infections. Ciprofloxacin is taken orally but it some cases intravenous administration is preferred (Campoli et al., 1988; Ciprofloxacin, 2018).

Fluconazole (Figure 8.13) is an antibiotic, frequently used in treating fungal and yeast infections (Fluconazole, 2018). This drug belongs to the triazole class of antifungal agents which works by inhibiting/reducing the fungal growth that causes infection in the human body. Substantial clinical studies demonstrated fluconazole's effectiveness, its favorable pharmacokinetics, and safety profile; all these have a large contribution to its widespread uses (Cha et al., 2004). This drug works well in fungal meningitis and as a preventive medicine for infections due to chemotherapy, radiation therapy, or bone marrow transplant (Fluconazole, 2018). It also demonstrated activity up to a certain extent for the treatment of fungal infections in HIV positive patients. Fluconazole is taken orally but it some cases intravenous administration is preferred (Zervos et al., 1993).

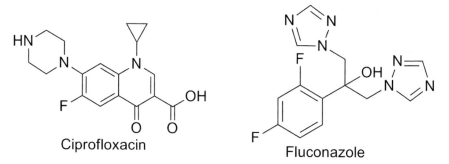

Ciprofloxacin

Fluconazole

FIGURE 8.13 Two important heterocyclic antibiotics: Ciprofloxacin and fluconazole.

8.6.1 ALPINUMISOFLAVONE

A pyran moiety is usually constructed of a six-membered heterocyclic, non-aromatic ring which has five carbon atoms and one oxygen atom as ring constituents along with two double bonds. Adding on to this isoflavones are a class of phytochemicals that are sort of phytoestrogen, a plant hormone, which closely resembles with human estrogen by chemical structure (The Healing Power of Soy's Isoflavones, 2004). Taken together, pyranoiso-flavones are a subclass of flavonoids which has had a pyran group attach itself. Alpinumisoflavone ($C_{20}H_{16}O_5$), a pyranoisoflavone (Figure 8.14) was extracted from the plant *Rinorea welwitschii, Helichrysum aureonitens, Erythrine senegalensis,* and *Derris eriocarpa.* This pyranoiso flavone has demonstrated its potency in several diseases (Lead Phytochemicals for Anti-cancer Drug Development, 2016).

Alpinimuso flavone along with seven other compounds were separated from the plant *Erythrine senegalensis.* Alpinimusoflavoneis a dose-dependent inhibitor of HIV-1 protease (Lee et al., 2009). The major source of alpinumisoflavone is the stem bark of *Erythrine senegalensis* but it was also isolated from the chloroform extract of the seeds of *Millettiathon ningii* (Togola et al., 2008).

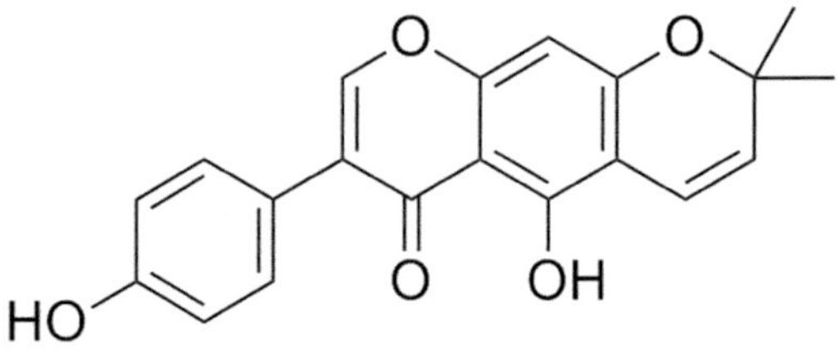

Alpinumisoflavone

FIGURE 8.14   Structure for alpinumisoflavone: An antimicrobial pyranoisoflavone.

8.6.2 ARTEMISININ FROM ARTEMISIA ANNUA

Terpenoids are an important naturally occurring organic chemicals. Terpenes are mostly volatile aromatic hydrocarbons, enriched in the essential oils plants like conifers and citrus trees. Many terpenoids are utilized in aroma-therapy (Terpenes and the "Entourage Effect," 2018). Holistic aromatherapy, sometimes called essential oil therapy, is the procedure of using naturally extracted phytochemicals to improve the body, mind, and spirit (Exploring

Aromartherapy, 2018). Aromatherapy works towards the unification of the physiological, psychological, and spiritual processes to further enhance the individual's innate healing process. Artemisinin (Figure 8.15) is a terpenoid more specifically a sesquiterpene lactone (Cowan, 1999). Artemisinin ($C_{15}H_{22}O_{15}$), was isolated from *Artemisia annua L.* (*Asteraceae* family), a plant indigenous to temperate Asia and other countries (Bilia et al., 2014; Wan-Su, 2015). The Chinese name is Qinghao, or Ching hao (means green herb). Adding on to this, it is also known as variations of wormwood. *A. annua* derived essential oil is abundant in monoterpenoids and sesquiterpenes and represents a byproduct with medicinal properties. This essential oil is extracted by hydrodistillation, another form of steam distillation (Bilia et al., 2014). In this form of distillation, the plant part is water-logged for a stipulated time and then heated. The heat of the distillation carries away the volatile materials (Zheljazkov et al., 2014). Some terpenoids are able to increase blood flow and to kill respiratory pathogens such as MRSA (methicillin-resistant staphylococcus aureus), the antibiotic-resistant bacteria responsible for many types of illness (Terpenes and the "Entourage Effect," 2018). Artemisinin demonstrated very high potential against some bacteria and fungi. Two examples are *C. albicans* and *A. fumigatus* (Bilia et al., 2014). Another such experiment was done and extract showed efficacy against some periodontopathic microorganisms. *A. actinomycetemcomitans, F. nucleatum* subsp. *animalis, F. nucleatum* subsp. *polymorphum,* and *P. intermedia* are some of the microorganisms wherethe extract showed potency (Wan-Su, 2015). Most importantly, artemisinin, and its semi-synthetic derivatives are highly active commercially available drugs in treating malignant malaria, caused by *Plasmodium falciparum.*

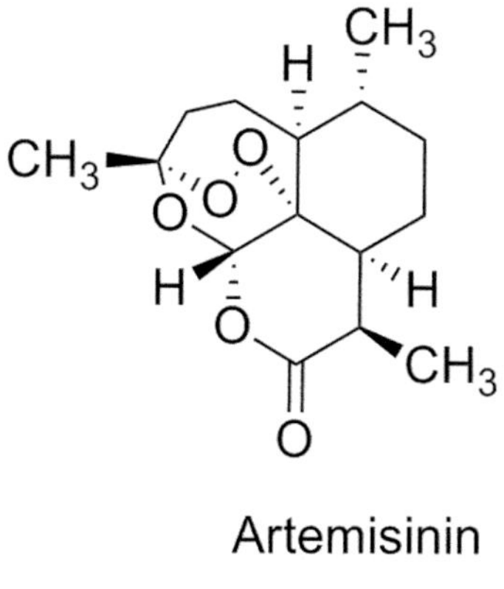

Artemisinin

FIGURE 8.15 Artemisinin: An antimicrobial drug for malignant malaria isolated from *Artemisia annua.*

8.7   HETEROCYCLIC CARDIOVASCULAR DRUGS FROM PLANTS

Cardiovascular diseases (CVDs) are heart and the vascular system related disorders. Atherosclerosis, cerebrovascular disease, coronary &peripheral heart diseases, rheumatic & congenital heart disease, deep vein thrombosis and pulmonary embolism, heart attacks, strokes, heart failure, arrhythmia, and heart valve problems are some popularly known CVDs (American Heart Association, 2017). Many of these can either be prevented and/or are controllable by healthy lifestyle and proper medication. *COPD* also belongs to CVD. In the US, the most common heart disease is coronary artery disease, abbreviated as CAD (Heart disease, 2018). If someone is diagnosed with CAD and left unattended the next step might be a heart attack. Figure 8.16 shows a few heterocyclic drugs for CVDs.

It is matter of serious concern that only in the US about 610,000 people on average die every year owing to CVDs. CVDs are the leading cause of death globally representing 31% of annual global deaths. (Heart Disease Facts & Statistics, 2017). Data on these cardiovascular diseases indicate one in every four premature deaths (under 70 years) is because of CVD. Furthermore, CVD-related deaths can vary by ethnicity and race. Non-Hispanic whites has the highest percentage of CVD deaths while American Indians has the lowest. Apart from diabetes and obesity, some other lifestyle choices can also put people at a higher risk of CVDs which include poor diet, physical inactivity, and intake of excessive alcohol. According to the CDC about 28.1 million adults (11.5% of the adult population) were diagnosed with CVD in 2016 in the US (Heart Disease Statistics, 2016).

The heart acts as a pump with a filter (lung) that controls blood circulation and purification in a definite pattern. It is well-known that human heart has four chambers; the right and left atrium and the right and left ventricle. Blood flow, through the systemic system, enters from the right side of the atrium and into the right ventricle. The systemic circuit is the part of the system which carries blood away from the heart and towards the organs until it once again reaches the heart. After leaving the right ventricle blood enters the pulmonary trunk, into the pulmonary circuit, and circulates until the blood once again enters the heart through the left atrium. The pulmonary circuit carries oxygen-poor blood to the lungs and returns oxygen rich blood to the left atrium (How the Heart Works, 2018). The blood that leaves the left atrium then enters the left ventricle which then enters the systemic circuit once again. Oral medication and nutrients are transported by blood leading from the digestive tract and into the liver. After absorption the medication is distributed throughout the body by blood to the targeted organ and other organs (Medicines by Design, 2018).

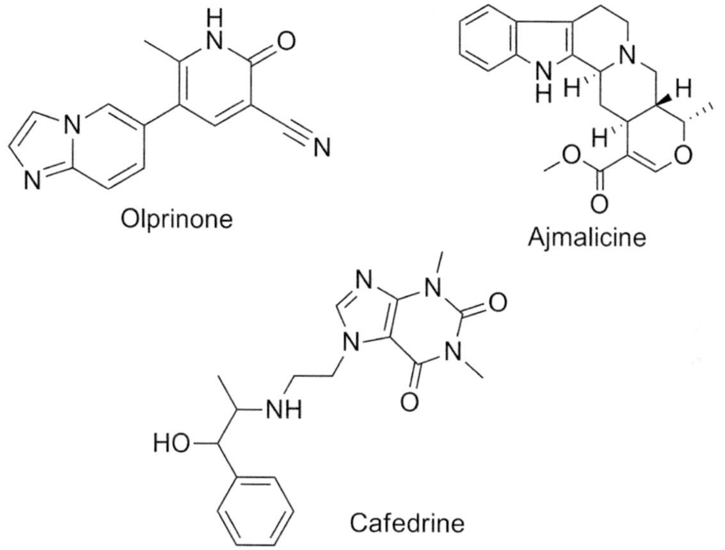

FIGURE 8.16   Frontline heterocyclic drugs for cardiovascular diseases.

8.7.1   *AMIODARONE FROM KHELLA (AMMIVISNAGA)*

In arrhythmia the heartbeats or rhythms become abnormal. Arrhythmia arises mainly owing to the heart related electrical impulses. If the electrical impulses do not work properly, the heart will become unable to coordinate its beats resulting in irregular rhythms. These impulses can be too fast, too slow, or too erratic causing the heart not to pump blood systematically (About Arrhythmia, 2016). If the heart does not pump blood effectively, the organs might be damaged or completely shut down. Arrhythmias have various types, e.g., tachycardia, bradycardia, and a trial fibrillation. The average heart rate is normally between 60 to 100 beats per minute under normal physical state (All About Heart Rate (Pulse), 2015). Tachycardia refers to a heart rate which is much faster, at or faster than 100 bpm (beats per minute), than the average heart rate (Tachycardia: Fast Heart Rate, 2016). Bradycardia refers to heart rate that is slower that 60 bpm (Bradycardia: Slow Heart Rate, 2016). Atrial fibrillation, also called AFib or AF, is a heartbeat which is irregular/ flutters (Atrial Fibrillation, 2016). Arrhythmia occurs if the heart's natural pacemaker develops an abnormal rhythm or rate, the conduction pathway is disrupted, or if another part takes on the role as the pacemaker.

Amiodarone (Figure 8.17), also calledNexterone, is an antiarrhythmic medication for arrhythmias, mainly for atrial fibrillation (Amiodarone: Guidelines for Use and Monitoring, 2003). It was extracted from *Ammi visnaga* (or khella), a flowering plant belongs to the carrot family having the active ingredient khellin (Bhagavathula, 2015). *Ammi visnaga* is an ancient Egyptian medicinal plant and the tea made from it was being used traditionally to treat several ailments, e.g., asthma, angina, and kidney stone (Bhagavathula, 2015). Khellin ($C_{14}H_{12}O_5$) itself is a vasodilator that also has bronchodilatory activity. Amiodarone regularizes the heart's electrical impulses (Amiodarone: Guidelines for Use and Monitoring, 2003). Many investigations were reported to validate amiodarone in several arrhythmias. The experiments concluded amiodarone as a safe and efficacious antiarrhythmic drug (Auer et al., 2002). In comparison to other antiarrhythmic drugs, amiodarone has higher level of tolerance, reduced noncardiac toxicity, and consequently fewer side effects.

The electrophysiological mechanism of actions of amiodarone are not completely understood. However, amiodarone prolongs repolarization by inhibition of outward potassium channels (Auer et al., 2002) and it is usually given to patients with an Implantable Cardioverter Defibrillator, abbreviated as ICD. An ICD is a battery-powered device which is placed under the skin to keep track of the patient's' heart rate (Implantable Cardioverter Defibrillator (ICD), 2016). If the device detects abnormal heartbeats, then the device will then deliver an electrical shock that will restore the normal heartbeat. By using amiodarone in ICD patients,' the shocks which the device gives out are reduced and amiodarone regulates the heartbeat.

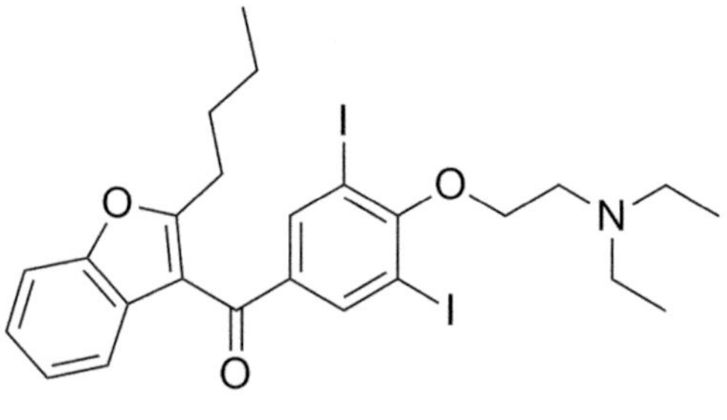

Amiodarone

FIGURE 8.17 Amiodarone: A heterocyclic drug for arrhythmias isolated from *Ammi visnaga*.

8.7.2 CONVALLATOXIN EXTRACTED FROM CONVALLARIA MAJALIS

Convallatoxin (Figure 8.18) is a cardiac glycoside, as well as a cardenolide, was extracted from the plant *Covallaria majalis*. (Schneider et al., 2017). A glycoside is a compound in which a sugar derivative is chemically bonded with a non-sugar group whereas cardenolide is a plant-derived steroid (Schneider et al., 2017). Since heart is the major site of action of convallatoxin (cardiac glycoside), it is simply recognized as cardenolide glycoside (Hi et al., 2010). Convallatoxin has analogs but all are not equally potent in CVDs. Cardiac glycosides are normally used for congestive heart failure and for ventricular rate control in atrial fibrillation (Hi et al., 2010). These compounds also inhibit cancer cell replication. Convallotoxin demonstrated anti-proliferative effects in many tumor cells at nano molar concentrations *in vitro*, and showed a low toxicity for healthy cells.

Cardiac glycosides inhibit the sodium-potassium ATPase pump that increases intracellular sodium (Schneider et al., 2017) which in turn accelerates the sodium-calcium exchange creating an elevation in intracellular calcium and an upgrading in cardiac contractility (Convallatoxin, 2003). The glycosides also increase cardiac vagal tone (an index of stress and stress vulnerability in mammals), which decreases cardiac sympathetic activity. Overdose of convallatoxin increases intracellular calcium which causes early after depolarization, cardiac irritability, and dysrhythmias. Although toxicity of cardiac glycoside is not common but based on other factors mild to severe toxicity may occur. In general, cardiac glycosides are administered orally and time to of action is about half an hour to two hours. The medication then peaks up between two to six hours and is excreted unchanged *via* kidney.

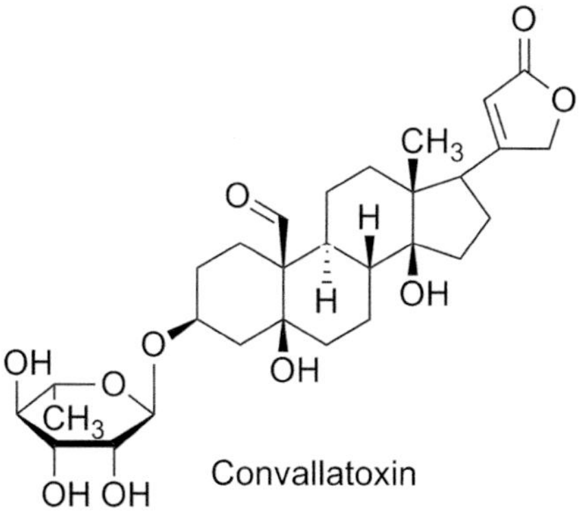

FIGURE 8.18 Convallatoxin: A cardiac glycoside ($C_{29}H_{42}O_{10}$) from the plant *Convallaria majalis*.

8.8   HETEROCYCLIC PLANT-DERIVED DRUGS FOR NEURODEGENERATIVE DISEASES

The human brain is the central command center for the CNS, comprises of six major sectors which are cerebrum, cerebellum, diencephalon, midbrain, pons, and medulla oblongata. Broadly, the CNS includes brain and spinal cord and regulates body and mind as a centralized computer by deciphering the information collected by the eye's to other organs, as needed (How the Spinal Cord Works, 2018). Spinal cord, the other part of CNS, serves like highway between the brain to other organs. Information collected by different organs reach to brain *via* spinal cord and subsequently brain interprets the collected information and creates instructions. Once the instructions are made the output is carried by spinal cord.

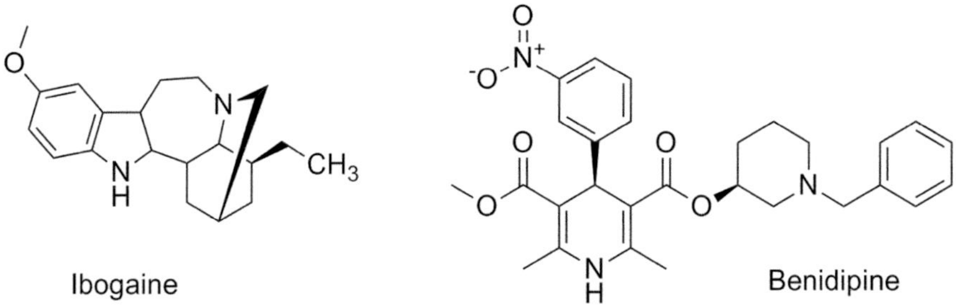

Ibogaine   Benidipine

FIGURE 8.19   Ibogaine and benidipine: Two widely used heterocyclic drugs for neurodegenerative diseases.

'Neurodegenerative disease' is an umbrella term to indicate a range of disorders that primarily affect the neurons (the building blocks of CNS) and consequently the brain functions begin to deteriorate. Once the neurons are damaged or died, they can't be regenerated or replaced unlike other cells which constantly replicate to produce themselves. Neurodegenerative diseases are incorrigible and debilitate the patient which results in the advanced degeneration and eventually the death of neurons (nerve cells) (JPND Research, 2018). The death of neurons can cause ataxias (movement problem), or mental dysfunction like dementias. Besides dementias some commonly known neurodegenerative diseases are Alzheimer's disease (AD) and other dementias, Parkinson's disease (PD) and PD-related disorders, Prion disease, Huntington's disease (HD), motor neuron disease (MND), spinocerebellar ataxia (SCA), spinal muscular atrophy (SMA), and amyotrophic lateral sclerosis (ALS) (Neurodegenerative Diseases, 2018). The genetics, age, and environmental factors are related to someone's susceptibility to one of the neurodegenerative diseases which are treatable up to a certain level but not completely curable. Although there are treatments for these diseases up to a certain level, but they are not completely curable. Two important heterocyclic drugs (Figure 8.19) are frequently prescribed in neurodegenerative disease.

8.8.1 L-DOPA FROM MUCUNAPRURIENS

Parkinson's disease (PD) is one of the most familiar neurodegenerative diseases which affects dopamine-producing (or dopaminergic) neurons of the midbrain, called substantia nigra (SN), a basal ganglia structure that dominates reward pathway (connected to behavior and memory) and movement (Parkinson's Foundation, 2018). PD is also known as paralysis agitans and/ or shaking palsy. The symptoms appear slowly, and the progression is based on genetic, environmental, and other relevant factors. PD begins to develop in the SN when the dopaminergic neurons begin to deteriorate (Young-onset Parkinson's, 2018). Once the dopaminergic neurons start to die, complication like difficulties with movement is seen. When someone experiences the symptoms of PD, it means the patient's nervous system has already experienced neurodegenerative damage and the patient begins to falter. The symptoms (body shaking including the trembling of hands, arms, legs, jaws, and face, stiffness of limbs, slow movement, etc.) begin on one side but with time the whole body (both sides) is affected. Since there is no universal biomarker for PD it is difficult to diagnose at the early stage. Although it is not fully curable but controllable through the enhancement of dopaminergic transmission by L-dihydroxyphenylalanine (L-Dopa) and dopamine agonists (Dorszewska et al., 2014). Levodopa (L-Dopa) (Figure 8.20) is the precursor to dopamine works to control Brady kinetic symptoms in PD. Studies showed that Levodopa can slow down the progression and provides beneficial effects even after drug administration has ended. Levodopa, unlike dopamine can cross the blood-brain barrier (BBB) and is transformed into dopamine in the CNS.

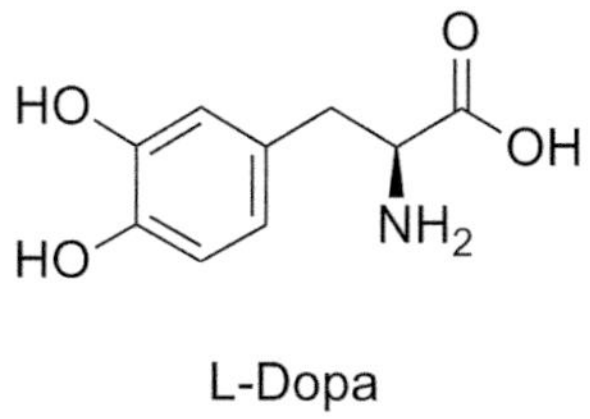

L-Dopa

FIGURE 8.20   L-Dopa ($C_9H_{11}NO_4$) isolated from the plant *Mucunapruriens*.

8.9 ANTI-INFLAMMATORY DRUGS

Inflammation is a symptom in which part of our body begins to turn red, irritated, swelled, heated, painful, and may lose its regular function (Punchard et al., 2004). These characterizations of inflammation were named by Celsius in ancient Rome (30-38 B.C.) and by Galen (A.D. 130-200). Inflammation is related to body's defense mechanism (Chronic Inflammation, 2018) to fight against the harmful agents including xenobiotics. Mostly inflammation occurs due to an injury or untreated infection. The latest definition of inflammation says that it begins due to the changes in a living tissue when it is injured but did not destroy its structure and vitality whereas cellular injury indicates interruption of living (Spector et al., 1963). Examples of what the body defends itself can be antigens, that prompt the immune system to produce antibodies (Antigen, 2018). Antibodies are proteins that recognize detrimental substances/ toxins called antigens (Antibody, 2018). When antigens damage the cells, the body releases chemicals which triggers a response by releasing/up-regulating both antibodies and proteins (Immune Response, 2018). The entire procedure coming from the commands of the immune system can last from hours today's in acute inflammation (Informed Health Online, 2018).

Inflammation has two major types: acute and chronic. Acute form refers to tissue damage owing to infliction of a wound, microbial invasion, or noxious compounds (Chronic inflammation, 2018). Acute inflammation starts quickly, and spreads rapidly. The period of changing an acute to chronic form is between two to six weeks. Chronic inflammation happens slowly in a long-term process and may last from several weeks to years. Chronic inflammation can begin from untreated wounds that might have acute inflammation and an autoimmune disorder. Chronic inflammation might be seen in diabetes, cardiovascular diseases, arthritis, allergies, COPD, or even in asthma. Furthermore, chronic inflammation can eventually start damaging healthy cells that can ultimately lead to DNA damage, tissue death, and internal scarring.

Turmeric, isolated from the rhizome of *Curcuma long* which belongs to the ginger family *Zingiberaceae*, is a golden-yellow colored spice with diverse health benefits, is very often used in the Indian subcontinent to induce a bright golden-yellow color, and amazing taste in the curries (Bandyopadhyay, 2014). The dry rhizomes are crushed to make turmeric powder. Curcumin is the most pharmacologically-privileged molecule (about 3% by weight) found in turmeric powder. Curcumin was first isolated about two centuries ago and it's structure was determined in 1910 (Bandyopadhyay, 2014) and the solution-phase structure was confirmed in 2007. Diverse therapeutic activities of curcumin ($C_{21}H_{20}O_6$) rather diferuloylmethane was explained in Ayurveda the traditional Indian medicines, which is the world's oldest holistic "whole-body" healing systems. Turmeric, with curcumin being the most active component is still being widely used in skin, pulmonary, and gastrointestinal systems, aches, wounds, sprains, and liver disorders. Systematic research showed multidimensional medicinal activities of curcumin (Aggarwal et al., 2007). Curcumin has successfully demonstrated several types of pharmacological activities such as anti-inflammatory, antioxidant, chemopreventive, antidiabetic among others (Hatcher et al., 2008). Apart from anti-inflammatory activity, curcumin also has executed anticancer functions through antioxidant mechanisms (Lead Phytochemicals for Anticancer Drug Development, 2016). Curcumin demonstrated these activities both *in vitro* and *in vivo*. These investigations paved the way for ongoing clinical trials.

Curcumin has demonstrated its efficacy as a chemotherapeutic agent against HNSCC also. Curcumin inhibits pro-inflammatory monocyte/macrophage-derived cytokines, monocyte inflammatory protein-1, monocyte chemotactic protein-1, interleukin-1β, and TNF–α in PMA- or LPS-stimulated peripheral blood monocytes and alveolar macrophages. PMA being a stimulant and LPS being a lipopolysaccharide which presents a sugar structure specific to its conformation. Studies were conducted to evaluate the toxicity of curcumin human body which concluded that use of 1125-2500 mg of curcumin per day had no symptom of toxicity. Furthermore, human trial was held, and the subjects used up to 8000 mg/day of curcumin for three months and found no toxicity from the usages of curcumin. In addition, many trials have been deemed safe for curcumin and it demonstrated anti-inflammatory action by inhibiting several different pain-causing biomolecules which use to play major roles in inflaming a person's body (Wilken et al., 2011) (Figure 8.21).

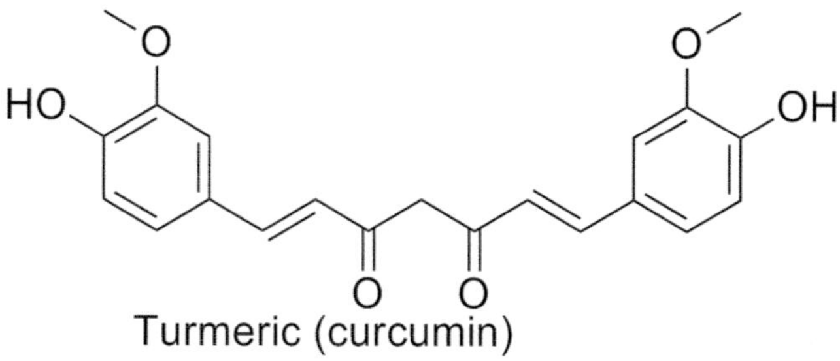

Turmeric (curcumin)

FIGURE 8.21 Curcumin: The major bioactive component of turmeric isolated from the genus *Curcuma*.

8.10   HETEROCYCLIC ANTI-MALARIAL DRUGS FROM PLANTS

Malaria is a life-threatening parasitic disease caused by *Plasmodium* parasites, that are transmitted through the bites of infected female *Anopheles* mosquitoes, called "malaria vectors" when they take a *Plasmodia*-bearing blood meal from an infected human. This disease, although it occurs mainly in travelers coming from an endemic region, is one of the most dominant infections which people must deal with worldwide (Suh et al., 2004). There are five different species that cause malaria in human, and two of these species – *P. falciparum* and *P. vivax* are the greatest threat. *P. falciparum* is prevalent malaria parasite in the African continent and counts the most malaria-related deaths globally. Furthermore, *P. vivax* is the dominant malaria parasite in most countries outside of sub-Saharan Africa.

Malaria is preventable and curable (Malaria, 2018). Malaria is still prevalent in tropical &subtropical countries (latitudes closer to the equator) (Koppen Climate Classification, 2018). The African region carry the lion's share of the global malaria burden. The WHO and other health-related organizations are working to reduce malaria by recommending insecticide-treated bed nets to protect people from infected mosquito bites in the endemic areas (Global Response to Malaria at Crossroads, 2017). Furthermore, many research groups are working to discover vaccine to prevent malaria (Malaria, 2018). Only in 2016, there were an estimated 216 million people of 91 countries diagnosed as malaria patients, which was an inflation of around 5 million cases over 2015. About 445,000 malaria patients died in 2016. In the year 2016, the African region held 90% of all malaria population and 91% of all malaria deaths. To fight this disease, funding for both malaria control (prevention) and treatment has reached an estimated US$2.7 billion in 2016. Malaria has been labeled as an acute febrile illness having an gestation period of one week or longer (Malaria, 2018). In a non-immune individual, symptoms (fever, headache, and chills) appear in 10–15 days after infective mosquito bite. Since the earlier symptoms are apparently common, it is difficult to diagnose at early stage. Furthermore, if malignant malaria is not treated timely, *P. falciparum* may progress to severe illness including death.

The transmission depends on parasitic factors and the environmental settings. The *Anopheles* mosquitoes lay their eggs in water, which hatch into larvae, eventually emerging as adult mosquitoes. The female mosquitoes seek blood meal to nurture their eggs. The female mosquitoes then proceed for blood meal from an infected host and planted the gametocyte of the parasite present in the blood. Each species of *Anopheles* mosquito has its own preferred aquatic habitat; for example, some prefer small, shallow collections of fresh water, such as puddles and hoof prints, which are abundant in tropical countries, especially in rainy season. Transmission is more intense where the mosquito can have a longer lifespan, the longer lifespan allows the parasite to have time of complete development. The long lifespan and strong human-biting habit of the African vectors are the primary reasons of why nearly 90% of the world's malaria cases are confined in Africa. Transmission also heavily depends upon the favorable climate like rainfall patterns, temperature, and humidity (Malaria, 2014) that extend the life span of the mosquitoes. Transmission may be seasonal (Schools & Health, 2018). The peak of transmission period is during or after the seasonal rain. The epidemics prefer when climate and related conditions suddenly favor transmission where people have no or negligible immunity to malaria (Kiszewski et al., 2004). An epidemic can also occur if people who have a low immunity move into an area which has an intense susceptibility of the disease (Doolan et al., 2009). People may also become partially immune to the sickness if they go through many years of exposure. While partial immunity never provides complete protection, it however, reduces the danger of becoming worse and severe illness. Since a person needs many years to attain the immunity, most deaths related to malaria, in Africa, occur in young children.

Simalikalactone D (or SkD) (Figure 8.22) is a quassinoid that is isolated from the plants *Quassia amara* and *Quassia africana* (Simalikalactone, 2018). This compound is an antimalarial, cytotoxic, and antiviral. As an antimalarial SkD attacks the life cycle of plasmodia if the host is human. *Quassia amara* is widely used for its antimalarial activity when it is even taken just as a tonic. This traditional antimalarial remedy showed excellent *in vitro* and *in vivo* activities (Bertani et al., 2006). Even the water extract of the freshly crushed leaves of the plant *Quassia amara* showed excellent antiplasmodial activity (Suh et al., 2004). The active principle of this extract, which is SkD, synergizes *in vitro* with atovaquone against *Plasmodium falciparum*. SkD (45 nM concentration) is noxious for mid-trophozoite *P. falciparum*, and the SkD and atovaquone combination acts upon the mitochondria of *P. falciparum* (Bertani et al., 2012). Another compound quassinoids malika lactone E (SkE), repressed the growth of *P. falciparum* culture *in vitro* by 50%, in the concentration range from 24 to 68 nM, independently of the strain sensitivity to chloroquine (Cachet et al., 2009). Both SkE and SkD showed promising efficacy against malignant malaria.

Bruceantin (Figure 8.22) is a triterpene quassinoid antineoplastic antibiotic, biosynthesized by the plant *Brucea antidysenterica* (*Simaroubaceae* family) (Bruceantin, 2018). A triterpene is a compound having six isoprene units (C_{30}) (Agra et al., 2015). Triterpenes like bruceantin have the ability to inhibit cell migration, cell proliferation, and also collagen deposition. Bruceantin has both antiamoebic and antimalarial activities. An antineoplastic antibiotic is a type of anticancer drug which blocks cell growth by disrupting the genetic material, DNA in the cell (Schools & Health, 2018).

The resistance of *P. falciparum* to commonly used antimalarial medications has begun a serious issue. As an example, quinine is one of the majorly known antimalarial drugs, however, it was then replaced by other synthetic quinolines because of drug-resistance (Antineoplastic Antibiotic, 2018). A more recent example would be the drug chloroquine which has developed a widespread resistance across Asia, America, and Africa. However, both simalikalactone D and bruceantin are known for showing to have high activity against *P. falciparum* making them a good treatment to use for this disease.

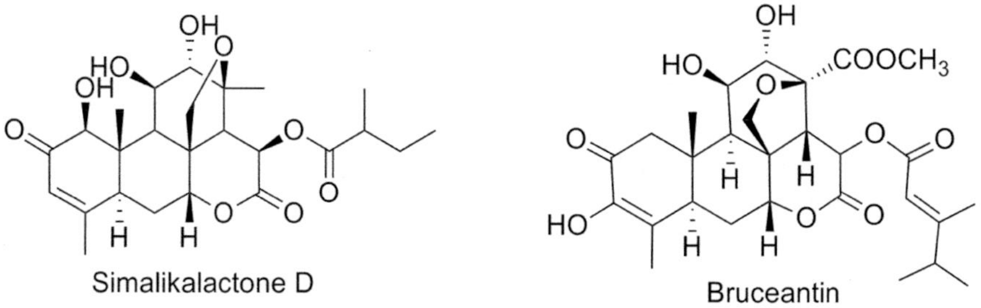

Simalikalactone D
Bruceantin

FIGURE 8.22 Bruceantin and simalikalactone D: Two plant-derived drugs for malignant malaria.

8.11 HETEROYCLIC ANTI-ASTHMATIC DRUGS FROM PLANTS

Asthma is a chronic (long-term) lung disease in which the airways or bronchial tubes, become narrower, swollen, and produce extra mucus that causes difficult to breathe (Asthma, 2018). Asthma patients have swollen bronchial tubes making them extremely sensitive. The sensitive bronchial tubes make the airways susceptible to reacting with certain inhaled substances (Asthma, 2018). While bronchial tubes interact with the inhaled substance, the muscles begin to tighten which narrows the airways. The narrowing down reduces the airflow into the lungs hence the patient feels breathing trouble. Higher swelling of the bronchial tubes worsens then condition as the airways start to shrink even more (Partners Healthcare Asthma Center, 2010). Movement of the bronchial tubes makes the cells react and generate more mucus than normal. The mucus makes the bronchial tubes even narrower making it even more difficult to get air into and out of the lungs. When breathing is made difficult, coughing, wheezing, and shortening of breaths start. This inflammatory disease affects over 300 million individuals worldwide (Kudo et al., 2013). Asthma is a minor nuisance for a certain population, nevertheless it can be a serious problem also that may interfere regular activities and induce fear of life-threatening asthma attack. It is non-curable but controllable and the symptoms are patient-dependent. However, because asthma can change over time, keeping a medical record is important (Mims, 2015). Symptoms of asthma arise if the airways are inflamed. Patients could also have symptoms only at specific times, for example when the patient is strenuously exercising or becomes scared/shocked, nonetheless the patient could also have symptoms at any time. Symptoms may depend on temperature fluctuation, sunlight, and/or seasonal change (Asthma Overview, 2017).

Asthma causes chest tightness or pain, trouble in sleeping (due to shortness of breath), coughing or wheezing, a whistling or wheezing sound when exhaling. The coughing or wheezing becomes worsened by the respiratory viral attack, e.g., cold or flu. Asthma can be worsened in cold and/or dry air. There could be occupational asthma, triggered by workplace hazards like chemical fumes, gases, or dust and allergy-induced asthma, triggered by airborne materials, e.g., pollen, mold spores, cockroach waste, or dried saliva shed by pets. Asthma can be controlled with appropriate medications and lifestyle.

Ginkgo biloba (maidenhair tree) is indigenous in Southern China and is cultivated around the globe. Its leaf extract is useful in asthma (Tang et al., 2007). This slow-growing ornamental tree can be seen in city parks or near commercial buildings, was investigated pharmacologically and the GBE or *Ginkgo biloba* extract came to be. Adding on to this, GBE had already been in use for hundreds of years to treat disorders such as asthma (Cybulska-Heinrich et al., 2012). The active ingredients (phytoceuticals) are flavonoids, ginkgolides, bilobalides (Figure 8.23), and terpenoids (Zuo et al., 2017). GBE is also labeled as a platelet-activating factor or PAF, an inflammatory pathophysiology mediator of asthma (Babayigit et al., 2009). However, despite use, controlled studies do not support the extract's efficacy for some of the indicated conditions.

This extract was used in a study with 75 asthma patients and the results demonstrated that GBE may decrease the activation of some inflammatory cells which decreases the production of sputum (Tang et al., 2007). Another experiment demonstrated that Gingko leaf extract effectively treated airway inflammation when administered orally (Li et al., 1997). This extract was not only beneficial for asthma but also for AD, and CVDs (Zuo et al., 2017). Another study to validate GBE as a PAF (platelet-activating factor) antagonist on lung histology was reported using mice model. The animals were divided into groups and received saline, GBE, and dexamethasone. GBE alleviated all established chronic histological changes which are done in the lungs except the breadth of smooth muscle in asthma (Haines et al., 2011).

Ginkgolide B or GKB, an ingredient of GBE, was tested by using a noninfectious mouse model to validate its anti-inflammatory activity through the mitogen-activated protein kinase pathway. The results showed that this active ingredient (GKB) could be useful and its efficacy depends on the suppression of extracellular regulating kinase/mitogen-activated protein kinase pathway.

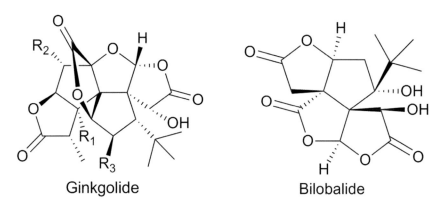

Ginkgolide   Bilobalide

FIGURE 8.23 Structures for the active ingredients of ginkgolides, and bilobalide.

8.12 HETEROCYCLIC PLANT-DERIVED DRUGS FOR PSYCHOTIC DISORDERS

Psychotic disorders are varying illnesses which affect the mind and cause mental disorders. These disorders cause abnormal thinking and perceptions. Psychosis is a mental disorder of feeling disconnected from reality (Early Psychosis and Psychosis, 2018). They make it hard to know what's real and what isn't. While the experience for everyone is different, many people say that psychosis is frightening and confusing. In general, "psychotic disorders" indicate schizophrenia, schizoaffective disorder, schizophreniform disorder, brief psychotic disorder, delusional disorder, shared psychotic disorder, etc.

Early or first-episode psychosis (FEP) refers beginning to lose contact with reality. Acting quickly to connect the patient with the right treatment during FEP can be life-changing and radically alter the future. FEP does not suddenly occur to someone, rather gradual, non-specific changes in the thoughts and perceptions are seen and the patient doesn't understand what exactly is happening. The early signs might be similar to typical teen or young adult behavior owing to the "phases" of this age. Even though such signs do not cause panic, however, they can assess by a doctor. Two main indications of psychosis are delusions and hallucinations (Psychotic Disorders, 2018). Delusions are false beliefs and hallucinations are false perceptions, such as hearing, seeing, or feeling something fantasy. People with bipolar disorder may also have psychotic symptoms. Psychosis may also be caused by consumption of alcohol and some drugs, brain tumors, brain infections, and stroke.

Treatment involves mental health professionals care and antipsychotic medicines in combination with other drugs. *Panax ginseng* root is a popular herbal remedy for psychosis (Kim et al., 2016). *Panax ginseng*, commonly known as ginseng or Korean ginseng, has the main active component (active principle), ginsenoside (Kim et al., 2013). Ginsenoside (Figure 8.24) is a steroid-like saponin which is unique to ginseng. Ginseng is often used for therapy to the patients suffering from central nervous system-related ailments. The therapeutic effects of ginseng were studied for Alzheimer's disease, Parkinson's disease, cerebral ischemia, depression, psychosis, and neurodevelopmental disorders. Ginsenosides showed anti-inflammatory activity by suppressing cytokines and regulating signaling pathways. When simply ginseng and its major bioactive component ginsenoside are administered, even in a single dosage, can create and effect on multiple sites of action. Accordingly, ginsenoside (ginsenoside and its analogs)have become the preferred candidates to develop multi-target drugs. In CNS diseases there are multiple etiological and pathological targets. Consequently, a multi-target drug-like ginsenoside might be the drug of choice.

Ginseng affects the neurotransmission of acetylcholine and γ-aminobutyric acid (GABA) through the mechanisms that involve neurotransmitter and its release and the related signaling pathways (Kim et al., 2013). Ginseng can also improve both learning and memory. This was observed through the usage of behavioral analysis by a mechanism involving the variation of synaptic plasticity and increase in neurogenesis which then affects the neuronal density in the hippocampus. Clinical trials were also done to investigate the efficacy of ginseng to the patients during a 12-weeks period, and prior and after the experiment the patients were requested to take a test to observe their mental capabilities to validate the beneficial effect of ginseng. The ginseng-treated group showed statistically significant development compared to the control (untreated) group.

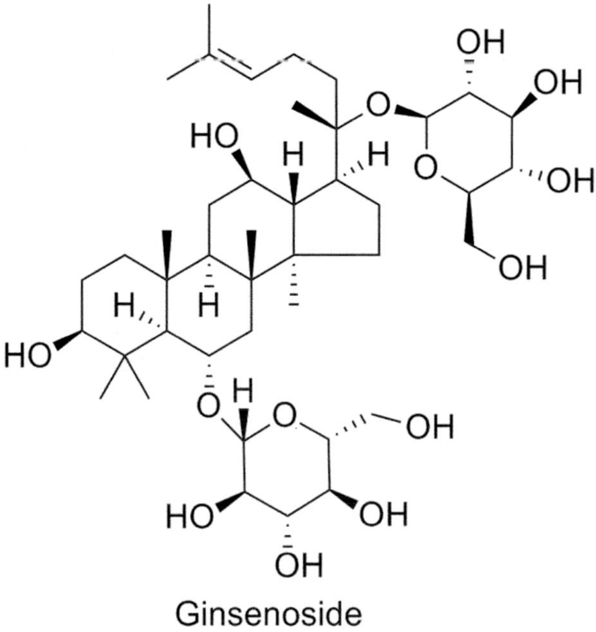

Ginsenoside

FIGURE 8.24 Ginsenoside from the root of *Panax ginseng*: A heterocyclic drug for psychotic disorders.

8.13 CONCLUSION

Heterocyclic chemistry deals with the extraction, isolation, chemical modification (semi-synthesis), total synthesis, and application of heterocyclic compounds in real world. Heterocycles are great stepping stool in the ladder of discovering modern drugs. The heterocycles most being searched for are coming from a major natural source, plants. For thousands of years, many kinds of plants have been used to treat several illnesses. A vast number of bio-privileged molecules including heterocyclic drugs from plants. Heterocyclic plant-based drugs can be anticancer, antidiabetic, anti-HIV, antimicrobial, anti-inflammatory, anti-asthmatic, and anti-psychotic, neuroprotective as well as drugs for cardiovascular diseases. Tannins, terpenoids, alkaloids, and flavonoids are some of the heterocyclic constituents which are secondary metabolites of plants. Presence of heterocyclic (either individual or hybrid) pharmacophore is essential for pharmacological activity of many commercial drugs. Medicinally privileged molecule(s) can be found at any part of a medicinal plant but the same molecule, in general, might not be found in every part of the plant. Still, today about 25% of commercial drugs are derived from the natural sources and majority (65–70%) of the natural drugs are present in terrestrial plants. Currently, about 61% of commercial drugs are either natural or semi-synthetic molecules. Depending on the type of disease the percentages are even higher. Interestingly, the majority of natural or chemically modified natural drugs contain at least one heterocyclic core which is responsible for the medicinal activity. Many natural plant species are still unexplored or less explored. Accordingly, more research should be carried out to discover new and novel natural drugs in the battle against dreadful diseases.

ACKNOWLEDGMENTS

We are thankful to the Department of Chemistry, University of Texas Rio Grande Valley for providing library and related facilities.

KEYWORDS

- **Anti-diabetic**
- **alkaloid**
- **Alzheimer's disease**
- **bioactive natural compound**
- **Anti-cancer**
- **heterocycles**

REFERENCES

About Arrhythmia, (2016). American Heart Association. http://www.heart.org/en/health-topics/arrhythmia/about-arrhythmia (accessed on 20 May 2020).
About Diabetes, (2017). *Centers for Disease Control and Prevention.* https://www.cdc.gov/diabetes/basics/diabetes.html (accessed on 20 May 2020).

About HIV/AIDS, (2018). *Centers for Disease Control and Prevention.* https://www.cdc.gov/ hiv/basics/whatishiv.html (accessed on 20 May 2020).

Aggarwal, B. B., Sundaram, C., Malani, N., & Ichikawa, H., (2007). Curcumin: The Indian solid gold. *Adv Exp Med Biol., 595*, 1–75.

Agra, L. C., Ferro, J. N., Barbosa, F. T., & Barreto, E., (2015). Triterpenes with healing activity: A systematic review. *J. Dermatolog Treat., 26*(5), 465–470.

Ahmed, A. S., Elgorashi, E. E., Moodley, N., Mcgraw, L. J., Naidoo, V., & Eloff, J. N., (2012). The antimicrobial, antioxidative, anti-inflammatory activity and cytotoxicity of different fractions of four South African Bauhinia species used traditionally to treat diarrhea. *J. Ethnopharmacol., 143*(3), 826–839.

All About Heart Rate (Pulse), (2015). American Heart Association. http://www.heart.org/ en/health-topics/high-blood-pressure/the-facts-about-high-blood-pressure/all-about-heart-rate-pulse (accessed on 20 May 2020).

Amar, M. B., (2006). Cannabinoids in medicine: A review of their therapeutic potential. *J Ethnopharmacol., 105*(1–2), 1–25.

American Cancer Society, (2015). *What is Cancer?* https://www.cancer.org/cancer/cancer-basics/what-is-cancer.html (accessed on 20 May 2020).

American Cancer Society, (2017). *What is Stomach Cancer?* https://www.cancer.org/cancer/ stomach-cancer/about/what-is-stomach-cancer.html (accessed on 20 May 2020).

American Heart Association, (2017). *What is Cardiovascular Disease?* http://www.heart.org/en/ health-topics/consumer-healthcare/what-is-cardiovascular-disease (accessed on 20 May 2020).

Amiodarone: Guidelines for Use and Monitoring, (2013). American Family Physician. https:// www.aafp.org/afp/2003/1201/p2189.html (accessed on 20 May 2020).

Antibody, (2018). *Medline Plus.* https://medlineplus.gov/ency/article/002223.htm (accessed on 20 May 2020).

Antigen, (2018). *Medline Plus.* https://medlineplus.gov/ency/article/002224.htm (accessed on 20 May 2020).

Antimicrobial Drugs, (2018). *Journal of Antimicrobial Drugs.* https://www.omicsonline.org/ scholarly/antimicrobial-drugs-journals-articles-ppts-list.php (accessed on 20 May 2020).

Aromatic Heterocycles: Pyridine and Pyrrole, (2016). *Chemistry LibreTexts.* https:// chem.libretexts.org/LibreTexts/Athabasca_University/Chemistry_350%3A_Organic_ Chemistry_I/Chapter_15%3A_Benzene_and_Aromaticity/15.05_Aromatic_ Heterocycles%3A_Pyridine_and_Pyrrole (accessed on 20 May 2020).

Arredondo, F., Echeverry, C., Blasina, F., Cassanello, L. V., Diaz, M., Rivera, F., Martinez, M., et al., (2015). Flavones and flavonols in brain and disease: Facts and pitfalls. In: Watson, R. R., & Preedy, V. C., (eds.), *Bioactive Nutraceuticals and Dietary Supplements in Neurological and Brain Disease* (pp. 229–236). San Diego.

Aslam, M. S., Wajid, M., Ahmad, M. S., & Mamat, A. S., (2015). A review on antidiabetic potential of genus Bauhinia. *IRJPS, 2*(3), 268–273.

Asthma Overview, (2017). *Informed Health Online.* https://www.ncbi.nlm.nih.gov/ pubmedhealth/PMH0072704/ (accessed on 20 May 2020).

Asthma, (2018). *American Academy of Allergy Asthma and Immunology.* https://www.aaaai. org/conditions-and-treatments/asthma (accessed on 20 May 2020).

Asthma, (2018). *Beth Israel Deaconess Medical Center.* https://www.bidmc.org/conditions-and-treatments/airway-breathing-and-lung/asthma (accessed on 20 May 2020).

Atrial Fibrillation, (2018). American Heart Association. http://www.heart.org/en/health-topics/atrial-fibrillation (accessed on 20 May 2020).

Auer, J., Berent, R., & Eber, B., (2002). Amiodarone in the prevention and treatment of arrhythmia. *Curr. Opin. Investig Drugs, 3*(7), 1037–1044.

Babayigit, A., Olmez, D., Karaman, O., Ozogul, C., Yilmaz, O., Kivkcak, B., Erbil, G., & Uzuner, N., (2009). Effects of Ginkgo biloba on airway histology in a mouse model of chronic asthma. *Allergy Asthma Proc., 30*(2), 186–191.

Bandyopadhyay, D., (2014). *Curcumin: A Folklore Remedy from Kitchen on the Way to Clinic as Cancer Drug in Horizons in Cancer Research* (pp. 1–42). Nova Science Publishers, Inc., Hauppauge, New York, USA.

Bandyopadhyay, D., (2014). Farmer to pharmacist: Curcumin as an anti-invasive and antimetastatic agent for the treatment of cancer. *Front Chem., 2*, 113.

Bertani, S., Houel, E., Jullian, V., Bourdy, G., Valentin, A., Stien, D., & Deharo, E., (2012). New findings on simalika lactone D, an antimalarial compound from *Quassia amara* L. (Simaroubaceae). *Exp. Parasitol., 130*(4), 341–347.

Bertani, S., Houel, E., Stien, D., Chevolot, L., Jullian, V., Garavito, G., Bourdy, G., & Deharo, E., (2006). Simalikalactone D is responsible for the antimalarial properties of an Amazonian traditional remedy made with *Quassia amara* L. (Simaroubaceae). *J. Ethnopharmacol., 108*(1), 155–157.

Bhagavathula, A. S., Al-khatib, A. J. M., Elnour, A. A., Al Kalbani, N. M. S., & Shehab, A., (2015). *Ammi Visnaga* in treatment of urolithiasis and hypertriglyceridemia. *Pharmacognosy Res., 7*(4), 397–400.

Biava, M., Porretta, G. C., Deidda, D., Pompei, R., Tafi, A., & Manetti, F., (2003). Importance of the thiomorpholine introduction in new pyrrole derivatives as antimycobacterial agents analogues of BM 212. *Bioorg. Med. Chem., 11*(4), 515–520.

Bilia, A. R., Satomauro, F., Sacco, C., Bergonzi, M. C., & Donato, R., (2014). Essential oil of *Artemisia annua* L.: An extraordinary component with numerous antimoicrobial properties. *Evid. Based Complement Alternat. Med., 2014*, 7.

Bonetti, F., Brombo, G., & Zuliani, G., (2017). Nootropics, functional foods, and dietary patterns for prevention of cognitive decline. In: Watson, R. R., (ed.), *Nutrition and Functional Foods for Healthy Aging* (pp. 211–232). San Diego.

Bradycardia: Slow Heart Rate, (2016). American Heart Association. http://www.heart.org/en/health-topics/arrhythmia/about-arrhythmia/bradycardia--slow-heart-rate (accessed on 20 May 2020).

Bruceantin, (2018). National Cancer Institute. https://www.cancer.gov/publications/dictionaries/cancer-drug/def/bruceantin (accessed on 20 May 2020).

Cachet, N., Hoakwie, F., Bertani, S., Bourdy, G., Deharo, E., Stien, D., Houel, E., et al., (2009). Antimalarial activity of simalikalactone E, a new quassinoid from *Quassia amara* L. (Simaroubaceae). *Antimicrob Agents Chemother., 53*(10), 4393–4398.

Camellia Sinensis, (2018). *Missouri Botanical Garden*. http://www.missouribotanicalgarden.org/PlantFinder/PlantFinderDetails.aspx?taxonid=287342 (accessed on 20 May 2020).

Campoli, D. M. R., Monk, J. P., Price, A., Benfield, P., Todd, P. A., & Ward, A., (1988). Ciprofloxacin: A review of its antibacterial activity, pharmacokinetic properties, and therapeutic use. *Drugs, 35*(4), 373–447.

Cha, R., & Sobel, J. D., (2004). Fluconazole for the treatment of candidiasis: 15 years experience. *Expert Rev. Anti. Infect. Ther., 2*(3), 357–366.

Chen, H., Gao, Y., Wu, J., Chen, Y., Chen, B., Hu, J., & Zhou, J., (2014). Exploring therapeutic potentials of baicalin and its aglycone baicalein for hematological malignancies. *Cancer Lett., 354*(1), 5–11.

Chen, S., Yong, T., Xiao, C., Su, J., Zhang, Y., Jiao, C., & Xie, Y., (2018). Pyrrole alkaloids and ergosterols from *Grifolafrondosa* exert anti-α-glucosidase and anti-proliferative activities. *J. Funct Foods., 43*, 196–205.

Choi, J., Ko, E., Chung, H. K., Lee, J. H., Ju, E. J., Lim, H. K., Park, I., et al., (2015). Nanoparticulated docetaxel exerts enhanced anticancer efficacy and overcomes existing limitations of traditional drugs. *Int. J. Nanomedicine, 10*, 6121–6132.

Chronic Inflammation, (2018). *StatPearls*. https://www.ncbi.nlm.nih.gov/books/NBK493173/ (accessed on 20 May 2020).

Ciprofloxacin, (2018). *MedlinePlus*. https://medlineplus.gov/druginfo/meds/a688016.html (accessed on 20 May 2020).

Cockrell, A. S., & Kafri, T., (2007). Gene delivery by lentivirus vectors. *Mol. Biotechnol., 36*(3), 184–204.

Convallatoxin, (2003). *Toxnet*. https://toxnet.nlm.nih.gov/cgi-bin/sis/search/a?dbs+hsdb:@term+@DOCNO+3475 (accessed on 20 May 2020).

Cowan, M. M., (1999). Plant products as antimicrobial agents. *Clin. Microbiol. Rev., 12*(4), 564–582.

Creagh, T., Ruckle, J. L., Tolbert, D. T., Giltner, J., Eiznhamer, D. A., Dutta, B., Flavin, M. T., & Xu, Z. Q., (2001). Safety and pharmacokinetics of single doses of (+)-calanolide A, a novel, naturally occurring nonnucleoside reverse transcriptase inhibitor, in healthy human immunodeficiency virus-negative human subjects. *Antimicrob Agents Chemother., 45*(5), 1379–1386.

Currens, M. J., Gulakowski, R. J., Mariner, J. M., Moran, R. A., Buckheit, R. W., Gustafson, K. R., McMahon, J. B., & Boyd, M. R., (1996). Antiviral activity and mechanism of action of calanolide A against the human immunodeficiency virus type-1. *J Pharmacol Exp Ther., 279*(2), 645–651.

Currens, M. J., Mariner, J. M., McMahon, J. B., & Boyd, M. R., (1996). Kinetic analysis of inhibition of human immunodeficiency virus type-1 reverse transcriptase by calanolide A. *J. Pharmacol Exp. Ther., 279*(2), 652–661.

Cybulska-Heinrich, A. K., Mozaffarieh, M., & Flammer, J., (2012). Ginkgo biloba: An adjuvant therapy for progressive normal and high-tension glaucoma. *Mol. Vis., 18*, 390–402.

Demain, A. L., & Fang, A., (2000). The natural functions of secondary metabolites. *Adv. Biochem. Eng. Biotechnol., 69*, 1–39.

Denny, W., & Kurosu, M., (2016). Advances in tuberculosis medicinal chemistry. In: Denny, W., & Kurosu, M., (eds.), *Advances in Tuberculosis Medicinal Chemistry* (pp. 2–5). Future Science: London.

Devi, M. L., Reddy, P. L., Yogeeswari, P., Sriram, D., Reddy, T. V., Reddy, B. V. S., & Narender, R., (2017). Design and synthesis of novel triazole linked pyrrole derivatives as potent mycobacterium tuberculosis inhibitors. *Med. Chem. Res., 26*, 2985–2999.

Diabetes Basics, (2018). American Diabetes Association. http://www.diabetes.org/diabetes-basics/?loc=db-slabnav (accessed on 20 May 2020).

Discovery and Development of Penicillin, (2018). ACS Chemistry for Life. https://www.acs.org/content/acs/en/education/whatischemistry/landmarks/flemingpenicillin.html#designation-acknowledgments (accessed on 20 May 2020).

Domagala, A., Jarosz, T., & Lapkowski, M., (2015). Living on pyrrolic foundations: Advances in natural and artificial bioactive pyrrole derivatives. *Eur. J. Med. Chem., 16*, 176–187.

Doolan, D. L., Dobano, C., & Baird, J. K., (2009). Acquired immunity to malaria. *Clin. Microbiol. Rev., 22*(1), 13–36.

Dorszewska, J., Prendecki, M., Lianeri, M., & Kozubski, W., (2014). Molecular effects of L-dopa therapy in Parkinson's disease. *Curr. Genomics., 15*(1), 11–7.

Du, G. J., Zhang, Z., Wen, X. D., Yu, C., Calway, T., Yuan, C. S., & Wang, C. Z., (2012). Epigallocatechin gallate (EGCG) is the most effective cancer chemopreventive polyphenol in green tea. *Nutrients, 4*(11), 1679–1691.

Early Psychosis and Psychosis, (2018). *National Alliance on Mental Illness.* https://www.nami.org/earlypsychosis (accessed on 20 May 2020).

Exploring Aromatherapy, (2018). *NAHA.* https://naha.org/explore-aromatherapy/about-aromatherapy/what-is-aromatherapy (accessed on 20 May 2020).

Fact Sheet Index-Bauhinia Variegate, (2016). Weeds of Australia. https://keyserver.lucidcentral.org/weeds/data/media/Html/bauhinia_variegata.htm (accessed on 20 May 2020).

Fernandes, E., Costa, D., Toste, S. A., Lima, J. L., & Reis, S., (2004). *In vitro* scavenging activity for reactive oxygen and nitrogen species by nonsteroidal anti-inflammatory indole, pyrrole, and oxazole derivative drugs. *Free Radic. Biol. Med., 37*(11), 1895–1905.

Florida Exotic Pest Plant Council Invasive Plant Lists. Florida Exotic Pest Plant Council. https://www.fleppc.org/list/list.htm (accessed on 20 May 2020).

Fluconazole, (2018). *MedlinePlus.* https://medlineplus.gov/druginfo/meds/a690002.html (accessed on 20 May 2020).

Frankish, N., De Sousa, M. F., Mills, C., & Sheridan, H., (2010). Enhancement of insulin release from the beta-cell line INS-1 by an ethanolic extract of Bauhinia variegate and its major constituent roseoside. *Planta Med., 76*(10), 995–997.

Fridlender, M., Kapulnik, Y., & Koltai, H., (2015). Plant derived substances with anti-cancer activity: From folklore to practice. *Front. Plant Sci., 6*, 799.

Gasiorowski, K., Lamer-Zarawaska, E., Leszek, J., Parvathaneni, K., Yendluri, B. B., Blach-Oszewska, Z., & Aliev, G., (2011). Flavones from root of Scutellaria baicalensis Georgi: Drugs of the future in neurodegenertation? *CNS Neurol. Disord. Drug Targets, 10*(2), 184–191.

General Cancer Information, (2016). Cancer Research UK. https://www.cancerresearchuk.org/about-cancer/cancer-in-general/treatment/cancer-drugs/drugs/docetaxel (accessed on 20 May 2020).

Genetics of Diabetes. (2017). American Diabetes Association. http://www.diabetes.org/diabetes-basics/genetics-of-diabetes.html?loc=db-slabnav (accessed on 20 May 2020).

Gholap, S. S., (2016). Pyrrole: An emerging scaffold for construction of valuable therapeutic agents. *Eur. J. Med. Chem., 3*, 13–31.

Global Response to Malaria at Crossroads, (2017). World Health Organization. http://www.who.int/en/news-room/detail/29-11-2017-global-response-to-malaria-at-crossroads (accessed on 20 May 2020).

Gomtsyan, A., (2012). Heterocycles in drugs and drug discovery. *Chem. Heterocycl. Compd., 48*(1), 7–10.

Guzman, M., (2018). Cannabis for the management of cancer symptoms: THC version 2.0? *Cannabis Cannabinoid Res., 3*(1), 117–119.

Haidari, F., Omidian, K., Rafiei, H., Zarei, M., & Shahi, M. M., (2013). Green tea (*Camellia sinensis*) supplementation to diabetic rats improves serum and hepatic oxidative stress markers. *Iran J. Pharm Res., 12*(1), 109–114.

Haines, D. D., Varga, B., Bak, I., Juhasz, B., Mahmoud, F. F., Kalantari, H., Gesztelyi, R., et al., (2011). Summative interaction between astaxanthin, Ginkgo biloba extract (EGb761) and vitamin C in suppression of respiratory inflammation: A comparison with ibuprofen. *Phytother. Res., 25*(1), 128–136.

Hanna, L., (1999). Calanolide A: A natural non-nucleoside reverse transcriptase inhibitor. *BETA, 12*(2), 8–9.

Hatcher, H., Planalp, R., Cho, J., Torti, F. M., & Torti, S. V., (2008). Curcumin: From ancient medicine to current clinical trials. *Cell Mol. Life Sci., 65*(11), 1631–1652.

Heart Disease Statistics, (2016). Centers for Disease Control and Prevention. https://www. cdc.gov/nchs/fastats/heart-disease.htm (accessed on 20 May 2020).

Heart Disease Facts and Statistics, (2017). Centers for Disease Control and Prevention. https://www.cdc.gov/heartdisease/facts.htm (accessed on 20 May 2020).

Heart Disease, (2018). Centers for Disease Control and Prevention. https://www.cdc.gov/ heartdisease/index.htm (accessed on 20 May 2020).

Hi, L. S., Liao, Y. R., Su, M. J., Lee, A. S., Kuo, P. C., Damu, A. G., Kuo, S. C., Sun, H. D., Lee, K. H., & Wu, T. S., (2010). Cardiac glycosides from Antiaristoxicaria with potent cardiotonic activity. *J. Nat. Prod., 73*(7), 1214–1222.

HIV Basics, (2018). Centers for Disease Control and Prevention. https://www.cdc.gov/hiv/ basics/ (accessed on 20 May 2020).

HIV Strains and Types, (2018). *Avert.* https://www.avert.org/professionals/hiv-science/types-strains (accessed on 20 May 2020).

HIV.gov, (2018). *What are HIV and AIDS?* https://www.hiv.gov/hiv-basics/overview/about-hiv-and-aids/what-are-hiv-and-aids (accessed on 20 May 2020).

How the Heart Works, (2018). National Heart, Lung, and Blood Institute. https://www.nhlbi. nih.gov/health-topics/how-heart-works (accessed on 20 May 2020).

How the Spinal Cord Works, (2018). Christopher & Dana Reeve Foundation. https://www. christopherreeve.org/living-with-paralysis/health/how-the-spinal-cord-works (accessed on 20 May 2020).

Hunter, P., (2012). The inflammation theory of disease. *EMBO Rep., 13*(11), 968–970.

Ichite, N., Chougule, M. B., Jackson, T., Fulzele, S. V., Safe, S., & Singh, M., (2009). Enhancement of docetaxel anticancer activity by a novel diindolylmethane compound in human non-small cell lung cancer. *Clin. Cancer Res., 15*(2), 543–552.

Immune Response, (2018). *MedlinePlus.* https://medlineplus.gov/ency/article/000821.htm (accessed on 20 May 2020).

Immunomodulation, (2018). National Cancer Institute. https://www.cancer.gov/publications/ dictionaries/cancer-terms/def/immunomodulation (accessed on 20 May 2020).

Implantable Cardioverter Defibrillator (ICD), (2016). American Heart Association. http:// www.heart.org/en/health-topics/arrhythmia/prevention--treatment-of-arrhythmia/ implantable-cardioverter-defibrillator-icd (accessed on 20 May 2020).

Informed Health Online, (2016). *What are Microbes?* https://www.ncbi.nlm.nih.gov/ pubmedhealth/PMH0072571/ (accessed on 20 May 2020).

Informed Health Online, (2018). *What is an Inflammation?* https://www.ncbi.nlm.nih.gov/ books/NBK279298/ (accessed on 20 May 2020).

Joray, M. B., Trucco, L. D., González, M. L., Napal, G. N. D., Palacios, S. M., Bocco, J. L., & Carpinella, M. C., (2015). Antibacterial and cytotoxic activity of compounds isolated from *Flourensiaoolepis. Evid. Based Complement Alternat. Med.*, 912484.

JPND Research, (2018). *What is Neurodegenerative Disease?* http://www. neurodegenerationresearch.eu/about/what/ (accessed on 20 May 2020).

Kamal, A., Hussaini, S. M. A., Faazil, S., Poornachandra, Y., Reddy, G. N., Kumar, C. G., Rajput, V. S., et al., (2013). Anti-tubercular agents. Part 8: Synthesis, antibacterial and antitubercular activity of 5-nitrofuran based 1,2,3-triazoles. *Bioorg. Med. Chem. Lett.*, *23*(24), 6842–6846.

Kashman, Y., Gustafson, K. R., Fuller, R. W., Cardellina, J. H., McMahon, J. B., Currens, M. J., Buckheit, R. W. Jr., et al., (1992). The calanolides, a novel HIV-inhibitory class of coumarin derivatives from the tropical rainforest tree, *Calophyllum lanigerum*. *J. Med. Chem.*, *35*(15), 2735–2743.

Kasim, L. S., Olatunde, A., Effedua, H. I., Adejumo, O. E., Ekor, M., & Fajemirokun, T. O., (2012). Antimicrobial activity of six selected plants against some strains of pathogenic organisms. *J. Microbiol. Antimicrob.*, *4*(3), 54–59.

Kaur, R., Rani, V., Abbot, V., Kapoor, Y., Konar, D., & Kumar, K., (2017). Recent synthetic and medicinal perspectives of pyrroles: An overview. *J. Pharm. Chem. Sci.*, *1*(1), 17–32.

Kenmotsu, H., & Tanigawara, Y., (2015). Pharmacokinetics, dynamics, and toxicity of docetaxel: Why the Japanese dose differs from the Western dose. *Cancer Sci.*, *106*(5), 497–504.

Key Statistics for Stomach Cancer, (2018). American Cancer Society. https://www.cancer.org/cancer/stomach-cancer/about/key-statistics.html (accessed on 20 May 2020).

Khazir, J., Hyder, I., Gayatri, J. L., Yandrati, L. P., Nalla, N., Chasoo, G., Mahajan, A., et al., (2014). Design and synthesis of novel 1,2,3-triazole derivatives of coronopilin as anti-cancer compounds. *Eur. J. Med. Chem.*, *82*, 255–262.

Kim, J. H., Kim, P., & Shin, C. Y., (2013). A comprehensive review of the therapeutic and pharmacological effects of ginseng and ginsenosides in central nervous system. *J. Ginseng Res.*, *37*(1), 8–29.

Kim, J. H., Yi, Y. S., Kim, M. Y., & Cho, J. Y., (2017). Role of ginsenosides, the main active components of *Panax ginseng*, in inflammatory responses and diseases. *J. Ginseng Res.*, *41*(4), 435–443.

Kiszewski, A. E., & Teklehaimanot, A., (2004). A review of the clinical and epidemiologic Burden of epidemic malaria. *Am. J. Trop. Med. Hyg.*, *71*(2), 128–135.

Kitamura, K., Honda, M., Yoshizaki, H., Yamamoto, S., Nakane, H., Fukushima, M., Ono, K., & Tokunaga, T., (1998). Baicalin, an inhibitor of HIV-1 production *in vitro*. *Antiviral Res.*, *37*(2), 131–140.

Koppen Climate Classification, (2018). ISC Audubon. http://www.thesustainabilitycouncil.org/the-koppen-climate-classification-system/the-temperate-climate/ (accessed on 20 May 2020).

Kudo, M., Ishigatsubo, Y., & Aoki, I., (2013). Pathology of asthma. *Front Microbiol.*, *4*, 263.

Kulkarni, Y. A., & Garud, M. S., (2016). Bauhinia variegate (Caesalpiniaceae) leaf extract: An effective treatment option in type I and type II diabetes. *Biomed Pharmacother.*, *83*, 122–129.

Kurapati, K. R. V., Atluri, V. S., Samikkannu, T., Garcia, G., & Nair, M. P. N., (2016). Natural products as anti-HIV agents and role in HIV-associated neurocognitive disorders (HAND): A brief overview. *Front Microbiol.*, *6*, 1444.

Lalitha, P., Veena, V., Vidhyapriya, P., Lakshmi, P., Krishna, R., & Sakthivel, N., (2016). Anticancer potential of pyrrole (1, 2, a) pyrazine 1, 4, dione, hexahydro 3-(2-Methyl Propyl) (PPDHMP) extracted from a new marine bacterium, *Staphylococcus* sp. Strain MB30. *Apoptosis, 21*(5), 566–577.

Lead Phytochemicals for Anticancer Drug Development, (2018). *Frontiers in Plant Science.* https://www.frontiersin.org/articles/10.3389/fpls.2016.01667/full (accessed on 20 May 2020).

Lebel, M., (1988). Ciprofloxacin: Chemistry, mechanism of action, resistance, antimicrobial spectrum, pharmacokinetics, clinical trials, and adverse reactions. *Pharmacotherapy, 8*(1), 3–33.

Lee, J., Oh, W. K., Ahn, J. S., Kim, Y. H., Mbafor, J. T., Wandiji, J., & Forum, Z. T., (2009). Prenylisoflavonoids from *Erythrina senegalensis* as novel HIV-1 protease inhibitors. *Planta Med., 75*(3), 268–270.

Lentivirus, (2018). *ViralZone.* https://viralzone.expasy.org/264?outline=all_by_species (accessed on 20 May 2020).

Li, B. Q., Fu, T., Dongyan, Y., Mikovits, J. A., Ruscetti, F. W., & Wang, J. M., (2000). Flavonoid baicalin inhibits HIV-1 infection at the level of viral entry. *Biochem. Biophys. Res. Commun., 276*(2), 534–538.

Li, M. H., Zhang, H. L., & Yang, B. Y., (1997). Effects of ginkgo leave concentrated oral liquor in treating asthma. *Chin. J. Integr. Med., 17*(4), 216–218.

Lin, W., Liu, A., Weng, C., Li, H., Sun, S., Song, A., & Zhu, H., (2018). Cloning and characterization of a novel phenylalanine ammonia-lyase gene from Inonotus baumii. *Enzyme Microb. Technol.*, 112, 52–58.

Lung Cancer, (2018). American Cancer Society. https://www.cancer.org/cancer/lung-cancer. html (accessed on 20 May 2020).

Ma, K., Han, J., Bao, L., Wei, T., & Liu, H., (2014). Two sarcoviolins with antioxidative and α-glucosidase inhibitory activity from the edible mushroom *Sarcodon leucopus* collected in tibet. *J. Nat. Prod., 77*(4), 942–947.

Mahmood, N., Pizza, C., Aquino, R., De Tommasi, N., Piacente, S., Colman, S., Burke, A., & Hay, A. J., (1993). Inhibition of HIV infection by flavonoids. *Antiviral Res., 22*(2–3), 189–199.

Mai, A., Valente, S., Rotili, D., Massa, S., Botta, G., Brosch, G., Miceli, M., et al., (2007). Novel pyrrole-containing histone deacetylase inhibitors endowed with cytodifferentiation activity. *Int. J. Biochem. Cell. Biol., 39*(7–8), 1510–1522.

Malaria, (2014). World Health Organization Thailand. http://www.searo.who.int/thailand/ factsheets/fs0007/en/ (accessed on 20 May 2020).

Malaria, (2018). *Mayo Clinic.* https://www.mayoclinic.org/diseases-conditions/malaria/ symptoms-causes/syc-20351184# (accessed on 20 May 2020).

Malaria, (2018). National Institute of Mental Health. http://www.who.int/news-room/fact-sheets/detail/malaria (accessed on 20 May 2020).

Malaria, (2018). Schools and Health. http://www.schoolsandhealth.org/Pages/Malaria.aspx (accessed on 20 May 2020).

Malaria, (2018). World Health Organization. http://www.who.int/malaria/en/ (accessed on 20 May 2020).

Mali, R. G., & Dhake, A. S., (2011). Evaluation of effects of *Bauhinia variegata* stem bark extracts against milk-induced eosinophilia in mice. *J. Adv. Pharm. Technol. Res., 2*(2), 132–134.

Marijuana and Cancer, (2017). American Cancer Society. https://www.cancer.org/treatment/ treatments-and-side-effects/complementary-and-alternative-medicine/marijuana-and-cancer.html (accessed on 20 May 2020).

Medical Marijuana and Cancer, (2018). Lungcancer.org. https://www.lungcancer.org/find_ information/publications/328-medical_marijuana_and_cancer (accessed on 20 May 2020).

Medicines by Design, (2018). National Institute of General Medical Sciences. https://publications.nigms.nih.gov/medbydesign/chapter1.html (accessed on 20 May 2020).

Micheli, F., Di Fabio, R., Bordi, F., Cavallini, P., Cavanni, P., Donati, D., Faedo, S., et al., (2003). Tarzia G. 2,4-dicarboxy-pyrroles as selective non-competitive mGluR1 antagonists: Further characterization of 3,5-dimethyl pyrrole-2,4-dicarboxylic acid 2-propyl ester 4-(1,2,2-trimethyl-propyl) ester and structure-activity relationships. *Bioorg. Med. Chem. Lett., 13,* 2113–2118.

Mims, J. W., (2015). Asthma: Definitions and pathophysiology. *Int Forum Allergy Rhinol., 1,* 2–6.

Mohamed, M. S., Ali, S. A., Abdelaziz, D. H. A., & Fathallah, S. S., (2014). Synthesis and evaluation of novel pyrroles and pyrrolopyrimidines as anti-hyperglycemic agents. *Bio. Med. Res. Int.,* 249780.

Murkovic, M., (2003). Phenolic compounds. In: Caballero, B., Finglas, P., & Toldra, D., (eds.), *Encyclopedia of Food Sciences and Nutrition* (2nd edn., pp. 4507–4514). Academic Press: San Diego, CA.

National Cancer Institute, (2018). *Antineoplastic Antibiotic.* https://www.cancer.gov/publications/dictionaries/cancer-terms/def/antineoplastic-antibiotic (accessed on 20 May 2020).

Neurodegenerative Diseases, (2018). National Institute of Environmental Health Sciences. https://www.niehs.nih.gov/research/supported/health/neurodegenerative/index.cfm (accessed on 20 May 2020).

Nishida, H., Fujimori, I., Arikawa, Y., Hirase, K., Ono, K., Nakai, K., et al., (2017). Exploration of pyrrole derivatives to find an effective potassium-competitive acid blocker with moderately long-lasting suppression of gastric acid secretion. *Bioorg. Med. Chem., 25*(13), 3447–3460.

Non-Small Cell Lung Cancer, (2016). American Cancer Society. https://www.cancer.org/cancer/non-small-cell-lung-cancer.html (accessed on 20 May 2020).

Nyamweya, S., Hegedus, A., Jaye, A., Rowland, S. J., Flanagan, K. L., & Macallan, D. C., (2013). Comparing HIV-1 and HIV-2 infection: Lessons for viral immunopathogenesis. *Rev. Med. Virol., 23*(4), 221–240.

Orhan, I. E., (2018). Phytochemical and pharmacological activity profile of Crataegus oxycantha L. (Hawthorn): A cardiotonic herb. *Curr. Med. Chem., 25*(37), 4854–4865.

Orzechowska, B., Chaber, R., Wiśniewska, A., Pajtasz-Piasecka, E., Jatczak, B., Siemieniec, I., Gulanowski, B., et al., (2014). Baicalin from the extract of *Scutellaria baicalensis* affects the innate immunity and apoptosis in leukocytes of children with acute lymphocytic leukemia. *Int. Immunopharmacol., 23*(2), 558–567.

Paclitaxel, (2018). National Cancer Institute. https://www.cancer.gov/about-cancer/treatment/drugs/paclitaxel (accessed on 20 May 2020).

Panche, A. N., Diwan, A. D., & Chandra, S. R., (2016). Flavonoids: An overview. *J. Nutr. Sci., 5,* e47.

Parkinson's Disease, (2018). *MedlinePlus.* https://medlineplus.gov/parkinsonsdisease.html (accessed on 20 May 2020).

Parkinson's Foundation, (2018). *What is Parkinson's?* http://www.parkinson.org/understanding-parkinsons/what-is-parkinsons (accessed on 20 May 2020).

Partners Healthcare Asthma Center, (2010). *What is Meant by "Inflammation" in Asthma?* http://www.asthma.partners.org/NewFiles/Inflammation.html (accessed on 20 May 2020).

Patel, D., Rathore, K. S., Mahatma, O. P., & Patel, T. *Herbs Used in Asthma Treatment: An Overview.* Pharmatutor. N.d.

Psychotic Disorders, (2018). *MedlinePlus.* https://medlineplus.gov/psychoticdisorders.html (accessed on 20 May 2020).

Punchard, N. A., Whelan, C. J., & Adcock, I., (2004). The journal of inflammation. *J. Inflamm., 1*, 1.

Rafieian-Kopaei, M., Motamedi, P., Vakili, L., Dehghani, N., Kiani, F., Taheri, Z., Torkamaneh, S., Nasri, P., & Nasri, H., (2014). Green tea and type 2 diabetes mellitus. *J. Nephropharmacol., 3*(1), 21–23.

Raimondi, M. V., Cascioferro, S., Schillaci, D., & Petruso, S., (2006). Synthesis and antimicrobial activity of new bromine-rich pyrrole derivatives related to monodeoxypyoluteorin. *Eur. J. Med. Chem., 41*(12), 1439–1445.

Regina, G. L., Bai, R., Coluccia, A., Famiglini, V., Pelliccia, S., Passacantilli, S., Mazzoccoli, C., et al., (2014). New pyrrole derivatives with potent tubulin polymerization inhibiting activity as anticancer agents including hedgehog-dependent cancer. *J. Med. Chem., 57*, 6531–6552.

Satterlund, T. D., Lee, J. P., & Morre, R. S., (2015). Stigma among California's medical marijuana patients. *J. Psychoactive Drugs, 47*(1), 10–17.

Schneider, N. F. Z., Silva, I. T., Persich, L., De Carvalho, A., Rocha, S. C., Marostica, L., Ramos, A. C. P., et al., (2017). Cytotoxic effects of the cardenolide convallatoxin and its Na, K-ATPase regulation. *Mol. Cell Biochem., 428*(1–2), 23–39.

Schneider, N. F., Geller, F. C., Persich, L., Marostica, L. L., Pádua, R. M., Kreis, W., Braga, F. C., & Simões, C. M., (2016). Inhibition of cell proliferation, invasion, and migration by the cardenolides digitoxigenin monodigitoxoside and convallatoxin in human lung cancer cell line. *Nat. Prod. Res., 30*(11), 1327–1331.

Sechi, M., Mura, A., Sannia, L., Orecchioni, M., & Paglietti, G., (2004). Synthesis of pyrrol-[1,2-a] indole-1,8(5H)-diones as new synthones for developing novel tricyclic compounds of pharmaceutical interest. *Arkivoc., 5*, 97–106.

SelecTree: Tree Detail, (2018). *Purple Orchid Tree.* https://selectree.calpoly.edu/tree-detail/bauhinia-variegata (accessed on 20 May 2020).

Shankar, E., Goel, A., Gupta, K., & Gupita, S., (2017). Plant flavone apigenin: An emerging anticancer agent. *Curr. Pharmacol. Rep., 3*(6), 423–446.

Simalikalactone. (2018). ChEBI.

Singh, S., Sharma, B., Kanwar, S. S., & Kumar, A., (2016). Lead phytochemicals for anticancer drug development. *Front Plant Sci.*, 7, 1667.

Śledziński, P., Zyeland, J., Slomski, R., & Nowak, A., (2018). The current state and future perspectives of cannabinoids in cancer biology. *Cancer Med., 7*(3), 765–775.

Spector, W., & Willoughby, D., (1963). The inflammatory response. *Bacteriol. Rev., 27*, 117–154.

Stark, L. A., (2010). Beneficial microorganisms: Countering microbephobia. *CBE Life Sciences Education, 9*, 387–389.

Statistics About Diabetes, (2018). American Cancer Association. http://www.diabetes.org/diabetes-basics/statistics/?loc=db-slabnav (accessed on 20 May 2020).

Suh, K. N., Kain, K. C., & Keystone, J. S., (2004). Malaria. *CMAJ, 170*(11), 1693–1702.

Tachycardia: Fast Heart Rate, (2016). American Heart Association. http://www.heart.org/en/health-topics/arrhythmia/about-arrhythmia/tachycardia--fast-heart-rate (accessed on 20 May 2020).

Tang, Y., Xu, Y., Xiong, S., Ni, W., Chen, S., Gao, B., Ye, T., et al., (2007). The effect of Ginkgo Biloba extract on the expression of PKCalpha in the inflammatory cells and the level of IL-5 in induced sputum of asthmatic patients. *J. Huazhong. Univ. Sci. Technolog. Med. Sci., 27*(4), 375–380.

Taxotere: Drug Information, (2016). Breastcancer.org. https://www.breastcancer.org/treatment/druglist/taxotere (accessed on 20 May 2020).

Terpenes and the "Entourage Effect," (2018). *Project CBD*. https://www.projectcbd.org/science/terpenes/terpenes-and-entourage-effect (accessed on 20 May 2020).

The Calophyllum Story, (2018). Sarawak Forest Department. https://archive.org/stream/SarawakForestDepartmentCalophyllumStory/Sarawak%20Forest%20Department%20-%20Calophyllum%20story#page/n0 (accessed on 20 May 2020).

The Healing Power of Soy's Isoflavones, (2004). *Women's Health Information*. https://www.fwhc.org/health/soy.htm (accessed on 20 May 2020).

Title 21 Code of Federal Regulations, (2018). U.S. Department of Justice Drug Enforcement Administration, Diversion Control Division. https://www.deadiversion.usdoj.gov/21cfr/cfr/2108cfrt.htm (accessed on 20 May 2020).

Togola, A., Austarheim, I., Theis, A., Diallo, D., & Paulsen, B. S., (2008). Ethnopharmacological uses of *Erythrina senegalensis*: A comparison of three areas in Mali, and a link between traditional knowledge and modern biological science. *J. Ethnobiol. Ethnomed., 4*, 6.

Unwalla, R., Mousseau, J. J., Fadeyi, O. O., Choi, C., Parris, K., Hu, B., Kenney, T., et al., (2017). Structure-based approach to identify 5-[4-Hydroxyphenyl]pyrrole-2-carbonitrile derivatives as potent and tissue selective androgen receptor modulators. *J. Med. Chem., 60*(14), 6451–6457.

Velasco, G., Sánchez, C., & Guzmán, M., (2016). Anticancer mechanisms of cannabinoids. *Curr. Oncol., 23*, 23–32.

Wang, S., Wan, N. C., Harrison, J., Miller, W., Chuckowree, I., Sohal, S., Hancox, T. C., et al., (2004). Design and synthesis of new templates derived from pyrrolopyrimidine as selective multidrug-resistance-associated protein inhibitors in multidrug resistance. *J. of Med. Chem., 47*, 1339–1350.

Wan-Su, K., Woo, J. C., Sunwoo, L., Woo, J. K., Dong, C. L., Uy, D. S., Hyoung, S. S., & Wonyong, K., (2015). Anti-inflammatory, antioxidant, and antimicrobial effects of artemisinin extracts from *Artemisia annua* L. *Korean J. Physiol. Pharmacol., 19*(1), 21–27.

Wavhale, R. D., Martis, E. A., Ambre, P. K., Wan, B., Franzblau, S. G., Iyer, K. R., Raikuvar, K., et al., (2017). Discovery of new leads against mycobacterium tuberculosis using scaffold hopping and shape based similarity. *Bioorg. Med. Chem., 25*(17), 4835–4844.

Wilken, R., Veena, M., Wang, M., & Srivatsan, E., (2011). Curcumin: A review of anti-cancer properties and therapeutic activity in head and neck squamous cell carcinoma. *Mol. Cancer, 10*, 12.

Wolfram, S., (2007). Effects of green tea and EGCG on cardiovascular and metabolic health. *J. Am. Coll. Nutr., 26*(4), 373–388.

Wolfram, S., Raederstorff, D., Preller, M., Wang, Y., Teixeira, S. R., Riegger, C., & Weber, P., (2006). Epigallocatechin gallate supplementation alleviates diabetes in rodents. *J. Nutr., 136*(10), 2512–2518.

Xu, L., Wang, J., Lei, M., Fu, Y., Wang, Z., Ao, M., & Li, Z., (2016). Transcriptome analysis of storage roots and fibrous roots of the traditional medicinal herb *Callerya speciose* (Champ.) ScHot. *PLoS One., 11*(8), e0160338.

Yoon, V., Fridiks-Hareli, M., Munisamy, S., Lee, J., Anastasiades, D., & Stevceva, L., (2010). The GP120 molecule of HIV-1 and its interaction with T cells. *Curr. Med. Chem., 17*(8), 741–749.

Young-Onset Parkinson's, (2018). Parkinson's Foundation. http://www.parkinson.org/ Understanding-Parkinsons/What-is-Parkinsons/Young-Onset-Parkinsons (accessed on 20 May 2020).

Zervos, M., & Meunier, F., (1993). Fluconazole (Diflucan): A review. *Int. J. Antimicrob. Agents, 3*(3), 147–170.

Zhao, P., & Astruc, D., (2012). Docetaxel nanotechnology in anticancer therapy. *Chem. Med. Chem., 7*(6), 952–972.

Zhao, Q., Chen, X. Y., & Martin, C., (2016). *Scutellaria baicalensis*, the golden herb from the garden of Chinese medicinal plants. *Sci. Bull. (Beijing), 61*(18), 1391–1398.

Zheljazkov, V. D., Astatkie, T., & Schlegel, V., (2014). Hydrodistillation extraction time effect on essential oil yield, composition, and bioactivity of coriander oil. *J. Oleo. Sci., 63*(9), 857–865.

Zuo, W., Yan, F., Zhang, B., Li, J., & Mei, D., (2017). Advances in the Studies of ginkgo biloba leaves extract on aging-related diseases. *Aging Dis., 8*(6), 812–826.

CHAPTER 9

Heterocyclic Drug Design and Development

GARIMA VERMA, MOHAMMAD SHAQUIQUZZAMAN, and
MOHAMMAD MUMTAZ ALAM

Department of Pharmaceutical Chemistry, School of Pharmaceutical Education and Research, Jamia Hamdard, New Delhi – 110062, India, Tel.: +91-7004101839, E-mail: drmmalam@gmail.com (M. M. Alam)

ABSTRACT

Heterocyclic compounds often termed as heterocycles are of great interest in our day-to-day life. These compounds form an integral part of nucleic acids, carrying genetic information. They are the main structural backbone of the majority of naturally occurring vitamins, pigments, etc. Dependence of modern-day society on synthetic heterocycles is such that the heterocyclic compounds are now known to be an indispensable part of human life.

Plants are known to have a long and impressive history in the treatment of a wide array of ailments. Heterocyclic drugs procured from these sources exhibit a broad spectrum of bioactivities like antimalarial, antimicrobial, anticancer, anti-inflammatory, CNS Stimulant, antidiabetic, and many more. They have proved their efficacy in several diseased conditions and have emerged as the boon to mankind. A few major breakthroughs that can be seen in the field of anticancer research are Vinblastine, Vincristine, Camptothecin, Taxol. World of antimalarial drug development has witnessed certain major candidates like Quinine and Artemisinin. Considering their efficacy and safety profile, they are for more superior to the commercially available synthetic agents intended for the treatment of these diseases.

9.1 INTRODUCTION

General Overview About Limitation of Current Allopathic Treatments and Role of Natural Agents in Overcoming These Challenges

9.1.1 ALLOPATHIC MEDICINE

The term 'allopathic medicine' was coined by Samuel Hahnemann (Figure 9.1) in the year 1810. In this system of medicine, doctors, and other healthcare professionals use drugs, radiation, or surgery for treating symptoms associated with the diseases. This is also known by other names like biomedicine, conventional medicine, mainstream medicine, orthodox medicine, evidence-based medicine, and Western medicine. This has proven to be a boon in case of infections, emergency, and several life-threatening conditions.

FIGURE 9.1   Samuel Hahnemann.

Despite its significant benefits in the health system, it suffers from numerous downsides. The most important one is that this system targets symptom and not the root cause Treatment in such a manner provides only a short term relief by chemical effects and does not alleviate the underlying cause. At present, the allopathic system is the most widely adopted one that utilizes synthetic chemical composition as drugs. Intake of these drugs for a considerably prolonged duration predisposes patients to a number of side effects. Invasive procedures can also pose serious consequences on human health (Drug-Free-Living.com; Allopathy-Strengths and Limitations). A few medical procedures are even associated with very serious adverse effects and even deaths in some cases. Indiscriminate use of antibiotics in the current scenario poses deleterious effects on the life of patients. Antibiotics are known to provide nothing more than momentary respite. Additionally, their use is also linked to serious side reactions. At times, owing to hectic schedules, physicians also end up with an inaccurate diagnosis. Administration of medications in such cases endangers the life of human beings. Also, in many instances, the consumption of powerful medications for not-that-serious health issues exposes patients to numerous health risks in the long run (Collins, 2016).

Apart from the allopathic system of medicine, other systems are also being adopted for the treatment of different ailments and medical conditions. Homeopathy is one of the most adopted methods. Homeopathic medicine focusses on the complete cure of the disease, i.e., from the root of the disease. The most important thing associated with these medicines is that they have either little or no side effects. However, the major demerit is that it is a bit more time-consuming. Thereby, this system is not suited for emergency conditions. Other systems of medicine are also there which involve Ayurveda, Siddha, Unani, Yoga, and Naturopathy. All these systems have their own pros and cons (CivilServiceIndia.com; Allopathy vs Homeopathy).

Natural products obtained from plants, animals, or micro-organisms have a human history of probably more than a thousand years but still, they continued to be in use (Ji et al., 2009). Herbal medicines, i.e., naturally occurring plant-derived substances are widely used for the treatment of illness. Roots of herbal medicine are spread in every culture across the globe. As per a report by the World Health Organization, approximately 80% of the population rely on these medicines for primary health care. The global market for these products is approaching $60 billion. Herbal medicines use either plants' seeds, roots, berries, flowers, bark, leaves, or flowers for medicinal use. Herbal medicines are known to offer several merits in terms of healthcare. Natural health care products provide patients with cost-effective options for the treatment of diseases. On the other hand, modern medicines come with a very high price and thus are inaccessible for people at times. This drawback is very effectively covered by the botanical medicines and thus has become a very popular option. It is very easier for patients to obtain herbal products like herbal extracts, essential oils, and herbal teas from the market. The majority of them are available in health food and grocery stores. One doesn't need any prescription for purchasing them. The most important fact associated with natural herbs is that these medicines target the underlying cause of the disease rather than just symptoms. Such herbs are routinely used for the treatment of different chronic and acute conditions. Their efficacy has been proven in the case of many ailments across the orb. Additionally, minimal or no side effects have been reported with the use of natural remedies (Ruggeri, 2017).

9.2 PLANTS AS A SOURCE FOR LIVING SYSTEM

The presence of plants on the planet Earth makes it a green planet. They constitute an indispensable part of the living system. They are known to be the first living organisms born on the earth. The sustenance of human life without them is impossible. They provide living beings with food, clean air, and water. Apart from these basic things, they greatly contribute to the economy. A great proportion of people rely on plant products for earning their livelihood (Ranga, 2018).

The significance of plants to the humankind can be understood through the following points:

- **Food:** Directly or indirectly, humans rely on plants for their nutrition. Major human nutrition comes in the form of grains, fruits, vegetables, seeds, mushrooms, leaves, seeds, etc. obtained from plants.
- **Clean Air and Regulation of Water Cycle:** Oxygen, a by-product of photosynthesis is provided by plants for the basic human use. Carbon dioxide produced on burning of fossil fuels is taken up in large amounts by the plants, which provides clean air. Water cycle is also regulated by these plants by the process of transpiration. Hence, these plants considered of prime importance.
- **Clothing:** Plants are considered as the largest source of textile and fabric material like cotton, jute, etc. These materials are less expensive and eco-friendly. Thereby, these can be used for making cloths and beeding material, depending on the requirements.
- **Furniture and Shelter:** Wood from plants is used for making houses and other furniture items. Wood obtained from trees like teak, neem, red sandal, etc. are used for making the furniture items.
- **Flowers:** Flowers obtained from plants are used by humans for varied reasons like beauty, fragrance, etc.

- **Medicines:** A number of medicines with excellent therapeutic worth intended for human use are obtained from plants. A few examples are given below.
- Digoxin from *Digitalis purpurea* for the treatment of congestive heart failure;
- Vincristine from *Vincarosea* for cancer;
- Theophylline from *Thea sinensis* for asthma; and
- Reserpine from *Rauvolfia serpentina* for combating high blood pressure.

- **Vitamins:** Plants serve as a great source of vitamins for human beings as the body can't synthesize its own vitamins for metabolism.
- **Rubber and Plastic:** Plants also provide rubber and plastic needed for different requirements like wiring, seats, tire, etc.
- **Natural Pesticides:** The use of natural pesticides is considered better in comparison to artificial pesticides. Artificial ones may get into food and cause harm to human health. However, natural pesticides are comparatively safe as they degrade with time and don't harm the soil, e.g., Neem, Pyrethrin.
- **Biofuels:** A few plants are deliberately grown for the purpose of biofuel. They are less toxic as they don't emit harmful gases in the environment.
- **Economic Contribution:** Plants contribute greatly to the nation's economy. Certain countries rely on agriculture for the source of revenue. Food, cosmetics, drugs, clothes, essential oils, etc. obtained from plant yield money and contribute greatly to economic growth.

Apart from all these roles, plants remain an integral part of the ecosystem. They form the starting point of the food chains. They occupy the position of every food chain. They also prevent soil erosion and keep the fertile top layer of soil intact. They act as environmental savvy as they reduce heat and prevent moisture from getting dried up. They also support plants and animals by providing those shelters to live (Botanical-online. The Importance of Plants. 2019; Jennifer, 2014).

9.3 HETEROCYCLIC DRUGS FROM PLANTS

Depending on the carbon framework, organic compounds can be classified into the open chain and closed chain or cyclic compounds. Cyclic compounds refer to those having atoms bound to each other in the form of a ring. These cyclic compounds can be sectioned into homocyclic and heterocyclic compounds (Figure 9.2). Homocyclic compounds, also known as carbocyclic or isocyclic compounds are the ones in which ring comprises of one type of atoms, mainly carbon. In heterocyclic compounds, the ring is formed by at least two different types of atoms (including a carbon). All atoms apart from carbon that are present in a ring are known as heteroatoms. However, in the majority of compounds, a major portion of the ring is composed of carbon. Most commonly witnessed heteroatoms include nitrogen, sulfur, and oxygen (Farlex. Heterocyclic Compounds. The Free Dictionary).

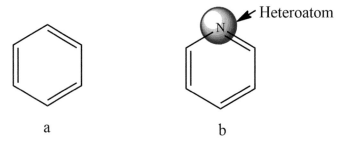

FIGURE 9.2   (a) Homocyclic ring and (b) Heterocyclic ring.

Heterocyclic compounds are of significant interest to humans. Such compounds are widely distributed in nature and are indispensable to life. They are known to play a vital role in metabolism of all living cells. They are integral parts of RNA and DNA. These compounds are also known to elicit a broad spectrum of applications. These are predominantly traceable in a number of pharmaceuticals, agrochemicals, and veterinary products. A number of natural products witnessing the presence of heterocyclic rings have reported to exhibit promising therapeutic effects. Some very common examples include antibiotics like penicillin, cephalosporin; alkaloids like vinblastine, morphine; antimalarial like cinchonine and many more drugs belonging to different categories. Different heterocyclic drugs obtained from plants have been mentioned in the subsequent section. These drugs have been classified in accordance with their therapeutic implications (Saini 2013; Maruthamuthu et al., 2016).

9.3.1   CARDIOTONIC AGENTS

Heart failure is an issue of serious health concern across the globe. The etiology behind heart failure is quite complex and multifactorial in nature.

Treatment of this problem aims at stimulating myocardial contractility by the administration of positive ionotropic agents, often termed as 'cardiotonic agents'

The word 'cardiotonic' can be understood by dividing it into two fragments 'cardio' and 'tonic.' Cardio refers to heart and tonic refers to have a tonic effect. So, the agents/compounds having a tonic effect on the heart are termed as 'cardiotonic agents.' Such agents are known to elicit a strengthening effect on the heart. They lead to an increased cardiac output.

On the whole, these agents exhibit the following effects:

- Increase the force of myocardial contraction (Positive inotropic effect)
- Increase cardiac output and renal perfusion;
- Increase urine output;
- Decrease blood volume;
- Slower down the heart rate;
- Decrease conduction velocity through the AV node; and
- Increase the movement of blood in the cells.

The list of certain commonly known cardiotonic agents includes deslanoside, digitoxin, digoxin, amrinone, inamrinone, cilostazol, milrinone, enoximone, etc. Those of natural origin have been mentioned in Table 9.1.

9.3.2   ANTI-INFLAMMATORY AGENTS

Whenever there is an infection or injury, the body of higher organisms responds in a protective manner. This defensive response of the body's immune system is known as inflammation. Such a response is mandatory for localization and the elimination of noxious agents. It is also important for the removal of damaged tissues so, that body can begin with the process of healing. The most common symptoms of inflammation include pain, redness, immobility, swelling, and heat. These symptoms are clearly noticeable on the skin. Commonly witnessed causes of inflammation include pathogens, external injuries, chemicals, or radiation. However, in some cases of chronic inflammation, symptoms are presented in a different manner like fatigue, abdominal pain, fever, rash, joint pain, chest pain, mouth sores, etc. (Nordqvist, 2019).

This process of inflammation is generally short-lived and subsides with the passage of time. However, non-steroidal anti-inflammatory drugs (NSAIDs) like aspirin control inflammation from further aggravation *via* inhibition of prostaglandin synthesis. Corticosteroids are also known to

TABLE 9.1   Natural Origin Drugs as Cardiotonic Agents

Acetyldigoxin (www.gobotany.com, *Digitalis lanata*; www.drugs.com. *Digitalis*)

Chemical Structure	

Biological Source	Leaves of *Digialis lanata*
Family	Scrophulariaceae
Other Chemical Constituents	Lanatosides A-E, Digitoxin, Digitoxose, Betaacetyldigoxin, Gluco-vatromonoside, Dioxigenin, Diginatigenin
Distribution	Found typically in Eastern Europe
Common Name(s)	Wooly Foxglove, Grecian Foxglove, Austrial Digitalis
Mechanism of Action	Slows atrioventricular conduction and slightly increases contraction power
Other Therapeutic Indications	Diuretic, Auricular fibrillation, Congestive heart failure

Convallatoxin (King's American Dispensatory, 1898. *Convallaria*; The Naturopathic Herbalist. *Convallaria majalis*)

Chemical Structure	

Biological Source	Leaves and flowers of *Convallaria majalis*
Family	Liliaceae
Other Chemical Constituents	Convallarin, Convallamarin, Majaloside, Convallosoid, Neoconvalloside, Convallotoxon, Corglycon, Glucoconnvalloside, Convallataxol
Distribution	Native of Europe

TABLE 9.1 *(Continued)*

Common Name(s)	Lilly of the Valley, May Bells, Our Lady's Tears, Mary's Tears
Mechanism of Action	Positive inotropic and negative chronotropic effects
Other Therapeutic Indications	Diuretic, Purgative, Emetic

Deslanoside (Manpreet, 2019; www.gobotany.com, *Digitalis lanata*; www.drugs.com. *Digitalis*)

Chemical Structure	
Biological Source	Leaves of *Digitalis lanata*
Family	Scrophulariaceae
Other Chemical Constituents	Lanatosides A-E, Digitoxin, Digitoxose, Betaacetyldigoxin, Gluco-vatromonoside, Dioxigenin, Diginatigenin
Distribution	Found predominantly in Eastern Europe
Common Name(s)	Wooly foxglove, Grecian foxglove
Mechanism of Action	Positive inotropic effects
Other Therapeutic Indications	Diuretic, Auricular fibrillation, Congestive heart failure

Digitoxin (www.abchomeopathy.com. *Digitalis purpurea*; www. alwaysayurveda.com. *Digitalis purpurea*)

Chemical Structure	
Biological Source	Dried leaves of *Digitalis purpurea*
Family	Scrophulariaceae
Other Chemical Constituents	Purpurea glycosides A & B, Glucogitaloxin, Digitalin, Digoxin, Ouabain, Digoxigenin, Gitoxin, Gitaloxin, Verodoxin
Distribution	Predominantly found throughout most of temperate Europe
Common Name(s)	Foxglove, Common foxglove, Purple foxglove, Lady's glove, Digitalis leaves

TABLE 9.1   *(Continued)*

Mechanism of Action	Inhibition of sodium-potassium ATPase
Other Therapeutic Indications	Congestive heart failure, Atrial flutter, Supraventricular tachycardia, Premature extra systoles

Ouabain (www.webmd.com. *Stropanthus*; www.tropical.theferns.info. *Stropanthus gratus*)

Chemical Structure	
Biological Source	Seeds of *Strophanthus gratus*
Family	Apocynaceae
Other Chemical Constituents	K-Strophanthin, Ouabagenin
Distribution	Native of Liberia, Burkina Faso, Ivory coast, Ghana, Nigeria, Cameroon, Equatorial Guinea, and Gabon
Common Name(s)	Ouabain tree
Mechanism of Action	Inhibits Na$^+$/K$^+$-ATPase ion pump
Other Therapeutic Indications	High blood pressure, Acute cardiac failure

Scillarin A (www.pacificbulbsociety.com. *Urginea*; Grieve, M. Squill. A Modern Herbal. www.botanical.com)

Chemical Structure	
Biological Source	Bulbs of *Urginea maritima*
Family	Asparagaceae
Other Chemical Constituents	Glucoscillaren A, Proscillaridin A, Scilliglaucoside, Scilliphaeoside, Scillaren A, Scilliphaeosidin

TABLE 9.1 *(Continued)*

Distribution	Native of Southern Europe, Western Asia, and Northern Africa
Common Name(s)	Squill, Sea squill, Sea onion, Maritime squill, Red squill
Mechanism of Action	Positive inotropic effects
Other Therapeutic Indications	Laxative, Expectorant, Jaundice, Convulsions, Asthma, Edema

control the inflammatory process. Anti-histaminic drugs are also used in certain cases. Anti-inflammatory drugs of plant origin are given in Table 9.2.

TABLE 9.2 Natural Origin Drugs as Anti-Inflammatory Agents

Aescin (Boericke, W. *Aesculus hippocastanum*. Homeoint.org)

Chemical Structure	

Biological Source	Leaves of *Aesculus hippocastanum*
Family	Sapindaceae
Other Chemical Constituents	Quercetin, Leucocyanidin, Leucodelphinidn, Procyanidin A2
Distribution	Native to the Pindus Mountains mixed forests and Balkan mixed forests of southeast Europe
Common Name	Horse Chestnut
Mechanism of Action	Induction of endothelial nitric oxide synthesis
Other Therapeutic Indications	Vasoconstriction and vasoprotection

9.3.3 ANTIDYSENTERY

An intestinal infection accompanied by severe diarrhea with blood is seen as a case of dysentery. In certain cases, mucus may be seen in the stool. Symptoms of dysentery include nausea, vomiting, abdominal cramps, high fever, and dehydration. Generally, it occurs due to poor hygiene. It may spread through contaminated food, contaminated water, improper handwashing, and physical contact. Treatment options for this issue include medications like bismuth subsalicylate, metronidazole, or tinidazole. Natural options for treating this problem are given in Table 9.3.

TABLE 9.3   Natural Origin Drugs for Dysentry

Neoandrographolide (Patel, 2018; The Herbal Resource. *Andrographis* – Health Benefits and Side Effects)

Chemical Structure	
Biological Source	Dried leaves and shoots of *Andrographis paniculata*
Family	Acanthaceae
Other Chemical Constituents	Andrographolide, Andrographine, Panicoline, Paniculide A-C
Distribution	Native of India and Sri Lanka
Common Name(s)	King of bitters, Kalmegh, Kirayat, Bhui-nimb
Mechanism of Action	–
Other Therapeutic Indications	Common cold, Jaundice, Gonorrhea, Malaria, Sinusitis, Bronchitis, Atherosclerosis, Diabetes

TABLE 9.3 *(Continued)*

Berberine (Boericke, W. *Beberis vulgaris.* Homeoint.org)

Chemical Structure	
Biological Source	Berries of *Berberis vulgaris*
Family	Berberidaceae
Other Chemical Constituents	Berberamine, Oxyacanthine, Isotetrandrine, Palmatine
Distribution	Native to Central and Southern Europe, Northwest Africa and Western Asia
Common Name(s)	Common Barberry, European Barberry
Mechanism of Action	–
Other Therapeutic Indications	Immune system stimulant, treatment of scurvy, treatment of coughs, treatment of eczema, Uterine stimulant, Improves digestion, improves blood flow

9.3.4 ANTHELMINTIC AGENTS

Helminths are the worm-like parasites which include flukes, roundworms, and tapeworms. These worms found in the intestine lead to a situation known as helminthiasis. Symptoms of helminthiasis include cough, wheeze, diarrhea, urticaria, etc. Drugs, which kill these worms are known as anthelmintics. Some commonly used anthelmintic agents include pyrantel, mebendazole, albendazole, ivermectin, and thiabendazole. A few anthelmintic agents are also obtained from plants. Such agents are mentioned in Table 9.4.

TABLE 9.4 Anthelmintic of Plant Origin

Arecoline (Useful Tropical Plants. *Areca catechu*; Palm Pediaa. *Areca catechu*)

Chemical Structure	

Biological Source	Dried ripe seeds of *Areca catechu*
Family	Piperaceae
Other Chemical Constituents	Arecatannin, Gallic acid, Terpineol, Arecaidine, Guvacine, Guvacoline
Distribution	Found predominantly in Tropical Pacific, Asia, and certain parts of East Africa
Common Name(s)	Areca palm, Areca nut palm, Betel palm, Indian nut, Pinang palm, Betel nut
Mechanism of Action	–
Other Therapeutic Indications	Sore throats, Arthritis, Tonic for exhaustion and debility, leucorrhoea, and vaginal laxity, sialogogue

Quisqualic Acid (Grant, *Quisqualis indica*; Weeds of Australia. *Quisqualis indica*)

Chemical Structure	

Biological Source	Root, seed, and fruit of *Quisqualis indica*
Family	Combretaceae
Other Chemical Constituents	Rutin, Trigonelline
Distribution	Found predominantly in Asia
Common Name(s)	Rangoon creeper, Drunken sailor, Chinese honeysuckle, Red Jasmine
Mechanism of Action	Agonist of AMPA receptor
Other Therapeutic Indications	Treatment of nephritis, rheumatism

9.3.5 CIRCULATORY DISORDERS

The circulatory system of the body includes the body's blood transport system (heart and blood vessels). The circulatory disorder is any state or condition that alters the circulatory system. Such conditions usually result in decreased blood flow and oxygen supply to different organs and tissues. Symptoms associated with these disorders generally include pain, cold intolerance, numbness, ulcers, which don't heal and color alterations at fingertips. Usual reasons behind these disorders are trauma, aneurysms, vascular malformations or Raynaud's disease. Losing weight, a balanced diet, regular exercise improves the condition of patients. Despite these lifestyle modifications, certain drugs of natural origin also relieve the patients. Drugs which can improve circulatory disorders are mentioned in Table 9.5.

TABLE 9.5 Plant Origin Drugs for Circulatory Disorders

Ajmalicine (Always Ayurveda, A Sister concern of Plant Ayurveda. *Rauwolfia serpentina*; Planet Ayurveda. Sarpgandha)

Chemical Structure	
Biological Source	Dried roots of *Rauvolfia serpentina*
Family	Apocynaceae
Other Chemical Constituents	Ajmaline, Deserpidine, Ajmalimine, Indobine, Indobinine, Reserpine, Reserpiline, Serpentine, Yohimbine, Rescinnamine
Distribution	Native to the Indian subcontinent and East Asia
Common Name(s)	Indian snakeroot, Devil pepper, Chhotachand, Serpentina root
Mechanism of Action	Lengthens the refractory period of heart by blockage of sodium channels
Other Therapeutic Indications	Antihypertensive, Insanity, Insomnia, Hysteria

9.3.6 VULNERARY

Vulnerary drugs refer to the agents which are used to heal a wound (Farlex. Vulnerary. The Free Dictionary). Drugs belonging to this category and those obtained from plants are mentioned in Table 9.6.

TABLE 9.6   Vulnerary Drugs of Plant Origin

Allantoin (Nelega, *Allantoin*. Naturalwellbeing.com)	
Chemical Structure	
Biological Source	Gel and latex of *Aloe barbadensis*
Family	Asphodelaceae
Other Chemical Constituents	Aloin, Barbaloin, Aloetic acid, Homoataloin, Choline, Choline salicylate, Coniferyl alcohol
Distribution	Native of South-West Arabian Peninsula
Common Name	Aloe Vera, Musabbar, Kumari
Mechanism of Action	-
Other Therapeutic Indications	Vaginal infections, Gynaecological disorders, Dietary supplement, Constipation, Genital herpes, Diabetes, Psoriasis, Weight loss

Asiaticoside (Murray, *Centella asiatica*; Patel, 2018. *Centella asiatica*)	
Chemical Structure	

TABLE 9.6 *(Continued)*

Biological Source	Leaves of *Centella asiatica*
Family	Apiaceae
Other Chemical Constituents	Asiatic acid, Madecassic acid, Madecassoside
Distribution	Found in many temperate and tropical swampy areas of the world.
Common Name	Centella, Asiatic pennywort, Gotu kola, Brahmi
Mechanism of Action	-
Other Therapeutic Indications	General tonic, Reduces arthritis, improves blood circulation, Hair tonic

9.3.7 MUSCLE RELAXANTS

Condition of muscle spasticity is associated with a number of clinical conditions like myositis, trauma, muscular, and ligamentous sprains and strains, tetanus, strychnine poisoning, neurological disorders and certain others as well (Fookes, 2018). In such cases, a category of drugs known as skeletal muscle relaxants is prescribed to the patients. Such drugs reduce tension in the muscles. They work either in the brain or spinal cord *via* blockage of excessively stimulated nerve pathways. Some commonly used muscle relaxants are baclofen, methocarbamol, tizanidine, and dantrolene. Those obtained from plants are given in Table 9.7.

TABLE 9.7 Plant Origin Drugs as Muscle Relaxant

Anabasine (Plants for a Future. *Anabasis aphylla*)	
Chemical Structure	
Biological Source	Aerial parts of *Anabasis aphylla*
Family	Amaranthaceae
Other Chemical Constituents	-
Distribution	Found majorly in Southern Europe, North Africa, and Asia

TABLE 9.7   *(Continued)*

Common Name	-
Mechanism of Action	Agonist of nicotinic acetylcholine receptor
Other Therapeutic Indications	Insecticide

Tubocurarine (Plants.usda.gov. Chondrodendron Ruiz & Pav. *Chondrodendron*)

Chemical Structure	

Biological Source	Leaves and roots of *Chondrodendron tomentosum*
Family	Menispermaceae
Other Chemical Constituents	Curine, Curarine, Chondrocurine, Isochnondroendrin, Tomentocurine
Distribution	Native of Central and South America
Common Name(s)	Curare vine, South American arrowroot poison
Mechanism of Action	Nicotinic acetylcholine receptor antagonist
Other Therapeutic Indications	Edema, Kidney stones, fever, dropsy

Cissampeline (Planet Ayurveda. Patha, Abuta)

Chemical Structure	

Biological Source	Roots and leaves of *Cissampelos pareira*
Family	Menispermaceae
Other Chemical Constituents	Arachidic acid, Berberine, Bulbocapnine, Cissamine, 4-methylcurine, Quercitol, Stearic acid, Tetrandrine
Distribution	Native of India

TABLE 9.7 *(Continued)*

Common Name(s)	Velvet Leaf
Mechanism of Action	-
Other Therapeutic Indications	Antispasmodic, Relieves menstrual cramps, Prevents miscarriage and uterine hemorrhages, Used in case of dysentery, piles, and urogenital disorders

Papaverine (Alwaysayurveda.com. *Papaver somniferum*; The Vaults of Erowid. Opium poppy)

Chemical Structure	
Biological Source	Latex of *Papaver somniferum*
Family	Papaveraceae
Other Chemical Constituents	Thebaine, Oripavine, Morphine, Codeine, Noscapine
Distribution	Native of Eastern Mediterranean area
Common Name(s)	Opium poppy, Breadseed poppy, Plant of joy, Mawseed
Mechanism of Action	Phosphodiesterase inhibitor
Other Therapeutic Indications	Insomnia, Dysentery, Anti-inflammatory, Febrifuge, Treatment of muscular pain

9.3.8 ANTICHOLINERGIC AGENTS

As the name suggests, anticholinergic agents are ones, which block the action of acetylcholine, a neurotransmitter. This chemical messenger is known to affect a number of body functions. Anticholinergic agents are used for treating a number of conditions like chronic obstructive pulmonary disorder, urinary incontinence, Parkinson's disease, overactive bladder, etc. (Cafasso, 2018). Some commonly used anticholinergic agents of natural origin are mentioned in Table 9.8.

TABLE 9.8   Plant Origin Drugs as Anticholinergic Agents

Anisodamine (Flora of China. *Anisodus tanguticus*; Revolvy.com, *Anisodus tanguticus*)

Chemical Structure	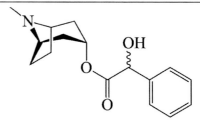

Biological Source	Roots of *Anisodus tanguticus*
Family	Solanaceae
Other Chemical Constituents	Anisodine, Hyoscyamine, Scopolamine
Distribution	Found in Qinghai-Tibetan Plateau
Common Name	-
Mechanism of Action	α_1 receptor antagonist
Other Therapeutic Indications	Epidemic meningitis, Rheumatoid arthritis, Pulmonary edema, Treatment of migraine, Noise-induced hearing loss

---

Atropine (Horticulture and soil science wiki. *Atropa belladona*)

Chemical Structure	

Biological Source	Leaves and other aerial parts of *Atropa belladonna*
Family	Solanaceae
Other Chemical Constituents	Atropine, Hyoscyamine, Scopolamine, Homatropine, Belladonine, Scopoletin
Distribution	Native to Europe, North Africa, and Western Asia
Common Name	Belladonna, Deadly nightshade
Mechanism of Action	Competitive, reversible antagonist of muscarinic acetylcholine receptor
Other Therapeutic Indications	Antispasmodic, Anodyne, Muscle relaxant, Anti-inflammatory, Mydriatic, Sedative, Antidepressant

TABLE 9.8 *(Continued)*

Anisodine (Horticulture and soil science wiki. *Atropa belladona*)

Chemical Structure	

Biological Source	Roots of *Anisodus tanguticus*
Family	Solanaceae
Other Chemical Constituents	Anisodamine, Hyoscyamine, Scopolamine
Distribution	Found mainly in Qinghai-Tibetan plateau
Common Name	-
Mechanism of Action	Muscarinic acetylcholine receptor antagonist, α_1-adrenergic receptor agonist
Other Therapeutic Indications	Glomerulonephritis, Rheumatoid arthritis, Hemorrhagic necrotic enteritis, Eclampsia, Pulmonary edema

Hyoscyamine (Truebotanica.com. *Hyoscyamus niger*; Medlineplus.gov. *Hyoscyamus niger*)

Chemical Structure	

Biological Source	Leaves, Seeds, and Roots of *Hyoscyamus niger*
Family	Solanaceae
Other Chemical Constituents	Scopolamine, Atropine, Hyoscine
Distribution	Native of Europe
Common Name(s)	Henbane, Black henbane, Stinking nightshade
Mechanism of Action	Antagonist of muscarinic acetylcholine receptor
Other Therapeutic Indications	Calmative, Anodyne, Antispasmodic, Expectorant, Antiasthmatic

9.3.9 ANTITUSSIVE AGENTS

Antitussives are the agents known to suppress cough. They are also termed as cough suppressants. They act by inhibition of a coordination region for coughing located in the brain stem, thereby disrupting the cough arc. Commonly used antitussive agents include dextromethorphan, benzonatate, etc. Those obtained from plants are given in Table 9.9.

TABLE 9.9   Plant Origin Drugs as Antitussive

Bergenin (Plants.ifas.ufl.edu. *Ardisia japonica*)	
Chemical Structure	
Biological Source	Stem bark of *Ardisia japonica*
Family	Primulaceae
Other Chemical Constituents	-
Distribution	Found predominantly in Eastern Asia, Eastern China, Japan, and Korea
Common Name	Marlberry
Mechanism of Action	-
Other Therapeutic Indications	Anticancer, Carminative, Diuretic, Depurative
Glaucine (Go Botany. *Glaucium falvum*; Pavlis, 2015)	
Chemical Structure	
Biological Source	Leaves and flowers of *Glaucium flavum*

TABLE 9.9   *(Continued)*

Family	Papaveraceae
Other Chemical Constituents	Protopine, Bocconoline
Distribution	Native of Europe, North Africa, Southwest, and Central Asia
Common Name(s)	Horned Poppy, Yellow Hornpoppy
Mechanism of Action	PDE4 Inhibitor
Other Therapeutic Indications	Anti-inflammatory effects, Calcium channel blocker, Bronchodilator

Codeine (The Vaults of Erowid. Opium poppy; Alwaysayurveda.com. *Papaver somniferum*)

Chemical Structure	
Biological Source	Air-dried milky exudation from excised unripe fruits of *Papaver somniferum*
Family	Papaveraceae
Other Chemical Constituents	Thebaine, Oripavine, Hydrocodone, Oxycodone, Noscapine, Oripavine, Narceine
Distribution	Native of Eastern Mediterranean region
Common Name	Opium poppy, Breadseed poppy, Plant of joy, Mawseed
Mechanism of Action	-
Other Therapeutic Indications	Sedative, Euphoriant, Relieve irritation and inflammation of stomach, Febrifuge

Noscapine (The Vaults of Erowid. Opium poppy; Alwaysayurveda.com. *Papaver somniferum*)

Chemical Structure	

TABLE 9.9   *(Continued)*

Biological Source	Air-dried milky exudation from excised unripe fruits of *Papaver somniferum*
Family	Papaveraceae
Other Chemical Constituents	Thebaine, Oripavine, Hydrocodone, Oxycodone, Codeine, Oripavine, Narceine
Distribution	Native of Eastern Mediterranean area
Common Name(s)	Opium poppy, Breadseed poppy, Plant of joy, Mawseed
Mechanism of Action	σ receptor agonistic effects
Other Therapeutic Indications	Sedative, Euphoriant, Relieve irritation and inflammation of stomach, Febrifuge

9.3.10   ANTICANCER

Cancer is referred to as the uncontrolled growth of abnormal cells in the body. The most common causes of cancer include chemical exposure, certain pathogens, ionizing radiation, and human genetics. Signs and symptoms associated with cancer include weight loss, fever, lumps, skin changes, alterations in bladder function, etc. (Patrick, C. Cancer. Medicinenet.com; Mayoclinic.org. Cancer). Treatment protocols depend on the type and stage of cancer. Surgery, chemotherapy, and radiation therapy are the viable options available for treatment. Chemotherapy is the most widely adopted option by the practitioners. Natural agents available for the treatment of cancer are documented in Table 9.10.

TABLE 9.10   Plant Origin Drugs as Anticancer

Camptothecin (Lucas, 2012; Lisa, 2014)	
Chemical Structure	
Biological Source	Bark and stem of *Camptotheca acuminata*
Family	Nyssaceae

TABLE 9.10 *(Continued)*

Other Chemical Constituents	10-Hydroxycamptothecin, Venoterpine, Syringic acid, Quercetin, Flavonoids, and Tannins
Distribution	Native to Southern China and Tibet
Common Name(s)	Happy tree, Cancer tree, Tree of life
Mechanism of Action	Topoisomerase-I inhibition
Other Therapeutic Indications	Common colds, Psoriasis, Liver problems, Digestive problems

Etoposide (Missouribotanicalgarden.org. *Podophyllum peltatum*; Homeopathicremediesblog. com. *Podophyllum peltatum*)

Chemical Structure	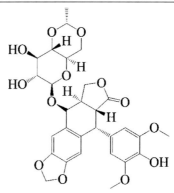
Biological Source	Root and rhizome of *Podophyllum peltatum*
Family	Berberidaceae
Other Chemical Constituents	Podophyllotoxin, Salicylic acid, Podophyllin, Quercetin, Astragallin
Distribution	Predominant occurrence in the Eastern United States and Southeastern Canada
Common Name(s)	Mayapple
Mechanism of Action	Topoisomerase II inhibitor
Other Therapeutic Indications	Emetic, Cathartic, Antihelmintic, Topical treatment of warts, Treatment of Crohn's disease

Irinotecan (Medicinalplantgenomics.msu.edu. *Camptotheca acuminata*; Sheffields.com. *Camptotheca acuminata*)

TABLE 9.10 *(Continued)*

Chemical Structure	

Biological Source	Stems and barks of *Camptotheca acuminata*
Family	Nyssaceae
Other Chemical Constituents	Camptothecin, Topotecan, Rubitecan, Trifolin, Hyperoside
Distribution	Native of Southern China and Tibet
Common Name(s)	Happy tree, Cancer tree, Tree of life
Mechanism of Action	Topoisomerase I inhibitor
Other Therapeutic Indications	Treatment of common colds, psoriasis, liver problems, and digestive problems

Podophyllotoxin (Missouribotanicalgarden.org. *Podophyllum peltatum*; Webmd.com. *Podophyllum*)

Chemical Structure	

Biological Source	Roots and rhizomes of *Podophyllum peltatum*
Family	Berberidaceae
Other Chemical Constituents	Podophyllin, Salicylic acid, Podophyllin, Quercetin, Astragallin
Distribution	Found mainly in the Eastern United States and Southeastern Canada
Common Name(s)	Mayapple, American mandrake, Wild mandrake, Ground lemon
Mechanism of Action	Antimitotic agent, Destabilizes microtubules and prevents cell division

TABLE 9.10 *(Continued)*

Other Therapeutic Indications	Purgative, Emetic, Cathartic, Anthelmintic, Used topically for treatment of warts

Taxol (Missouribotanicalgarden.org. *Podophyllum peltatum*; Webmd.com. *Podophyllum*)

Chemical Structure	
Biological Source	Dried leaves, bark, and roots of *Taxus brevifolia*
Family	Taxaceae
Other Chemical Constituents	Cephalomannine, 10-deacetyl baccatin, Baccatin
Distribution	Native to Pacific Northwest of North America
Common Name(s)	Pacific yew, Western yew, Mountain mahogany
Mechanism of Action	Tubulin targeting agent: Inhibits the formation of microtubules
Other Therapeutic Indications	-

Teniposide (Missouribotanicalgarden.org. *Podophyllum peltatum*; Webmd.com. *Podophyllum*)

Chemical Structure	
Biological Source	Roots and rhizomes of *Podophyllum peltatum*
Family	Berberidaceae
Other Chemical Constituents	Podophyllin, Salicylic acid, Podophyllin, Quercetin, Astragallin

TABLE 9.10   *(Continued)*

Distribution	Found predominantly in the Eastern United States and Southeastern Canada
Common Name(s)	Mayapple, American mandrake, Wild mandrake, Ground lemon
Mechanism of Action	Topoisomerase II inhibitor
Other Therapeutic Indications	Purgative, Emetic, Cathartic, Anthelmintic, Used topically for treatment of warts

Vinblastine (Muller, 2015; Sandoval, 2016)

Chemical Structure	
Biological Source	Dried whole plants of *Catharanthus roseus*
Family	Apocynaceae
Other Chemical Constituents	Vincristine, Rosinidin, Ajmalicine, Lochnerine, Serpentine, Tetrahydroalstonine
Distribution	Native of Madagascar
Common Name(s)	Madagascar periwinkle, Rose periwinkle, Rosy periwinkle
Mechanism of Action	Disruption of microtubules
Other Therapeutic Indications	-

Vincristine (Muller, 2015; Sandoval, 2016)

Chemical Structure	
Biological Source	Dried whole plants of *Catharanthus roseus*
Family	Apocynaceae

TABLE 9.10 *(Continued)*

Other Chemical Constituents	Vinblastine, Rosinidin, Ajmalicine, Lochnerine, Serpentine, Tetrahydroalstonine
Distribution	Native of Madagascar
Common Name(s)	Madagascar periwinkle, Rose periwinkle, Rosy periwinkle
Mechanism of Action	Disruption of microtubules
Other Therapeutic Indications	—

9.3.11 CNS STIMULANTS

The drug which acts in activation of the brain ultimately exciting body function, particularly act as a mood elevator. CNS Stimulants derived from plants are given in Table 9.11.

TABLE 9.11 Plant Origin Drugs as CNS Stimulants

Caffeine (Teaclass.com. *Camellia sinensis*; Herbpathy.com. *Camellia sinensis*)

Chemical Structure	
Biological Source	Leaves and leaf buds of *Camellia sinensis*
Family	Theaceae
Other Chemical Constituents	Epicatechin, Theobromine, Gallic acid, Catechin
Distribution	Native of East Asia, Indian subcontinent and Southeast Asia
Common Name(s)	Tea plant, Tea shrub, Tea tree
Mechanism of Action	Adenosine receptor antagonist
Other Therapeutic Indications	Prevention of cancer, Lowers cholesterol, Prevents/delays Parkinson's disease

Strychnine (Webmd.com. *Nux vomica*; Flowersofindia.net. *Nux vomica*)

TABLE 9.11 *(Continued)*

Chemical Structure	

Biological Source	Seeds of *Strychnosnux-vomica*
Family	Loganiaceae
Other Chemical Constituents	Brucine, Vomicine, Pseudostrychnine, α-colubrine
Distribution	Native of India and Southeast Asia
Common Name(s)	Strychnine tree, Nux vomica, Poison nut, Semen strychnos, Quaker buttons
Mechanism of Action	Antagonist of glycine and acetylcholine receptors
Other Therapeutic Indications	Treatment of different diseases of digestive tract, eye, lungs, heart, and circulatory system; Management of depression, migraine headache, symptoms of menopause, Raynaud's disease; Treatment of erectile dysfunction

9.3.12 HEMOSTATIC AGENTS

Hemostasis is known as a stoppage of bleeding from any blood vessel or body part. Agents which bring about hemostasis are known as hemostatic agents (Levy, 2009). Hemostasis is brought about by stimulation of fibrin formation or inhibition of fibrinolysis. Hemostatic agents derived from plants are given in Table 9.12.

TABLE 9.12 Plant Origin Drugs as Hemostatic Agents

(+)-Catechin (Tomczyk and Latte, 2009; Plants for a Future. *Potentilla fragarioides*)

Chemical Structure	

TABLE 9.12 *(Continued)*

Biological Source	Stems of *Potentilla fragarioides*
Family	Rosaceae
Other Chemical Constituents	–
Distribution	Native to China, Japan, Korea, Mongolia, and Russia
Common Name	Tannins, Triterpenes
Mechanism of Action	–
Other Therapeutic Indications	Astringent properties, Treating gynecological bleeding, Antiulcerogenic, Antimicrobial, Anti-inflammatory, Hepatoprotective, Antiviral

Hydrastine (Boericke, *Hydrastis canadensis*; Alwaysayurveda.com. *Hydrastis canadensis*)

Chemical Structure	
Biological Source	Roots and rhizomes of *Hydrastis canadensis*
Family	Ranunculaceae
Other Chemical Constituents	Berberine, Berberastine, Hydrastinine, Tetrahydroberberastine, Canadine, Canalidien
Distribution	Native of Southeastern Canada and the Eastern United States
Common Name(s)	Goldenseal, Orangeroot, Yellow pucoon
Mechanism of Action	
Other Therapeutic Indications	Anti-inflammatory, Anti-diarrheal, Antibacterial, Treatment of cancer, control muscle spasms, management of painful and heavy menstruation

9.3.13 *LOCAL ANESTHETIC AGENTS*

Local anesthetic agents are the ones which are used clinically for producing reversible loss of sensation in a circumscribed area of the body.

These agents are used in case of acute pain, chronic pain, and surgery. They bring about the reversible interruption of impulse conduction. A few local anesthetic agents naturally obtained from plants are discussed in Table 9.13.

TABLE 9.13 Plant Origin Drugs as Local Anesthetic Agents

Cocaine (Boericke, *Aesculus hippocastanum*; Natural Centre for Homeopathy. *Aesculus hippocastanum*)

Chemical Structure	
Biological Source	Leaves of *Erythroxylum coca*
Family	Erythroxylaceae
Other Chemical Constituents	Methyl ecgonine Cinnamate, Benzoylecgonine, Hydroxy tropacocaine, Tropacocaine, Ecgonine, Cuscohygrine, Dihydro cuscohygrine, Nicotine, Hygrine
Distribution	Native of Western South America
Common Name(s)	Coca Plant
Mechanism of Action	Affects serotonin (5-HT) receptors
Other Therapeutic Indications	Used for treating altitude sickness, Alleviates pain of headache, wounds, sores, rheumatism; Aphrodisiac

9.3.14 CHOLINESTERASE INHIBITORS

Acetylcholinesterase agents are those which prevent the breakdown of acetylcholine, a neurotransmitter. These inhibitors inhibit the enzyme cholinesterase which is actually responsible for the lysis of acetylcholine. They are used for the treatment of dementia encountered in the patients suffering from Alzheimer's disease (Ogbru, 2017). Some frequently used cholinesterase inhibitors are tacrine, donepezil, galantamine, rivastigmine, and memantine. Inhibitors obtained from plants are given in Table 9.14.

TABLE 9.14 Plant Origin Drugs as Local Cholinesterase Inhibitors

Galantamine (Missouribotanicalgarden.org. *Lycoris squamigera*; Badgett, 2018)

Chemical Structure

Biological Source	Bulb of *Lycoris squamigera*
Family	Amaryllidaceae
Other Chemical Constituents	Ducasterol, *o*-methyl corenine, Lycorenine, Lycorine, Lycoricidine
Distribution	Native of Japan
Common Name(s)	Magic Lily, Resurrection Lily, Naked lady
Mechanism of Action	Competitive and reversible cholinesterase inhibitor
Other Therapeutic Indications	Antibacterial, Antiviral

Physostigmine (Zhao et al., 2005)

Chemical Structure

Biological Source	Seeds of *Physostigma venenosum*
Family	Fabaceae
Other Chemical Constituents	Atropine, Eserine, Calabarine
Distribution	Native of Tropical Asia
Common Name(s)	Calabar bean, Ordeal bean
Mechanism of Action	Reversible cholinesterase inhibitor
Other Therapeutic Indications	Treatment of astigmatism, cholera, epilepsy, glaucoma, herpes, insomnia, leucorrhoea, paraplegia, hysteria

9.3.15 AMOEBICIDE

Any agent that leads to destruction of amoeba is called amoebicide. Such agents are employed in the treatment of amoebazoa infections. Metronidazole, tinidazole, secnidazole, and ornidazole are commonly employed agents. A few such agents are obtained from plant sources as well. They have been documented in Table 9.15.

TABLE 9.15   Plant Origin Drugs as Amoebicide

Glaucarubin (The Tropical Plant database. *Simaouba*; Revalgo, 2016)

Chemical Structure	
Biological Source	Bark of *Simarouba glauca*
Family	Simaroubaceae
Other Chemical Constituents	Ailanthinone, Glaucorubinone, Holacanthone
Distribution	Native of Florida, South America, and Lesser Antilles
Common Name(s)	Paradise tree, Dysentery bark, Bitterwood
Mechanism of Action	-
Other Therapeutic Indications	Good remedy for viruses, Treatment of intestinal worms and parasites, Astringent effects, Anticancer

9.3.16   ADDISON'S DISEASE

Addison's disease is a long term endocrine disorder characterized by insufficient production of steroid hormones. Symptoms of the disease include fatigue, sores in mouth, weight loss, darkened skin color, depression, fainting spells, decreased heart rate, etc. (Macon and Yu, 2016). Treatment includes administration of glucocorticoids. Drugs obtained from plants intended for the treatment of Addison's disease are given in Table 9.16.

9.3.17   CAPILLARY FRAGILITY

The condition of capillary fragility deals with bleeding beneath the skin. Increased intake of vitamin C is known to provide beneficial effects in such cases. A few natural sources are discussed in Table 9.17.

TABLE 9.16 Plant Origin Drugs for Treatment of Addison's Disease

Glycyrrhizin (Alwaysayurveda.com. *Glycyrrhiza glabra*; Himalayawellness.com. Licorice)

Chemical Structure	
Biological Source	Root of *Glycyrrhiza galbra*
Family	Fabaceae
Other Chemical Constituents	Glucose, Sucrose, Mannite, Starch, Asparagine
Distribution	Native of Southern Europe, some parts of Asia including India
Common Name(s)	Licorice
Mechanism of Action	-
Other Therapeutic Indications	Relieves heartburn, Prevents ulcer formation, Beneficial effects on liver, Relieves symptoms of premenstrual syndrome, Anti-allergy effects, Fights dermatitis, eczema, and psoriasis

TABLE 9.17 Plant Origin Drugs for Treatment of Capillary Fragility

Hesperidin (Eol.org. Naranjo Dulce; eFlora of India. *Citrus sinensis*)

Chemical Structure	
Biological Source	Citrus species, e.g., Oranges Fruit of *Citrus sinensis*
Family	Rutaceae

TABLE 9.17   *(Continued)*

Other Chemical Constituents	Rutin, Iso-hesperidin, Neohesperidin, Vitamin C, Pectin
Distribution	Found predominantly in subtropical areas
Common Name(s)	Orange
Mechanism of Action	-
Other Therapeutic Indications	Anti-inflammatory, Stomachic, Carminative

Rutin (Eol.org. Naranjo Dulce; eFlora of India. *Citrus sinensis*)

Chemical Structure	

Biological Source	Fruit of *Citrus sinensis*
Family	Rutaceae
Other Chemical Constituents	Hesperidin, Iso-hesperidin, Neohesperidin, Vitamin C, Pectin
Distribution	Found predominantly in subtropical areas
Common Name(s)	Orange
Mechanism of Action	-
Other Therapeutic Indications	Anti-inflammatory, Stomachic, Carminative

9.3.18   TRANQUILIZERS

Tranquilizers are the drugs used for treating anxiety, fear, tension, agitation, and certain other states of mental disturbance. They are of two types: major and minor ones. Minor tranquilizers are referred to as anxiolytics whereas minor ones are known as antipsychotics. A few tranquilizers are obtained from plants and are mentioned in Table 9.18.

TABLE 9.18 Plant Origin Drugs as Tranquilizers

Kawain (Entheology.com. *Piper methysticum*: kava; Glosbe.com. *Piper methysticum*)

Chemical Structure	

Biological Source	Roots of *Piper methysticum*
Family	Piperaceae
Other Chemical Constituents	Kavain, Dihydrokavain, Methysticin, Dihydromethysticin, Yangonin, Desmethoxyyangonin, Flavokavain A, B, and C
Distribution	Native of Pacific islands
Common Name(s)	Kava kava
Mechanism of Action	Potentiates activity of $GABA_A$ receptor
Other Therapeutic Indications	Treatment of liver issues, Diuretic, Relieves pain

9.3.19 BRONCHODILATORS

Bronchodilators are the drugs, which dilate bronchi and bronchioles, and decrease resistance in the airway, resulting in increased airflow to the lungs. Such medications are prescribed for the treatment of various breathing difficulties. Common bronchodilators are salbutamol, salmeterol, terbutaline, bambuterol, formoterol, clenbuterol, etc. (Drugs.com. Bronchodilators). Those obtained from natural origin are given in Table 9.19.

TABLE 9.19 Plant Origin Drugs as Bronchodilators

Khellin (Wong, 2017. The Benefits of *Ammi visnaga*; Sarahraven.com. *Ammi visnaga*)

Chemical Structure

TABLE 9.19   *(Continued)*

Biological Source	Fruits of *Ammi visnaga*
Family	Apiaceae
Other Chemical Constituents	Visnagin, Visnadine
Distribution	Native of Egypt
Common Name(s)	Visnaga, Khelle, Picktooth, Bisnaga, Bishop's weed
Mechanism of Action	-
Other Therapeutic Indications	Angina, Asthma, Atherosclerosis, Bronchitis, Diabetes, Heart disease, High blood pressure, Hypercholesterolemia, Premenstrual syndrome, Vitiligo,

Vasicine (Ayurtimes.com. *Adhatoda vasica*)

Chemical Structure	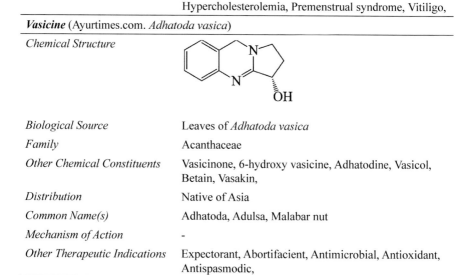

Biological Source	Leaves of *Adhatoda vasica*
Family	Acanthaceae
Other Chemical Constituents	Vasicinone, 6-hydroxy vasicine, Adhatodine, Vasicol, Betain, Vasakin,
Distribution	Native of Asia
Common Name(s)	Adhatoda, Adulsa, Malabar nut
Mechanism of Action	-
Other Therapeutic Indications	Expectorant, Abortifacient, Antimicrobial, Antioxidant, Antispasmodic,

9.3.20   SMOKING DETERRENTS

Smoking deterrents are the agents which prevent an individual from smoking. Bupropion is most commonly used for smoking cessation (Medicinenet. com. Bupropion sustained-release). Smoking deterrents obtained from plants are mentioned in Table 9.20.

TABLE 9.20   Plant Origin Drugs as Smoking Deterrents

α-Lobeline (Horne, 2010. *Lobelia inflata*)

Chemical Structure	

TABLE 9.20   *(Continued)*

Biological Source	Seeds of *Lobelia inflata*
Family	Campanulaceae
Other Chemical Constituents	Lobelanine, Lobelanidine
Distribution	Native of Eastern North America, Southeastern Canada, and the Eastern United States
Common Name(s)	Indian tobacco, Puke weed
Mechanism of Action	Agonist-antagonist at nicotinic acetylcholine receptors, μ-opioid receptor antagonist
Other Therapeutic Indications	Diaphoretic, Resuscitation after asphyxia

9.3.21   ANALGESICS

Pain is a distressing condition produced in response to some intense or damaging stimuli. In order to get relief from pain, a group of drugs known as analgesics is employed. They produce analgesia by acting on peripheral and central nervous systems. Nonsteroidal anti-inflammatory drugs (NSAIDs) are most frequently used. Certain analgesics procured from plants are mentioned in Table 9.21.

TABLE 9.21   Plant Origin Drugs as Analgesics

Morphine (Alwaysayurveda.com. *Papaver somniferum*; Pappas, 2017 Massive Poppy Bust)

Chemical Structure	
Biological Source	Latex of *Papaver somniferum*
Family	Papaveraceae
Other Chemical Constituents	Thebaine, Oripavine, Narcotine, Narceine, Papaverine, Codeine
Distribution	Native of Eastern Mediterranean region
Common Name(s)	Opium poppy, Bread seed poppy
Mechanism of Action	Opioid receptor antagonist
Other Therapeutic Indications	Relieves irritation and inflammation of stomach, Febrifuge, Tonic.

TABLE 9.21   *(Continued)*

Salicin (Woodlandtrust.org.uk. Willow white; Go Botany. *Salix alba*)

Chemical Structure

Biological Source	Bark of *Salix alba*
Family	Saliaceae
Other Chemical Constituents	Salicortin, Populin, Fragilin, Tremulacin, Saligenin
Distribution	Native of Europe, Western, and Central Asia
Common Name(s)	White willow
Mechanism of Action	-
Other Therapeutic Indications	Anticancer, Antioxidant, Anti-inflammatory

9.3.22   INSECTICIDES

Insecticides are the substances intended to kill, repel, harm, or mitigate one or more species of insects. Different insecticides exhibit insecticidal properties through different modes of action. A few insecticides mentioned in Table 9.22 are obtained from natural sources.

TABLE 9.22   Plant Origin Drugs as Insecticides

Nicotine (Datiles and Rodríguez, 2014)

Chemical Structure

Biological Source	Leaves of *Nicotiana tabacum*
Family	Solanaceae
Other Chemical Constituents	Nor-nicotine, Anabesine
Distribution	Native of tropical and subtropical America
Common Name(s)	Tobacco
Mechanism of Action	-
Other Therapeutic Indications	Seed oil is used as a substitute for groundnut oil, Treatment of different skin ailments

TABLE 9.22   *(Continued)*

Rotenone (Tropical.theferns.info. *Lonchocarpus nicou*)

Chemical Structure	

Biological Source	Roots of *Pachyrhizus erosus*
Family	Fabaceae
Other Chemical Constituents	Butein, Medicarpin, Cycloeucalenol, Maackiain, Pongaglabol
Distribution	Native of South America
Common Name(s)	Barbasco
Mechanism of Action	Interferes with electron transport chain in mitochondria
Other Therapeutic Indications	Pesticide, Non selective piscicide

9.3.23   ANTIPYRETIC

Reduction in body temperature i.e. treatment of fever is achieved with the use of certain substances known as antipyretics. In this case also, NSAIDs are widely used. Antipyretic agents procured from plants are given in Table 9.23.

TABLE 9.23   Plant Origin Drugs as Antipyretic

Palmatine (Natural medicines herbs.net. *Coptis japonica*; Chineseherbshealing.com. Coptis root)

Chemical Structure	

TABLE 9.23   *(Continued)*

Biological Source	Roots of *Coptis japonica*
Family	Ranunculaceae
Other Chemical Constituents	Berberine
Distribution	Native of Asia and North America
Common Name(s)	Goldthread, Canker root
Other Therapeutic Indications	Anti-inflammatory, Stomachic, Local analgesic and anesthetic, Treatment of intestinal catarrh, dysentery, enteritis, high fevers, conjunctivitis

9.3.24   SWEETENERS

Sugar substitutes which impart sweet taste like that of sugar are known as sweeteners. However, their food energy is less in comparison to sugar. Some sweeteners are also obtained from plant sources mentioned in Table 9.24.

TABLE 9.24   Plant Origin Drugs as Sweeteners

Phyllodulcin (Gardenia.net. *Hydrangea macrophylla*)	
Chemical Structure	
Biological Source	Leaves of *Hydrangea macrophylla*
Family	Hydrangeaceae
Other Chemical Constituents	Hydrangeol, Hydrangeic acid
Distribution	Native of Japan
Common Name(s)	Bigleaf hydrangea, French hydrangea, Iacecap hydrangea, Mophead hyudrangea, Penny mac, Hortensia
Mechanism of Action	-
Other Therapeutic Indications	Antiallergic, Antimicrobial, Antimalarial

Stevioside (Plantsguru.com. *Stevia rebaudiana*)	

TABLE 9.24 *(Continued)*

Chemical Structure	

Biological Source	Leaves of *Stevia rebaudiana*
Family	Asteraceae
Other Chemical Constituents	Rebaudiioside
Distribution	Native of Brazil and Paraguay
Common Name(s)	Candyleaf, Sweetleaf, Sugar leaf
Mechanism of Action	Interaction with protein channel, TRPM5
Other Therapeutic Indications	-

9.3.25 ANALEPTICS

Analeptics are the agents meant for stimulation of the central nervous system. These drugs are used for combating a number of diseases like depression, attention deficit hyperactivity disorder, and respiratory depression. Certain analeptics obtained from plants as well. These have been mentioned in Table 9.25.

TABLE 9.25 Plant Origin Drugs as Analeptics

Picrotoxin (Tropical.theferns.info. *Anamirta cocculus*)	
Chemical Structure	

Biological Source	Fruit of *Anamirta cocculus*

TABLE 9.25 *(Continued)*

Family	Menispermaceae
Other Chemical Constituents	Berberine, Palmatine, Magnoflorine, Menispermine, Paramenispermine
Distribution	Native of Southeast Asia
Common Name(s)	Indian berry, Fishberry, Levant nut, Poisonberry
Mechanism of Action	Antagonist of GABA
Other Therapeutic Indications	Antidote against barbiturate and morphine poisoning, Nervine tonic in schizophrenia and epilepsy, treatment of dyspepsia and menstrual problems

9.3.26   PARASYMPATHOMIMETICS

Parasympathomimetic drugs, also termed as cholinomimetic drugs are known to stimulate the parasympathetic nervous system. A few parasympathomimetic drugs obtained from plants are given in Table 9.26.

TABLE 9.26   Plant Origin Drugs as Parasympathomimetics

Pilocarpine (The Tropical Plant database. *Jaborandi*; Felter, 1922. Pilocarpus)

Chemical Structure	

Biological Source	Leaves and roots of *Pilocarpus jaborandi*
Family	Rutaceae
Other Chemical Constituents	Isopilocarpine. Pilocarpidine
Distribution	Native of Neotropics of South America
Common Name(s)	Jaborandi, Indian hemp
Mechanism of Action	Acts on muscarinic receptor (M_3)
Other Therapeutic Indications	Diaphoretic, Sialagogue, Myotic, Sedative, Antispasmodic, Diuretic

9.3.27 ANTIMALARIAL

Malaria is a mosquito-borne disease affecting humans caused by *Plasmodium*. Symptoms of this disease include vomiting, fever, tiredness, and headaches. Antimalarial drugs are used for curing malaria. Those procured from the natural origin are documented in Table 9.27.

TABLE 9.27 Plant Origin Drugs as Antimalarial

Quinine (Tropical.theferns.info. *Cinchona calisaya*)	
Chemical Structure	
Biological Source	Bark of *Cinchona calisaya*
Family	Rubiaceae
Other Chemical Constituents	Quinidine, Cinchonine, Cinchonidine, Quinicine, Hyrocinchonidine, Homocinchonidine
Distribution	Native of South America
Common Name(s)	Peruvian bark
Mechanism of Action	Interferes with malaria parasite's ability to dissolve and metabolize the hemoglobin
Other Therapeutic Indications	Treatment of arrhythmias, increases appetite, treatment of blood vessel disorders like hemorrhoids, varicose veins and leg cramps, cancer, enlarged spleen, muscle cramps
Artemisinin (Ayurtimes.com. *Artemisia annua*; Mskcc.org. *Artemisia annua*)	
Chemical Structure	
Biological Source	Aerial parts and leaves of *Artemisia annua*

TABLE 9.27   *(Continued)*

Family	Asteraceae
Other Chemical Constituents	Artemetin, Friedelin, Artemisinic acid, Deoxyartemisinin, Stigmasterol, Arteannuin B
Distribution	Native of temperate Asia
Common Name(s)	Sweet wormwood, Sweet annie, Sweet sagewort, Annual mugwort, Annual wormwood
Mechanism of Action	Creation of oxidative stress in the cells of the parasite
Other Therapeutic Indications	Antipyretic, Anti-inflammatory

9.3.28   ANTIHYPERTENSIVES

Hypertension, well known as high blood pressure is a matter of serious concern across the globe. Present lifestyle is one of the biggest reasons behind this. Its treatment is achieved by a class of drugs known as antihypertensive drugs. Antihypertensive agents of plant origin are mentioned in Table 9.28.

TABLE 9.28   Plant Origin Drugs as Antihypertensives

Rescinnamine (Muller, 2015. *Rauwolfia serpentina*)

Chemical Structure	
Biological Source	Dried roots and bark of *Rauvolfia serpentine*
Family	Apocynaceae
Other Chemical Constituents	Ajmaline, Ajmalicine, Deserpidine, Serpentine, Yohimbine, Reserpine
Distribution	Native of Indian subcontinent and East Asia
Common Name(s)	Indian snakeroot, Devil pepper
Mechanism of Action	Angiotensin converting enzyme (ACE) inhibitor
Other Therapeutic Indications	Treatment of insomnia, cataract, schizophrenia, anxiety, psychosis, hysteria, urticaria, stimulates uterine contraction, Tranquilizer

TABLE 9.28 *(Continued)*

Reserpine (Muller, 2015. *Rauwolfia serpentina*)

Chemical Structure	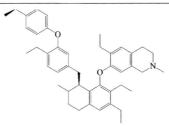

Biological Source	Dried roots and bark of *Rauvolfia serpentina*
Family	Apocynaceae
Other Chemical Constituents	Ajmaline, Ajmalicine, Deserpidine, Serpentine, Yohimbine, Rescinnamine
Distribution	Native of the Indian subcontinent and East Asia
Common Name(s)	Indian snakeroot, Devil pepper
Mechanism of Action	Irreversible blocker of vesicular monoamine transporter (VMAT)
Other Therapeutic Indications	Treatment of insomnia, cataract, schizophrenia, anxiety, psychosis, hysteria, urticaria, stimulates uterine contraction, Tranquilizer

Tetrandrine (Herbpathy.com. *Stephania tetrandra*; Chineseherbshealing.com. *Stephania tetrandra*)

Chemical Structure	

Biological Source	Roots of *Stephania tetrandra*
Family	Menispermaceae
Other Chemical Constituents	Fangchinoline, Cyclanoline, Menisine, Menisidine, Oxofangchirine, Stephenanthrine, Stepholidine
Distribution	Native of China and Taiwan
Common Name(s)	--
Mechanism of Action	Calcium channel blocker
Other Therapeutic Indications	Analgesic, Antiallergic, Antiarthritic, Antiasthmatic, Anticancer, Anti-inflammatory

9.3.29 SEDATIVES

Sedatives are compounds which induce sedation *via* reduction of irritability or excitement. They cause CNS depression. The majority of such agents affect γ-aminobutyric acid (GABA) and bring about the desired effects. These are generally used along with analgesics prior to surgery. Sedatives procured from plants are mentioned in Table 9.29.

TABLE 9.29 Plant Origin Drugs as Sedatives

Rotundine (Adooq.com. Rotundine; Tropical.theferns.info. *Stephania sinica*)

Chemical Structure	
Biological Source	Roots of *Stephania sinica*
Family	Menispermaceae
Other Chemical Constituents	Thicanine*N*-oxide (4-hydroxycorynoxidine), Thicanine 4-*O*-beta-D-glucoside
Distribution	Native of Eastern, Southern Asia, and Australasia
Common Name(s)	--
Mechanism of Action	Selective dopamine (D1) receptor antagonist
Other Therapeutic Indications	Neural anodyne, Treatment of asthma, dysentery, acute stomach trouble, and sore throat, Analgesic, Tranquilizer

Tetrahydropalmatine (Cham. & Schltdl. *Corydalis ambigua*; Sobejko, 2014. *Corydalis ambigua*)

Chemical Structure	

TABLE 9.29 *(Continued)*

Biological Source	Roots of *Corydalis ambigua*
Family	Papaveraceae
Other Chemical Constituents	Corynoline, Acetylcorynoline, Protopine, Tetrahydrocoptisine, Allocryptopine
Distribution	Native of Northern Hemisphere and high mountains of Tropical Eastern
Common Name(s)	-
Mechanism of Action	Antagonist of dopamine receptors (D1, D2 and D3)
Other Therapeutic Indications	Astringent, Antiperiodic, Diuretic, Deobstruent, Tonic, Antispasmodic, Analgesic

Scopolamine (Gyanunlimited.com *Datura stramonium*; Meriam-webster.com. *Datura*)

Chemical Structure	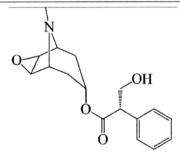
Biological Source	Seeds, flowers, and leaves of *Datura stramonium*
Family	Solanaceae
Other Chemical Constituents	Atropine, Hyoscyamine, Mallic acid
Distribution	Native of North America
Common Name(s)	Jimsonweed, Devil's snare, Thornapple, Moon flower
Mechanism of Action	Nonspecific antimuscarinic effects
Other Therapeutic Indications	Treatment of asthma, analgesic

9.3.30 DENTAL PLAQUE INHIBITORS

Dental plaque is a sticky, clear film which is deposited on teeth, between teeth, both above and below the gum line. Removal of this film is important as it may lead to numerous dental problems. Dental plaque inhibitors are available which prevent the formation of this plaque. Inhibitors obtained from plants are mentioned in Table 9.30.

TABLE 9.30   Plant Origin Drugs as Dental Plaque Inhibitors

Sanguinarine (Southard et al., 1984)	
Chemical Structure	
Biological Source	Rhizomes of *Sanguinaria canadensis*
Family	Papaveraceae
Other Chemical Constituents	Protopine, Sanguilutine, Chelirubine, Sanguirubine,
Distribution	Native of Eastern North America
Common Name(s)	Bloodroot, Canada puccoon, Bloodwort, Redroot, Tetterwort
Mechanism of Action	-
Other Therapeutic Indications	Respiratory aid, Antibacterial, anti-inflammatory

9.3.31   ASCARICIDES

Drugs intended for the treatment of roundworm infections are called ascaricides. Mebendazole, pyrantel, albendazole, thiabendazole, hexylresorcinol, santonin are well known ascaricides. Drugs of plant origin are given in Table 9.31.

TABLE 9.31   Plant Origin Drugs as Ascaricides

Santonin (Chopra and Chandler, 1924)	
Chemical Structure	

TABLE 9.31 *(Continued)*

Biological Source	Leaves of *Artemisia maritima*
Family	Asteraceae
Other Chemical Constituents	Artemisin, Cineole, Pinene, Resin, Santonic acid
Distribution	Native of France, the United Kingdom, Italy, Belgium, Germany, Denmark, Sweeden, Bulgaria, and Russia
Common Name(s)	Sea wormwood, Old woman
Mechanism of Action	–
Other Therapeutic Indications	Tonic for digestive system, Antiseptic, Antispasmodic, Carminative, Emmenogue, Cholagogue

9.3.32 APHRODISIAC AGENTS

Aphrodisiac agents, also known as love drugs are the ones which lead to an increased libido on consumption. Amphetamine, Methamphetamine, and phenylethylene derivatives are commonly used but are associated with a panel of side effects. Agents procured from plants are mentioned in Table 9.32.

TABLE 9.32 Plant Origin Drugs as Aphrodisiac Agents

Yohimbine (Webmd.com. *Yohimbe*; Medicinenet.com. *Yohimbe*)

Chemical Structure	
Biological Source	Bark extracts of *Pausinystalia johimbe*
Family	Rubiaceae
Other Chemical Constituents	Corynanthine, Raubasine
Distribution	Native of Western and Central Africa
Common Name(s)	Yohimbe
Mechanism of Action	α_1 and α_2 receptors antagonist
Other Therapeutic Indications	Erectile dysfunction

9.3.33   LAXATIVES

Laxatives are the drugs which relieve constipation by increasing stool motility, bulk, and frequency. Magnesium citrate, lactulose, docusate, polycarbophil, sodium phosphate, bisacodyl, etc. are commonly used. Laxatives obtained from natural origin are mentioned in Table 9.33.

TABLE 9.33   Plant Origin Drugs as Laxatives

Sennonsides A, B (Sharma, 2017 *Cassia angustifolia*)

Chemical Structure	
Biological Source	Leaves of *Cassia angustifolia*
Family	Fabaceae
Other Chemical Constituents	Kaempferol, Rhein, Aloe-emodin, Chrysophanic acid, Salicylic acid, Myricyl alcohol
Distribution	Found majorly in Egypt
Common Name(s)	Senna, Alexandrian Senna, Sennae folium
Mechanism of Action	Act as stimulant laxatives
Other Therapeutic Indications	Irritable bowel syndrome, Anal fissures, Hemorrhoids, Fungicide

9.3.34   ANTIHEPATOTOXIC AGENTS

Antihepatotoxic agents are the ones used for the cure or prevention of liver damage. They are also termed as hepatoprotective drugs. Hepatoprotective herbs are mentioned in Table 9.34.

TABLE 9.34 Plant Origin Drugs as Antihepatotoxic Agents

Silymarin (Popay, 2013; Sahelian, 2017)

Chemical Structure	
Biological Source	Seeds of *Silybum marianum*
Family	Asteraceae
Other Chemical Constituents	Silibnin, Silybin, Silydianin, Silychrisin
Distribution	Native of Southern Europe and Western Asia
Common Name(s)	Cardus marianus, Milk thistle, Blessed milkthistle, Marian thistle, Mary thistle, Saint Mary's thistle, Mediterranean milk thistle, Variegated thistle, Scotch thistle
Mechanism of Action	Promotes synthesis of liver cell protein and reduces oxidation of glutathione
Other Therapeutic Indications	Anticancer, Treatment of mushroom poisoning, Cerebral edema

9.3.35 OXYTOCIC DRUGS

Oxytocic drugs are employed for a number of fields in obstetrics like induction of labor, treatment of uterine atony, management of labor, and many more. A number of synthetic oxytocic agents are available in the market. However, a few are obtained from plants as well. Certain ones obtained from plants are mentioned in Table 9.35.

TABLE 9.35 Plant Origin Drugs as Oxytocic Agents

Sparteine (Sandoval, 2016; *Cytisus scoparius*)	

Chemical Structure	
Biological Source	*Cytisus scoparius*
Family	Fabaceae
Other Chemical Constituents	Scoparin, Spiraeoside, Genistin, Lupaine, Scoparoside
Distribution	Native of Western and Central Europe
Common Name(s)	Common broom, Scotch broom
Mechanism of Action	–
Other Therapeutic Indications	Diuretic, Cardiac stimulant, Treatment of gout

9.3.36   ANTIEMETICS

Antiemetic refers to an agent used for combating vomiting and nausea. Such drugs are used primarily for the treatment of motion sickness. Antiemetics of natural origin are given in Table 9.36.

TABLE 9.36 Plant Origin Drugs as Antiemetics

Tetrahydrocannabinol (Cannabissativa.com. *Cannabis sativa*; Alwaysayurveda.com. *Cannabis sativa*)	

Chemical Structure	
Biological Source	Flowers of *Cannabis sativa*
Family	Cannabinaceae
Other Chemical Constituents	Cannabidiol, Cannabinol, Cannabidiolic acid, Cannabigerol

TABLE 9.36 *(Continued)*

Distribution	Indigenous to Eastern Asia
Common Name(s)	Marijuana, Marihuana, Mary Jane
Mechanism of Action	Partial agonistic activity at cannabinoid (CB) receptor, particularly, CB_1 and CB_2
Other Therapeutic Indications	Management of cancer, Analgesic, Reduces anxiety,

9.3.37 DIURETICS

Any compound or agent that leads to an increased production of urine is termed as diuretic. Diuretics are of immense use in treating heart failure, influenza, water poisoning, kidney diseases, and many more. A few examples include acetazolamide, bumetanide, ethacrynic acid, furosemide, etc. Diuretics of natural origin are discussed in Table 9.37.

TABLE 9.37 Plant Origin Drugs as Diuretics

Theobromine (Audioenglish.org. *Theobroma cacao*)

Chemical Structure	

Biological Source	Seeds of *Theobroma cacao*
Family	Sterculiaceae
Other Chemical Constituents	Theophylline, Caffeine, Polyphenols
Distribution	Native of deep tropical areas of America
Common Name(s)	Cacao tree, Cocoa tree
Mechanism of Action	PDE Inhibitor
Other Therapeutic Indications	Vasodilator

Theophylline (Audioenglish.org. *Theobroma cacao*)

TABLE 9.37 *(Continued)*

Chemical Structure

Biological Source	Seeds of *Theobroma cacao*
Family	Malvaceae
Other Chemical Constituents	Theobromine, Caffeine, Polyphenols
Distribution	Native of deep tropical areas of America
Common Name(s)	Cacao tree, Cocoa tree
Mechanism of Action	Competitive nonselective phosphodiesterase (PDE) inhibitor, particularly PDE III and IV
Other Therapeutic Indications	Bronchodilator

9.4 CONCLUSION

Heterocyclic moieties are the main constituent of naturally occurring as well as synthetically derived drugs, dyes, and pigments. Various naturally occurring plant products have been covered under headings like cardiotonic, anti-inflammatory, muscle relaxants, anticancer, anticholinergic agents, antitussive, hemostatic agents, amoebicide, etc. The detailed description of plant products containing heterocycles is discussed here by exploring their source, structure, family, chemical constituent, and therapeutic applications.

KEYWORDS

- **hemostatic agents**
- **heterocycles**
- **mechanism of action**
- **natural drugs**
- **non-steroidal anti-inflammatory drugs**
- **γ-aminobutyric acid**

REFERENCES

Abchomeopathy.com. *Digitalis Purpurea*. https://abchomeopathy.com/r.php/Dig (accessed on 20 May 2020).

Abchomeopathy.com. *Hydrastis Canadensis*. https://abchomeopathy.com/r.php/Hydr (accessed on 20 May 2020).

Abchomeopathy.com. *Hyoscyamus Niger*. https://abchomeopathy.com/r.php/Hyos (accessed on 20 May 2020).

Abchomeopathy.com. *Podophyllum Peltatum*. https://abchomeopathy.com/r.php/Podo (accessed on 20 May 2020).

Adooq.com. Rotundine www.adooq.com/rotundine.html (accessed on 20 May 2020).

Always Ayurveda, a sister concern of plant Ayurveda. *Rauwolfia serpentina*. www.alwaysayurveda.com/rauwolfia-serpentina/ (accessed on 20 May 2020).

Alwaysayurveda.com. *Cannabis Sativa*. www.alwaysayurveda.com/cannabis-sativa/(accessed on 20 May 2020).

Alwaysayurveda.com. *Digitalis Purpurea*. www.alwaysayurveda.com/digitalis-purpurea/ (accessed on 20 May 2020).

Alwaysayurveda.com. *Glycyrrhiza Glabra*. http://www.alwaysayurveda.com/glycyrrhiza-glabra/ (accessed on 20 May 2020).

Alwaysayurveda.com. *Hydrastis Canadensis*. http://www.alwaysayurveda.com/hydrastis-canadensis/ (accessed on 20 May 2020).

Alwaysayurveda.com. *Papaver Somniferum*. www.alwaysayurveda.com/papaver-somniferum/ (accessed on 20 May 2020).

Alwaysayurveda.com. *Rauwolfia Serpentina*. www.alwaysayurveda.com/rauwolfia-serpentina/ (accessed on 20 May 2020).

Arora, P., Arora, V., Lamba, H. S., & Wadhwa, D., (2012). Importance of heterocyclic chemistry: A review. *Inter. J. Pharm. Sci. Res., 3*, 2947–2954.

Audioenglish.org. *Theobroma Cacao*. https://www.audioenglish.org/dictionary/theobroma_cacao.htm (accessed on 20 May 2020).

Ayurtimes.com. *Adhatoda Vasica*. https://www.ayurtimes.com/adhatoda-vasica-vasaka/ (accessed on 20 May 2020).

Ayurtimes.com. *Artemisia Annua*. https://www.ayurtimes.com/sweet-wormwood-artemisia-annua-sweet-annie/ (accessed on 20 May 2020).

Badgett, B., (2018). *Lycoris Care-How to grow the Lycoris Flower in the Garden*. Gardeningknowhow.com. https://www.gardeningknowhow.com/ornamental/bulbs/lycoris-lily/lycoris-lily.htm (accessed 20 May 2020)

Boericke, W. *Aesculus Hippocastanum*. Homeoint.org. www.homeoint.org/books/boericmm/a/aesc.htm (accessed on 20 May 2020).

Boericke, W. *Beberis Vulgaris*. Homeoint.org. http://homeoint.org/books/boericmm/b/berb.htm (accessed on 20 May 2020).

Boericke, W. *Hydrastis Canadensis*. Homeoint.org. http://www.homeoint.org/books/boericmm/h/hydr.htm (accessed on 20 May 2020).

Botanical-online, (2019). *The Importance of Plants*. https://www.botanical-online.com/theimportanceofplants.htm (accessed on 20 May 2020).

Cafasso, J., (2018). *Anticholinergic*. Healthline.com. https://www.healthline.com/health/anticholinergics#drug-list (accessed on 20 May 2020).

Cannabissativa.com. *Cannabis Sativa.* cannabissativa.com/cannabis/ (accessed on 20 May 2020).

Cham. & Schltdl. *Corydalis ambigua. Plants for a Future.* https://pfaf.org/User/Plant.aspx?LatinName=Corydalis+ambigua (accessed on 20 May 2020).

Chineseherbshealing.com. Coptis root www.chineseherbshealing.com/coptis-root/ (accessed on 20 May 2020).

Chineseherbshealing.com. *Stephania Tetrandra* www.chineseherbshealing.com/stephania-tetrandra-han-fang-ji/ (accessed on 20 May 2020).

Chopra, R. N., & Chandler, (1924). ASAC, Indian Santonin. *The Indian Medical Gazette,* 537–540

Civilserviceindia.com. *Allopathy vs. Homeopathy, Discuss.* https://www.civilserviceindia.com/subject/Essay/allopathy-vs-homeopathy3.html (accessed on 20 May 2020).

Collins, S., (2016). *Drawbacks of Allopathy-Facts You Cannot Ignore: 1800 Remedies.* https://www.1800remedies.com/drawbacks-of-allopathy/ (accessed on 20 May 2020).

Datiles, M. J., & Rodríguez, R. A., (2014). *Nicotiana tabacum.* Cabi.org. https://www.cabi.org/isc/datasheet/36326 (accessed on 20 May 2020).

Divineessence.com. *Ammi Visnaga* (Khella). https://divineessence.com/en/product/ammi-visnaga-khella/ (accessed on 20 May 2020).

Drug-Free-Living.com. *Allopathy-Strengths and Limitations.* http://www.drug-free-living.com/allopathy-limitations.html (accessed on 20 May 2020).

Drugs.com. *Bronchodilators.* https://www.drugs.com/drug-class/bronchodilators.html (accessed on 20 May 2020).

Drugs.com. *Digitalis.* https://www.drugs.com/npp/digitalis.html (accessed on 20 May 2020).

Drugs.com. *Lobelia.* https://www.drugs.com/npp/lobelia.html (accessed on 20 May 2020).

Drugs.com. *Pilocarpine.* https://www.drugs.com/pro/pilocarpine.html (accessed on 20 May 2020).

Drugs.com. *Rescinnamine.* https://www.drugs.com/international/rescinnamine.html (accessed on 20 May 2020).

Drugs.com. *Reserpine.* https://www.drugs.com/cdi/reserpine.html (accessed on 20 May 2020).

Drugs.com. *Willow Bark.* https://www.drugs.com/npc/willow-bark.html (accessed on 20 May 2020).

Dxnhealthramanis.com. *Simarouba Glauca for Cancer.* www.dxnhealthramanis.com/simarouba-glauca-for-cancer.asp (accessed on 20 May 2020).

eFlora of India. *Citrus Sinensis.* https://sites.google.com/site/efloraofindia/species/m---z/r/rutaceae/citrus/citrus-sinensis (accessed on 20 May 2020).

Entheology.com. *Piper Methysticum-kava Kava.* https://entheology.com/plants/piper-methysticum-kava-kava/(accessed on 20 May 2020).

Eol.org. Naranjo Dulce (*Citrus sinensis*). https://eol.org/pages/582206/overview/ (accessed on 20 May 2020).

Farlex. *Heterocyclic Compounds.* The Free Dictionary. https://encyclopedia2.thefreedictionary.com/Heterocyclic+Compounds (accessed on 20 May 2020).

Farlex. *Vulnerary.* The Free Dictionary. https://medical-dictionary.thefreedictionary.com/vulnerary (accessed on 20 May 2020).

Felter, H. W., (1922). *Pilocarpus.* Henriettes-herb.com. https://www.henriettes-herb.com/eclectic/felter/pilocarpus.html (accessed on 20 May 2020).

Flora of China. *Anisodus Tanguticus.* efloras.org.http://www.efloras.org/florataxon. aspx?flora_id=2&taxon_id=200020508 (accessed on 20 May 2020).

Flowersofindia.net. *Nux Vomica.* http://www.flowersofindia.net/catalog/slides/Nux%20 Vomica.html (accessed on 20 May 2020).

Fookes, C., (2018). *Skeletal Muscle Relaxants.* Drugs.com. https://www.drugs.com/drug-class/skeletal-muscle-relaxants.html (accessed on 20 May 2020).

Gardenia.net. *Hydrangea Macrophylla.* https://www.gardenia.net/plant-variety/hydrangea-macrophylla (accessed on 20 May 2020).

Glosbe.com. *Piper Methysticum.* https://glosbe.com/en/hi/piper%20methysticum (accessed on 20 May 2020).

Go Botany. *Digitalis lanata.* https://gobotany.newenglandwild.org/species/digitalis/lanata/ (accessed on 20 May 2020).

Go Botany. *Digitalis Purpurea.* https://gobotany.newenglandwild.org/species/digitalis/purpurea/(accessed on 20 May 2020).

Go Botany. *Glaucium Falvum.* https://gobotany.newenglandwild.org/species/glaucium/flavum/(accessed on 20 May 2020).

Go Botany. *Salix Alba.* https://gobotany.newenglandwild.org/species/salix/alba/ (accessed on 20 May 2020).

Grant, A. *Quisqualis Indica: Care-Information About Rangoon Creeper Vine.* Gardeningknowhow.com https://www.gardeningknowhow.com/ornamental/vines/rangoon-creeper/rangoon-creeper-vine.htm (accessed on 20 May 2020).

Grieve, M. *Squill: A Modern Herbal.* Botanical.com https://www.botanical.com/botanical/mgmh/s/squill86.html (accessed on 20 May 2020).

Gyanunlimited.com *Datura Stramonium (Common Name: Jimson Weed) Medicinal Uses, Side Effects and Benefits.* www.gyanunlimited.com/health/datura-stramonium-medicinal-uses-side-effects-and-benefits/9184/ (accessed on 20 May 2020).

Herbpathy.com. *Camellia Sinensis.* https://herbpathy.com/Uses-and-Benefits-of-Camellia-Sinensis-Cid2543(accessed on 20 May 2020).

Herbpathy.com. *Stephania Tetrandra.* https://herbpathy.com/Uses-and-Benefits-of-Stephania-Tetrandra-Cid1332 (accessed on 20 May 2020).

Himalayawellness.com. *Licorice, Licorice.* www.himalayawellness.com/herbfinder/glycyrrhiza-glabra.htm (accessed on 20 May 2020).

Homeopathicremediesblog.com.*PodophyllumPeltatum*https://www.homeopathicremediesblog.com/remedies/podophyllum-peltatum/ (accessed on 20 May 2020).

Horne, S., (2010). *Lobelia Inflata.* Treelite.com http://www.treelite.com/articles/articles/lobelia-(lobelia-inflata).html (accessed on 20 May 2020).

Horticulture and soil science wiki. *Atropa Belladone.* Horticultureandsoilscience, fandom. com. http://horticultureandsoilscience.wikia.com/wiki/Atropa_belladonna (accessed on 20 May 2020).

Jennifer, C., (2014). *The Importance of Plants to Life on Earth.* Blog.udemy.com. https://blog.udemy.com/importance-of-plants/ (accessed on 20 May 2020).

Ji, H. F., Li, X. J., Zhang, H. Y., & Zhang, (2009). Natural products and drug discovery. Can thousands of years of ancient medical knowledge lead us to new and powerful drug combinations in the fight against cancer and dementia? *EMBO Reports, 10*(3), 194–200.

King's American Dispensatory, (1898). *Convallaria.* Henriettes-herb.com. https://www.henriettes-herb.com/eclectic/kings/convallaria.html (accessed on 20 May 2020).

Levy, J. H., (2009). Hemostatic agents. *Transfusion, 44*, 58S–62S.

Lisa, M. S., (2014). *Camptotheca Aka Cancer Tree*. Medicinalplants101.blogspot.com. https://medicinalplants101.blogspot.com/2014/01/camptotheca-aka-cancer-tree.html (accessed on 20 May 2020).

Lucas, J. W., (2012). *Why is this the "Happy Tree?"* (*Camptotheca acuminata*) (Xi Shu). Davesgarden.com. https://davesgarden.com/guides/articles/view/206 (accessed on 20 May 2020).

Macon, L., & Yu, W., (2016). *Addison's Disease*. Healthline.com. https://www.healthline.com/health/addisons-disease#the-long-term-outlook (accessed on 20 May 2020).

Manpreet, K. *Digitalis Lanata: Sources, Macroscopical Character and Uses* (With Diagram). Yourarticlelibrary.com. www.yourarticlelibrary.com/biology/alkaloid/digitalis-lanata-sources-macroscopical-character-and-uses-with-diagram/49763 (accessed on 20 May 2020).

Maruthamuthu, S., Rajam, C. R., Stella, P., Dileepan, A. G., & Ranjith, R., (2016). The chemistry and biological significance of imidazole, benzimidazole, benzoxazole, tetrazole and quinazoline nucleus. *J. Chem. Pharm. Res., 8*, 505–526.

Mayoclinic.org. *Cancer*. https://www.mayoclinic.org/diseases-conditions/cancer/symptoms-causes/syc-20370588 (accessed on 20 May 2020).

Medicinalplantgenomics.msu.edu. *Camptotheca Acuminata: Overview Page*. https://medicinalplantgenomics.msu.edu/16922.shtml (accessed on 20 May 2020).

Medicinenet.com. *Bupropion Sustained-Release (Smoking Deterrent)-Oral*. https://www.medicinenet.com/bupropion_sustained_release-oral_smoking_deter/article.htm (accessed on 20 May 2020).

Medicinenet.com. *Camellia Sinensis*. https://www.medicinenet.com/green_tea_camellia_sinensis-oral/article.htm (accessed on 20 May 2020).

Medicinenet.com. *Valeriana Officinalis-Oral*. https://www.medicinenet.com/valerian_valeriana_officinalis-oral/article.htm (accessed on 20 May 2020).

Medicinenet.com. *Yohimbe (PAUSINYSTALIA YOHIMBE)-Oral*. https://www.medicinenet.com/yohimbe_pausinystalia_yohimbe-oral/article.htm (accessed on 20 May 2020).

Medlineplus.gov. *Hyoscyamus niger*. https://medlineplus.gov/druginfo/meds/a684010.html (accessed on 20 May 2020).

Meriam-webster.com. *Datura*. https://www.merriam-webster.com/dictionary/datura (accessed on 20 May 2020).

Missouribotanicalgarden.org. *Ardisia Japonica*. https://www.missouribotanicalgarden.org/PlantFinder/PlantFinderDetails.aspx?taxonid=282842 (accessed on 20 May 2020).

Missouribotanicalgarden.org. *Lycoris Squamigera*. www.missouribotanicalgarden.org/PlantFinder/PlantFinderDetails.aspx?kempercode=a464 (accessed on 20 May 2020).

Missouribotanicalgarden.org. *Podophyllum Peltatum*. www.missouribotanicalgarden.org/PlantFinder/PlantFinderDetails.aspx?kempercode=l800 (accessed on 20 May 2020).

Moroz, A. *Treatment of Pain and Inflammation*. Merckmanual.com. https://www.merckmanuals.com/home/fundamentals/rehabilitation/treatment-of-pain-and-inflammation (accessed on 20 May 2020).

Mskcc.org. *Artemisia Annua*. https://www.mskcc.org/cancer-care/integrative-medicine/herbs/artemisia-annua (accessed on 20 May 2020).

Muller, M., (2015). *Catharanthus Roseus (Vinca) Medicinal Plant Benefits and Images*. Homeremedies.com. https://www.homeremediess.com/medicinal-plant-catharanthus-roseus-vinca-benefits-and-images/(accessed on 20 May 2020).

Muller, M., (2015). *Rauwolfia Serpentina (Sarpagandha) Medicinal Herb Uses and Pictures*. Homeremedies.com. https://www.homeremediess.com/medicinal-herb-sarpagandha-uses/ (accessed on 20 May 2020).

Murray, M. T. *Centella Asiatica*. musculoskeletalkey.com. https://musculoskeletalkey.com/ centella-asiatica-gotu-kola/(accessed on 20 May 2020).

Natural medicines herbs.net. *Coptis Japonica*. www.naturalmedicinalherbs.net/herbs/c/ coptis-japonica.php (accessed on 20 May 2020).

Nelega, P. *Allantoin*. Naturalwellbeing.com. www.naturalwellbeing.com/learning-center/ Allantoin/ (accessed on 20 May 2020).

Nordqvist, C. *Everything you Need to Know About Inflammation*. Medicalnewstoday.com https://www.medicalnewstoday.com/articles/248423.php (accessed on 20 May 2020).

Ogbru, O., (2017). *Cholinesterase Inhibitors Side Effects, Uses, and Drug Interactions*. Medicinenet.com. https://www.medicinenet.com/cholinesterase_inhibitors/article.htm#what_ are_cholinesterase_inhibitors_cheis_how_do_they_work (accessed on 20 May 2020).

Pacific Island Ecosystems at Risk (PIER). *Quisqualis Indica*. www.hear.org/pier/species/ quisqualis_indica.htm (accessed on 20 May 2020).

Pacificbulbsociety.com. *Urginea*. https://www.pacificbulbsociety.org/pbswiki/index.php/ Urginea (accessed on 20 May 2020).

Palm Pediaa. *Areca Catechu*. www.palmpedia.net/wiki/Areca_catechu (accessed on 20 May 2020).

Pappas, S., (2017). *Massive Poppy Bust: Why Home-Grown Opium is Rare*. Livescience.com https://www.livescience.com/59452-why-opium-is-grown-outside-us.html (accessed on 20 May 2020).

Patel, K., (2018). *Andrographis Paniculata*. Examine.com. https://examine.com/supplements/ andrographis-paniculata/(accessed on 20 May 2020).

Patel, K., (2018). *Centella Asiatica*. Examine.com. https://examine.com/supplements/ centella-asiatica/ (accessed on 20 May 2020).

Patel, K., (2018). *Valerina Officinalis*. Examine.com. https://examine.com/supplements/ valeriana-officinalis/ (accessed on 20 May 2020).

Patrick, C. *Cancer*. Medicinenet.com. https://www.medicinenet.com/cancer/article. htm#what_is_the_prognosis_for_cancer (accessed on 20 May 2020).

Pavlis, R., (2015). *Glaucium Falvum*. Garden fundamentals.com. www.gardenfundamentals. com/glaucium-flavum/ (accessed on 20 May 2020).

Planet Ayurveda. *Patha, Abuta (Cissampleos Pareira)*. www.planetayurveda.com/library/ patha-cissampelos-pareira/(accessed on 20 May 2020).

Planet Ayurveda. *Sarpgandha (Rauwolfia Serpentine)*. http://www.planetayurveda.com/ library/sarpagandha-rauwolfia-serpentina (accessed on 20 May 2020).

Plants for a Future. *Anabasis Aphylla*. https://pfaf.org/User/Plant.aspx?LatinName=Anabasis +aphyllaen.hortipedia.com/wiki/Anabasis_aphylla (accessed on 20 May 2020).

Plants for a Future. *Potentilla Fragarioides*. https://pfaf.org/user/Plant.aspx?LatinName=Pot entilla+fragarioides (accessed on 20 May 2020).

Plants.ifas.ufl.edu. *Ardisia Japonica*. http://plants.ifas.ufl.edu/plant-directory/ardisia- japonica/ (accessed on 20 May 2020).

Plants.usda.gov. *Chondrodendron Ruiz & Pav. Chondrodendron* https://plants.usda.gov/core/ profile?symbol=CHOND2 (accessed on 20 May 2020).

Plantsguru.com. *Stevia Rebaudiana-Seeds*. www.plantsguru.com/stevia-rebaudiana-seeds (accessed on 20 May 2020).

Popay, I., (2013). *Silybum Marianum.* Cabi.org. https://www.cabi.org/isc/datasheet/50304 (accessed on 20 May 2020).

Ranga, N. R., (2018). Importance of plants in our life: Their role on the earth. *Study Read.* https://www.studyread.com/importance-plants-life-earth/ (accessed on 20 May 2020).

Revalgo, (2016). *Lakshmi Taru (Simarouba Glauca) Cancer Cure.* Revalgo.com. www.revalgo.com/blog/lakshmi-taru-simarouba-glauca-cancer-cure/ (accessed on 20 May 2020).

Revolvy.com. *Anisodus Tanguticus.* https://www.revolvy.com/page/Anisodus-tanguticus (accessed on 20 May 2020).

Ruggeri, C., (2017). *Herbal Medicine Benefits and the Top Medicinal Herbs More People are Using.* Draxe.com. https://draxe.com/herbal-medicine/ (accessed on 20 May 2020).

Sahelian, R., (2017). *Silymarin from Milk Thistle Research, Dosage, Benefit and Side Effects, Use for Liver Health.* Raysahelian.com. www.raysahelian.com/silymarin.html (accessed on 20 May 2020).

Saini, M. S., Kumar, A., Dwivedi, J., & Singh, R., (2013). A review: Biological significances of heterocyclic compounds. *Inter. J. Pharma Sci. Res., 4*, 66–77.

Sandoval R. J., (2016). *Catharanthus Roseus.* Cabi.org. https://www.cabi.org/isc/datasheet/16884 (accessed on 20 May 2020).

Sandoval R. J., (2016). *Cytisus Scoparius.* Cabi.org. https://www.cabi.org/isc/datasheet/17610 (accessed on 20 May 2020).

Sarahraven.com. *Ammi Visnaga.* https://www.sarahraven.com/flowers/seeds/annuals/ammi_visnaga.htm (accessed on 20 May 2020).

Sharma, A., (2017). *Cassia Angustifolia (Senna Plant) Benefits, Uses and Side Effects.* Medicinalplantsanduses.com. https://www.medicinalplantsanduses.com/cassia-angustifolia-benefits-uses (accessed on 20 May 2020).

Sheffields.com. *Camptotheca Acuminate.* https://sheffields.com/seeds/Camptotheca/acuminata/ (accessed on 20 May 2020).

Sobejko, C., (2014). *Corydalis Ambigua.* Herbalmateriamedica.wordpress.com. https://herbalmateriamedica.wordpress.com/2014/08/08/corydalis-ambigua/ (accessed on 20 May 2020).

Southard, G. L., Boulware, R. T., Walborn, D. R., Groznik, W. J., Thorne, E. E., & Yankell, S. L., (1984). Sanguinarine, a new antiplaque agent: Retention and plaque specificity, *J. Amer. Dent. Assoc., 108*(3), 338–341

Teaclass.com. *Camellia Sinensis.* https://www.teaclass.com/lesson_0111.html (accessed on 20 May 2020).

The Herbal Resource. *Andrographis-Health Benefits and Side Effects.* https://www.herbal-supplement-resource.com/andrographis-herb.html (accessed on 20 May 2020).

The Naturopathic Herbalist. *Convallaria Majalis.* https://thenaturopathicherbalist.com/herbs/c-2/convallaria-majalis/ (accessed on 20 May 2020).

The Tropical Plant Database. *Jaborandi.* Raintree.com. www.rain-tree.com/jaborand.htm#.W9as6vZuLOY (accessed on 20 May 2020).

The Tropical Plant Database. *Simarouba.* Raintree.com. www.rain-tree.com/simaruba.htm#.W9HMO_ZuI2x (accessed on 20 May 2020).

The Vaults of Erowid. *Opium Poppy.* Erowid.org https://erowid.org/plants/poppy/poppy.shtml (accessed on 20 May 2020).

Tomczyk, M., & Latte, K. P., (2009). Potentilla: A review of its phytochemical and pharmacological profile. *J. Ethnopharmacology, 122*(2), 184–204.

Tropical.theferns.info. *Anamirta Cocculus*. https://tropical.theferns.info/viewtropical. php?id=Anamirta+cocculus (accessed on 20 May 2020).

Tropical.theferns.info. *Cinchona Calisaya*. http://tropical.theferns.info/viewtropical. php?id=Cinchona+calisaya (accessed on 20 May 2020).

Tropical.theferns.info. *Lonchocarpus Nicou*. tropical.theferns.info/viewtropical.php?id= Lonchocarpus+nicou (accessed on 20 May 2020).

Tropical.theferns.info. *Stephania Sinica*. http://tropical.theferns.info/viewtropical.php?id= Stephania+sinica (accessed on 20 May 2020).

Tropical.theferns.info. *Stropanthus Gratus*. https://tropical.theferns.info/viewtropical. php?id=Strophanthus+gratus (accessed on 20 May 2020).

Truebotanica.com. *Hyoscyamus Niger*. www.truebotanica.com/hyoscyamus-niger/ (accessed on 20 May 2020).

Useful Tropical Plants. *Areca Catechu*. tropical.theferns.info/viewtropical.php?id=Areca+catechu (accessed on 20 May 2020).

Webmd.com. *Lobelia*. https://www.webmd.com/vitamins/ai/ingredientmono-231/lobelia (accessed on 20 May 2020).

Webmd.com. *Nux Vomica*. https://www.webmd.com/vitamins/ai/ingredientmono-58/ nux-vomica (accessed on 20 May 2020).

Webmd.com. *Piper Methysticum*. https://www.webmd.com/drugs/2/drug-17314/kava-piper-methysticum-oral/details (accessed on 20 May 2020).

Webmd.com. *Podophyllum*. https://www.webmd.com/vitamins/ai/ingredientmono-806/ podophyllum (accessed on 20 May 2020).

Webmd.com. *Stropanthus*. https://www.webmd.com/vitamins/ai/ingredientmono-223/ strophanthus (accessed on 20 May 2020).

Webmd.com. *Valerian*. https://www.webmd.com/vitamins/ai/ingredientmono-870/valerian (accessed on 20 May 2020).

Webmd.com. *Yohimbe*. https://www.webmd.com/vitamins/ai/ingredientmono-759/yohimbe (accessed on 20 May 2020).

Weeds of Australia. *Quisqualis Indica*. https://keyserver.lucidcentral.org/weeds/data/media/ Html/quisqualis_indica.htm (accessed on 20 May 2020).

Wong, C., (2017). *The Benefits of Ammi Visnaga*. Health benefits, uses, and more. Verywellhealth.com. https://www.verywellhealth.com/the-benefits-of-ammi-visnaga-89405 (accessed on 20 May 2020).

Woodlandtrust.org.uk. *Willow White (Salix Alba)*. https://www.woodlandtrust.org.uk/visiting-woods/trees-woods-and-wildlife/british-trees/native-trees/white-willow/ (accessed on 20 May 2020).

Zhao, B., Moochhala, S., & Tham, S. Y., (2005). Biologically active components of *Physostigma venenosum*. *J. Chrom. B, Analytical Technologies in the Biomedical, and Life Sciences, 812*, 183–192.

Part IV
Diverse Applications of Herbal Medicines

CHAPTER 10

Role of Natural Agents in the Management of Diabetes

MONIKA ELŻBIETA JACH[1] and ANNA SEREFKO[2]

[1]Department of Molecular Biology, The John Paul II Catholic University of Lublin, Konstantynów Street 1I, Lublin – 20–708, Poland, E-mail: monijach@kul.lublin.pl

[2]Department of Applied and Social Pharmacy, Laboratory of Pre-clinical Testing, Medical University of Lublin, Chodźki Street 4a, Lublin – 20–093, Poland

ABSTRACT

According to the World Health Organization (WHO), almost 422 million adults across the world suffer from diabetes mellitus. When not treated properly, diabetes generates various serious complications that increase the cost of treatment and impair a patient's quality of life. Apart from synthetic medicinal products, over 1200 traditional herbs have been used by diabetic patients. Efficacy and tolerability of certain natural substances have been confirmed in numerous preclinical experiments and clinical trials. This review focuses on the antidiabetic properties and the mechanism of action of the most common natural agents used in type 2 diabetes, including acacia gum, aloe vera, chromium-enriched *Saccharomyces cerevisiae* yeasts, cinnamon, *Ficus carica*, *Glycyrrhiza glabra*, *Momordica charantia*, *Punica granatum*, *Vitis vinifera*, and others. Natural substances as the safe, easily available, and quite cheap ones seem to be beneficial as a complementary therapy in the management of diabetes, especially in its type 2. Therefore, further studies on the identification of new anti-hyperglycemic herbal compounds are highly recommended.

10.1   INTRODUCTION

According to the Global Report on Diabetes from 2016 issued by the World Health Organization (WHO), almost 422 million adults across the world have diabetes mellitus, and 1.6 million deaths are directly attributed to diabetes each year (Global Report on Diabetes, 2016). Therefore, diabetes is one of the fastest increasing chronic metabolic diseases and one of the most common causes of death in the world. It is estimated that the number of diabetes patients will grow to 642 million worldwide by 2040. The greatest growth in the incidence and prevalence of diabetes is noted in the African population, which is attributed to changes in the eating habits related to fast urbanization and westernization (Mennen et al., 2001; Babiker et al., 2017). Therefore, it is so important to take steps to reverse the diabetes epidemic in the world.

The American Diabetes Association (ADA) classified diabetes mellitus into the following general categories: (1) Type 1 diabetes mellitus (T1DM) (due to autoimmune β-cell destruction, usually leading to absolute insulin deficiency); (2) Type 2 diabetes mellitus (T2DM) (due to an increasing loss of β-cell insulin secretion often associated with insulin resistance); (3) Gestational diabetes mellitus (GDM) (diabetes diagnosed in the second or third trimester of pregnancy but not clearly overt prior to gestation); (4) Specific types of diabetes-related to other causes such as monogenic diabetes syndromes (e.g., neonatal diabetes and maturity-onset diabetes of the young (MODY)), disorders of the exocrine pancreas (e.g., cystic fibrosis and pancreatitis), and drug- or chemical-induced diabetes (e.g., by the use of glucocorticoids in the treatment of HIV/AIDS, or after organ transplantation) (ADA, 2018).

Diabetes is usually caused by a combination both of genetically determined and environmental factors (Patel et al., 2012; Bobiş et al., 2018). As a metabolic disorder, diabetes occurs with multifactorial and heterogeneous etiologies. The direct and best-known "symptom" of diabetes is a high blood sugar level; however, other symptoms must also be taken into account: increased thirst and hunger, unexplained fatigue, enhanced urination, blurred vision, and unexpected weight loss or weight gain (Bobiş et al., 2018). T1DM is well-known as juvenile diabetes or insulin-dependent diabetes. It is a chronic syndrome where insulin is produced by the pancreas in a small amount or not at all. T1DM patients feel the symptoms of diabetes generally very quickly, i.e., usually within a few weeks (Bobiş et al., 2018; Rios et al., 2015). For type 1 diabetics, access to easy and affordable treatment, including insulin, is essential to life. However, type 2 is the most common form of diabetes in the world. In developed countries, even 90–95% of all cases of diabetes are type 2. The risk of T2DM increases after 30 years of age (Rios et al., 2015). This type of disease is also a chronic metabolic condition characterized by permanent hyperglycemia resulting in incorrectly high blood sugar levels above the normal value (ADA, 2018). T2DM usually occurs in adults when the organism becomes resistant to insulin or has insulin deficiency (Rios et al., 2015). T2DM develops for many years, and its symptoms are not noticeable. T2DM is most often associated with overweight or obesity (Bobiş et al., 2018). Long-term hyperglycemia leads to serious complications and emerging risk factors for several organs such as the eyes, blood vessels, nerves, and kidneys. It also increases the risk of development of heart diseases (Danesh et al., 2007).

In the case of T2DM, including individual cases, the most common condition is insulin resistance usually related to insulin deficiency. Insulin resistance is an impaired response of peripheral tissues to insulin causing a gradual transition in which changes in the pancreatic β-cell mass and function occur. At the beginning, these cases do not usually require using exogenous insulin and, sometimes weight loss; proper diet, exercise, and/or oral glucose-lowering treatment are only required to achieve adequate glycemic status (ADA, 2013). Therefore, such up to 30% of diabetic patients are interested in complementary strategies as alternatives to mainstream western medical treatment (Cefalu et al., 2011; Raman et al., 2012). Nevertheless, diabetics with severe β-cell destruction and, consequently, no residual secretion of insulin, need insulin treatment (Rios et al., 2015).

T2DM patients that do not require insulin can choose to supplement their therapeutic regimen with dietary supplementation in many pharmaceutical forms, such as vitamin and/or mineral mixtures, but the most popular agents are natural products originating from herbal or botanical sources (Cefalu et al., 2011). Medicinal plants and their products as well as naturally obtained substances have long been used as an important therapeutic support for alleviating human ailments (Joseph and Raj, 2010b; Rios et al., 2015; Sign et al., 2012). The use of herbs for diabetes treatment is especially in the focus of scientific interest. Since ancient times, herbs, and herb extracts have been used to combat T2DM. Many traditional natural medicinal substances on the world market are obtained from medicinal plants, minerals, microorganisms, or organic matter. The WHO has made a list of 21,000 plants that are used for pharmacological aims around the world. In particular, 150 species are used commonly on a particularly large scale (Cefalu et al., 2011; Rios et al., 2015; Zohary et al., 2000). Noteworthy, it has been reported that over 1200 traditional herbs have been used for diabetes treatment with real or perceptible medicinal support benefit (Jung et al., 2006). Moreover, an additional option to reduce the risk of serious T2DM complications is to improve dietary habits to enhance the intake of glucose-lowering nutritional products, e.g., dietary fibers (DF) (Danesh et al., 2007).

10.2 PROPOSED MECHANISMS OF THE ACTION OF NATURAL SUBSTANCES AS ANTIDIABETIC AGENTS

A majority of described plants and their extracts used to treat T2DM and related symptoms have a long-term tradition of application, but one of the major concerns related to natural agents is the deficiency of definitive and compatible data on the efficacy and, more importantly, the insufficient knowledge about the precise mechanism(s) of the action of natural agents. These very important limitations may explain why there is noticeable skepticism as regards the effectiveness of herbal agents in conventional medicine. Nevertheless, there is increasing evidence of the effectiveness of herbs, which will provide a rationale for further definitive research on a given mechanism (Cefalu et al., 2011). Thus, most authors have focused their research on the mechanism of action of natural substances with antidiabetic properties (Rios et al., 2015). Researchers are considering the physiological parameters of the regulation of glucose metabolism and pathophysiological alterations occurring in diabetes. These involve the interaction and function of multiple peripheral tissues, including liver, muscle, and fat tissue. To have a therapeutic effect, natural plant substances and herb extracts must theoretically affect glucose at a few different levels in tissues such as adipocytes (in adipogenesis, lipolysis, and modulation of adipocyte secretion of leptin and resistin) and liver (in hepatic metabolism, modulation of gluconeogenesis, enhancement of hepatic sensitivity) (Cefalu et al., 2011). Therefore, based on the anomalies accompanying T2DM, the application of herbs can be proposed as a way to modulate the whole-body metabolism by affecting adipocyte function and, next, regulating endocrine secretion involved in the extension of insulin action in the skeletal muscle. Additionally, plants can regulate hepatic gluconeogenesis and modulate glucose levels in the whole organism. For instance, a specific substance named "biguanide" or "metformin" originating from plant sources improves hyperglycemia by regulating hepatic processes.

It is well known that T2DM is related to defects in insulin secretion; thus, enhancement of the function of pancreatic β-cells may be another pathway by which natural substances may theoretically act. Increasing insulin secretory function may not yield a satisfactory effect. However, as shown in many preclinical and clinical trials, some botanical agents can increase proliferation and/or affect apoptosis of pancreatic islet tissue. This may significantly influence the natural diabetes control. There is no evidence to date that any of these proposed effects are consistently observed with any plant supplement available currently. Therefore, another pathway by which plants may act is the direct regulation of insulin action in multiple peripheral tissues (e.g., skeletal muscle and fat tissue) (Cefalu et al., 2011). Tabatabaei-Malazy et al., (2012) analyzed the effects of the primary component of antidiabetic botanicals which show direct action on the secretion of insulin by the pancreas. They found that amelioration of pancreas β-cell function and secretion of insulin is possible in connection with antioxidants by suppression of oxidative stress. In addition, there are other mechanisms that have a significant impact on improving health in diabetes; these include suppression of NF-κB, cytokine-induced impairment, uncoupling protein 2 activation, insulin-like activity, as well as enhancing intracellular calcium.

Other potential mechanisms of the antidiabetic effects of natural substances may involve inhibition of glucose absorption in the gut, increased glucose uptake and upregulation of glucose transport, alteration of glycogen metabolism, activation of nuclear receptors, insulin-mimetic, and -tropic effects, increased adiponectin release, increase in D-chiro-inositol, incretin mimetics and incretin enhancers, the action of endogenous opioids on glucose homeostasis, and antioxidation. Several herb substances such as antagonists of cannabinoid receptor type 1 (CB1), inhibitors of DP-4, and GLP-1analogs could work via these mechanisms (El-Abhar and Schaalam, 2014).

10.3 NATURAL AGENTS FOR TYPE 2 DIABETES

10.3.1 *ACACIA GUM (COMMON NAME: GUM ARABIC)*

Acacia gum is a water-soluble DF indigestible to both humans and animals. Although the DF intake is low in the human diet, the impact of fiber on the prevention of T2DM and its complications has been documented well (Danesh et al., 2007). Gum Arabic has nutritional value and some impact on the metabolism of glucose and lipids (Babiker et al., 2017).

Gum Arabic consists of a mixture of polysaccharides, oligosaccharides, and glycoproteins. It is an exudate of *Acacia Senegal* trees with interesting properties (Godrum et al., 2000). Acacia gum dissolves in water, forming a gel-like fluid with a viscous texture. It is able to absorb and digest carbohydrates slowly due to its viscosity effect (Marciani et al., 2000), which can change colonic microbial fermentation to produce short-chain fatty acids (SCFA), especially butyrate, and thereby possibly reduce the risk of metabolic syndrome (Babiker et al., 2017; Ulven, 2012). Gum Arabic improves bowel movements in patients with glycemic control (Min et al. 2012; Nasir, 2013).

The effect of Gum Arabic on the serum level of glucose and lipids, and on the *Body Mass Index* (BMI) in T2DM patients was examined in a double-blind randomized placebo-controlled trial. A significant reduction in fasting plasma glucose (FPG) and glycated hemoglobin (HbA1c) was reported in a group of patients who consumed a dietary supplement with 30 g of *Acacia Senegal*. Furthermore, Gum Arabic supplementation improved lipid profiles, decreased low-density lipoprotein (LDL-) cholesterol, total cholesterol, and triglycerides. In turn, *high*-density lipoprotein cholesterol (HDL-) cholesterol showed a significant enhancement in patients receiving Gum Arabic supplementation. These patients also lost weight and their BMI was reduced significantly. Acacia Gum can be consumed for improving the nutrition of T2DM patients (Babiker et al., 2017).

10.3.2 AEGLE MARMELOS (COMMON NAME: BAEL)

Beal is a popular herb in Ayurveda, Unani, Siddha, and folk systems of medicine used for the treatment of multifarious ailments. Different parts of this herb such as seeds, leaves, and fruit have hypoglycemic, hypolipidemic, and anti-hypertensive properties (Lambole et al., 2010). Supplementation with *A. marmelos* fruit pulp powder lowered rapidly postprandial plasma glucose (PPG) levels, HbA1c, total cholesterol, LDL-cholesterol, VLDL-cholesterol, and total triglycerides, and there was no impact on the HDL-cholesterol parameter due to the short period of supplementation (Sharma et al., 2016). *A. marmelos* fruit extract is rich in various active substances such as alkaloids, carbohydrates, coumarins, flavonoids, glycosides, sterols, tannins, and triterpenoids. The fruit extract exhibits hypoglycemic activity (500 mg/kg) and this effect is probably associated with its antioxidant activity (Abdullah et al., 2017; Saha et al., 2016).

In turn, *A. marmelos* Correa leaf juice supplementation (20 g/100 ml, per 60 days) caused improvement in all biochemical parameters of T2DM, with increased efficacy and irrelevant side effects. The leaf juice or extract of beal can be employed in the management of T2DM, possibly due to its content of active substances such as aegelin 2, scopoletin, and sitosterol, (Nigam and Nambiar, 2016; Sharma et al., 2016). *A. marmelos* alleviates the symptoms of T2DM in a natural way and exerts no adverse effects on health (Sharma et al., 2016).

10.3.3 *AGROPYRON REPENS (COMMON NAME: COUGH GRASS)*

This medicinal herb is applied as whole, cut herb, or dried rhizomes. Couch grass contains 3–8% triticin, which is similar to inulin and yields fructose after hydrolysis. Therefore, couch grass is referred to as fructosan (EMA, 2011; El-Snafi, 2015).

An aqueous extract of couch grass rhizome is traditionally used in Morocco as an antidiabetic agent. In rat studies, after oral application of a single dose (20 mg/kg) of a daily prepared and lyophilized aqueous extract (1 g/100 ml of water), significant enhancement of blood glucose levels was noticed in the diabetic animals. The levels of blood sugar were normalized after 2 weeks of daily oral application (20 mg/kg) of the water extract in diabetic animals and significantly reduction in normal (non-diabetic) rats after acute and chronic treatment. No modifications were detected in the concentrations of basal plasma insulin after the administration in either normal or diabetic rats (El-Snafi, 2015; Eddouks et al., 2005).

10.3.4 *ALOE VERA (COMMON NAME: BITTER ALOE)*

Aloe vera is used as dried sap in traditional medicine for the treatment of diabetes in the Arabian Peninsula and India (Cefalu et al., 2011). The pulp, gel, or juice obtained from the inner part of the leaves can contain glucomannan. Glucomannan is a water-soluble DF with hypoglycemic and insulin-sensitizing properties (Vuksan et al., 2000). Preclinical studies in rats have reported that bitter aloe leaf pulp (10 ml/kg) significantly decreased serum glucose and enhanced serum insulin levels, compared to control diabetic rats (Abo-Youssef et al., 2013). It is believed that one tablespoonful of aloe vera juice administered orally twice a day for 42 days (in the morning and at bedtime) shows anti-hyperglycemic activity in T2DM patients (Kochhar et al., 2018). However, other preclinical studies have shown incompatible results (Yeh et al., 2003).

On the other hand, small-scale clinical trials suggested that the consumption of aloe extract improved fasting glucose levels (Bunyapraphatsara et al., 1996; Yeh et al., 2003). Other studies have shown that *A. vera* gel (15 g) consumed by diabetic women caused a significant reduction in the blood sugar level. A similar impact was observed when *A. vera* gel was administrated in combination with a standard oral antidiabetic medication (2×5 mg glibenclamide) (Vogler and Ernst, 1999). A double-blind, placebo-controlled pilot study of two Aloe products (UP780 and AC952) in patients with prediabetes over an 8-week period demonstrated that only AC952 *A. vera* inner leaf gel powder significantly lowered levels of glucose and fructosamine. In the case of another product, i.e., *A. vera* inner leaf gel powder UP780 standardized with a 2% aloesin, there were significant reductions in fasting glucose, fructosamine, HbA1c, insulin, the homeostatic model assessment (HOMA) score, and in F2-isoprostanes, in comparison with the placebo group (Devaraj et al., 2013).

Another clinical study with application of a complex of *A. vera* gel (147 mg) and aloesin powder (3 mg) as an Aloe QDM complex in obese patients with prediabetes or early T2DM has shown the impact of this complex on fasting blood glucose (FBG), fasting serum insulin, and the score of HOMA of insulin resistance (HOMA-IR). After both 4 and 8 weeks, both the serum insulin level and HOMA-IR were significantly decreased in the intervention group. The Aloe QDM complex also lowered other parameters, e.g., body fat mass and body weight in obese patients with prediabetes or untreated early T2DM (Choi et al., 2013).

10.3.5 AMORPHOPHALLUS KONJAC (COMMON NAME: KONJAC)

Traditionally, konjac is used in China, Japan, and Southeast Asia as a food and medicinal product. The main component of crude konjac is glucomannan, which is a 100% water-soluble DF derived mainly from the tuberous roots of this plant (Cheang et al., 2017). The glucomannan fiber cannot be hydrolyzed by salivary and pancreatic amylase (Chua et al., 2010).

A clinical study has shown that supplementation of konjac-derived glucomannan (10 g/day) plus plant sterols (1.8 g/day) reduced plasma LDL-cholesterol concentrations in patients with T2DM after 21 days of intervention (Yoshida et al., 2006). In another study, konjac glucomannan supplementation (3.6 g/day, per 28 days) reduced lipid and glucose levels in the blood by increasing fecal excretion of neutral sterol and bile acid and alleviated enhanced glucose levels in hyperglycemic diabetic patients (Chen et al., 2003). In a double-blinded randomized crossover controlled trial, 75% of the T2DM patients had significantly reduced body weight and waist circumference, BMI, high-sensitivity C-reactive protein (hs-CRP), and HbA1c after consumption of konjac-derived glucomannan noodles (400 g, per 4 weeks). The other parameters such as FBG, insulin, lipid profile, and HOMA-IR of the subjects did not show any significant difference (Cheang et al., 2017).

10.3.6 ANDROGRAPHIS PANICULATA (COMMON NAME: KING OF BITTERNESS)

A. paniculata is mostly used in Ayurvedic and Unani medicine. The most important active substance in *A. paniculata* is andrographolide. This substance possesses an α-amylase inhibitor activity with *in vivo* effectiveness (10 mg/kg) in diabetic rats (Rios et al., 2015). Moreover, andrographolide (50 mg/kg) effectively lowered blood glucose levels, stimulated glucose transporter type 4 (GLUT4) translocation, and improved pancreatic islets in diabetic rats and the function of pancreatic β-cell (Nugroho et al., 2014, Zang et al., 2009). Andrographolide efficiently prevented the attack of insulitis in a dose-dependent manner and thus delayed the attack and suppressed the progression of diabetes in 30-week-old non-obese diabetic (NOD) mice. Andrographolide also regulates the homeostasis of T-helper-cells (Th1/Th2/Th17) and can thereby prevent *β*-cells from death and inhibit infiltration of T-cells into pancreatic islets and, hence, prevent the development of T1DM (Zang et al., 2013).

    A. paniculata water extracts administered orally prevented hyperglycemia in non-diabetic rats without affecting epinephrine-induced hyperglycemia (Borhannudin et al., 1994). In turn, *A. paniculata* ethanol extracts administered orally significantly lessened the amount of the FBG in humans (Hossain et al., 2014).

    A. paniculata can decrease the blood glucose level by augmenting glucose oxidation and utilization, restoration of insulin signaling molecules in liver, and lowering the serum lipid levels in high fat and T2DM rats without a hypoglycemic effect (Augustine et al. 2014). Moreover, a mixture of an n-hexane insoluble fraction of *A. paniculata* and a curcuminoid fraction of

Curcuma xanthorrhiza rhizome exerted a considerable antihyperglycemic effect in high-fructose-fat-fed rats (Nugroho et al., 2014). Another bioactive compound from *A. peniculata*, namely 14-deoxy-11, 12-didehydroandrographolide, also showed anti-hyperglycemic activity (Lee et al., 2010).

10.3.7 *ARTEMISIA DRACUNCULUS (COMMON NAME: RUSSIAN TARRAGON)*

The herb is commonly used worldwide. Several preclinical studies have shown that an ethanolic extract of Russian Tarragon has antidiabetic properties in animal models (Ribnicky et al., 2006; Vandanmagsar et al., 2014). This extract contains many active substances such as essential oils, coumarins, flavonoids (Cefalu et al., 2011; Obolskiy et al., 2011). Both *in vitro* and preclinical studies strongly suggest that the main effect of the extract consists of a beneficial impact on insulin signaling in the muscle (Wang et al., 2010). The improved cellular signaling was related to increased insulin sensitivity of the whole organism (Vandanmagsar et al., 2014).

In turn, a clinical trial in insulin-resistant patients suggested that an ethanolic extract of *A. dracunculus* increased insulin sensitivity in these subjects, but it was not significantly modified. Additionally, no modifications in the components of body weight or fat were observed (Cefalu et al., 2011). In another study, murine, and human pancreatic β-cells were treated with a standardized ethanol extract of *A. dracunculus*. This extract elevated release of insulin from pancreatic β-cells (Aggarwal et al., 2015**)**.

10.3.8 *ASTRAGALUS MEMBRANACEUS (COMMON NAME: MILK VETCH)*

A. membranaceus root is one of the most important herbs in traditional Chinese medicine (Guo et al., 2017). Recently, it has been included in European Pharmacopoeias as a medicinal plant. The milk vetch root has many pharmacological activities, e.g., multiple organ protection as well as antioxidant and hypoglycemic effects (Cui et al., 2016; Li et al., 2018; Kim et al., 2013; Wang et al., 2018).

The water extract is one of the primary active preparations obtained from the milk vetch root (Cui et al., 2018). *The root contains many* isoflavones, other polyphenolic compounds, and other bioactive metabolites (Matkowski et al., 2003). The most important compound is astragaloside IV (ASIV). It is a small molecular weight saponin with many pharmacological activities such as antihypertensive, antidiabetic, and myocardial protective properties (Gui et al., 2013). The ASIV treatment(10 mg/kg/day, per 8 weeks) can improve the structural and functional abnormalities observed in diabetic nephropathy rats, with the renoprotective activity mediated via suppression of endoplasmic reticulum stress (Barky et al., 2017; Wang et al., 2015). Furthermore, another research in diabetic mice showed synergic effects and potential mechanisms of the action of Astragalus polysaccharides (APS) administered as a mixture with Crataegus flavonoids (CF) in the treatment of T1DM and T2DM. Therapeutic agents e.g., CF (200 mg/kg/day), APS (200 mg/kg/day), APS + CF (200 mg/kg/day of each), were administrated orally to T2DM mice for 4 weeks. The results proved that the APS + CF mixture significantly reduced FBG as well as water and food consumption by the diabetic mice. Simultaneous administration of APS + CF elevated serum insulin levels and restored islet cell function. The APS + CF combination exerted greater antidiabetic effects than treatments with Astragalus polysaccharides or CF alone (Cui. et al., 2016).

Small-scale clinical studies showed that a mixture of *A. membranaceus* and *Lycii fructus* extracts exerted positive effects on immunomodulators and prevention of T2DM and eye disease complications (Jeon et al., 2010).

10.3.9   *AZADIRACHTA INDICA (COMMON NAME: NEEM)*

A. indica leaves have hypoglycemic, hypolipidemic, hepatoprotective, and immunostimulant properties (Khosla et al., 2000). Neem leaves contain many bioactive compounds such as azadirachtin, flavonoids, and several other bioactives (Bhat et al., 2011a). Neem inhibits the action of epinephrine on glucose metabolism, thereby enhancing the utilization of peripheral glucose (Kochhar et al., 2018).

In a preclinical study, a mixture (1:1) of an aqueous extract of dried powder of root and leaves (200 mg/kg body wt., per 8 weeks) of two strains *Abroma augusta* and *A. indica*, respectively, caused a significant diminution of FBG and serum lipids. The water extract also lowered the formation of lipid peroxides (LPO) referred to as thiobarbituric acid reactive substance (TBARS) and enhanced antioxidants such as catalase (CAT), glutathione peroxidase (GPX), glutathione transferase and superoxide dismutase (SOD) in erythrocytes. Reduction in the levels of LPO as TBARS was observed in the heart, kidney, liver, and muscles. The aqueous extracts of *A. augusta* roots and *A. indica* leaves when served together had better hypoglycemic and antidiabetic effects than when used alone (Helim, 2003).

A combination of *A. indica* and *Bougainvillea spectabilis* leaf extracts exhibited glucosidase inhibitory activity against murine pancreatic and intestinal glucosidase (Bhat et al., 2011b). It was detected that *B. spectabilis* leaves contained D-pinitol (3-O-methylchiroinositol), which is believed to exert an insulin-like effect (Adebayo et al., 2005). In another preclinical study, *A. indica* chloroform extracts and *B. spectabilis* methanolic and aqueous extracts induced high oral glucose tolerance and significantly lowered the intestinal activity of glucosidase. In immunohistochemical analysis, regeneration of insulin-producing cells and a corresponding augmentation in the plasma insulin and c-peptide levels were observed after treatment with these extracts (Bhat et al., 2011b).

In preclinical study treatment with *A. indica* oil (1.2 mL/day) and azadirachtin (1.0 mg/mL/day) during pregnancy had no hypoglycemic and anti-hyperglycemic effects on all animal groups, but affected the oral *glucose tolerance test* (*OGTT*) glycemic levels in the diabetic rats. The oil lowered the rate of total cholesterol and non-esterified fatty acids (NEFA) in diabetic rats. Both *A. indica* oil and azadirachtin enhanced lipoperoxidation via increasing levels of malondialdehyde (MDA) in the non-diabetic rats, impaired intrauterine development, and changed the antioxidant/oxidative status during pregnancy (Dallaqua et al., 2012).

10.3.10 *BOERHAVIA DIFFUSA (COMMON NAMES: PUNARNAVA OR SPIDERLING)*

B. diffusa is one of the famous traditional medicinal herbs used to treat a large number of human ailments, as mentioned in the Ayurveda, Charaka Samhita, Sushrita Samhita, and Unani medical systems. With their numerous medicinal properties, the entire *B. diffusa* plant or its peculiar parts (aerial parts and roots) are used by the populations of India and Arab countries (Mahesh et al., 2012). The aerial *B. diffusa* parts, especially leaves, are rich in alkaloids and sterols (Sharma et al., 2000).

A *B. diffusa* leaf extract with various solvents exhibited hypoglycemic activity in normal animals and anti-hyperglycemic activity in diabetes models. After 4 weeks, this extract (200 mg/kg of body weight) significantly limited free fatty acids, phospholipids, serum, and tissue cholesterol, and triglycerides. Furthermore, the leaf extract supplementation was more efficacious than glibenclamide in the treatment of diabetic rats (Nayak and Thirunavoukkarasu, 2016; Pari and Satheesh, 2004). The Spiderling leaf extract can also improve antioxidation in diabetic rats (Satheesh and Pari, 2004). Moreover, a water extract of *B. diffusa* leaves (200 mg/kg for 4 weeks) induced significant modifications in the accumulation of blood glucose and hepatic gluconeogenic enzyme activity in all groups of rats (Nayak and Thirunavoukkarasu, 2016). Furthermore, a chloroform extract of Spiderling leaves induced a dose-dependent diminution (50, 100, and 200 mg/kg body weight, for 4 weeks) in the level of blood glucose of diabetic rats in comparison with glibenclamide (Nalamolu et al., 2004).

In turn, *B. diffusa* root extracts may have potential activity against corticosteroid-induced diabetes in rats, because antiperoxidative factors, hypoglycemia, and cortisol decreased the activities of *B. diffusa* root extracts (150 mg/kg) in dexamethasone-induced diabetic rats (Nayak and Thirunavoukkarasu, 2016).

10.3.11 CAMELIA SINENSIS (COMMON NAME: TEA)

C. sinensis is commonly known as green or black tea and is the most popular beverage consumed all over the world. Green tea is a rich source of polyphenols, especially catechins. Its epigallocatechin gallate-enriched extract ameliorated glucose homeostasis and enhanced the expression of peroxisome proliferator-activated receptorγ (PPARγ) in a fructose-fed insulin-resistant hamster model (Rios et al., 2015). Various studies report that the polyphenolic compounds prevalent in both green and black tea are connected with favorable effects in the prophylaxis of cardiovascular disease and exhibit anti-diabetic properties (Khan and Mukhtar, 2013). Preliminary research on healthy women demonstrated that green or black tea drinkers (drinking more than 4 cups of tea a day) had a 30% lower risk of developing T2DM in comparison with non-tea drinking subjects (Song et al., 2005). However, in the Japanese population, where more than 4 cups of green tea are consumed per day, the T2DM risk was reduced by 33%, but no diminution was detected for the consumption of black or red teas (Iso et al., 2006). Nevertheless, black tea exerts a positive impact on other serum biomarkers of T2DM patients such as C-reactive protein, glutathione levels, MDA, and total antioxidant capacity (Neyestani et al., 2010).

It seems that the beneficial impact of green tea is connected with the constant consumption of catechin-enriched beverages. This was confirmed in a double-blind controlled study where T2DM patients without insulin remedy received green tea (582.8 mg catechins or 96.3 mg of catechins per day). After 12 weeks, enhanced insulin concentration and lessened the levels of HbA1cwere observed in the catechin T2DM group, in comparison with the control (Nagao et al., 2009).

10.3.12 CHROMIUM-ENRICHED SACCHAROMYCES CEREVISIAE YEASTS

Severe chromium (Cr) deficiency may lead to increased fasting hyperglycemia, glucose intolerance, elevated circulating insulin, insulin resistance, or even in diabetes (Balk et al. 2007). It has been shown that regular intake of Cr-enriched *Saccharomyces cerevisiae* (brewer's yeasts) is beneficial in T2DM patients, improving their glycemia, though Cr supplementation did not affect lipid or glucose metabolism in healthy people (Balk et al., 2007). Hosseinzadeh et al. (2013) also demonstrated that the consumption of brewer's yeasts by the diabetic population can reduce diastolic and systolic blood pressures. The decrease in 2-hour post glucose load, fasting glucose, and fructosamine were reported in clinical trials (Bahijiri et al., 2000; Yin and Phung, 2015), though the values of HbA1c remained unchanged. Offenbacher and Pi-Sunyer (1980) revealed that the ingestion of brewer's yeasts rich in Cr can increase both glucose tolerance and insulin sensitivity. In a study by Sharma et al. (2011), a 3-month intake of the Cr-yeasts improved the values of fasting glucose and HbA1c. Unfortunately, the outcomes from clinical trials are divergent. Racek and colleagues (2006) reported that 12-week treatment with Cr-yeasts containing a daily dose of 400 µg of Cr reduced insulin and serum glucose levels, whereas Król et al. (2011) did not observe considerable changes in blood glucose in patients suffering from T2DM receiving for 8 weeks a yeast preparation that contained even a higher dose (500 µg/day) of Cr. Another scientific group conducted by Hunt (1985) also did not demonstrate any significant alterations of the blood parameters in subjects with diabetes who had taken brewer's yeast for 3 months. Similarly, elderly patients presenting stable impaired glucose tolerance did not benefit from the supplementation of Cr-yeasts (Uusitupa and et al., 1983). In trials by Rabinowitz et al. (1985), Cr-rich preparation of brewer's yeasts did not influence FPG or glucose response to meal or tolbutamide, but it significantly increased the postprandial insulin level in the ketosis-resistant subgroups.

10.3.13 COMMIPHORA MUCUL (COMMON NAMES:GUM-GUGGUL, INDIAN BDELLIUM TREE)

Commiphora mucul, as an herbal tree with a resinous secretion, is one of the most valuable remedies in the traditional Ayurvedic medicine, especially for its antidiabetic, anti-oxidant, and anti-inflammatory activities (Kumar et al., 2016, Sudhakara et al., 2012). It has not well-defined chemical composition. Phyto-chemical studies of *C. mucul* indicate its content of sugars such as fructose or sucrose, amino acids, oil, and several steroids. However, only guggulsterone, as a bioactive sterol, has been purified from an ethyl-acetate extract of the gum guggul (Cornick et al., 2009). The essential oil of guggulsterone and guggul was found to be an efficient antioxidant (Siddiqui et al., 2013).

Guggul possesses antidiabetic activity, resulting in its action on PPAR α and γ (Cornick et al., 2009). Additionally, some preclinical studies with diabetic rats proved that *C. mucul* had hepatoprotective activity against oxidative stress associated with insulin deficiency and insulin resistance (Bellamkonda et al., 2011; Ramesh and Saralakumari, 2012). Moreover, gum guggul had a neuroprotective effect against diabetes-induced oxidative stress in the brain of diabetic rats (Sudhakara et al., 2012). This is important, as diabetes impairs learning and memory. The administration of a gum guggul extract significantly increased the time spent in a dark chamber and the number of crossing in the retention test (after 24 and 48 hours). There was no significant distinction in the bodyweight of the diabetic + *C. mucul* extract and diabetic control groups, but the serum glucose in the group receiving the *C. mucul* extract was significantly lessened in comparison with the diabetic control group (Salehi et al., 2015).

In another preclinical study with diabetic rats, *C. mucul* (50 and 100 mg/kg) was estimated in diabetic neuropathy in comparison with ramipril (0.2 and 2.3 mg/kg). After 8 weeks, gum guggul and ramipril individually significantly protected all the biochemical, behavioral, histopathological, and electrophysiological anomalies against diabetes-induced neuropathic pain. Administration of the gum guggul with ramipril mixture significantly improved the protective impact of *C. mucul* (Kumar et al., 2016).

10.3.14 ELEUTHEROCOCCUS SENTICOSUS (COMMON NAME: SIBERIAN GINSENG)

Eleutherococcus senticosus is widely exported from China as spices and healthy food. It is known as an adaptogenic agent and remedy for various complaints including diabetes (Ahn et al., 2013). Siberian ginseng contains several active components such as lignans, glycans, triterpene saponins, steroid glycosides, hydroxycoumarins, phenyl acrylic acid derivatives, especially syringin, and flavones (Sun et al., 2011).

In a preclinical study, a Siberian ginseng methanol extract (1.0 mg/kg) enhanced glucose utilization and lessened the levels of serum glucose in diabetic rats suffering from insulin deficiency. Additionally, Siberian ginseng at the effectual dose (1.0 mg/kg) significantly reduced serum glucose in non-diabetic rats, too. Syringin (0.01 to 10.0 micromol/L) induced glucose uptake in soleus muscle and enhanced glycogen synthesis in a dose-dependent manner (Niu et al., 2008). However, which bioactive component mediates the antidiabetic effect of Siberian ginseng remains unknown (Ahn et al., 2013).

In another preclinical study, the hypoglycemic effect via suppression of α-amylase and α-glucosidase activity was evaluated using leaf extracts of several *Eleutherococcus* species, e.g., *E. senticosus*, *E. gracilistylus*, *E. sieboldianus*, and *E. sessiliflorus*. Among the species, suppression of α-glucosidase activity was revealed by *E.gracilistylus* and *E.senticosus* extracts. All the extracts from *Eleutherococcus* spp exerted a hypoglycemic effect and improved fat metabolism in the diabetic rats (Lim et al., 2010).

Clinical studies with *E. senticosus* root extracts have demonstrated better utilization of glucose with a reduction of elevated glucose plasma levels in T2DM patients (Ivorra et al., 1988; Sui et al., 1994). In a double-blinded randomized clinical study, in contrast to the placebo and Panax ginseng, Siberian ginseng consumption (480 mg/day for 3 months) appeared in a highly significant decline in the level of the FPG and PPG. Moreover, Siberian ginseng significantly decreased HbA1c, total cholesterol, and total triglyceride levels. Both of the herbs had a wide margin of safety with no side effects.*E. senticosus* extract contributed to some recovery of the sensory stimulus, which became an electrical stimulus again and reduced symptoms of peripheral polyneuropathy (Freye et al., 2013).

10.3.15   EMBLICA OFFICINALIS (COMMON NAME: AMLA)

Emblica officinalis is used in traditional Ayurveda medical systems as a potent Rasayana. The rind is the commonly used part of amla fruit. It is a herb-derived substance promoting health and longevity by enhancement of defense against diseases (Akhtar et al., 2011; D'souza et al., 2014; Walia and Boolchandani, 2015). *E. officinalis* is a rich source of vitamin C, which has strong antioxidation activity. Additionally, it has other important constituents, i.e., amino acids, alkaloids, carbohydrates, flavonoids, gallotanin, phenolic, pectin, and tannins. Amla exerts antidiabetic effects via antioxidant and free radical scavenging properties and reduces oxidative stress (D'souza et al., 2014; Nain et al., 2012; Walia and Boolchandani, 2015). Amla fruit ash contains minerals such as chromium, zinc, and copper. Cr had significant antidiabetic activity and improved impaired lipid metabolism in various experimental studies of T1DM and T2DM (Staniek, 2018; Walia and Boolchandani, 2015).

Tannoids contained in amla fruit are inhibitors of aldose reductase (AR). AR acts an indispensible role in the enlargement of diabetes complications as cataract. Amla fruit activates antioxidant enzymes and prevents aggregation and insolubilization of lens proteins caused by hyperglycemia. Other substances with antidiabetic and antihyperlipidemic activity are polyphenolic constituents. It has been revealed that polyphenol-enrichedamla fruit juice (541.3 mg gallic acid equivalent in 1 g amla extract) can reduce glucose and lipids and protect from cardiac complications (Walia and Boolchandani, 2015).

The 4-week administration of amla water extract improved oral glucose tolerance and, after 8 weeks, it induced a significant diminution in the FPG level compared to day 0 in T2DM rats. Triglycerides were reduced (14%), however, there was no significant modification in serum alanine aminotransferase (ALT or *SGPT*)), creatinine, cholesterol, and insulin in any group. In turn, the glutathione content was reduced (Ansari et al., 2014). Oral administration of a hydro-methanolic (20:80) extract of *E. officinalis* leaves (100, 200, 300, and 400 mg/kg b.w. daily for 45 days) significantly lessened the amount of FBG and enhanced insulin levels, compared with diabetic rats. Furthermore, the extract significantly decreased all biochemical parameters, i.e., ALT, aspartate aminotransferase (AST or *SGOT*), lipid profile, serum creatinine, and serum urea. The treatment of the leaf extract also significantly increased the reduction of CAT, glutathione, GPX, SOD, and decreased the LPO levels in the kidney and liver of diabetic rats (Nain et al., 2012). Furthermore, in another preclinical study, an aqueous extract of amla seeds exhibited hypoglycemic potential and antidiabetic activity (Koshy et al., 2012).

Amla fruit powder (1, 2, or 3 g) significantly decreased FPG and 2-h PPG levels on the 21st day in all group of patients, compared with their baseline values (Akhtar et al., 2011). A 2-month intake (a tablespoon per day) of a mixture of amla fruit juice with bitter gourd (*Momordica charantia*) stimulated the pancreas to secrete insulin, thereby declining the blood glucose in T2DM. This mixture also prevented eye complications in T2DM (Patel and Goyal, 2011).

10.3.16   *EUCALYPTUS GLOBULUS (COMMON NAME: EUCALYPTUS)*

Eucalyptus globules intake as herbal tea has been described in Aboriginal, European, and British Pharmacopoeias. Hot waterleaf decoctions of eucalyptus have been used in traditional medicine as antidiabetic remedy (Dey and Mitra, 2013, Konoshima, andTakasaki, 2002). Thus, different preclinical studies in diabetic animals confirmed its anti-hyperglycemic properties, and the antioxidant activity of eucalyptus leaves (Dey and Mitra, 2013, Nakhaee et al., 2009). Eucalyptus contains many phytochemicals, both volatile and non-volatile fractions. Among them, terpenoids are one of the major bioactive compounds comprising most of the essential oil (eucalyptus oil) with its characteristic aroma (Dey and Mitra, 2013).

In preclinical studies, 4-week oral administration of eucalyptus leaves (20 and 62.5 g/kg of eucalyptus in diet and 2.5 g/L of aqueous eucalyptus extract in drinking water) significantly reduced the weight loss and enhanced water and food intake in the treated diabetic groups in comparison to the non-diabetic group. Eucalyptus improved diabetic status in a dose-dependent manner by partial restoration of pancreatic β-cells and repair of the streptozotocin-induced damage in rats (Mahmoudzadeh-Sagheb et al., 2010). At the end of the treatment period (4 weeks), the administration of eucalyptus leaves caused a significant reduction of plasma glucose, plasma, and liver MDA, protein carbonyl (PC), and HbA1c levels with a concomitant increase in the levels of reduction of ferric reducing antioxidant power (FRAP) in the treated diabetic rats (Nakhaee et al., 2009).

The administration of eucalyptus in normal and diabetic models with infection of *Candida albicans* significantly improved hyperglycemia, polydipsia, polyphagia, and compensated the weight loss in diabetic rats. Moreover, *E. globules* induced a significant diminution in the *C. albicans* amount in kidney and liver homogenates (Bokaeian et al., 2012). An aqueous eucalyptus leaf extracts significantly lessened the levels of FBG, liver MDA, HCO_3^-, and liver enzymes, and enhanced levels of xanthine oxidase and serum CAT in diabetic rats in a similar manner as glibenclamide (Saka et al., 2017). In a dose-dependent manner, eucalyptus leaf extracts (10 g/kg diet) inhibited absorption of intestinal fructose and suppressed adiposity related to dietary sucrose and simultaneously reduced plasma and hepatic triacylglycerol concentrations in rats (Sugimoto et al., 2005). Furthermore, *E. globulus* leaf extracts simultaneously inhibited the activities of fructokinase, which metabolizes fructose in the liver, and *glucose-6-phosphate dehydrogenase* (G6PD or G6PDH), preventing the activation of fructose and sugar metabolism as well as the synthesis of fatty acid-induced by sucrose consumption (Sugimoto et al., 2005; Sugimoto et al., 2010).

Besides *E. globules,* other *Eucalyptus* species (*E. citriodora, E. camaldulensis,* and *E. tereticornis*) seem to possess antidiabetic properties. Since ancient times, the bark and leaves of different *Eucalyptus* species have been used as folk and alternative medicine for the treatment of ailments (Dey and Mitra, 2013).

10.3.17 EUGENIA JAMBOLANA (COMMON NAME: JAMBOLAN)

Eugenia jambolana is commonly used in traditional Ayurveda medicine as a medicinal herb for various ailments. Various parts of *E. jambolana, e.g.,* the bark, fruits, kernel, leaves, and septum, are frequently used for diabetes treatment. *E. jambolana* is administered in various pharmaceutical preparations such as tinctures, aqueous extracts, infusions, simple aqueous extracts prepared with hot water but without boiling, and decoctions as boiled infusions to lessen the levels of blood glucose in diabetic animals. Kernel decoctions are widely drunk as an antidiabetic remedy, especially in India (Kochhar and Pathak, 2018; Petato et al., 2001; Sridhar et al., 2005)*. E. jambolana* seed contains gallic acid, oxalic acid, tri-terpenoids, and tannins (Sridhar et al., 2005).

E. jambolana seed powder (500 or 1000 mg/kg for 15 days) enhanced animals' body weight on day 20 in relation to day 5 in diabetic female rats. It also reduced FBG, differences in post-treatment fasting and peak blood glucose, and differences in liver glycogen. However, the best anti-diabetic effect was obtained when 500 mg/kg were applied (Sridhar et al., 2005). In turn, administration of an ethanolic or methanolic extract of *E. jambolana* kernels (100 mg/kg or 150 mg/kg of body weight, respectively) significantly lessened the levels of blood glucose, blood urea, and cholesterol, enhanced glucose tolerance and concentration of total proteins and liver glycogen, and lowered the activities of glutamate oxaloacetate transaminase (GOT) and serum glutamate pyruvate transaminase (SGPT) in diabetic rats (Jasmine and Daisy, 2007; Ravi et al., 2004). Furthermore, subacute toxicity studies with a single administration dose (2.5 and 5.0 g/kg)of *E. jambolana* seed powder detected no mortality or abnormality (Sridhar et al., 2005). Additionally, *E. jambolana* seems to exert an ulcer protective effect due to its antidiabetic and gastric antisecretory properties (Chaturvedi et al., 2009). Moreover, analyses of water and alcoholic extracts, as well as a lyophilized powder (200mg/kg per day) of E. jambolana in hyperglycemic rats, showed a reduction of blood glucose by the water seed extract (2.5 and 5.0g/kg body weight p. o. for 6 weeks) along with an enhancement of total hemoglobin and antioxidant activity in the diabetic group (Kochhar and Pathak, 2018; Sharma et al., 2003).

The antidiabetic effect of an ethyl acetate fraction of *E. jambolana* seeds (200 mg/kg of body weight/day) was evaluated in the diabetic male rat at the genomic level. After 35 days, the seed extract, in comparison to glibenclamide, exhibited a significant anti-hyperglycemic action, improved serum insulin, and HbA1c levels, and contributed to the regeneration of pancreatic β-cells. The seed fraction also contributed to the normalization of expression of the Bax, i.e., Bcl-2, and Hex-I genes (members of the B-cell lymphoma 2 called Bcl-2 gene family) in hepatic tissue (Jana et al., 2014). An *E. jambolana* ethanolic seed extract significantly lessened the levels of blood glucose in a dose-dependent manner. Additionally, the seed extract significantly restored the bodyweight loss, reduced kidney weight, HbA1c, blood factors such as urea, uric acid, urea nitrogen, and creatinine, the volume of urine, and the level of urine microalbumin. In turn, a histopathological study of rat kidney tissues confirmed the protective impact of *E. jambolana* seed extracts in diabetic nephropathy (Esther and Manonman, 2014). However, oral administration of an *E. jambolana* leaf decoction (15%, w/v) as a substitute for water to all groups of rats had no hypoglycemic impact and no antidiabetic activity (Petato et al., 2001).

10.3.18　FICUS CARICA (COMMON NAME: FIG)

Ficus carica is a well-known medicinal plant used as a diabetic remedy in Spain and south-western Europe and in traditional medicine systems such as Ayurveda, Siddha, and Unani (Prasad et al., 2006; Tian et al., 2014). Decoctions of *F. carica* leaves are traditionally used for diabetes treatment in folk medicine (Irudayaraj et al., 2017). *F. carica* fruits are normally used as an edible food. *The* fruits contain active compounds, e.g., polysaccharides and polyphenols (Tian et al., 2014; Mopuri et al., 2018). Furthermore, different parts of *F. carica* such as leaves, fruits, and roots also contain amino acids, aliphatic alcohols, anthocyanins, fatty acids, hydrocarbons, organic acid, phenolic components, phytosterols, volatile components, and a few other classes of secondary metabolites. The most of the phytochemicals have been detected in latex of fit leaves, roots, and fruits (Badgujar et al., 2014; Nadeem and Zeb, 2018).

F. carica leaves showed antidiabetic, antihyperlipidemic, and antioxidative activities in diabetic rats (Asadi et al., 2006; Belguith-Hadriche et al., 2016; Perez et al., 2003). The antidiabetic activity of fig methanolic leaf extracts has also been reported in diabetic rats (Stalin et al., 2012). The fig ethyl acetate leaf extract (250 and 500 mg/kg) contributed to significant improvement of concentration of blood glucose, hepatic glycogen, total cholesterol, and triglycerides as well as body weight. Additionally, the extract changed the activities of key carbohydrate-metabolizing enzymes such as fructose-1,6-bisphosphatase, glucose-6-phosphatase, and hexokinase in the liver tissue of diabetic rats to revert to near-normal levels. The extract of *F. carica* leaves showed a cytoprotective effect on islets of pancreatic β-cells (Irudayaraj et al., 2017).

Comparative *in vitro* research of various parts of *F. carica* (fruits, leaves, and stem bark extracts) showed the impact of these extracts on antioxidative, antidiabetic (e.g., inhibition of α-amylase and α-glucosidase enzymes), and antiobesogenic activities (e.g., pancreatic lipase suppression). The fruit ethanolic extract consisted of a higher quantity of polyphenols and flavonoids in comparison with leaves and stem bark extracts. The activity of the fruit ethanolic extract was significantly higher than other extracts and parts of the fig plant in terms of antioxidative, antidiabetic, and antiobesogenic effects (Mopuri et al., 2018). Furthermore, treatment with *F. carica* fruits appeared in an increment in the survival index that correlated with enhanced plasma insulin levels (El-Shobaki et al., 2010).

In a clinical study, the supplementation of a decoction of *F. carica* leaves to the diet of ten insulin-dependent T1DM patients helped to control PPG (Serraclara et al., 1998). A 1-month addition of the fig leaf decoction to oral hypoglycemic drugs significantly reduced PPG concentration in T2DM patients ($n = 14$) (Mazhim et al., 2016).

10.3.19   GALEGA OFFICINALIS (COMMON NAME: GALEGA)

Galega officinalis is a leguminous plant, which aerial parts have long been used in traditional and folk medicine to treat diabetes in Chile, Japan as well as Europe (Bailey and Day, 2004; Gunn and Farnsworth, 2013; Rios et al., 2015). This herb consists of two nitrogen guanidine constituents: galegin (syn. galegine) as isoamylene guanidine and hydroxygalegine prevalent in all parts during flowering and forming fruits. These bioactive substances possess pharmacological features as hypoglycemic and galactogenic factors. However, guanidine is excessively toxic for clinical treat; hence, the study focused on galegine, which turned out to be less toxic as an extract of *G. officinalis*. In the 1920s, the extract was specified as an antidiabetic formulation (Bailey and Day, 2004; Martínez-Larrañaga and Martínez, 2018).

Preliminary research in an animal model established that administration of *G. officinal is* extracts (600 mg/kg a day) exerted a marked hypoglycemic impact on the diabetic condition. These studies revealed the mobilization of antioxidant and antiradical protection mechanisms. Simultaneously, the extract had a corrective influence on the leukocyte differential count. *G. officinalis* extracts had an inhibitory effect on the genetically programmed cell death through normalization of the quantity of white blood cells containing regulatory proteins of apoptosis (p53 and Bcl-2) and poly-(ADP)-rybosylated proteins in leukocytes of rats (Khokhla et al., 2010; Khokhla et al. 2012; Khokhla et al., 2013). Furthermore, in another preclinical study with T1DM subjects, the administration of a chloroform fraction of a *G. officinalis* extract promoted restoration of the neutrophil bone marrow pool and reduction of the number of lymphoblasts. It also caused inhibition of the lymphocyte apoptosis process (Nagalievska et al., 2018).

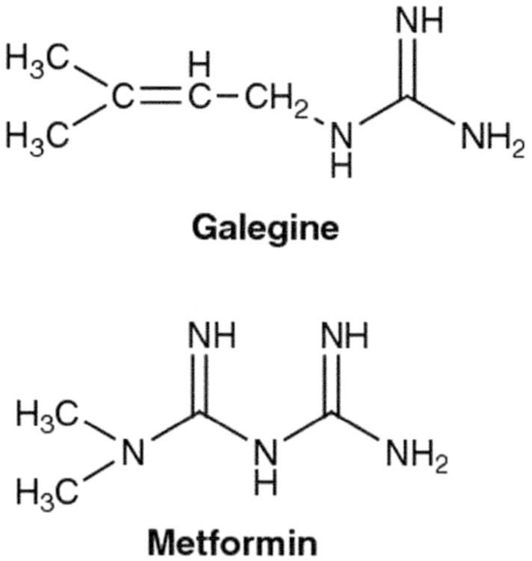

Galegine

Metformin

FIGURE 10.1   Chemical structure of galegine and metformin (Bailey and Day, 2004).

The synthetic analog of galegine, metformin (dimethylbiguanide) is better tolerated and more effective than the extract (Bailey and Day, 2004; Gunn and Farnsworth, 2013). Numerous studies found that an early intake of the synthetic analog of galegine reduced several diabetic complaints e.g., cardio-vascular mortality and enhanced survival in overweight and obese T2DM patients beyond that expected for the prevailing control of the glycaemic level (Bailey and Day, 2004). The chemical structure of metformin as the antidiabetic medicinal product is really similar to galegine, which is responsible for reduction of blood glucose caused by galega extract. The structure of both compounds is shown in Figure 10.1 (Rios et al., 2015).

10.3.20   GLOBULARIA ALYPUM (COMMON NAME: TASSELGHA)

Globularia alypum is commonly used in Algeria as a medical herb showing some benefits in the traditional treatment of diabetes, cardiovascular, and renal diseases, and a number of other complaints (Djeridane et al., 2006). Phytochemical analysis of hydromethanolic extracts from the aerial parts of *G. alypum* appeared the content of high amounts of phenolic acids, which possess strong hypoglycaemic and antioxidative properties. The extract also contains terpenoids and flavonoids (Djellouli et al., 2014; Es-Safi et al., 2007).

In a preclinical study with rats fed a high-fructose diet, supplementation of the water extract of *G. alypum* revealed improvement of plasma triglycerides and lipid peroxidation (Taleb-Dida et al., 2011). A lyophilized ethanolic extract of Tasselgha (1 g/kg a day) reduced glycaemia, values of liver and plasma lipids, levels of VLDL-LDL- and HDL-cholesterol, and lipid peroxidation and ameliorated the reverse transport of cholesterol in the diabetic group (Djellouli et al., 2014). In a comparative study, the impact of oral administration of water extracts of *Rubus fructicosis* (0–6 g/kg) and *G. alypum* leaves (2–10.5 g/kg a day) on blood glucose concentration was analyzed in normal and diabetic models. In non-diabetic rats, single, and repeated intake of *R. fructicosis* reduced significantly concentration of blood glucose, while the therapy with *G. alypum* did not modify this blood parameter. In diabetic rats, single, and repeated intake of both *R. fructicosis* and *G. alypum* caused a significant lessened the levels of blood glucose. The plant treatments did not have an influence on insulin secretion in all groups of rats, showing that the lowering mechanism of blood glucose concentration by these plants is extra-pancreatic, at least for the doses used. Additionally, an acute toxicity study revealed that the water extract of Tasselgha water extract might be considered relatively safe, since the LD50 value was over 14.5 g/kg (Jouad et al., 2002).

10.3.21   GLYCYRRHIZA GLABRA (COMMON NAME: LICORICE)

Glycyrrhiza glabra root is traditionally used in Chinese medicine in the therapy of diabetes and its complications (Li et al., 2004; Kataria et al.,

2013; Tang et al., 2006). The primary bioactive compound in *G. glabra* root is an isoflavonoid glabridin (isoflavane). It is believed that it possesses antioxidative activity, inhibits LPO, prevents low-density lipoprotein oxidation, protects mitochondrial functions from oxidative stress, and helps in other complications. Licorice also possesses hypoglycaemic activities and its consumption helps in decreasing blood glucose or sugar levels (Abo et al., 2012; Choi, 2005; Kataria et al., 2013; Shang et al., 2009).

In a preclinical study with normal mice, seven bioactives isolated from rhizomes revealed a significant blood glucose-reduction effect, especially 2′,4′-dimethoxy-4-hydroxychalcone, liquiritigenin-7,4′-dibenzoate, and isoliquiritigenin (50, 50 and 200 mg/kg body weight, respectively). The structure-activity relationship signified that the ether and ester groups contained in these bioactive compounds are essential for the antidiabetic activity. The antidiabetic activity of chalcone derivatives is mediated by stimulation of PPARγ, and these flavonoids modify the activity of intracellular enzymes e.g., glucosidases (Gaur et al., 2014). The licorice extract inhibits α-glucosidase and α-amylase enzyme activity, leading to diminution of disaccharide hydrolysis (Karthikeson and Lakshmi, 2017).

In several preclinical studies with animal models, extracts of glabridin (25 to 100 mg/kg) exerted a beneficial effect on hyperglycemia, hyperlipidemia, and liver dysfunction (El-Ghffar, 2016). *G. glabra* augmented insulin-mediated glucose removal in muscles and increased transport of GLUT1 from inside the cell to the plasma membrane (Gupta et al., 2011). Furthermore, the consumption of glabridin extracts (25 and 50 mg/kg) had antidiabetic effects (especially the higher dose). It was associated with significantly reduced body weight gain, enzymic/non-enzymic antioxidants, and HDL-cholesterol, and significantly lowered relative organ weight, lipid profiles, LPO, serum glucose levels, pro-inflammatory cytokines, and kidney and liver functions (El-Ghffar, 2016). The oral administration of glabridin (25 and 50 mg/kg) improved learning and memory in non-diabetic rats and reversed learning and memory deficits in diabetic rats (Hasanein, 2011).

10.3.22 GYMNEMA SYLVESTRE (COMMON NAME: GURMAR)

Gymnema sylvestre leaves or their extracts are traditionally used to treat T1DM and T2DM diabetes, especially in Ayurvedic and Western medicines (Grover et al., 2002; Kochhar et al., 2018; Rios et al., 2015; Tiwari et al.,

2014). Gurmar contains potential antidiabetic compounds, e.g., saponins, triterpenoid saponins, and polypeptides (Porchezhian and Dobriyal, 2003; Rios et al., 2015). Gurmar intake exerts positive effects on blood glucose homeostasis and controls sugar thirst through the suppression of sweet taste sensation (Cefalu et al., 2011; Tiwari et al., 2014). In animal model studies, the impact of *G. sylvestre* extracts on sugar metabolism suggested to be secondary to improving glucose uptake in peripheral tissues. Gurmar extract increases the secretion of insulin from pancreatic β cells, the quantity of β cells in the pancreas, and enhances the activity of enzymes responsible for the utilization of glucose via insulin-dependent pathways. It causes enhancement of phosphorylase action, reduction of the effect of the action of sorbitol dehydrogenase, and gluconeogenic enzymes, finally inhibiting glucose absorption from the small intestine (Dey et al., 2002; Di Fabio et al., 2013; Kochhar et al., 2018). The hypoglycemic activity of gurmar is also connected with cooperation with glyceraldehyde-3-phosphate dehydrogenase (GADPH), an indispensable enzyme in the glycolysis pathway (Tiwari et al., 2014). However, *G. sylvestre* action most likely is not related to its ability to inhibit DPP-4. Gurmar has also been revealed to have a limited effect on GLP-1 levels (Kosarayu et al., 2014).

Preclinical studies in dogs without a pancreas showed that administration of *G. sylvestre* extract, a saponin fraction, or isolated triterpene glycosides caused enhanced hypoglycemic action of exogenous insulin, intensification of insulin effects, and prolonged duration of lowered glucose levels (Alqahtani et al., 2013; Di Fabio et al., 2013).

Various small-scale clinical trials with a patented preparation based on a standardized extract (400 mg/day) indicated that the extract increased the endogenous insulin levels, presumably due to pancreas regeneration in T1DM patients (Di Fabio et al., 2013). Furthermore, it was shown in a small-scale open-label trial that 18- to 20-month administration of an ethanolic extract (200 mg a day) significantly ameliorated levels of FBG and HbA1c in T2DM patients. Another uncontrolled clinical trial demonstrated that a comparable extract (800 mg a day) also reduced concentration of FBG and HbA1c in both T1DM and T2DM patients after 3 months of application (Nahas et al., 2009). A clinical study with T2DM patients appeared that, after 3 months, the extract of gurmar (500 mg a day) decreased both FPG and PPG and HbA1c (Kumar et al., 2010). Generally, 6- to 12-month intake of gurmar liquid extracts (25 to 75 ml per week) is recommended (Saneja et al., 2010).

10.3.23   GYNOSTEMMA PENTAPHYLLUM (COMMON NAME: MAKINO)

Gynostemma pentaphyllum is a plant used as a tea in traditional and folk medicines. For centuries, it has played a primary role in the diabetes prophylaxis and treatment in Chinese, Vietnam, and many other Asian countries, especially Southeast Asia (Hoa et al., 2009; Norberg et al., 2004; Yeh et al., 2003). *G. pentaphyllum* extract contains a large group of saponins, also called gypenosides (Le et al., 2010).

In preclinical studies with rats, crude saponin fractions of *G. pentaphyllum* extracts were shown to exert hypoglycemic and hypolipidemic activity (Jang et al., 2001). After treatment for 4 days, the whole extract of *G. pentaphyllum* (250 mg/kg) reduced both hyperglycemia and hyperlipidemia in diabetic rats. These effects were maintained for at least 5 weeks (Megalli et al., 2006). Furthermore, a ethanol extract of Makino extract manufactured in Vietnam suppressed the activity of protein tyrosine phosphatase 1B (PTP-1B), which may have led to increased insulin sensitivity hence improved tolerance of glucose (Hung et al., 2009).

In randomized clinical studies, *G. pentaphyllum* tea (6 g per day) exhibited an antidiabetic effect both as an independent therapy and as an addition to therapy of sulfonylureas (SU) showing good safety in recently diagnosed T2DM patients. It was also shown to have an effect on insulin sensitivity (Huyen et al., 2010). After an 8-week SU treatment with the *G. pentaphyllum* extract (6 g a day per 4 weeks), the concentration of FPG and HbA1c were reduced significantly *G. pentaphyllum* extracts also lowered post load values in a 30- and 120-minute oral glucose tolerance test (OGTT). The HbA1c levels were lessened significantly in comparison to the placebo group. Addition of the Makino extract to SU enhanced the cure of T2DM patients (Huyen et al., 2012). In another randomized clinical study, *G. pentaphyllum* tea (6 g per day) significantly reduced fasting and steady-state plasma glucose (SIGIT mean) after 4 weeks. The effect on FPG was diverted after changing the treatment on placebo. The glycometabolic amelioration were attained without any considerable modification of circulating insulin concentration. The *G. pentaphyllum* tea exerted an antidiabetic impact by enhancing the sensitivity of insulin (Huyen et al., 2013).

10.3.24   HEMPSEED OIL FROM CANNABIS SATIVA (COMMON NAME: HEMP)

Cannabis sativa is traditionally used as an indigenous medicinal agent in South Africa for the treatment of such disorders as diabetes and for the early treatment of snakebite. Worldwide, *C. sativa* is used as a recreational drug, mostly for the psychoactive effect of the primary active compound alkaloid Δ9-tetrahydrocannabinol (Δ9-THC). *C. sativa* contains more than 500 components, of which 104 cannabinoids (cannabidiol (CBD) and cannabinol) and 200 to 250 non-cannabinoid compounds were identified. For this reason, it is important to transform this plant from an illegal drug into a medicinal herbal product with therapeutic benefits (Levendal and Frost, 2006; Lafaye et al., 2017; van Wyk and Gericke, 1997).

C. sativa alkaloids (endocannabinoids) are well-known exogenous agonists of cannabinoid receptors (CB$_1$R and CB$_2$R) belonging to the group of membrane receptors coupled to protein G (Simonienko et al., 2018). Cannabinoid receptors and their endogenous ligands (endocannabinoids) represent the endocannabinoid system, which plays an indispensable role in regulating energy balance, appetite, insulin sensitivity, the function of pancreatic β-cells, and lipid metabolism (Kim et al., 2012; Liu et al., 2012; Pacher et al., 2006; Pagotto et al., 2006). Endogenous ligands (e.g., CB$_1$Rs) occurring in peripheral tissues are involved in energy homeostasis, e.g., in adipose tissue, liver, pancreas, and skeletal muscle. Activation of CB$_1$R attenuates insulin function and secretion from the pancreas, reduces insulin responsiveness in skeletal muscles, and promotes lipogenesis in the liver and adipose tissue. Peripheral and central CB$_1$Rs are activated via Δ9-THC (Kim et al., 2012; Liu et al., 2012; Osei-Hyiaman et al., 2005; Pacher et al., 2006; Pagotto et al., 2006). Hence, the antagonism of CB$_1$R reduces body weight, improves dyslipidemia, and attenuates insulin resistance in humans (Després et al., 2005; Triay et al., 2012).

Permanent use of cannabis with a high THC concentration leading to activation of CB$_1$R may negatively affect the metabolic actions of insulin and facilitate hepatic steatosis (Hézode et al., 2008; Muniyappa et al., 2013). Even short-term (13 days) experimental smoking of marijuana enhanced appetite, food ingestion, body weight, and induced glucose intolerance in healthy men (Foltin et al., 1988; Hollister et al., 1974). Therefore, two primary types of *C. sativa* must be differentiated: the narcotic and non-narcotic types. The former is also known as marijuana, hashish, or Cannabis tincture and contains Δ9-THC, in concentrations between 1–20%, exhibiting psychotropic properties. The latter type of *C. sativa* is industrial hemp with THC concentrations < 0.3%, which has no psychoactive activities (Ross et al., 2000; Holler et al., 2008). Hempseed and hempseed oil are enriched in linoleic acid (LA) and gamma-linolenic acid (*GLA*). In preclinical studies, orally supplemented GLA-LA treatments (20 mg/kg) corrected a diabetes-related diminution in the gastrocnemius muscle low-molecular-weight decreased thiol content and caused a mild curtailment in FBG concentration after ten days (Khamaisi et al., 1999). Additionally, an antinociceptive effect of a controlled cannabis extract (eCBD) was effective in attenuation of diabetic neuropathic pain in diabetic rats. Furthermore, after the repeated eCBD treatment, the nerve growth factor level in the sciatic nerve of diabetic rats was restored to normal (Comelli et al., 2009).

10.3.25   JUNIPERUS COMMUNIS (COMMON NAME: JUNIPER)

Juniperus communis is a coniferous plant whose bark is commonly used in traditional and folk medicine, especially in the Northern Hemisphere, for management of diabetes. Besides bark, its aerial parts, berries, and fruits are widely used for alleviation of various complaints (Bais et al., 2014; Banerjee et al., 2013). Juniper contains various active compounds including flavonoids, volatile oil, coumarins, and several labdane diterpenes and diterpenoids (Bais et al., 2014).

*In a preclinical study with d*iabetic rats, significantly lessened blood glucose concentration and enhanced HDL levels were noted after 21-day oral administration of juniper methanolic extracts (100 and 200 mg/kg) (Banerjee et al., 2013). In another preclinical study a decoction from juniper "berries" (250 mg/kg) caused the diminution of blood glucose concentration in normoglycaemic rats. This effect was associated with increased consumption of peripheral glucose and intensification of glucose-induced insulin secretion. After 24 days, the administration of this decoction (125 mg total "berries"/kg) to diabetic rats contributed to a significant diminution of both blood glucose concentration and the mortality index and prevented body weight loss (de Medina et al., 1994).

10.3.26   LAGERSTROEMIA SPECIOSA (COMMON NAME: BANABA)

Lagerstroemia speciosa is a plant whose leaves are used in folk medicine as a remedy for diabetes in many parts of the world, especially in Southeast Asia. The major bioactive compounds of banaba leaves are corosolic acid and ellagitannins. Corosolic acid has also been isolated from several other plants. Banaba also contains tannins and terpene acids (e.g., oleanolic acid) (Klein et al., 2007; Miura et al., 2012).

The hypoglycemic effect of hot water and methanol extracts of *L. speciosa* has been revealed in both numerous animal models and several human studies. Some studies are focused on corosolic acid originated from banaba leaves as a main bioactive substance responsible for antidiabetic properties. In these studies, purified corosolic acid is used as an isolate with an organic solvent (10 mg/30 days), and corosolic acid as a constituent of standardized banaba extracts (60 mg comprising 10 mg corosolic acid/2 weeks) (Park et al., 2011; Ulbricht et al., 2007). In other studies, there are ellagitannins in water-soluble fractions, which might have an activity similar to insulin, and tannins which have the antioxidant and glucose regulatory properties (Klein et al., 2007).

In preclinical studies, an aqueous extract(equivalent to 1–2 g of dried leaves or 150 mg/kg) or a spray-dried extract of banaba (100 mg/kg) lessened the concentration of blood glucose, whereas banaba aqueous decoctions did not affect blood glucose levels (Matsuura et al., 2004; Miura et al., 2012; Saumya et al., 2011). Different animal studies showed that banaba extracts standardized to corosolic acid and highly purified corosolic acid improved glucose and lipid regulation. After 5 weeks, oral administration of the leaf extracts as a mixture of hot water extract (5%) and methanol extract (2%) to genetically diabetic mice caused significant suppression of blood glucose concentration and reduction of plasma insulin levels, total cholesterol, and the quantity of urinary glucose (Kakuda et al., 1996; Miura et al., 2012). Banaba extract is effective in a dose-dependent manner, while a low dose of banaba extract (0.8 mg/kg) had a poor anti-diabetic effect (Hong et al., 2004). In another study with genetically diabetic mice, 12-week supplementation of a water-soluble banaba extract (0.5%) significantly lessened the concentration of serum glucose, insulin, HbA1c, and triglycerides. Therefore, the banaba extract elevated insulin sensitivity and glucose control by regulating PPAR-mediated lipid metabolism (Park et al., 2005). The hot water (75–90°C) leaf extract of *L. speciosa* has hypoglycemic activity via limitation of gluconeo-genesis and indication of glucose oxidation through the pentose phosphate pathway, augmenting the *G6PD*activity, concentration of glutathione and lowering the action of gluconeogenic enzymes (glucose-6-phosphatase and fructose-1,6-diphosphatase) (Saha et al., 2009).

In other preclinical studies, the hot water extract of *L. speciosa* leaves inhibited the enlargement of blood glucose concentration after starch consumption but not after glucose intake, and limited the activities of α-amylase, glucoamylase, isomaltase, maltase, and sucrase (Suzuki et al., 2001). A 9-week administration of isolated corosolic acid (0.023%) was efficient in the reduction of FPG concentration, levels of insulin and triglyc-erides, and body weight. Additionally, corosolic acid in the diet enhanced the expression of PPAR-α in the liver and PPAR-γ in white adipose tissues, thereby elucidating the mechanism of body weight loss and reduction of hepatic steatosis in mice (Yamada et al., 2008a). Moreover, corosolic acid (20–100μm) reduced gluconeogenesis in a dose-dependent manner by augmenting of fructose-2,6 diphosphate formation, decreasing levels of cyclic AMP and inhibiting an activity of protein kinase A, and enhanced glucokinase action without affecting glucose-6-phosphatase activity, pointing to an intensification of glycolysis (Yamada et al., 2008b). A single dose of corosolic acid (10 mg/kg) is sufficient for a significant diminution of blood sugar concentration. This impact was associated with enhancing the translocation of GLUT4 in the muscle (Miura et al., 2004). In another study, 2-week administration of corosolic acid (2 mg/kg) lessened the concentration of blood sugar, supporting the hypothesis that this acid ameliorates glucose metabolism by suppressing insulin resistance (Miura et al., 2006). Corosolic acid (10 mg/kg) suspended in water also suppressed the intestinal hydrolysis of sucrose but not lactose or maltose (Takagi et al., 2008).

In another preclinical study with diabetic rats, another active substance of banaba, i.e., oleanolic acid, isolated from leaves exhibited α-glucosidase activity (Hou et al., 2009). Daily intake of oleanolic acid significantly lessened the concentration of blood glucose in diabetic animals with simultaneous regeneration of hepatic and muscle glycogen stocks to almost normal amounts and, in combination with insulin therapy, caused the potentiation of antihyperglycemic activity (Miura et al., 2012).

In human clinical trials, two products containing a water-soluble banaba extract and other ingredients lessened the concentration of blood glucose in T2DM patients without causing any adverse reactions. Besides banaba, the first product manufactured in a form of tablets comprised extracts from green tea, green coffee, and *Garcinia* (3 tablets 3 times a day). Other products in this form contained such bioactive compounds as extracts from banaba (16 mg), bitter melon (100 mg), *Gymnema* (133 mg), and *Garcinia cambogia* which extract was standardized in 60% hydroxycitric acid (1500 mg), and yet Bioperine from black pepper (2.6 mg), wheat amylase inhibitors (10 mg), and addition 3 elements: chromium (167 mcg), vanadium (50 mcg), and magnesium (50 mg). Nevertheless, which of the active substances in the products was responsible for the antidiabetic effect has not been determined (Ikeda et al., 1999, Lieberman et al., 2005). Moreover, a banaba extract standardized to 1% corosolic acid (0.32 and 0.48 mg corosolic acid in a soft gel capsule taken daily for 2 weeks) lessened the concentration of blood glucose in T2DM patients (Judy et al., 2002).

In other clinical studies with non-diabetic subjects, after 2 weeks of daily oral administration of one soft gel capsule comprising corosolic acid (10 mg) as a banaba extract standardized to 18% corosolic acid, lessened levels of FPG and 60-min PPG with weight loss and without adverse reactions were reported. This dose of this extract contributed to the mitigation of diabetic symptoms, as it reduced drowsiness, thirst, and hunger. Furthermore, no adverse reactions were noted with no variation in blood pressure, kidney or liver function, and numbers of blood cells or hemoglobin (Miura et al., 2012, Tsuchibe et al., 2006). In this beneficial effect of *L. speciosa* and corosolic acid with related to glucose metabolism is involved many mechanisms, including elevation of cellular glucose uptake, impairment of sucrose and starch hydrolysis, and suppression of gluconeogenesis (Stohs et al., 2012).

10.3.27 MICROBIAL ORIGIN COMPOUNDS

Substances of microbial origin such as acarbose, miglitol, and voglibose are frequently used as natural agents against diabetes (Table 10.1). These are the most commonly used digestive enzyme inhibitors to treat T2DM. Acarbose acts on α-glucosidase, sucrose, α-amylase, and maltase, but it does not possess insulinotropic properties (Bedekar et al., 2010; Wehmeier, 2003; Wehmeier, 2004). Acarbose is used as an antidiabetic drug with IC_{50} value of 9.04 μg/ml. Inhibition of α-glucosidase activity allows to control of glycemia and by dint of glucosidase inhibitors are generally used in T2DM therapy (Moelands et al., 2018; Rios et al., 2015). However, acarbose is currently restricted because of its poor efficacy in lessening glycemia and causing uncomfortable side effects, which are not entirely acceptable by both diabetics and physicians (Rios et al., 2015).

TABLE 10.1 The Origin of Anti-Diabetic Microbiological Agents

Antidiabetic Microbiological Agents	Origin
Acarbose	*Actinoplanes* sp.
Miglitol	*Bacillus* and *Streptomyces*
Voglibose	*Streptomyces*

Miglitol, as an oral antihyperglycemic agent in the T2DM treatment, is pseudo monosaccharide α-glucosidase inhibitor, which makes improvements to glycemic control. It lessens the levels of FPG, PPG and HbA1c. Miglitol is usually well-tolerated and is not correlated with body weight gain or hypoglycemia when it is taken as monotherapy. The substance is absorbed systemically, however, it is not metabolized and is fast excreted by kidneys. Clinical studies with miglitol (predominately 50 or 100 mg, 3 times a day) in T2DM patients revealed a significant amelioration in glycemic control for periods of 6 to 12 months. In the comparative studies, lower miglitol therapeutic dose had comparable efficacy to a higher acarbose dose (50 and 100 mg, 3 times a day, respectively). In turn, in combination with insulin therapy or other oral antidiabetic substances, miglitol intensified glycemic control in individuals whose metabolic control was suboptimal despite the use of diet and concomitant medications. Most adverse reactions after miglitol uptake are involved in disturbances of the gastrointestinal tract such as abdominal pain, diarrhea, and flatulence. These complaints are generally dependent on the miglitol dose, mild to moderate in intensity, initially transpire and diminish with time. Attenuation of side effects is resolved quickly on miglitol discontinuation or dosage adjustment. Miglitol also proved no significant influence on cardiovascular, renal, respiratory, or hematological function in long term studies (Scott and Spencer, 2000).

Voglibose is also an α-glucosidase inhibitor which has been found to significantly decrease the rapid rise in PPG. Voglibose reduced blood glucose concentration after uptake of maltose, sucrose, and starch but not upon administration of glucose, fructose, and lactose, something that is in accordance with its mechanism of action, wherein it binds to enzymes which break down complex starch and disaccharides into simple sugars. Therefore, voglibose inhibits the hydrolysis of oligo-, di-, and trisaccharides to glucose and other monosaccharides, thus lessening the postprandial elevation of blood glucose. Voglibose also demonstrated the diminution of serum insulin and serum glucose concentration (Bedekar et al., 2010; Ritz et al., 2013; Kim et al., 2018). In a double-blind randomized clinical trial, voglibose intake (0.2 mg three times a day) as an addition to lifestyle modification delayed the development of T2DM in individuals with attenuated glucose tolerance (Kawamori et al., 2009; Moelands et al., 2018).

10.3.28 MOMORDICA CHARANTIA (COMMON NAME: BITTER MELON)

Based on the pre-clinical and clinical studies, *Momordica charantia* fruit extracts, leaves extracts, and seed extracts may exert a hypoglycemic activity. Charantin, vicine, and polypeptide-p are the active substances mostly responsible for the observed effects. Following Zhu and colleagues (2012), the administration of *M. charantia* extracts results in an increased cellular glucose uptake and protects the pancreas (including the islet β-cells), while Yibchok-anun et al. (2006) confirmed its secretagogue activity as well as stimulating effects on glycogen storage. Most probably, the extracts up-regulate GLUT4, PI3K, and PPAR-γ and influence the cellular insulin signaling pathways (Zhu et al., 2012). However, not all clinical trials confirm the results of the pre-clinical experiments. 4-week administration of the dry bitter melon extract (given at a dose of 6 g) did not improve FPG, elevated PPG, and fructosamine levels in T2DM patients (John et al., 2003). Similarly, there was no significant difference in the HbA1c and FPG in T2DM after a 3-month treatment with *M. charantia* capsules (Dans et al., 2007). Though the 4-week administration of the 2000 mg/day of bitter melon dried fruit pulp significantly reduced fructosamine levels in T2DM patients, it did not significantly influence the FPG and the 2-h PPG levels (Fuangchan et al., 2011). On the other hand, scientific teams conducted by Ahmad (1999) and Welihinda (1986) observed the beneficial effects of the fruit juices of *M. charantia* given to diabetic patients. Significant reduction in the FPG and/or the PPG levels were reported (Ahmad et al., 1999; Welihinda et al., 1986). Bitter melon extracts are generally regarded as safe and well-tolerated by patients, though several cases of abdominal pain and diarrhea have been reported (Dans et al., 2007).

10.3.29 *PANAX GINSENG(COMMON NAME: GINSENG)*

Amongst several commercially available ginseng varieties, there are two particularly widely consumed: *Panax ginseng* and *Panax quinquefolius*. It has been found out that root extracts, leaves, and berries of ginseng influence blood glucose. Ginsenosides are particularly responsible for the observed effects. Antihyperglycemic activity of *P. ginseng* berry extract and ginsenoside Re was confirmed in the experiments by Attele et al. (2002). After 12 days of intraperitoneal treatment with 150 mg/kg of the above-mentioned extract in diabetic mice, a significant improvement of glucose tolerance, a decrease in serum insulin levels, and increased insulin-stimulated glucose disposal were recorded. Waki et al. (1982) showed that components of ginseng radix can stimulate the biosynthesis of insulin. According to findings by several research teams, administration of ginseng therapy for 4–12 weeks to patients with diabetes mellitus or glucose intolerance may reduce the serum and whole blood glucose (Bang et al., 2014), including FPG (Zhang et al., 2007; Park et al., 2014) and PPG (Oh et al., 2014), and may moderately improve HbA1c levels (Yoon et al., 2012). However, it seems that ginseng preparations do not improve the β-cell function and insulin sensitivity (Reeds et al., 2011; Zhang et al., 2007). Following Gui et al. (2016) administration of ginseng extracts is more effective in patients who are not subjected to insulin treatment and who do not take antidiabetic drugs. The molecular mechanism of the hypo-glycemic potential of ginseng preparations has not been fully understood yet, but literature data give evidences that the active substances of ginseng extracts may affect insulin production and secretion, glucose metabolism and uptake, and may modulate the inflammatory pathways. Moreover, ginsenosides may activate the AMP-activated protein kinase (AMPK) signaling, which is most probably responsible for the suppression of hepatic steatosis and gluconeogen-esis (Bai et al., 2018). Toxicological studies over ginseng preparations shown that they are safe and well-tolerated, with the recommended daily dose of 3 g (Dharmananda, 2002). No adverse reactions were observed in hemostatic, kidney, or liver functions after long-term ginseng consumption by T2DM patients (Mucalo et al., 2014).

10.3.30 *PUNICA GRANATUM (COMMON NAME: POMEGRANATE)*

The antihyperglycemic effects of pomegranate are well-known in traditional medicine in the Middle East. The beneficial influence of the pomegranate juice, pomegranate extracts as well as their active components on the pathologic markers of T2DM has been demonstrated in the laboratory, pre-clinical, and clinical studies. Therefore, the pomegranate products seem to be an effective and safe treatment option for T2DM patients. Moreover, due to their anti-resist in activity (Makino-Wakagi et al., 2012), they may be also beneficial in preventing the obesity-induced diabetes. Though the exact mechanism of the antidiabetic activity of pomegranate has not been fully understood yet, it is known that the pomegranate products neutralize the reactive oxygen species (ROS) (Tzulker et al., 2007), reduce lipid peroxi-dation and oxidative stress (Hunet et al., 1990; Bierhaus et al., 1997) and activation of the nuclear factor κB (Schubert et al., 2002), and increase the activity of PPAR-γ (Huang et al., 2005), as well as the activity of the antioxi-dant enzymes, including CAT, paraoxonase 1 (PON1) and SOD (Rock et al., 2008; Mohan et al., 2010). All the above-mentioned factors are involved in the pathophysiology of diabetes.

The results of the cell culture experiments have revealed that different kinds of pomegranate products (such as pomegranate wine, pomegranate juice, extracts of pomegranate flowers, or peel) produce the antioxidant effects and affect the metabolism of carbohydrates. After exposure to the pomegranate preparations, an inhibition of the nuclear factor κB (NF-κB) in the vascular-endothelial cells (Schubert et al., 2010), enhancement of the PPAR-γ in the human acute monocytic leukemia cell line-1-differentiated macrophages (Huang et al., 2005a), inhibition of the H_2O_2-induced lipid peroxidation in the rat red blood cells (Parmar et al., 2008), decrease in secretion and intracellular levels of resistin in the differentiated murine 3T3-L1 adipocytes (Makino-Wakagi et al., 2012), and enhancement of insulin release from the β-tumor cell line (Koren-Gluzer et al., 2011) were observed. Most probably, punicic acid, punicalagin, and ellagic acid were responsible for the majority of these effects (Hontecillas et al., 2009; Koren-Gluzer et al., 2011; Makino-Wakagi et al., 2012). Similarly, encouraging results have been obtained in the studies performed in rodents. It was shown that the pomegranate peel extract given at a dose of 200 mg/kg reduces the serum level of glucose (Parmar et al., 2007; Parmar et al., 2008), increases the concentration of insulin (Parmar et al., 2007) and exerts an anti-lipid peroxidation activity in several organs (i.e., in the heart, liver, and kidneys) (Parmar et al., 2007; Parmar et al., 2008). Moreover, the pomegranate flower extract administered orally to the diabetic rats for several weeks at a dose of 400–500 mg/kg produces a hypoglycemic activity (Jafri et al., 2000), reduces FPG (Bagri et al., 2009), the cardiac, serum, and total levels of triglycerides, the total values of cholesterol and fatty acids (Huang et al., 2005b; Bagri et al., 2009) it prevents the enhance of the cardiac fatty acids uptake and their oxidation (Huang et al., 2005b), as well as it elevates the concentration of high-density lipoprotein, glutathione, glutathione reductase, GPX, glutathione-S-transferase, SOD, and CAT (Bagri et al., 2009). According to Huang and colleagues (2005a), Gallic acid is mainly responsible for the antidiabetic properties of the pomegranate flower extract, and the observed antidiabetic effects may be due to elevated sensitivity of the insulin receptor. Additionally, the pomegranate flower extract modulates the NF-κB pathway, cardiac endothelin-1 (Huang et al., 2005c), and the hepatic PPAR-α, acyl-CoA oxidase, and carnitine palmitoyltransferase-1 (Xu et al., 2009). The hypoglycemic effects were also demonstrated for the seed extract of pomegranate, when given at the high doses. Following Das et al. (2001), rats with streptozotocin-induced diabetes presented significantly decreased levels of blood glucose after oral administration of the pomegranate seed extract (300 and 600 mg/kg). In the experiments by McFarlin (2009), the pomegranate seed oil improved the insulin sensitivity and thus, it reduced the risk of T2DM in mice. Pomegranate juice, rich in variety of bioactive compounds, including coumaric acids, phenolic acids, non-phenolic acids (e.g., ascorbic acid), tannins (ellagitannins and gallotannins), flavonoids (e.g., catechin and quercetin), flavanols (e.g., catechin, epicatechin, and epigallocatechin), and anthocyanins (Banihani et al., 2013). According to literature data, oral administration of the pomegranate juice to diabetic mice or diabetic rats increased the cellular glutathione level (Rozenberg et al., 2006) and the plasma concentrations of stable endogenous nitric oxide (de Nigris et al., 2007), reduced oxidative stress (Mohan et al., 2010), decreased the total peroxide level in the peritoneal macrophages, the activity of macrophage paraoxonase 2 (Rozenberg et al., 2006), and the serum values of resistin (Makino-Wakagi et al., 2012). As was demonstrated by Vroegrijk et al. (2011), the pomegranate seed oil given at a dose of 1 g/kg, improved insulin resistance in the experimentally obese mice. These observations were in line with studies by Hontecillas and colleagues (2009), who had found out that 30-day supplementation of the main ingredient of the pomegranate seed oil, i.e., punicic acid (consumed at a dose of 1 g per 100 g of the feed) reduced the values of FPG in obese mice. Hontecillas et al. (2009) also reported that punicic acid inhibits the activation of NF-κB, reduces the expression of tumor necrosis factor-α (TNF-α) and up-regulates PPAR-α- and γ-responsive genes in the adipose tissue and the skeletal muscles.

As for the human studies focused on the antidiabetic effects of pomegranates, a great majority of them assessed the activity of the pomegranate juice. It was demonstrated that consumption of the pomegranate juice for several weeks by diabetic patients reduces the serum values of total cholesterol, decreases the level of lipid peroxidation, and antagonizes the oxidative stress induced by hyperglycemia (Basu et al., 2009; Esmaillzadeh et al., 2004; Fenercioglu et al., 2010). Moreover, Rock et al. (2008) suggested that the regular intake of the pomegranate juice may hold up the development of atherosclerosis in patients suffering from T2DM. Interestingly enough, sugars contained in the pomegranate juice have no negative effect on T2DM variables (Rozenberg et al., 2006).

10.3.31 *SILYBUM MARIANUM (COMMON NAME: MILK THISTLE)*

Silybum marianum contains flavonolignans. Amongst them, silibinin is the mostly studied one. Silibin belongs to the inhibitors of AR, i.e., an enzyme partially responsible for the development of complications of diabetes. *In vitro* studies showed that silibinin had positive effects on diabetic neuropathy (Di Giulio et al., 1999), reduced both gluconeogenesis and glycolysis (Guigas et al., 2007), and increased viability of pancreatic β-cells (Cheng et al., 2012). Moreover, it has been demonstrated that isosylibin A acts as an agonist of PPARγ (Pferschy-Wenzig et al., 2014). Silibin significantly reduced glucose levels in serum of diabetic rats and restored HbA1c, insulin values, with an effect significantly more pronounced when it was used in a form of nanoparticles (Soto et al., 2003; Das et al., 2014). This flavonolignan exerts a potent antioxidant effect (Soto et al., 1998; Soto et al., 2004) and according to Zhang et al. (2014), has a potential to improve diabetic retinopathy. Human studies confirmed observations from the animal ones. Silymarin, i.e., a dry mixture of flavonolignans from milk thistle, given for 6 months at a dose of 600 mg/day reduced FPG and values of mean daily glucose, starting from the second month of therapy. It also decreased HbA1c levels and insulin resistance (Velussi et al., 1993). Similar results were obtained by Huseini et al. (2006), after a 2-month therapy with the same dose of silymarin. The authors suggested that silymarin effects are not dependent on insulin production. Silymarin seems to be safe and well-tolerated, since no adverse reactions were observed after the dose of 2100 mg taken for 7 days (Hawke et al., 2010). Other clinical studies did not report any alarming side effects related to the prolonged (up to 223 months) silymarin therapy (Gordon et al., 2006; Angulo et al., 2008). Isolated cases of headache, dizziness, symptoms from the alimentary system, or transient elevation of bilirubin value have been recorded (Gordon et al., 2006; Ferenci and Beinhardt, 2013).

10.3.32 SPICES

According to the available literature data, a variety of spices and their extracts may be beneficial in the management of T2DM. Amongst them, aniseed, black pepper, cinnamon, cloves, coriander, cumin, curry leaves, fenugreek, garlic, ginger, onion, mustard, and turmeric are mentioned the most frequently. The antidiabetic effects of species are principally attributed to the presence of phenolic compounds, including flavonoids, terpenes, vanilloids, and organosulfur substances. All of them possess an antioxidant activity and in consequence, they exert the protective effects against diabetes mellitus.

Cinnamon is one of the most popular spices worldwide. There are several species used as spices such as *Cinnamomum cassia, C. aromaticum,* or

C. zeylanicum. Historically, cinnamon has been used for diabetes treatment. The active substances of cinnamon responsible for the antidiabetic effects are not known precisely, but polyphenol type-A polymers are considered to represent some of the bioactive compounds that may have insulin-mimetic effects (Cefalu et al., 2011).

In preclinical studies with murine diabetes models, antidiabetic properties of cinnamon were observed. It was claimed that cinnamon was a natural insulin sensitizer in adipocytes and in different animal models (Allen et al., 2013). On the other hand, the results of cinnamon action reported in clinical studies are not entirely consistent (Dugoua et al., 2007). However, several studies have suggested a positive impact in some settings. Crawford et al. (2009) evaluated T2DM patients who had been previously treated with diet and exercise. After 90 days, orally administered cinnamon (1 g per day) seemed to be effective and significantly decreased glycemia, as estimated with HbA1c. Other clinical studies also suggested positive reducing effects of cinnamon (1 to 6 g/day) on the levels of glucose and lipids (Anderson, 2008; Khan et al., 2003; Mang et al., 2006), while other human studies failed to reveal an impact of cinnamon on glycemia or lipids (Altschuler et al., 2007; Blevins et al., 2007). Vanschoonbeek et al. (2006) evaluated oral supplementation of cinnamon (1.5 g/day) in postmenopausal T2DM patients and found that it did not ameliorate whole-body insulin sensitivity or oral glucose tolerance. In other clinical randomized controlled trials comparing the effects of cinnamon intake (2 g/daily for a period 4 to 16 weeks) vs. a placebo, active medication, or no treatment, the impact of cinnamon on FPG level was not conclusive (Leach and Kumar, 2012).

However, a meta-analysis has revealed that consumption of about half a teaspoon of cinnamon per day (120 mg/daily to 6 g/daily for 4 to 18 weeks) can result in significant improvement in blood glucose, cholesterol, and triglyceride concentration, and sensitivity to insulin in T2DM patients. Cinnamon has no significant effect on HbA1c (Allen et al., 2013). However, a meta-analysis evaluating the impact of cinnamon on glycemic control in T2DM patients (1 to 6 g a day for 40 days to 4 months) demonstrated a significant diminution of HbA1c and FPG levels (Akilen et al., 2012).

The therapeutic potential of different species of cinnamon (*C. verum* and *C. cassia,* 1 to 10 g/daily for 1 to 3 months) was evaluated in T2DM patients and insulin resistance, especially the ability of these species to lessen the levels of blood glucose and suppress protein glycation. In clinical studies, significant diminution of FPG and in glucose response using OGTT was noted (Kirkham et al., 2009; Kadan et al., 2013).

The hypoglycemic potential of *Trigonella foenum-graecum* has been confirmed in diabetic animals and humans. It seems that fenugreek increases glucose uptake from the blood and reduces its plasma level. Most probably, galactomannan soluble fiber and saponins are responsible for this activity. It has been suggested that saponins increase the activity of SOD, GPX, and CAT in the liver as well as reduce hepatic lipid peroxidation, and thus contribute to the antidiabetic effect (Lu et al., 2008). Moreover, according to Srichamroen et al. (2009), fenugreek may increase viscosity in the gastrointestinal tract, leading to reduced activity of carbohydrates and reduced mobility of glucose. In consequence, both decreased hydrolysis of carbohydrates and decreased glucose absorption by the intestine may be observed. It also has been reported that fenugreek seed extract normal-izes the altered levels of the enzymes crucial for glucose homeostasis (i.e., pyruvate kinase and phosphoenolpyruvate carboxykinase) as well as normalizes the changed distribution of GLUT4 in the skeletal muscles in diabetic rats (Mohammad et al., 2006). The experiments by Vijayakumar et al. (Vijaykumar et al., 2005) carried out in diabetic mice showed that the fenugreek seed extract induces tyrosine phosphorylation of different proteins, including the insulin receptor and stimulates the insulin signaling pathway in adipocytes and liver cells. According to the experiments by Ajabnoor and Tilmisany (1997), 200 mg/kg of *Trigenolla foenum graceum* seed extract given to diabetic and normal mice had the same hypoglycemic effect as 200 mg/kg of tolbutamide, and 15 mg/kg of dry fenugreek extract reduced blood sugar to the similar point as insulin in alloxan-induced diabetic rats (Vijayakumar et al., 2008). T2DM subjects receiving 15 g of powdered fenugreek seed with a meal presented the reduced subsequent levels of PPG (Madar et al., 1998). Administration of the fenugreek seeds extract at a dose of 1 g/day for 2 months as an adjunct treatment to SU or a biguanide improved both glycemic control and insulin resistance in patients with a mild T2DM and significant decrease in FPG (Gupta et al., 2001). Similarly, patients receiving the SU treatment in combination with *T. foenum-graecum*, total saponins presented, reduced levels of FBG and HbA1c (Lu et al., 2008). Kassaian et al. (2009) found out that the administration method of the fenugreek seed extract may be of relevance. They showed that T2DM patients who received 10 g/day of powdered fenugreek seeds soaked in hot water for 8 weeks presented decreased levels of fasting blood sugar (FBS), triglycerides, and VLDL-cholesterol, whereas these parameters did not change in patients who were given equal treatment with powdered fenugreek seeds mixed with yogurt. Ingestion of fenugreek preparations may lead to mild gastrointestinal problems, including abdominal bloating, diarrhea, dyspepsia, or flatulence. Amongst the other side effects, hypoglycemia, hypokalemia, dizziness, or increased frequency of urination may be reported (Haber and Keonavong, 2013; Madar et al., 1998).

The main bioactive compounds of *Zingiber officinale* are gingerol, shogaol, and zerumbone. Whereas shogaol, and zerumbone exert the antioxidant effects, gingerol is particularly responsible for the antidiabetic potential of ginger (Rahman et al., 2014). It was demonstrated that ginger inhibits α-amylase, α-glucosidase, and enzymes involved in gluconeo-genesis and glycogenolysis, enhances insulin sensitivity and its release, increases glucose uptake through translocation of the GLUT4 to the cell surface and improves the lipid profile (Li. et al., 2012). Mozaffari-Khosravi et al. (2014) have shown that daily administration of ginger powder (3 g for 8 weeks) may improve insulin resistance and reduce both FBG and HbA1c values.

Curcumin is responsible for the antidiabetic properties of turmeric. It has been reported that this active substance may reduce the risk of diabetes in pre-diabetes patients (Chuengsamarn et al., 2012). In the laboratory experiments, curcumin increased insulin secretion and elevated glucose intake (Best et al., 2007; Kim et al., 2010). Lekshmi and colleagues (2012) found out that the dried extracts of turmeric volatile oil inhibited α-amylase and α-glucosidase. Moreover, turmeric increases an antioxidant glutathione, reduces inflammation and ROS, and thus, it may exert the beneficial effects on blood glucose. In the clinical study of Selvi et al. (2015), patients with diabetes who received metformin with turmeric supplementation (at a dose of 2 g) presented better FBG and HbA1c values than patients receiving only metformin.

The antidiabetic potential of cumin seed was demonstrated both in the preclinical (Iyer et al., 2009) and clinical studies (Ismail et., 2010). Cumin protects the surviving pancreatic β-cells, increases insulin secretion and glycogen storage. The main insulinotropic active substances of cumin are cumin aldehyde and cuminol. Ismail and colleagues reported that the administration of cumin was more effective than glibenclamide, whereas Bamosa et al. (2010) reported that daily consumption of this species (at a dose of 2 g) with their regular drugs for 12 weeks can significantly decrease the blood sugar levels.

Coriander seeds increase insulin secretion, elevate the glucose transport with glucose incorporation into muscle glycogen, decrease serum lipids, and possess an antioxidant activity (Gray and Flatt, 1999; Rajeshwari et al., 2011). Most probably, this pharmacological potential of coriander seeds is due to the synergistic effect of its numerous active components. The antidiabetic effects of coriander seeds were observed in diabetic mice (Gray and Flatt, 1999) and patients (Rajeshwari et al., 2011). Daily administration of coriander seed powder at a dose of 5 g exerted a significant antihyperglycemic effect in T2DM patients.

Antidiabetic potential of sulfur-containing species, such as garlic and onion, has been confirmed in preclinical and clinical studies (Kochhar et al., 2018). Jelodar and colleagues (2005) demonstrated that the administration of garlic lowered the levels of blood glucose in diabetic rats, whereas Sobenin et al. (2008) reported that the time-released garlic powder tablets lessened the number of both FBG values and triglyceride levels more effectively than a placebo in diabetic patients. As for onion extracts, they stimulated glucose uptake in the muscle cell line L6 (Noipha et al., 2010) and reduced the glycemic response and FBG levels in rats with induced diabetes (Jevas, 2011). Furthermore, the onion peel extract elevated glycogen levels in liver and muscle as well as glucose uptake in diabetic rats. In humans, onion reduces inflammation, stimulates GLUT4, and expression of the insulin receptor, and thus improves glucose intake. Eldin and colleagues (2010) found out that the ingestion of crude *Allium cepa* (100 g) resulted in a significant decrease in FBG concentration as well as induced hyperglycemia in T2DM patients.

Dehydrodieugenol, dehydrodieugenol B, and oleanolic acid are primarily responsible for the hypoglycemic effects of cloves. They activate PPAR-γ and modulate the expression of the gluconeogenesis enzymes (Kuroda et al., 2012). 14-day administration of the clove powder to diabetic rats resulted in reduced FBG values (Prasad et al., 2005). On the other hand, the team led by Khan (2006) reported that 30-day supplementation with capsules containing clove extract (1–3 g) was able to lower serum glucose, total cholesterol, and triglycerides in diabetic patients.

10.3.33 VITIS VINIFERA (COMMON NAME: GRAPES)

Based on literature data, grapes contain a variety of polyphenols (like resveratrol, quercetin, catechins, and anthocyanins) that are able to reduce hyperglycemia, improve functioning of the β-cells, and protect against β-cell loss. Al-Awwadi et al. (2004) demonstrated that administration of a polyphenol extract from a red wine at a dose of 200 mg/kg for 6 weeks to diabetic mice resulted in reduced glycemia. Similar observations were made by Pinent and colleagues (2004) who investigated the antihyperglycemic effects of the grape seed-derived procyanidins. The authors found out that the extract of grape seed procyanidins possesses the insulin-like potential in the insulin-sensitive cells, stimulating both glucose uptake and translocation of the GLUT4 to the plasma membrane. Furthermore, grape seed proanthocyanidins seem to produce protective effects towards pancreatic tissue, and when administered at a dose of 50–100 mg/kg for 72 hours to diabetic rats significantly elevates the pancreatic glutathione levels as well as exerts a potent antioxidant activity (El-Alfy et al., 2005). In the human population, the consumption of anthocyanins and flavonoids is associated with the reduced risk of T2DM, which was confirmed by Wedick et al. (2012) and Liu et al. (2014), respectively. It has been shown that regular consumption of the grape seeds extracts rich in proanthocyanidin or red wine polyphenols leads to a significant increase in the amount of fecal *Bifidobacteria*. High level of fecal *Bifidobacteria* is associated with improved tolerance of glucose and lessened values of inflammatory markers (Yamakoshi et al., 2001; Queipo-Ortuno et al., 2012). According to clinical data, grape polyphenols may have protective effects on fructose-induced oxidative stress as well as on insulin resistance (Morrow, 2005). Consumption of red wine grape pomace at a daily dose of 20 mg for 16 weeks reduced postprandial insulin and FPG (Urquiaga et al., 2015), whereas a 4-week intake of dealcoholized grape wine reduced fasting insulin values in blood and increased FBG:insulin ratio (Banini et al., 2006) in T2DM subjects.

Resveratrol given to the diabetic patients at a dose of 250 mg/day for 3 months had a positive impact on the mean HbA1c levels (Bhatt et al., 2012).

10.4 CONCLUSIONS

Natural substances being easily available and low cost, can be used as a complementary therapy in the management of diabetes, possibly due to the presence of bioactives. A lot of natural substances could be the potential sources in the development of antidiabetic agents. Therefore, the identification of more antihyperglycemic compounds of herbs and combinations of these herbs with other medicinal plants should be a center of researchers' attention. This would allow for better treatment options for diabetic patients, especially type 2. The therapeutic value of natural agents that people can incorporate into everyday life may be an effective approach in the management of diabetic complications.

KEYWORDS

- **antidiabetic agents**
- **antidiabetic medical plants**
- **antihyperglycemic compounds**
- **diabetes mellitus**
- **herbs**
- **natural agents**

REFERENCES

Abdallah, I. Z. A., Salem, I. S., & Abd El-Salam, N. A. S., (2017). Evaluation of antidiabetic and antioxidant activity of *Aegle marmelos* L. Correa fruit extract in diabetic rats. *Egypt J. Hosp. Med., 67*(2), 731–741.

Abo, G. M. M., & Hasan, N. M., (2012). Effect of glabridin on the structure of ileum and pancreas in diabetic rats: A histological, immunohistochemical and ultrastructural study. *Nat. Sci., 10*, 78–90.

Abo-Youssef, A. M. H., & Mesiha, B. A. S., (2013). Beneficial effects of *Aloe Vera* in treatment of diabetes: Comparative *in vivo* and *in vitro* studies. *Bull. Fac. Pharm. Cairo. Univ., 1*(1), 7–11.

Adebayo, J. O., Adesokan, A. A., Olatunji, L. A., Buoro, D. O., & Soladoye, A. O., (2005). "Effect of ethanolic extract of *Bougainvillea spectabilis* leaves on hematological and serum lipid variables in rats. *Biochemistry, 17*, 45–50.

Aggarwal, S., Shailendra, G., Ribnicky, D. M., Burk, D., Karki, N., & Qingxia, W. M. S., (2015). An extract of *Artemisia dracunculus* L. stimulates insulin secretion from β cells, activates AMPK, and suppresses inflammation. *J. Ethnopharmacol., 170*, 98–105.

Ahmad, N., Hassan, M. R., Halder, H., & Bennoor, K. S., (1999). Effect of Momordica charantia (Karolla) extracts on fasting and postprandial serum glucose levels in NIDDM patients. *Bangladesh Med. Res. Counc. Bull., 25*(1), 11–13.

Ahn, J., Um, M. Y., Lee, H., Jung, C. H., Heo, S. H., & Ha, T. Y., (2013). Eleutheroside E, an active component of *Eleutherococcus senticosus*, ameliorates insulin resistance in type 2 diabetic db/db mice. *eCAM, 934183*, 1–9.

Ajabnoor, M. A., & Tilmisany, A. K., (1988). Effect of *Trigonella foenumgraecum* on blood glucose levels in normal and alloxandiabetic mice. *J. Ethnopharmacol., 22*, 45–49.

Akhtar, M. S., Ramzan, A., Ali, A., & Ahmad, M., (2011). Effect of Amla fruit (*Emblica officinalis* Gaertn.) on blood glucose and lipid profile of normal subjects and type 2 diabetic patients. *Int. J. Food Sci. Nutr., 62*, 609–616.

Akilen, R., Tsiami, A., Devendra, D., & Robinson, N., (2012). Cinnamon in glycaemic control. Systematic review and meta analysis. *Clin. Nutr., 31*, 609–615

Al-Awwadi, N., Azay, J., Poucheret, P., Cassanas, G., Krosniak, M., Auger, C., Gasc, F., et al., (2004). Antidiabetic activity of red wine polyphenolic extract, ethanol, or both in streptozotocin-treated rats. *J. Agric. Food Chem., 52*, 1008–1016.

Allen, R. W., Schwartzman, E., Baker, W. L., Coleman, C. I., & Phung, O. J., (2013). Cinnamon use in type 2 diabetes: An updated systematic review and meta-analysis. *Ann. Fam. Med., 11*(5), 452–9.

Alqahtani, A., Hamid, K., Kam, A., Wong, K. H., Abdelhak, Z., Razmovski-Naumovski, V., Chan, K., et al., (2013). The pentacyclic triterpenoids in herbal medicines and tier pharmacological activities in diabetes and diabetic complication. *Curr. Med. Chem., 20*, 908–931.

Altschuler, J. A., Casella, S. J., MacKenzie, T. A., & Curtis, K. M., (2007). The effect of cinnamon on A1C among adolescents with type 1 diabetes. *Diab Care., 30*, 813–816

American Diabetes Association (ADA), (2013). Diagnosis and classification of diabetes mellitus. *Diab Care, 36*, S67–S79.

American Diabetes Association (ADA), (2018). Classification and diagnosis of diabetes: Standards of medical care in diabetes 2018. *Diab. Care, 41*(1), S13–S27.

Anderson, R. A., (2008). Chromium and polyphenols from cinnamon improve insulin sensitivity. *Proc. Nutr. So., 67*, 48–53.

Angulo, P., Jorgensen, R. A., Kowdley, K. V., & Lindor, K. D., (2008). Silymarin in the treatment of patients with primary sclerosing cholangitis: An open-label pilot study. *Dig. Dis. Sci.*, 53, 1716–1720.

Ansari, A., Shahriar, S. Z., Hassan, M., Das, S. R., Rokeya, B., Haque, A., Haque, E., et al., (2014). *Emblica officinalis* improves glycemic status and oxidative stress in STZ induced type 2 diabetic model rat. *Asian Pac. J. Trop. Med., 7*(1), 21–25.

Asadi, F., Pourkabir, M., Maclaren, R., & Shahriari, A., (2006). Alterations to lipid parameters in response to fig tree (Ficus carica) leaf extract in chicken liver slices. *Turk J. Vet. Anim. Sci., 30*, 315–318.

Attele, A. S., Zhou, Y. P., Xie, J. T., Wu, J. A., Zhang, L., Dey, L., Pugh, W., et al., (2002). Antidiabetic effects of *Panax ginseng* berry extract and the identification of an effective component. *Diabetes, 51*, 1851–1858.

Augustine, A. W., Narasimhan, A., Vishwanathan, M., & Karundevi, B., (2014). Evaluation of antidiabetic property of *Andrographis paniculata* powder in high fat and sucrose-induced type-2 diabetic adult male rat. *Asian Pac. J. Trop. Dis., 4*(1), S140–S147.

Babiker, R., Elmusharaf, K., Keogh, M. B., Banaga, A. S. I., & Saeed, A. M., (2017). Metabolic effect of Gum Arabic (*Acacia Senegal*) in patients with Type 2 diabetes mellitus (T2DM): Randomized, placebo-controlled double-blind trial. *FFHD, 7*(3), 219–231.

Badgujar, S. B., Patel, V. V., Bandivdekar, A. H., & Mahajan, R. T., (2014). Traditional uses phytochemistry and pharmacology of Ficus carica: A review. *Pharm. Biol., 52*, 1487–1503.

Bagri, P., Ali, M., Aeri, V., Bhowmik, M., & Sultana, S., (2009). Antidiabetic effect of *Punica granatum* flowers: Effect on hyperlipidemia, pancreatic cells lipid peroxidation, and antioxidant enzymes in experimental diabetes. *Food Chem. Toxicol., 47*, 50–54.

Bahijiri, S. M., Mira, S. A., Mufti, A. M., & Ajabnoor, M. A., (2000). The effects of inorganic chromium and brewer's yeast supplementation on glucose tolerance, serum lipids and drug dosage in individuals with type 2 diabetes. *Saudi Med. J., 21*, 831–837.

Bai, L., Gao, J., Wei, F., Zhao, J., Wang, D., & Wei, J., (2018). Therapeutic potential of Ginsenosides as an adjuvant treatment for diabetes. *Front Pharmacol., 9*, 423.

Bailey, C. J., & Day, C., (2004). Metformin: Its botanical background. *Pract. Diab. Int., 21*(3), 115–117.

Bais, S., Gill, N. S., Rana, N., & Shandil S., (2014). A phytopharmacological review on a medicinal plant: *Juniperus communis*. *Int. Sch. Res. Notices, 634723*, 1–6.

Balk, E. M., Tatsioni, A., Lichtenstein, A. H., Lau, J., & Pittas, A. G., (2007). Effect of chromium supplementation on glucose metabolism and lipids: A systematic review of randomized controlled trials. *Diabetes Care, 30*, 2154–2163.

Bamosa, A. O., Kaatabi, H., Lebda, F. M., Elq, S. M., & Al-Sultanb, A., (2010). Effect of Nigella Sativa seeds on the glycemic control of patients with type 2 diabetes mellitus. *Indian J. Physiol. Pharmacol., 54*, 344–354.

Banerjee, S., Singh, H. O., & Chatterjee, T. K., (2013). Evaluation of anti-diabetic and anti-hyperlipidemic potential of methanolic extract of *Juniperus communis* (L.) in *Streptozotocin nicotinamide* induced diabetic rats. *Int. J. Pharma. Bio. Sci., 4*(3), 10–17.

Bang, H., Kwak, J. H., Ahn, H. Y., Shin, D. Y., & Lee, J. H., (2014). Korean red ginseng improves glucose control in subjects with impaired fasting glucose, impaired glucose tolerance, or newly diagnosed type 2 diabetes mellitus. *J. Med. Food, 17*, 128–134.

Banihani, S., Swedana, S., & Alguraanb, Z., (2013). Pomegranate and type 2 diabetes. *Nutr. Res., 33*, 341–348.

Banini, A. E., Boyd, L. C., Allen, J. C., Allen, H. G., & Sauls, D. L., (2006). Muscadine grape products intake, diet and blood constituents of non-diabetic and type 2 diabetic subjects. *Nutrition, 22*, 1137–1145.

Basu, A., & Penugond, K., (2009). Pomegranate juice: A heart-healthy fruit juice. *Nutr. Rev., 67*, 49–56.

Bedekar, A., Shah, K., & Koffas, M., (2010). Natural products for type II diabetes treatment. *Adv. Appl. Microbiol., 71*, 21–73

Belguith-Hadriche, O., Ammar, S., Contreras, M. M., Turki, M., Sequra-Carretero, A., Ei Feki, A., Makni-Ayedi, F., & Bouaziz, M., (2016). Antihyperlipidemic and antioxidant activates of edible Tunisian Ficus carica L. Fruits in high fat diet induced hyperlipidemic rats. *Plant Foods Hum. Nutr., 71*, 183–189.

Bellamkonda, R., Rasineni, K., Singareddy, S. R., Kasetti, R., Pasurla, R., Chippada, A., & Desireddy, S., (2011). Antihyperglycemic and antioxidant activities of alcoholic extract of *Commiphora mukul* gum resin in streptozotocin-induced diabetic rats. *Pathophysiology, 18*, 255–261.

Best, L., Elliott, A. C., & Brown, P. D., (2007). Curcumin induces electrical activity in rat pancreatic b-cells by activating the volume-regulated anion channel. *Biochem. Pharmacol., 73*, 1768–1775.

Bhat, M., Kothiwale, S. K., Tirmale, A. R., Bhargava, S. Y., & Joshi, B. N., (2011b). Antidiabetic properties of *Azardiracta indica* and *Bougainvillea spectabilis*: *In vivo* studies in murine diabetes model. *Evid. Based Complement Alternat. Med., 561625*, 1–9.

Bhat, M., Zinjarde, S. S., Bhargava, S. Y., Kumar, A. R., & Joshi, B. N., (2011a). Antidiabetic Indian plants: A good source of potent amylase inhibitors. *Evid. Based Complement Alternat. Med.*, 810207, 1–6.

Bhatt, J. K., Thomas, S., & Nanjan, M. J., (2012). Resveratrol supplementation improves glycemic control in type 2 diabetes mellitus. *Nutr. Res., 32*, 537–541.

Bierhaus, A., Chevion, S., Chevion, M., Hofmann, M., Quehenberger, P., Illmer, T., Luther, T., Berentshtein, E., et al., (1997). Advanced glycation end product-induced activation of NF-kappaB is suppressed by alpha-lipoic acid in cultured endothelial cells. *Diabetes, 46*, 1481–1490.

Bobiş, O., Dezmirean, D. S., & Moise, A. R., (2018). Honey and diabetes: The importance of natural simple sugars in diet for preventing and treating different type of diabetes. *Oxid. Med. Cell Longev., 4757893*, 1–12.

Bokaeian, M., Nakhaee, A., & Moodi, B., (2012). Effect of *E. globulus* upon Candida colonization in normal and diabetic rats. *ZJRMS, 13*(9), 21–26.

Bunyapraphatsara, N., Yongchaiyudha, S., Rungpitarangsi, V., & Chokechaijaroenporn, O., (1996). Antidiabetic activity of *Aloe Vera* L juice. II. Clinical trial in diabetes mellitus patients in combination with glibenclamide. *Phytomed., 3*, 245–254.

Cefalu, W. T., Stephens, J. M., & Ribnicky, D. M., (2011). Chapter 19. Diabetes and herbal (Botanical) medicine. In: Benzie, I. F. F., & Wachtel-Galor, S., (eds.), *Herbal Medicine: Biomolecular and Clinical Aspects* (2nd edn.). CRC Press. Chapter 19.

Chander, R., Khanna, A. K., & Pratap, R., (2002). Antioxidant activity of guggulsterone, the active principle of guggulipid from *Commiphora* mukul. *J. Med. Arom. Plant Sci., 24*, 1142–1148.

Chaturvedi, A., Bhawani, G., Agarwal, P. K., Goel, S., Singh, A., & Goel, R. K., (2009). Antidiabetic and antiulcer effects of extract of *Eugenia jambolana* seed in mild diabetic rats: Study on gastric mucosal offensive acid-pepsin secretion. *Indian J. Physiol. Pharmacol., 53*(2), 137–146.

Cheang, K. U., Chen, C. M., Chen, C. Y. O., Liang, F. Y., Shih, C. K., & Li, S. C., (**2017**). Effects of glucomannan noodle on diabetes risk factors in patients with metabolic syndrome: A double-blinded, randomized crossover controlled trial. *J. Food Nutr. Res., 5*(8), 622–628.

Chen, H. L., Sheu, W. H., Tai, T. S., Liaw, Y. P., & Chen, Y. C., (2003). Konjac supplement alleviated hypercholesterolemia and hyperglycemia in type 2 diabetic subjects-a randomized double-blind trial. *J. Am. Coll. Nutr., 22* (1), 36–42.

Chen, X., Zheng, Y., & Shen, Y., (2006). Voglibose (Basen, AO-128), one of the most important α-glucosidase inhibitors. *Curr. Med. Chem., 13*, 109–116.

Cheng, B., Gong, H., Li, X., Sun, Y., Zhang, X., Chen, H., Liu, X., et al., (2012). Silibinin inhibits the toxic aggregation of human islet amyloid polypeptide. *Biochem. Biophys. Res. Commun., 419*, 495–499.

Choi, E. M., (2005). The licorice root derived isoflavan glabridin increases the function of osteoblastic MC3T3-E1 cells. *Biochem. Pharm, 70*, 363–368.

Choi, H. C., Kim, S. J., Son, K. Y., Oh, B. J., & Cho, B. L., (2013). Metabolic effects of Aloe vera gel complex in obese prediabetes and early non-treated diabetic patients: Randomized controlled trial. *Nutrition, 29*, 1110–1114

Chua, M., Baldwin, T. C., Hocking, T. J., & Chan, K., (2010). Traditional uses and potential health benefits of *Amorphophallus konjac* K. Koch ex N.E.Br. *J, Ethnopharmacol., 128*(2), 268–278.

Chuengsamarn, S., Rattanamongkolgul, S., Luechapudiporn, R., Phisalaphong, C., & Jirawatnotai, S., (2012). Curcumin extract for prevention of type 2 diabetes. *Diab. Care., 35*, 2121–2127.

Comelli, F., Bettoni, I., Colleoni, M., Giagnoni, G., & Costa, B., (2009). Beneficial effects of a Cannabis sativa extract treatment on diabetes-induced neuropathy and oxidative stress. *Phytother. Res., 23*(12), 1678–1684.

Cornick, C. L., Strongitharm, B. H., Sassano, G., Rawlins, C., Mayes, A. E., Joseph, A. N., O'Dowd, J., Stocker, C., Wargent, E., Cawthorne, M. A., Brown, A. L., & Arch, J. R., (2009). Identification of a novel agonist of peroxisome proliferator-activated receptors alpha and gamma that may contribute to the anti-diabetic activity of guggulipid in Lep(ob)/Lep(ob) mice. *J. Nutr. Biochem., 20*(10), 806–815.

Cui, K., Zhang, S., Jiang, X., & Xie, W., (2016). Novel synergic antidiabetic effects of Astragalus polysaccharides combined with Crataegus flavonoids via improvement of islet function and liver metabolism. *Mol. Med. Rep., 13*(6), 4737–4744.

Cui, Y., Wang, Q., Sun, R., Guo, L., Wang, M., Jia, J., et al., (2018). *Astragalus membranaceus* (Fisch.) Bunge repairs intestinal mucosal injury induced by LPS in mice. *BMC Complement Altern. Med., 18*, 230.

D'souza, J. J., D'souza, P. P., Fazal, F., Kumar, A., Bhat, H. P., & Baliga, M. S., (2014). Anti-diabetic effects of the Indian indigenous fruit *Emblica officinalis* Gaertn: Active constituents and modes of action. *Food Funct., 5*(4), 635–644.

Dallaqua, B., Saito, F. H., Rodrigues, T., Calderon, I. M. P., Rudge, M. V. C., Herrera, E., & Damasceno, D. C., (2012). Treatment with *Azadirachta indica* in diabetic pregnant rats: Negative effects on maternal outcome. *J. Ethnopharmacol., 143*(3), 805–811.

Danesh, J., Erqou, S., Walker, M., & Thompson, S. G., (2007). The emerging risk factors collaboration: Analysis of individual data on lipid, inflammatory and other markers in over 1.1 million participants in 104 prospective studies of cardiovascular diseases. *Eur. J. Epidemiol., 22*(12), 839–869.

Dans, A. M., Villarruz, M. V., Jimeno, C. A., Javelosa, M. A., Chua, J., Bautista, R., & Velez, G. G., (2007). The effect of *Momordica charantia* capsule preparation on glycemic control in type 2 diabetes mellitus needs further studies. *J. Clin. Epidemiol., 60*(6), 554–559.

Das, A. K., Mandal, S. C., Banerjee, S. K., Sinha, S., Saha, B. P., & Pal, M., (2001). Studies on the hypoglycaemic activity of *Punica granatum* seed in streptozotocin induced diabetic rats. *Phytother. Res., 15*, 628–629.

Das, S., Roy, P., Pal, R., Auddy, R. G., Chakraborti, A. S., & Mukherjee, A., (2014). Engineered silybin nanoparticles reduce efficient control in experimental diabetes. *Plos One., 9*, e101818.

De Medina, S. F., Gamez, M. J., Jiménez, I., Jiménez, J., Osuna, J. I., & Zarzuelo, A., (1994). Hypoglycemic activity of juniper "berries." *Planta Med., 60*(3), 197–200.

De Nigris, F., Balestrieri, M. L., Williams-Ignarro, S., D'Armiento, F. P., Fiorito, C., Ignarro, L. J., & Napoli, C., (2007). The influence of pomegranate fruit extract in comparison to regular pomegranate juice and seed oil on nitric oxide and arterial function in obese zucker rats. *Nitric Oxide, 17*, 50–54.

Després, J. P., Golay, A., & Sjöström, L., (2005). Rimonabant in obesity-lipids study group. Effects of rimonabant on metabolic risk factors in overweight patients with dyslipidemia. *N Engl. J. Med., 353*, 2121–2134.

Dey, B., & Mitra, A., (2013). Chemo-profiling of eucalyptus and study of its hypoglycemic potential. *World J. Diabetes., 4*(5), 170–176.

Dey, L., Attele, A. S., & Yuan, C. S., (2002). Alternative therapies for type 2 diabetes. *Altern. Med. Rev., 7*, 45–58.

Dharmananda, S., (2002). The nature of ginseng. Traditional use, modern research, and the question of dosage. *Herbal Gram, 54*, 34–51.

Di Fabio, G., Romanucci, V., Zarrelli, M., Giordano, M., & Zarrelli, A., (2013). C-4-gem-dimethylated oleanes of *Gymnema sylvestre* and their pharmacological activities. *Molecules, 18*, 14892–14919.

Di Giulio, A. M., Lesma, E., Germani, E., & Gorio, A., (1999). Inhibition of high glucose-induced protein mono-ADP ribosylation restores neuritogenesis and sodium-pump activity in SY5Y neuroblastoma cells. *J. Neurosci. Res., 46*, 565–571.

Djellouli, F., Krouf, D., Bouchenak, M., & Dubois, M. A. L., (2014–15). Favorable effects of *Globularia alypum* L. lyophilized methanolic extracton the reverse cholesterol transport and lipoprotein peroxidation in streptozotocin-induced diabetic rats. *J. Pharmacogn. Phytochem., 6*(4), 758–765.

Dugoua, J. J., Seely, D., Perri, D., Cooley, K., Forelli, T., Mills, E., & Koren, G., (2007). From type 2 diabetes to antioxidant activity: A systematic review of the safety and efficacy of common and cassia cinnamon bark. *Can J. Physiol. Pharmacol., 85*, 837–847,

Eddouks, M., Maghrani, M., & Michel, J. B., (2005). Hypoglycaemic effect of *Triticum repens* P. Beauv. In normal and diabetic rats. *J. Ethnopharmacol., 102*(2), 228–232.

El-Abhar, H. S., & Schaalam, M. F., (2014). Phytotherapy in diabetes: Review on potential mechanistic perspectives. *World J. Diabetes, 5*, 176–197.

El-Alfy, A. T., Ahmed, A. A. E., & Fatani, A. J., (2005). Protective effect of red grape seeds proanthocyanidins against induction of diabetes by alloxan in rats. *Pharmacol. Res., 52*, 264–270.

Eldin, I. M. T., Ahmed, E. M., & Abd, E. H. M., (2010). Preliminary study of the clinical hypoglycemic effects of *Allium cepa* (red onion) in Type 1 and type 2 diabetic patients. *Environ Health Insights, 4*, 71–77.

El-Ghffar, E. A. A., (2016). Ameliorative effect of glabridin, a main component of *Glycyrrhiza glabra* L. roots in *Streptozotocin* induced Type 1 diabetes in male albino rats. *Indian J. Tradit. Know., 15*(4), 570–579.

El-Shobaki, F. A., El-Bahay, A. M., Esmail, R. S. A., Abd-El-Megeid, A. A., & Esmail, N. S., (2010). Effect of figs fruit (*Ficus carica* L.) and its leaves on hyperglycemia in alloxan diabetic rats. *World J. Dairy Food Sci., 5*, 47–57.

El-Snafi, A. E., (2015). Chemical constituents and pharmacological importance of *Agropyron repens*-A review. *J. Toxicol. Pharmacol., 01*(02), 37–41.

EMA, (2011). *European Medicinal Agency Committee on Herbal Medicinal Products (HMPC).* Assessment report on *Agropyron repen*s (L.) P. Beauv., rhizome. EMA/ HMPC/563395/2010.

Esmaillzadeh, A., Tahbaz, F., Gaieni, I., Alavi-Majd, H., & Azadbakht, L., (2004). Concentrated pomegranate juice improves lipid profiles in diabetic patients with hyperlipidemia. *J. Med. Food., 7*, 305–308.

Es-Safi, N., Kollman, A., Khlifi, S., & Ducrot, P. H., (2007). Antioxidative effect of compounds isolated from *Globularia alypum* L. structure-activity relationship. LWT-food. *Sci. Technol., 40*, 1246–1252.

Esther, G. S., & Manonmani, A. J., (2014). Effect of Eugenia *Jambolana* on streptozotocin-nicotinamide induced type-2 diabetic nephropathy in rats. *Int. J. Drug Dev. Res., 6*(1), 175–187.

Fenercioglu, A. K., Saler, T., Genc, E., Sabuncu, H., & Altuntas, Y., (2010). The effects of polyphenol-containing antioxidants on oxidative stress and lipid peroxidation in type 2 diabetes mellitus without complications. *J. Endocrinol. Invest., 33*, 118–124.

Ferenci, P., & Beinhardt, S., (2013). Silibinin: An old drug in the high tech era of liver transplantation. *J. Hepatol., 58*, 409–411.

Foltin, R. W., Fischman, M. W., & Byrne, M. F., (1988). Effects of smoked marijuana on food intake and body weight of humans living in a residential laboratory. *Appetite, 11*, 1–14.

Freye, E., & Gleske, J., (2013). Siberian ginseng results in beneficial effects on glucose metabolism in diabetes type 2 patients: A double blind placebo-controlled study in comparison to panax ginseng. *Int. Jof. Clin. Nutr., 1*(1), 11–17.

Fu, Z., Gilbert, E. R., & Liu, D., (2013). Regulation of insulin synthesis and secretion and pancreatic beta-cell dysfunction in diabetes. *Curr. Diabetes Rev., 9*(1), 25–53.

Fuangchan, A., Sonthisombat, P., Seubnukarn, T., Chanouan, R., Chotchaisuwat, P., Sirigulsatien, V., Ingkaninan, K., et al., (2011). Hypoglycemic effect of bitter melon compared with metformin in newly diagnosed type 2 diabetes patients. *J. Ethnopharmacol., 134*(2), 422–428.

Gaur, R., Yadav, K. S., Verma, R. K., Yadav, N. P., & Bhakuni, R. S., (2014). *In vivo* anti-diabetic activity of derivatives of isoliquiritigenin and liquiritigenin. *Phytomedicine, 21*, 415–422.

Goodrum, L. J., Patel, A., Leykam, J. F., & Kieliszewski, M. J., (2000). Gum arabic glycoprotein contains glycomodules of both extensin and arabinogalactan-glycoproteins. *Phytochemistry, 54*(1), 99–106.

Gordon, A., Hobbs, D. A., Bowden, D. S., Bailey, M. J., Mitchell, J., Francis, A. J., & Roberts, S. K., (2006). Effects of *Silybum marianum* on serum hepatitis C virus RNA, alanine aminotransferase levels and well-being in patients with chronic hepatitis C. *J. Gastroenterol. Hepatol., 21*, 275–280.

Gray, A. M., & Flatt, P. R., (1999). Insulin releasing and insulin-like activity of the traditional antidiabetic plant *Coriandrum sativum* (coriander). *Br. J. Nutr., 81*, 203–209.

Grover, J. K., Yadav, S., & Vats, V., (2002). Medicinal plants of India with anti-diabetic potential. *J. Ethnopharmacol., 81*, 81–100.

Gui, D., Huang, J., Guo, Y., Chen, J., Chen, Y., Xiao, W., Liu, X., & Wang, N., (2013). Astragaloside IV ameliorates renal injury in streptozotocin-induced diabetic rats through inhibiting NF kappaB mediated inflammatory genes expression. *Cytokine, 61*(3), 970–977.

Gui, Q. F., Xu, Z. R., Xu, K. Y., & Yang, Y. M., (2016). The efficacy of ginseng-related therapies in type 2 diabetes mellitus: An updated systematic review and meta-analysis. *Medicine (Baltimore), 95*, e2584.

Guigas, B., Naboulsi, R., Villanueva, G. R., Taleux, N., Lopez-Novoa, J. M., Leverve, X. M., & El-Mir, L. Y., (2007). The flavonoid silibinin decreases glucose-6-phosphate hydrolysis in perifused rat hepatocytes by an inhibitory effect on glucose-6-phosphatase. *Cell Physiol. Biochem., 20*, 925–934.

Gunn, J., & Farnsworth, C. T. N., (2013). Chapter 33-diabetes and natural products. In: Watson, R., & Preedy, B., (eds.), *Bioactive Food as Dietary Interventions for Diabetes* (pp. 381–394).

Guo, K., He, X., Lu, D., Zhang, Y., Li, X., Yan, Z., & Qin, B., (2017). Cycloartane-type triterpenoids from *Astragalus hoantchy* French. *Nat. Prod. Res., 31*(3), 314–319.

Gupta, A., Gupta, R., & Lal, B., (2001). Effect of *Trigonella foenum-graecum* (fenugreek) seeds on glycaemic control and insulin resistance in type 2 diabetes mellitus: A double blind placebo controlled study. *J. Assoc. Physicians India, 49*, 1057–1061.

Gupta, N., Belmkar, S., Gupta, P. K., & Jain, A., (2011). Study of *Glycyrrhiza glabra* on glucose uptake mechanism in rats. *IJDDHR, 1*(2), 50–55.

Haber, S. L., & Keonavong, J., (2013). Fenugreek use in patients with diabetes mellitus. *Am. J. Health-Sysyt. Pharm., 70*, 1198–2013.

Hasanein, P., (2011). Glabridin as a major active isoflavan from *Glycyrrhiza glabra* (licorice) reverses learning and memory deficits in diabetic rats. *Acta Physiol. Hung., 98*(2), 221–230.

Hawke, R. L., Schrieber, S. J., Soule, T. A., Wen, Z., Smith, P. C., Reddy, K. R., et al., (2010). SyNCH trial group. Silymarin ascending multiple oral dosing phase I study in noncirrhotic patients with chronic hepatitis C. *J. Clin. Pharmacol., 50*, 434–449.

Helim, E. M., (2003). Lowering of blood sugar by water extract of *Azadirachta indica* and *Abroma augusta* in diabetes rats. *Indian J. Exp. Biol., 41*(6), 636–640.

Hézode, C., Zafrani, E. S., Roudot-Thoraval, F., Costentin, C., Hessami, A., Bouvier-Alias, M., Medkour, F., et al., (2008). Daily cannabis use: A novel risk factor of steatosis severity in patients with chronic hepatitis C. *Gastroenterology., 134,* 432–439.

Hoa, N. K., Phan, D. V., Thuan, N. D., & Östenson, C. G., (2009). Screening of the hypoglycemic effect of eight Vietnamese herbal drugs. *Methods Find Exp. Clin. Pharmacol., 31*(3), 165–169.

Holler, J. M., Bosy, T. Z., Dunkley, C. S., Levine, B., Past, M. R., & Jacobs, A., (2008). Delta9-tetrahydrocannabinol content of commercially available hemp products. *J. Anal. Toxicol., 32,* 428–432.

Hollister, L. E., & Reaven, G. M., (1974). Delta-9-tetrahydrocannabinol and glucose tolerance. *Clin. Pharmacol. Ther., 16,* 297–302.

Hong, H., & Won, J. M., (2004). Effects of malted barley extract and banaba extract on blood glucose levels in genetically diabetic mice. *J. Med. Food., 7*(4), 487–490.

Hontecillas, R., O'Shea, M., Einerhand, A., Diguardo, M., & Bassaganya-Riera, J., (2009). Activation of PPAR gamma and alpha by punicic acid ameliorates glucose tolerance and suppresses obesity-related inflammation. *J. Am. Coll. Nutr., 28,* 184–195.

Hossain, M. S., Urbi, Z., Sule, A., & Rahman, K. M. H., (2014). *Andrographis paniculata* (Burm. f.) Wall. Ex nees: A review of ethnobotany, phytochemistry, and pharmacology. *Sci. World J., 274905,* 1–28.

Hosseinzadeh, P., Djazayery, A., Mostafavi, S. A., Javanbakht, M. H., Derakhshanian, H., Rahimiforoushani, A., & Djalali, M., (2013). Brewer's yeast improves blood pressure in type 2 diabetes mellitus. *Iran J. Public Health, 42,* 602–609.

Hou, W., Li, Y., Zhang, Q., Wei, X., Peng, A., Chen, L., & Wei, Y., (2009). Triterpene acids isolated from *Lagerstroemia speciosa* leaves as α-glucosidase inhibitors. *Phytother. Res., 23*(5), 614–618.

Hu, C. K., Lee, Y. J., Colitz, C. M., Chang, C. J., & Lin, C. T., (2012). The protective effects of *Lycium barbarum* and *Chrysanthemum morifolum* on diabetic retinopathies in rats. *Vet. Ophthalmol., 2,* 65–71.

Huang, T. H., Peng, G., Kota, B. P., Li, G. Q., Yamahara, J., Roufogalis, B. D., & Li, Y., (2005a). Anti-diabetic action of *Punica granatum* flower extract: Activation of PPAR-gamma and identification of an active component. *Toxicol. Appl. Pharmacol., 207,* 160–169.

Huang, T. H., Peng, G., Kota, B. P., Li, G. Q., Yamahara, J., Roufogalis, B. D., & Li Y., (2005b). Pomegranate flower improves cardiac lipid metabolism in a diabetic rat model: Role of lowering circulating lipids. *Br. J. Pharmacol., 145,* 767–774.

Huang, T. H., Yang, Q., Harada, M., Li, G. Q., Yamahara, J., Roufogalis, B. D., & Li, Y., (2005c). Pomegranate flower extract diminishes cardiac fibrosis in Zucker diabetic fatty rats: Modulation of cardiac endothelin-1 and nuclear factor-kappaB pathways. *J. Cardiovasc. Pharmacol., 46,* 856–862.

Hung, T. M., Hoang, D. M., Kim, J. C., Jang, H. S., Ahn, J. S., & Min, B. S., (2009). Protein tyrosine phosphatase 1B inhibitory by dammaranes from Vietnamese Giao-Co-Lam tea. *J. Ethnopharmacol., 124*(2), 240–245.

Hunt, J. V., Smith, C. C., & Wolff, S. P., (1990). Autoxidative glycosylation and possible involvement of peroxides and free radicals in LDL modification by glucose. *Diabetes, 39,* 1420–1424.

Huseini, H. F., Larijani, B., Heshmat, R., Fakhrzadeh, H., Radjabipour, B., Toliat, T., & Raza, M., (2006). The efficacy of *Silybum marianum* (L.) Gaertn. (silymarin) in the treatment of type II diabetes: A randomized, double-blind, placebo controlled, clinical trial. *Phytother Res., 20*, 1036–1039.

Huyen, V. T. T., Phan, D. V., Thang, P., Hoa, N. K., & Östenson, C. G., (2010). Antidiabetic effect of *Gynostemma pentaphyllum* tea in randomly assigned type 2 diabetic patients. *Horm. Metab. Res., 42*(5), 353–357.

Huyen, V. T. T., Phan, D. V., Thang, P., Hoa, N. K., & Östenson, C. G., (2012). Antidiabetic effects of add-on *Gynostemma pentaphyllum* extract therapy with sulfonylureas in type 2 diabetic patients. *Evid. Based Complement Alternat. Med.,* 452313, 1–7.

Huyen, V. T. T., Phan, D. V., Thang, P., Hoa, N. K., & Östenson, C. G., (2013). *Gynostemma pentaphyllum* tea improves insulin sensitivity in type 2 diabetic patients. *J. Nutr. Metab.,* 765383, 1–7.

Ikeda, Y., Chen, J. T., & Matsuda, T., (1999). Effectiveness and safety of banabamin tablet containing extract from banaba in patients with mild type 2 diabetes. *Jpn. Pharmacol. Ther.,* 27(5), 72–73.

Irudayaraj, S. S., Sunil, C., Stalin, A., Veeramuthu, D., Al-Dhabi, N. A., & Savarimuthu, I., (2017). Protective effects of *Ficus carica* leaves on glucose and lipids levels, carbohydrate metabolism enzymes, and β-cells in type 2 diabetic rats. *Pharm. Biol., 55*(1), 1074–1081.

Iso, H., Date, C., Wakai, K., Fukui, M., & Tamakoshi, A., (2006). The relationship between green tea and total caffeine intake and risk for self-reported type 2 diabetes among Japanese adults. *Ann. Intern. Med., 144*, 554–562

Ivorra, M. D., D'Ocon, M. P., Paya, M., & Villar, A., (1988). Antihyperglycemic and insulin-releasing effects of beta-sitosterol 3-beta-D-glucoside and its aglycone, beta-sitosterol. *Arch. Int. Pharmacodyn., 296*, 224–231.

Iyer, A., Panchal, S., Poudyal, H., & Brown, L., (2009). Potential health benefits of Indian spices in the symptoms of metabolic syndrome: A review. *Indian J. Biochem. Biophys., 46*, 467–481.

Jafri, M. A., Aslam, M., Javed, K., & Singh, S., (2000). Effect of *Punica granatum* Linn. (flowers) on blood glucose level in normal and alloxan-induced diabetic rats. *J Ethnopharmacol., 70*, 309–314.

Jana, K., Ghosh, A., Chatterjee, K., & Ghosh, D., (2014). Antidiabetic activity of seed of *Eugenia jambolana* in streptozotocin induced diabetic male albino rat: An apoptotic and genomic approach. *Int. J. Pharm. Pharm. Sci., 6*(11), 407–412.

Jang, Y. J., Kim, J. K., Lee, M. S., Ham, I. H., Whang, W. K., Kim, K. H., & Kim, H. J., (2001). Hypoglycemic and hypolipidemic effects of crude saponin fractions from *Panax ginseng* and *Gynostemma pentaphyllum. Yakhak Hoechi., 45*, 545–556.

Jasmine, R., & Daisy, P., (2007). Hypoglycemic and hypolipidemic activity of *Eugenia jambolana* in streptozotocin-diabetic rats. *Asian J. Biochem., 2*(4), 269–273.

Jelodar, G. A., Maleki, M., Motadayen, M. H., & Sirus, S., (2005). Effect of fenugreek, onion and garlic on blood glucose and histopathology of pancreas of alloxan-induced diabetic rats. *Indian J. Med. Sci., 59*, 64–69.

Jeon, Y. H., Moon, J. W., Kweon, H. J., Jeoung, Y. J., An, C. S., Jin, H. L., Hur, S. J., & Lim, B. O., (2010). Effects of *Lycii fructus* and *Astragalus membranaceus* mixed extracts on immunomodulators and prevention of diabetic cataract and retinopathy in streptozotocin-induced diabetes rat model. *Korean J. Medicinal Crop Sci., 18*(1), 15–21.

Jevas, C., (2011). Anti-diabetic effects of *Allium cepa* (onions) aqueous extracts on alloxan-induced diabetic *Rattus novergicus*. *J. Med. Plants Res., 5*, 1134–1139.

John, A. J., Cherian, R., Subhash, H. S., & Cheria, A. M., (2003). Evaluation of the efficacy of bitter gourd (*Momordica charantia*) as an oral hypoglycemic agent: A randomized controlled clinical trial. *Indian J. Physiol. Pharmacol., 47*(3), 363–365.

Joseph, B., & Jini, D., (2011). Insight into the hypoglycaemic effect of traditional Indian herbs used in the treatment of diabetes. *Res. J. Med. Plant., 5*(4), 352–376.

Joseph, B., & Raj, S. J., (2010). Phytopharmacological properties of *Ficus racemosa* Linn.: An overview. *Int. J. Pharm. Sci. Rev. Res., 3*(2), 134–138.

Jouad, H., Maghrani, M., & Eddouks, M., (2002). Hypoglycaemic effect of *Rubus fructicosis* L. and *Globularia alypum* L. in normal and streptozotocin-induced diabetic rats. *J. Ethnopharmacol.*, 81, 351–356.

Jung, M., Park, M., Lee, H. C., Kang, Y. H., Kang, E. S., & Kim, S. K., (2006). Antidiabetic agents from medicinal plants. *Curr. Med. Chem., 13*, 1203–1218.

Kadan, S., Saad, B., Sasson, Y., & Zaid, H., (2013). *In vitro* evaluation as of cytotoxicity of eight antidiabetic medicinal plants and their effect on GLUT4 translocation. *Evid. Based Complement Alternat. Med., 549345*, 1–9.

Kakuda, T., Sakane, I., Takihara, T., Ozaki, Y., Takeuchi, H., & Kuroyanagi, M., (1996). Hypoglycemic effect of extracts from *Lagerstroemia speciosa* L. leaves in genetically diabetic KK-AY mice. *Biosci. Biotechnol. Biochem., 60*(2), 204–208.

Karthikeson, P. S., & Lakshmi, T., (2017). Anti-diabetic activity of *Glycyrrhiza glabra*: An *in vitro* study. *Int. J. Pharm. Sci. Rev. Res., 44*(1), 80–81.

Kassaian, N., Azadbakht, L., Forghani, B., & Amini, M., (2009). Effect of fenugreek seeds on blood glucose and lipid profiles in type 2 diabetic patients. *Int. J. Vitam. Nutr. Res., 79*, 34–39.

Kataria, R., Hemraj, S. G., Gupta, A., Jalhan, S., Jindal, A., & Jindal, A., (2013). Pharmacological activities on *Glycyrrhiza Glabra*: A review. *Asian J. Pharm. Clin. Res., 6*(1), 5–7.

Kawamori, R., Tajima, N., Iwamoto, Y., Kashiwagi, A., Shimamoto, K., & Kaku, K., (2009). Voglibose Ph-3 study group. Voglibose for prevention of type 2 diabetes mellitus: A randomized, double-blind trial in Japanese individuals with impaired glucose tolerance. *Lancet, 9, 373*(9675), 1607–1614.

Khamaisi, M., Rudich, A., Beeri, I., Pessler, D., Friger, M., Gavrilov, V., Tritschler, H., & Bashan, N., (1999). Metabolic effects of gamma-linolenic acid-alpha-lipoic acid conjugate in streptozotocin diabetic rats. *Antioxid Redox Signal, 1*(4), 523–535.

Khan, A., Qadir, S. S., & Khattak, K. N., (2006). Cloves improve glucose, cholesterol and triglycerides of people with type 2 diabetes mellitus. *FASEB J., 20*, A990.

Khan, A., Safdar, M., Ali, K. M. M., Khattak, K. N., & Anderson, R. A., (2003). Cinnamon improves glucose and lipids of people with type 2 diabetes. *Diab Care., 26*, 3215–3218.

Khan, N., & Mukhtar, H., (2013). Tea and health: Studies in humans. *Curr. Pharm. Des., 19*, 6141–6147.

Khokhla, M. R., HIa, K., IaP, C., Skybitska, M. I., & Sybirna, N. O., (2012). Cytological and biochemical characteristics of rats' peripheral blood under the condition of experimental diabetes mellitus type 1 and *Galega officinalis* admission. *Studia Biologica., 6*(1), 37–46.

Khokhla, M., Kleveta, G., Kotyk, A., Skybitska, M., Ya, C., & Sybirna, N., (2010). Sugar-lowering effects of *Galega officinalis* L. *Ann. Univ. Mariae Curie Sklodowska., 4*(30), 177–182.

Khokhla, M., Kleveta, G., Lupak, M., Skybitska, M., Chajka, Y., & Sybirna, N., (2013). The inhibition of rat leukocytes apoptosis under the condition of experimental diabetes mellitus type 1 by *Galega officinalis* L. extract. *Curr. Issues Pharm. Med. Sci., 26*(4), 393–397.

Khosla, P. S., Bhanwara, J., Singh, S., Seth, S., & Srivastava, R. K., (2000). A study of hyperglycemia effects of *A. indica* (Neem) in normal and alloxan diabetic rabbits. *Indian J Physiol. Pharmacol., 44*, 69–74.

Kim, G. D., Oh, J., Park, H. J., Bae, K., & Lee, S. K., (2013). Magnolol inhibits angiogenesis by regulating ROS-mediated apoptosis and the PI3K/AKT/mTOR signaling pathway in mES/EB-derived endothelial-like cells. *Int. J. Oncol., 43*(2), 600–610.

Kim, J. H., Park, J. M., Kim, E. K., Lee, J. O., Lee, S. K., Jung, J. H., & Kim, H. S., (2010). Curcumin stimulates glucose uptake through AMPK-p38 MAPK pathways in L6 myotube cells. *J. Cell Physiol., 223*, 771–778.

Kim, J. W., Lee, Y. J., You, Y. H., Moon, M. K., Yoon, K. H., Ahn, Y. B., & Ko, S. H., (2018). Effect of sodium-glucose cotransporter 2 inhibitor, empagliflozin, and α-glucosidase inhibitor, voglibose, on hepatic steatosis in an animal model of type 2 diabetes. *J. Cell Biochem.*, 26. https://doi.org/10.1002/jcb.28141 (accessed on 20 May 2020).

Kim, W., Lao, Q., Shin, Y. K., Carlson, O. D., Lee, E. K., Gorospe, M., Kulkarni, R. N., & Egan, J. M., (2012). Cannabinoids induce pancreatic β-cell death by directly inhibiting insulin receptor activation. *Sci. Signal., 5(*216), ra23.

Kirkham, S., Akilen, R., Sharma, S., & Tsiami, A., (2009). The potential of cinnamon to reduce blood glucose levels in patients with type 2 diabetes and insulin resistance. *Diabetes Obes. Metab., 11*, 1100–1113.

Klein, G., Kim, J., Himmeldirk, K., Cao, Y., & Chen, X., (2007). Antidiabetes and anti-obesity activity of *Lagerstroemia speciosa. J. Evid. Based Complementary Altern. Med., 4*(4), 401–407.

Klonof, D. C., & Schwartz, D. M., (2000). An economic analysis of interventions for diabetes. *Diab. Care., 23*, 390–404.

Kochhar, A., & Pathak, N., (2018). Nutraceutical based approach to combat diabetes mellitus. *Curr. Res. Diabetes Obes. J., 8*(2), CRDOJ.MS.ID.555739.

Konoshima, T., & Takasaki, M., (2002). *Chemistry and bioactivity of the non-volatile constituents of eucalyptus. In: Biological and End Use Aspects (*pp. 1–23*).* CRC Press: Taylor and Francis group.

Koren-Gluzer, M., Aviram, M., Meilin, E., & Hayek, T., (2011). The antioxidant HDL-associated paraoxonase-1 (PON1) attenuates diabetes development and stimulates beta-cell insulin release. *Atherosclerosis, 219*, 510–518.

Koshy, S. M., Bobby, Z., Hariharan, A. P., & Gopalakrishna, S. M., (2012). Amla (*Emblica officinalis*) extract is effective in preventing high fructose diet-induced insulin resistance and atherogenic dyslipidemia profile in ovariectomized female albino rats/*NAMS, 19*(10), 1146–1155.

Król, E., Krejpcio, Z., Byks, H., Bogdański, P., & Pupek-Musialik, D., (2011). Effects of chromium brewer's yeast supplementation on body mass, blood carbohydrates, and lipids and minerals in type 2 diabetic patients. *Biol. Trace Elem. Res., 143*, 726–737.

Kumar, A., Kaur, A., Mittal, R., & Kaur, H., (2016). Protective effect of *Commiphora mukul* in experimental paradigm in streptozotocin-induced diabetic neuropathy and nephropathy." *EC Pharmacol. Toxicol., 2*(3), 129–147.

Kumar, S. N., Mani, U. V., & Mani, I., (2010). An open label study on the supplementation of *Gymnema sylvestre* in type 2 diabetes. *J. Diet Suppl., 7*, 273–282.

Kuroda, M., Mimaki, Y., Ohtomo, T., Yamada, J., Nishiyama, T., Mae, T., & Kawada, T., (2012). Hypoglycemic effects of clove (*Syzygium aromaticum* flower buds) on genetically diabetic KK-Ay mice and identification of the active ingredients. *J. Nat. Med., 66*, 394–399.

Lafaye, G., Karila, L., Blecha, L., & Benyamina, A., (2017). Cannabis, cannabinoids, and health. *Dialogues Clin. Neurosci., 19*(3), 309–316.

Lambole, V. B., Murti, K., Kumar, U., Bhatt, S. P., & Gajera, V., (2010). Phyto pharmacological properties of *Aegles marmelos* as a potential medicinal tree: An overview. *Int. J. Pharm. Sci. Rev. Res., 5*(2), 67–72.

Leach, M. J., & Kumar, S., (2012). Cinnamon for diabetes mellitus. *Cochrane Database Syst. Rev., 9*, CD007170.

Lee, M. J., Rao, Y. K., Chen, K., Lee, Y. C., Chung, Y. S., & Tzeng, Y. M., (2010). Andrographolide and 14-deoxy-11,12-didehydroandrographolide from *Andrographis paniculata* attenuate high glucose-induced fibrosis and apoptosis in murine renal mesangeal cell lines. *J. Ethnopharmacol., 132*(2), 497–505.

Lekshmi, P. C., Arimboor, R., Indulekha, P. S., & Menon, A. N., (2012). Turmeric (*Curcuma longa* L.) volatile oil inhibits key enzymes linked to type 2 diabetes. *Int. J. Food Sci. Nutr., 63*, 832–834.

Levendal, R. A., & Frost, C. L., (2006). *In vivo* effects of *Cannabis sativa* L. extract on blood coagulation, fat and glucose metabolism in normal and streptozocin-induced diabetic rats. *Afr. J. Trad. CAM, 3*(4), 1–12.

Levins, S. M., Leyva, M. J., Brown, J., Wright, J., Scofield, R. H., & Aston, C. E., (2007). Effect of cinnamon on glucose and lipid levels in non insulin-dependent type 2 diabetes. *Diab. Care, 30*, 2236–2237.

Li, H., Wang, P., Huang, F., Jin, J., Wu, H., Zhang, B., Wang, Z., Shi, H., & Wu, X., (2018). Astragaloside IV protects blood-brain barrier integrity from LPS-induced disruption via activating Nrf2 antioxidant signaling pathway in mice. *Toxicol. Appl. Pharmacol., 340*, 58–66.

Li, W. L., Zheng, H. C., Bukuru, J., & De Kimpe, N., (2004). Natural medicines used in the traditional Chinese medical system for therapy of diabetes mellitus. *J. Ethnopharmacol., 92*, 1–2.

Li, Y., Tran, V. H., Duke, C. C., & Roufogalis, B. D., (2012). Preventive and protective properties of *Zingiber officinale* (Ginger) in diabetes mellitus, diabetic complications, and associated lipid and other metabolic disorders: A brief review. *J. Evid. Based Complementary Altern. Med., 516870*, 1–10.

Lieberman, S., Spahrs, R., Stanton, A., Martinez, L., & Grinder, M., (2005). Weight loss, body measurements, and compliance: A 12-week total lifestyle intervention pilot study. *Altern. Complement Ther., 11*(6), 307–313.

Lim, S. H., Park, Y. H., Kwon, C. J., Ham, H. J., Jeong, H. N., Kim, H. K., & Ahn, Y. S., (2010). Anti-diabetic and hypoglycemic effect of *Eleutherococcus* spp. *J. Korean Soc. Food Sci. Nutr., 39*(12), 1761–1768.

Liu, J., Zhou, L., Xiong, K., Godlewski, G., Mukhopadhyay, B., Tam, J., Yin, S., et al., (2012). Hepatic cannabinoid receptor-1 mediates diet-induced insulin resistance via inhibition of insulin signaling and clearance in mice. *Gastroenterology, 142,* 1218–1228.

Liu, Y. J., Zhan, J., Liu, X. L., Wang, Y., Ji, I., & He, Q. Q., (2014). Dietary flavonoids intake and risk of type 2 diabetes: A meta-analysis of prospective cohort studies. *Clin. Nutr., 33,* 59–63.

Lu, F. R., Shen, L., Qin, Y., Gao, L., Li, H., & Dai, Y., (2008). Clinical observation on trigonella *Foenum graecum* L. total saponins in combination with sulfonylureas in the treatment of type 2 diabetes mellitus. *Chin. J. Integr. Med., 14,* 56–60.

Lu, K. W., Chen, J. C., Lai, T. Y., Yang, J. S., Weng, S. W., Ma, Y. S., Tang, N. Y., Lu, P. J., Weng, J. R., & Chung, J. P., (2010). Gypenosides causes DNA damage and inhibits expression of DNA repair genes of human oral cancer SAS cells. *In vivo, 24*(3), 287–291.

Madar, Z., Abel, R., Samish, S., & Arad, J., (1988). Glucose-lowering effect of fenugreek in non-insulin dependent diabetics. *Eur. J. Clin. Nutr., 42,* 51–54.

Mahesh, A. R., Kumar, H., & Devkar, R. A., (2012). Detail study on *Boerhaavia diffusa* plant for its medicinal importance: A review. *Res. J. Pharmaceutical Sci., 1*(1), 28–36.

Mahmoudzadeh-Sagheb, H., Heidari, Z., Bokaeian, M., & Moudi, B., (2010). Antidiabetic effects of eucalyptus globulus on pancreatic islets: A stereological study. *Folia Morphol. (Warsz)., 69*(2), 112–118.

Makino-Wakagi, Y., Yoshimura, Y., Uzawa, Y., Zaima, N., Moriyama, T., & Kawamura, Y., (2012). Ellagic acid in pomegranate suppresses resistin secretion by a novel regulatory mechanism involving the degradation of intracellular resistin protein in adipocytes. *Biochem. Biophys. Res. Commun., 417,* 880–885.

Mang, B., Wolters, M., Schmitt, B., Kelb, K., Lichtinghagen, R., Stichtenoth, D. O., & Hahn, A., (2006). Effects of a cinnamon extract on plasma glucose, HbA, and serum lipids in diabetes mellitus type 2. *Eur. J. Clin. Invest., 36,* 340–344.

Marciani, L., Gowland, P., Spiller, R. C., Manoj, P., Moore, R. J., Young, P., Al-Sahab, S., et al., (2000). Gastric response to increased meal viscosity assessed by echo-planar magnetic resonance imaging in humans. *J. Nutr., 130*(1), 122–127.

Martínez-Larrañaga, A. A. M. R., & Martínez, I. A. M. A., (2018). Chapter 62-poisonous plants of Europe. Basic and clinical principles. In: *Veterinary Toxocology* (3rd edn., pp. 891–909).

Mary, N. K., Babu, B. H., & Padikkala, J., (2003). Anti-atherogenic effect of Caps HT2, a herbal Ayurvedic medicine formulation. *Phyto-Medicine., 10,* 474–482.

Matkowski, A., Woźniak, D., Lamer-Zarawska, E., Oszmański, J., & Leszczyńske, A., (2003). Flavonoids and phenol carboxylic acids in the oriental medicinal plant *Astragalus membranaceus* acclimated in Poland. *Z Naturforsch C., 58*(7–8), 602–604.

Matsuo, T., Odaka, H., & Ikeda, H., (1992). Effect of an intestinal disaccharidase inhibitor (AO-128) on obesity and diabetes. *Am. J. Clin Nutr., 55,* 314S–317S.

Mazhin, S. A., Zaker, M. A., Shahbazian, H. B., Azemi, M. E., & Madanchi, N., (2016). *Ficus carica* leaves decoction on glycemic factors of patients with Type 2 diabetes mellitus: A double-blind clinical trial. *Jundishapur J. Nat. Pharm. Prod., 11*(1), e25814.

McFarlin, B. K., Strohacker, K. A., & Kueht, M. L., (2009). Pomegranate seed oil consumption during a period of high-fat feeding reduces weight gain and reduces type 2 diabetes risk in CD-1 mice. *Br. J. Nutr., 102,* 54–59.

Megalli, S., Davies, N. M., & Roufogalis, B. D., (2006). Anti-hyperlipidemic and hypoglycemic effects of *Gynostemma pentaphyllum* in the Zucker fatty rat. *J. Pharm. Sci. Exp. Pharmacol., 9*(3), 281–291.

Mennen, L. I., Jackson, M., Sharma, S., Mbanya, J. C., Cade, J., Walker, S., Riste, L., et al., (2001). Habitual diet in four populations of African origin: A descriptive paper on nutrient intakes in rural and urban Cameroon, Jamaica, and Caribbean migrants in Britain. *Public Health Nutr., 4*(3), 765–772.

Min, Y. W., Park, S. U., Jang, Y. S., Kim, Y. H., Rhee, P. L., Ko, S. H., Joo, N., et al., (2012). Effect of composite yogurt enriched with acacia fiber and *Bifidobacterium lactis*. *World J. Gastroenterol., 18*(33), 4563–4569.

Miura, T., Itoh, Y., Kaneko, T., Ueda, N., Ishida, T., Fukushima, M., Matsuyama, F., & Seino, Y., (2004). Corosolic acid induces GLUT4 translocation in genetically type 2 diabetic mice. *Biol. Pharm. Bull., 27*(7), 1103–1105.

Miura, T., Takagi, S., & Ishida, T., (2012). Management of diabetes and its complications with Banaba (*Lagerstroemia speciosa* L.) and Corosolic acid. *Evid. Based Complement Alternat. Med., 871495*, 1–8.

Miura, T., Ueda, N., Yamada, K., Fukushima, M., Ishida, T., Kaneko, T., Matsuyama, F., & Seino, Y., (2006). Antidiabetic effects of corosolic acid in KK-Ay diabetic mice. *Biol. Pharm. Bull., 29*(3), 585–587.

Moelands, S. V., Lucassen, P. L., Akkermans, R. P., De Grauw, W. J., & Van, D. L. F. A., (2018). Alpha-glucosidase inhibitors for prevention or delay of type 2 diabetes mellitus and its associated complications in people at increased risk of developing type 2 diabetes mellitus. *Cochrane Database Syst. Rev., 12*(CD005061), 1–4.

Mohammad, S., Taha, A., Akhtar, K., Bamezai, R. N., & Baquer, N. Z., (2006). *In vivo* effect of *Trigonella foenum graecum* on the expression of pyruvate kinase, phosphoenolpyruvate carboxykinase, and distribution of glucose transporter (GLUT4) in alloxan-diabetic rats. *Can J. Physiol. Pharmacol., 84*(6), 647–654.

Mohan, M., Waghulde, H., & Kasture, S., (2010). Effect of pomegranate juice on Angiotensin II-induced hypertension in diabetic wistar rats. *Phytother. Res., 24*(2), S196–203.

Mopuri, R., Ganjayi, M., Meriga, B., Koorbanally, N. A., & Islam, M. S., (2018). The effects of *Ficus carica* on the activity of enzymes related to metabolic syndrome. *J. Food Drug An., 26*(1), 201–210.

Morrow, J. D., (2005). Quantification of isoprostanes as indices of oxidant stress and the risk of atherosclerosis in humans. *Arterioscler. Thromb. Vasc Biol., 25*, 279–286.

Mozaffari-Khosravi, H., Talaei, B., Jalali, B. A., Najarzadeh, A., & Mozayan, M. R., (2014). The effect of ginger powder supplementation on insulin resistance and glycemic indices in patients with type 2 diabetes: A randomized, double-blind, placebo-controlled trial. *Complement Ther. Med., 22*, 9–16.

Mucalo, I., Jovanovski, E., Vuksan, V., Božikov, V., Romić, Ž., & Rahelić, D., (2014). American ginseng extract (*Panax quinquefolius* L.) is safe in long-term use in type 2 diabetic patients. *Evid. Based Complement Alternat. Med., 969168*, 1–6.

Muniyappa, R., Sable, S., Ouwerkerk, R., Mari, A., Gharib, A. M., Walter, M., & Courville, A., et al., (2013). Metabolic effects of chronic cannabis smoking. *Diab Care., 36*(8), 2415–2422.

Nadeem, M., & Zeb, A., (2018). Impact of maturity on phenolic composition and antioxidant activity of medicinally important leaves of *Ficus carica* L. *Physiol. Mol. Biol. Plants., 24*(5), 881–887.

Nagalievska, M., Sabadashka, M., Hachkova, H., & Sybirna, N., (2018). *Galega officinalis* extract regulate the diabetes mellitus related violations of proliferation, functions, and apoptosis of leukocytes. *BMC Complement Altern. Med., 18*, 4.

Nagao, T., Meguro, S., Hase, T., Otsuka, K., Komikado, M., Tokimitsu, I., Yamamoto, T., & Yamamoto, K., (2009). A catechin-rich beverage improves obesity and blood glucose control in patient with type 2 diabetes. *Obesity, 17,* 310–317.

Nahas, R., & Moher, M., (2009). Complementary and alternative medicine for the treatment of type 2 diabetes. *Can Fam. Physician, 55,* 591–596.

Nain, P., Saini, V., Sharma, S., & Nain, J., (2012). Antidiabetic and antioxidant potential of *Emblica officinalis gaertn.* leaves extract in streptozotocin-induced type-2 diabetes mellitus (T2DM) rats. *J. Ethnopharmacol., 142,* 65–67.

Nakhaee, A., Bokaeian, M., Saravani, M., Farhangi, A., & Akbarzadeh, A., (2009). Attenuation of oxidative stress in streptozotocin-induced diabetic rats by *Eucalyptus globulus. Indian J. Clin. Biochem., 24*(4), 419–425.

Nalamolu, R. K., Boini, K. M., & Nammi, S., (2004). Effect of chronic administration of *Boerhaavia diffusa* Linn. leaf extract on experimental diabetes in rats. *Trop. J. Pharm. Res., 3*(1). http://www.tjpr.freehosting.net (accessed on 20 May 2020).

Nasir, O., (2013). Renal and extrarenal effects of gum arabic (Acacia senegal), What can be learned from animal experiments? *Kidney Blood Press R., 37*(4–5), 269–279.

Nayak, P., & Thirunavoukkarasu, M., (2016). A review of the plant *Boerhaavia diffusa*: Its chemistry, pharmacology and therapeutical potential. *J. Phytopharmacol., 5*(2), 83–92.

Nigam, V., & Nambiar, V. S., (2018). *Aegle marmelos* leaf juice as a complementary therapy to control type 2 diabetes—Randomized controlled trial in Gujarat. *India Adv. Integr. Med.* https://doi.org/10.1016/j.aimed.2018.03.002 (accessed on 20 May 2020).

Niu, H. S., Liu, I. M., Cheng, J. T., Lin, C. L., & Hsu, F. L., (2008). Hypoglycemic effect of syringin from *Eleutherococcus senticosus* in streptozotocin-induced diabetic rats. *Planta Med., 74*(2), 109–113.

Noipha, K., Ratanachaiyavong, S., & Ninla-Aesong, P., (2010). Enhancement of glucose transport by selected plant foods in muscle cell line L6. *Diabetes Res. Clin. Pract., 89,* e22–e26.

Norberg, A., Nguyen, K. H., Liepinsh, E., Phan, D. V., Thuan, N. D., Jornvall, H., Sillard, R., & Ostenson, C. G., (2004). A novel insulin-releasing substance, phanoside, from the plant *Gynostemma pentaphyllum. J. Biol. Chem., 279*(40), 41361–41367.

Nugroho, A. E., Kusumaramdani, G., Widyaninggar, A., Anggoro, D. P., & Pramono, S., (2014). Antidiabetic effect of combinations of n-hexane insoluble fraction of ethanolic extract of *Andrographis paniculata* with other traditional medicines. *Int. Food Res. J., 21*(2), 785–789.

Nugroho, A. E., Rais, I. R., Setiawan, I., Pratiwi, P. Y., Hadibarata, T., Tegar, M., & Pramono, S., (2014). Pancreatic effect of andrographolide isolated from *Andrographis paniculata* (Burm. f.) Nees. *Pak. J. Biol. Sci., 17*(1), 22–31.

Obolskiy, D., Pischel, I., Feistel, B., Glotov, N., & Heinrich, M., (2011). *Artemisia dracunculus* L. (Tarragon): A critical review of its traditional use, chemical composition, pharmacology, and safety. *J. Agric. Food Chem., 59*(21), 11367–11384.

Offenbacher, E. G., & Pi-Sunyer, F. X., (1980). Beneficial effect of chromium-rich yeast on glucose tolerance and blood lipids in elderly subjects. *Diabetes, 29,* 919–925.

Oh, M. R., Park, S. H., Kim, S. Y., Back, H. I., Kim, M. G., Jeon, J. Y., et al., (2014). Postprandial glucose-lowering effects of fermented red ginseng in subjects with impaired fasting glucose or type 2 diabetes: A randomized, double-blind, placebo controlled clinical trial. *BMC Complement Altern. Med., 14,* 237.

Osei-Hyiaman, D., De Petrillo, M., Pacher, P., Liu, J., Radaeva, S., Bátkai, S., Harvey-White, J., et al., (2005). Endocannabinoid activation at hepatic CB1 receptors stimulates fatty acid synthesis and contributes to diet-induced obesity. *J. Clin. Invest, 115,* 1298–1305.

Ozougwu, J. C., (2011). Anti-diabetic effects of *Allium cepa* (ONIONS) aqueous extracts on alloxan-induced diabetic *Rattus novergicus*. *J. Med. Plants Res., 5*(7), 1134–1139

Pacher, P., Bátkai, S., & Kunos, G., (2006). The endocannabinoid system as an emerging target of pharmacotherapy. *Pharmacol. Rev., 58,* 389–462.

Pagotto, U., Marsicano, G., Cota, D., Lutz, B., & Pasquali, R., (2006). The emerging role of the endocannabinoid system in endocrine regulation and energy balance. *Endocr. Rev., 27,* 73.

Pari, L., & Satcheesh, M. A., (2004). Antidiabetic effect of *Boerhavia diffusa*: Effect on serum and tissue lipids in experimental diabetes. *Med. Food., 7*(4), 472–476.

Park, C., & Lee, J. S., (2011). Banaba: The natural remedy as antidiabetic drug. *Biomedical Res., 22,* 127–131.

Park, M. Y., Lee, K. S., & Sung, M. K., (2005). Effects of dietary mulberry, Korean red ginseng, and banaba on glucose homeostasis in relation to PPAR-α, PPAR-γ, and LPL mRNA expressions. *Life Sci., 77*(26), 3344–3354.

Park, S. H., Oh, M. R., Choi, E. K., Kim, M. G., Ha, K. C., Lee, S. K., Kim, Y. G., et al., (2014). An 8-wk, randomized, double blind, placebo-controlled clinical trial for the antidiabetic effects of hydrolyzed ginseng extract. *J. Ginseng Res., 38,* 39243.

Parmar, H. S., & Kar, A., (2007). Antidiabetic potential of *Citrus sinensis* and *Punica granatum* peel extracts in alloxan treated male mice. *Biofactors, 31,* 17–24.

Parmar, H. S., & Kar, A., (2008). Medicinal values of fruit peels from *Citrus sinensis, Punica granatum,* and *Musa paradisiaca* with respect to alterations in tissue lipid peroxidation and serum concentration of glucose, insulin, and thyroid hormones. *J. Med. Food, 11,* 376–381.

Patel, D. K., Prasad, S. K., Kumar, R., & Hemelatha, S., (2012). An overview on antidiabetic medicinal plants having insulin mimetic property. *Asian Pac. J. Trop. Biomed., 2,* 320–330.

Patel, S. S., & Goyal, R. K., (2011). Prevention of diabetes-induced myocardial dysfunction in rats using the juice of the *Emblica officinalis* fruit. *Exp. Clin. Cardiol., 16*(3), 87–91.

Pepato, M. T., Folgado, V. B. B., Kettelhut, I. C., & Brunetti, I. L., (2001). Lack of antidiabetic effect of a *Eugenia jambolana* leaf decoction on rat streptozotocin diabetes. *Braz. J. Med. Biol. Res., 34*(3), 389–395.

Perez, C., Canal, J. R., & Torres, M. D., (2003). Experimental diabetes treated with Ficus carica extract: Effect on oxidative stress parameters. *Acta Diabetol., 40,* 3–8.

Pferschy-Wenzig, E. M., Atanasov, A. G., Malainer, C., Noha, S. M., Kunert, O., Schuster, D., Heiss, E. H., et al., (2014). Identification of isosilybin a from milk thistle seeds as an agonist of peroxisome proliferator-activated receptor gamma. *J. Nat. Prod., 77,* 842–847.

Pinent, M., Blay, M., Bladé, M. C., Salvado, M. J., Arola, L., & Ardévol, A., (2004). Grape seed-derived procyanidins have an antihyperglycemic effect in streptozotocin-induced diabetic rats and insulinomimetic activity in insulin-sensitive cell lines. *Endocrinology, 145,* 4985–4990.

Porchezhian, E., & Dobriyal, R. M., (2003). An overview on the advances of *Gymnema sylvestre*: Chemistry, pharmacology and patents. *Pharmazie., 58,* 5–12.

Prasad, P. V., Subhaktha, P. K., Narayana, A., & Rao, M. M., (2006). Medico-historical study of "asvattha" (sacred fig tree). *Bull. Indian Inst. History Med., 36,* 1–20.

Prasad, R. C., Herzog, B., Boone, B., Sims, L., & Waltner-Law, M., (2005). An extract of *Syzygium aromaticum* represses genes encoding hepatic gluconeogenic enzymes. *J. Ethnopharmacol., 96,* 295–301.

Queipo-Ortuno, M. I., Boto-Ordonez, M., Murri, M., Gomez-Zumaquero, J. M., Clemente-Postigo, M., Estruch, R., et al., (2012). Influence of red wine polyphenols and ethanol on the gut microbiota ecology and biochemical biomarkers. *Am. J. Clin. Nutr., 95*, 1323–1334.

Rabinowitz, M. B., Gonick, H. C., Levin, S. R., & Davidson, M. B., (1983). Effects of chromium and yeast supplements on carbohydrate and lipid metabolism in diabetic men. *Diab. Care, 6*, 319–327.

Racek, J., Trefil, L., Rajdl, D., Mudrová, V., Hunter, D., & Senft, V., (2006). Influence of chromium-enriched yeast on blood glucose and insulin variables, blood lipids, and markers of oxidative stress in subjects with type 2 diabetes mellitus. *Biol. Trace Elem. Res., 109*, 215–230.

Rahman, H. S., Rasedee, A., Yeap, S. K., Othman, H. H., Chartrand, M. S., Namvar, F., & How, C. W., (2014). Biomedical properties of a natural dietary plant metabolite, zerumbone, in cancer therapy and chemoprevention trials. *Bio. Med. Research International*, p. 920742.

Rajeshwari, U., Shobha, I., & Andallu, B., (2011). Comparison of aniseeds and coriander seeds for antidiabetic, hypolipidemic and antioxidant activities. *Spatula DD., 1*, 9–16.

Raman, B. V., Krishna, N. V., Rao, N. B., Saradhi, P. M., & Rao, B. M. V., (2012). Plants with antidiabetic activities and their medicinal values. *Int. Res. J. Pharm., 3*(3), 11–15.

Ramesh, B., & Saralakumari, D., (2012). Antihyperglycemic, hypolipidemic and anti-oxidant activities of ethanolic extract of *Commiphora mukul gum* resin in fructose-fed male wistar rats. *J. Physiol. Biochem.* doi: 10.1007/s13105–012–0175.

Ravi, K., Sivagnanam, K., & Subramanian, S., (2004). Anti-diabetic activity of *Eugenia jambolana* seed kernels on streptozotocin-induced diabetic rats. *J. Med. Food, 7*(2), 187–191.

Reeds, D. N., Patterson, B. W., Okunade, A., Holloszy, J. O., Polonsky, K. S., & Klein, S., (2011). Ginseng and ginsenoside Re do not improve beta-cell function or insulin sensitivity in overweight and obese subjects with impaired glucose tolerance or diabetes. *Diab. Care, 34*, 1071–1076.

Ribnicky, D. M., Poulev, A., Watford, M., Cefalu, W. T., & Raskin, I., (2006). Antihyperglycemic activity of Tarralin, an ethanolic extract of *Artemisia dracunculus* L. *Phytomed., 13*, 550–557.

Rios, J. L., Francini, F., & Schinella, R. G., (2015). Natural products for the treatment of type 2 diabetes mellitus. *Planta Med., 81*, 975–994.

Ritz, E., Adamczak, M., & Wiecek, A., (2013). Chapter 2-carbohydrate metabolism in kidney disease and kidney failure. In: Kopple, J. D., Massry, S. G., & Kalantar-Zadeh K., (eds.), *Nutritional Management of Renal Disease* (3rd edn., pp. 17–30); Academic Press.

Rock, W., Rosenblat, M., Miller-Lotan, R., Levy, A. P., Elias, M., & Aviram, M., (2008). Consumption of wonderful variety pomegranate juice and extract by diabetic patients increases paraoxonase 1 association with high-density lipoprotein and stimulates its catalytic activities. *J. Agric. Food Chem., 56*, 8704–8713.

Ross, S. A., Mehmedic, Z., Murphy, T. P., & Elsohly, M. A., (2000). GC-MS analysis of the total Δ9-THC content of both drug-and fiber-type cannabis seeds. *J. Anal. Toxicol., 24*, 715–717.

Rozenberg, O., Howell, A., & Aviram, M., (2006). Pomegranate juice sugar fraction reduces macrophage oxidative state, whereas white grape juice sugar fraction increases it. *Atherosclerosis, 188*, 68–76.

Saha, B. K., Bhuiyan, M. N. H., Mazumder, K., & Haque, K. M. F., (2009). Hypoglycemic activity of *Lagerstroemia speciosa* L. extract on streptozotocin-induced diabetic rat: Underlying mechanism of action. *Bangl. J. Pharmacol., 4*(2), 79–83.

Saha, R. K., Nesa, A., Nahar, K., & Akter, M., (2016). Anti-diabetic activities of the fruit *Aegle mamelos*. *J. Mol. Biomark. Diagn., 7*, 272.

Saka, W. A., Akhigbe, R. E., Ajayi, A. F., Ajayi, L. O., & Nwabuzor, O. E., (2017). Anti-diabetic and antioxidant potentials of aqueous extract of *Eucalyptus globulus* in experimentally-induced diabetic rats. *Afr J. Tradit. Complement Altern. Med., 14*(6), 20–26.

Salehi, Y., Taheraslani, Z., & Moradkhanin, S., (2015). Hydro-alcoholic extract of *Commiphora mukul* gum resin may improve cognitive impairments in diabetic rats. *Avicenna J Med. Biochem., 2*(2), e24906.

Satheesh, M. A., & Pari, L., (2004). Antioxidant effect of *Boerhavia diffusa* L. in tissues of alloxan induced diabetic rats. *Indian J. Exp. Biol., 42*, 989–992.

Saumya, S. M., & Basha, P. M., (2011). Antioxidant effect of *Lagerstroemia speciosa* Pers (Banaba) leaf extract in streptozotocin-induced diabetic mice. *Indian J. Exp. Biol., 49*(2), 125–131.

Schubert, S. Y., Neeman, I., & Resnick, N., (2002). A novel mechanism for the inhibition of NF-kappaB activation in vascular endothelial cells by natural antioxidants. *FASEB J., 16*, 1931–1933.

Scott, L. J., & Spencer, C. M., (2000). Miglitol: A review of its therapeutic potential in type 2 diabetes mellitus. *Drugs, 59*(3), 521–549.

Selvi, N. M. K., Sridhar, M. G., Swaminathan, R. P., & Sripradha, R., (2015). Efficacy of turmeric as adjuvant therapy in type 2 diabetic patients. *Indian J. Clin. Biochem., 30*, 180–186.

Serraclara, A., Hawkins, F., Perez, C., Dominguez, E., Campillo, J. E., & Torres, M. D., (1998). Hypoglycemic action of an oral fig-leaf decoction in type-I diabetic patients. *Diabetes Res. Clin. Pract., 39*(1), 19–22.

Shang, H., Cao, S., Wang, J., Zheng, H., & Putheti, R., (2009). Glabridin from Chinese herb licorice inhibits fatigue in mice. *Afr. J. Tradit. Comple. Altern. Med., 7*, 17–23.

Sharma, K., Shukla, S., & Chauhan, E. S., (2016). Evaluation of *Aegle marmelos* (Bael) as hyperglycemic and hyperlipidemic diminuting agent in type II diabetes mellitus subjects. *Pharma Innovation., 5*(5), 43–46.

Sharma, P. C., Yelne, M. B., & Dennis, T. J., (2000). *Database on Medicinal Plants Used in Ayurveda* (Vol. 1, No. XIII, p. 528). In Central Council for Research in Ayurveda & Sidha; New Delhi.

Sharma, S. B., Nasir, A., Prabhu, K. M., Murthy, P. S., & Dev, G., (2003). Hypoglycaemic and hypolipidemic effect of ethanolic extract of seeds of *Eugenia jambolana* in alloxan-induced diabetic rabbits. *J. Ethnopharmacol., 85*(2–3), 201–206.

Sharma, S., Agrawal, R. P., Choudhary, M., Jain, S., Goyal, S., & Agarwal, V., (2011). Beneficial effect of chromium supplementation on glucose, HbA1C, and lipid variables in individuals with newly onset type-2 diabetes. *J. Trace Elem. Med. Biol., 25*, 149–153.

Shaw, J. E., Sicree, R. A., & Zimmet, P. Z., (2010). Global estimates of the prevalence of diabetes for 2010 and 2030. *Diabetes Res. Clin. Pract., 87*, 4–14.

Siddiqui, M. Z., Tomas, M., & Prasad, N., (2013). physicochemical characterization and antioxidant activity of essential oils of *Guggul* (*Commiphora wightii*) collected from Madhya Pradesh. *Indian J. Pharm. Sci., 75*(3), 368–372.

Simonienko, K., Wygnał, N., Cwalina, U., Kwiatkowski, M., Szulc, A., & Waszkiewicz, N., (2018). The reasons for use of cannabinoids and stimulants in patients with schizophrenia. *Psychiatr. Pol., 52*(2), 261–273.

Singh, U., Singh, S., & Kochhar, A., (2012). Therapeutic potential of antidiabetic neutraceuticals. *Phytopharmacol., 2*(1), 144–169.

Sobenin, I. A., Nedosugova, L. V., Filatova, L. V., Balabolkin, M. I., Gorchakova, T. V., & Orekhov, A. N., (2008). Metabolic effects of time-released garlic powder tablets in type 2 diabetes mellitus: The results of double-blinded placebo-controlled study. *Acta Diabetol., 45*, 1–6.

Song, Y., Manson, J. E., Buring, J. E., Sesso, H. D., & Liu, S., (2005). Associations of dietary flavonoids with risk of type 2 dibetes, and markers of insulin resistance and systematic inflammation in women: A prospective study and cross-section analysis. *J. Am Coll. Nutr., 24*, 376–384.

Soto, C. P., Perez, B. L., Favari, L. P., & Reyes, J. L., (1998). Prevention of alloxan-induced diabetes mellitus in the rat by silymarin. *Comp. Biochem. Physiol. C. Pharmacol. Toxicol. Endocrinol., 119*, 125–129.

Soto, C., Mena, R., Luna, J., Cerbon, M., Larrieta, E., Vital, P., Uria, E., et al., (2004). Silymarin induces recovery of pancreatic function after alloxan damage in rats. *Life Sci., 75*, 2167–2180.

Soto, C., Recoba, R., Barron, H., Alvarez, C., & Favari, L., (2003). Silymarin increases antioxidant enzymes in alloxan-induced diabetes in rat pancreas. *Comp. Biochem. Physiol. Part C Toxicol. Pharmacol., 136*, 205–212.

Srichamroen, A., Thomson, A. B. R., Field, C. J., & Basu, T. K., (2009). *In vitro* intestinal glucose uptake is inhibited by galactomannan from Canadian fenugreek seed (*Trigonella foenum graecum* L) in genetically lean and obese rats. *Nutr Res., 29*, 49–54.

Sridhar, S. B., Sheetal, U. D., Pai, M. R. S. M., & Shastri, M. S., (2005). Preclinical evaluation of the antidiabetic effect of *Eugenia jambolana* seed powder in streptozotocin-diabetic rats. *Braz. J. Med. Biol. Res., 38*(3), 463–468.

Stalin, C., Dineshkumar, P., & Nithiyananthan, K., (2012). Evaluation of antidiabetic activity of methanolic leaf extract of *Ficus carica* in alloxan induced diabetic rats. *Asian J. Pharm. Clin. Res., 5*, 85–87.

Staniek, H., (2018). The combined effects of Cr(III) supplementation and iron deficiency on the copper and zinc status in wistar rats. *Biol. Trace Elem. Res.* doi: 10.1007/s12011-018-1568-7.

Stohs, S. J., Miller, H., & Kaats, G. R., (2012). A review of the efficacy and safety of banaba (*Lagerstoemia spinosa* L.) and corosolic acid. *Phytother Res., 26*, 317–324.

Sudhakara, G., Ramesh, B., Mallaiah, P., Sreenivasulu, N., & Saralakumari, D., (2012). Protective effect of ethanolic extract of *Commiphora mukul* gum resin against oxidative stress in the brain of streptozotocin induced diabetic wistar male rats. *EXCLI Journal, 11*, 576–592.

Sugimoto, K., Hosotani, T., Kawasaki, T., Nakagawa, K., Hayashi, S., Nakano, Y., Inui, H., & Yamanouchi, T., (2010). Eucalyptus leaf extract suppresses the postprandial elevation of portal, cardiac and peripheral fructose concentrations after sucrose ingestion in rats. *J. Clin. Biochem. Nutr., 46*, 205–211.

Sugimoto, K., Suzuki, J., Nakagawa, K., Hayashi, S., Enomoto, T., Fujita, T., Yamaji, R., Inui, H., & Nakano, Y., (2005). Eucalyptus leaf extract inhibits intestinal fructose absorption, and suppresses adiposity due to dietary sucrose in rats. *Br. J. Nutr., 93*, 957–963.

Sui, D. Y., Lu, Z. Z., Li, S. H., & Cai, Y., (1994). Hypoglycemic effect of saponin isolated from leaves of *Acanthopanax senticosus* (Rupr. et Maxin.) Harms. *Zhongguo Zhong Yao Za Zhi., 19*, 683–685.

Sun, Y. L., Liu, L. D., & Hong, S. K., (2011). *Eleutherococcus senticosus* as a crude medicine: Review of biological and pharmacological effects. *J. Med. Plants Res., 5*(25), 7.

Suzuki, Y., Hayashi, K., Sukabe, I., & Kakuda, T., (2001). Effects and mode of action of banaba (*Lagerstroemia speciosa* L.) leaf extracts on postprandial blood glucose in rats. *Jpn. Soc. Nutr. and Food Sci., 54*, 131–137.

Tabatabaei-Malazy, O., Larijami, B., & Abdollahi, M., (2012). A systematic review of in vitro studies conducted on effect of herbal products on secretion of insulin from langerhans islets. *J. Pharm. Pharm. Sci., 15*, 447–466.

Takagi, S., Miura, T., Ishibashi, C., Kawata, T., Ishihara, E., Gu, Y., & Ishida, T., (2008). Effect of corosolic acid on the hydrolysis of disaccharides. *J. Nutr. Sci. Vitaminol., 54*(3), 266–268.

Taleb-Dida, N., Krouf, D., & Bouchenak, M., (2011). *Globularia alypum* aqueous extract decreases hypertriglyceridemia and ameliorates oxidative status of the muscle, kidney, and heart in rats fed a high-fructose diet. *Nutr. Res., 31*(6), 488–495.

Tang, L. Q., Wei, W., Chen, L. M., & Liu, S., (2006). Effects of berberine on diabetes induced by alloxan and a high-fat/high-cholesterol diet in rats. *J. Ethnopharmacol., 108*, 109–115.

Tian, J., Zhang, Y., Yang, X., Rui, K., Tang, X., Ma, J., Chen, J., Xu, H., Lu, L., & Wang, S., (2014). *Ficus carica* polysaccharides promote the maturation and function of dendritic cells. *Int. J. Mol Sci., 15*, 12469–12479.

Tiwari, P., Mishra, B. N., & Sangwan, N. S., (2014). Phytochemical and pharmacological properties of *Gymnema sylvestre*: An important medicinal plant. *Biomed. Res. Int.*, 830285.

Triay, J., Mundi, M., Klein, S., Toledo, F. G., Smith, S. R., Abu-Lebdeh, H., & Jensen, M., (2012). Does rimonabant independently affect free fatty acid and glucose metabolism? *J. Clin. Endocrinol. Metab., 97,* 819–827.

Tsuchibe, S., Kataumi, S., Mori, M., & Mori, H., (2006). An inhibitory effect on the increase in the postprandial glucose by banaba extract capsule enriched corosolic acid. *J. Integrated Study Diet Hab., 17*, 255–259.

Tzulker, R., Glazer, I., Bar-Ilan, I., Holland, D., Aviram, M., & Amir, R., (2007). Antioxidant activity, polyphenol content, and related compounds in different fruit juices and homogenates prepared from 29 different pomegranate accessions. *J. Agric. Food Chem., 55*, 9559–9570.

Ulbricht, C., Dam, C., Milkin, T., Seamon, E., Weissner, W., & Woods, J., (2007). Banaba, (*Lagerstroemia speciosa* L.): An evidence-based systematic review by the natural standard research collaboration. *J. Herb Pharmacother., 7*(1), 99–113.

Ulven, T., (2012). Short-chain free fatty acid receptors FFA2/GPR43 and FFA3/GPR41 as new potential therapeutic targets. *Front Endocrinol., 3*, 111.

Urquiaga, I., D'Acuna, S., Perez, D., Dicenta, S., Echeverria, G., Rigotti, A., & Leighton, F., (2015). Wine grape pomace flour improves blood pressure, fasting glucose and protein damage in humans: A randomized controlled trial. *Biol. Res., 48*, 49.

Uusitupa, M. I., Kumpulainen, J. T., Voutilainen, E., Hersio, K., Sarlund, H., Pyörälä, K. P., Koivistoinen, P. E., & Lehto, J. T., (1983). Effect of inorganic chromium supplementation on glucose tolerance, insulin response, and serum lipids in noninsulin-dependent diabetics. *Am. J. Clin. Nutr., 38*, 404–410.

Van, D. R. M., (2012). Dietary flavonoid intakes and risk of type 2 diabetes in US men and women. *Am. J. Clin. Nutr., 95*, 925–933.

Van, W. B., Van, O. B., & Gericke, N., (1997). *Medicinal Plants of South Africa* (p. 66). Briza Publications, Pretoria, South Africa.

Vandanmagsar, B., Haynie, K. R., Wicks, S. E., Bermudez, E. M., Mendoza, T. M., Ribnicky, D., Cefalu, W. T., & Mynatt, R. L., (2014). *Artemisia dracunculus* L. extract ameliorates insulin sensitivity by attenuating inflammatory signaling in human skeletal muscle culture. *Diabetes Obes. Metab., 16*(8), 728–738.

Vanschoonbeek, K., Thomassen, B. J., Senden, J. M., Wodzig, W. K., & Van, L. L. J., (2006). Cinnamon supplementation does not improve glycemic control in postmenopausal type 2 diabetes patients. *J. Nutr., 136*, 977–980.

Velussi, M., Cernigoi, A. M., Viezzoli, L., Dapas, F., Caffau, C., & Zilli, M., (1993). Silymarin reduces hyperinsulinemia, malondialdehyde levels, and daily insulin need in cirrhotic diabetic patients. *Curr. Therap. Res., 53*, 533–545.

Vijayakumar, M. V., & Bhat, M. K., (2008). Hypoglycemic effect of a novel dialyzed fenugreek seeds extract is sustainable and is mediated, in part, by the activation of hepatic enzymes. *Phytother. Res., 22*, 500–505.

Vijaykumar, M. V., Singh, S., Chhipa, R. R., Chhipa, R. R., & Bhat, M. K., (2005). The hypoglycaemic activity of fenugreek seed extract is mediated through the stimulation of an insulin signaling pathway. *Br. J. Pharmcol., 146*, 41–48.

Vogler, B. K., & Ernst, E., (1999). Aloe Vera: A systematic review of its clinical effectiveness. *Br. J. Gen. Pract., 49*, 823–828.

Vroegrijk, I. O., Van, D. J. A., Van, D. B. S., Westbroek, I., Keizer, H., Gambelli, L., et al., (2011). Pomegranate seed oil, a rich source of punicic acid, prevents diet-induced obesity and insulin resistance in mice. *Food Chem. Toxicol., 49*, 1426–1430.

Vuksan, V., Sievenpiper, J. L., & Owen, R., (2000). Beneficial effects of viscous dietary fiber from Konjac-Mannan in subjects with the insulin resistance syndrome: Results of a controlled metabolic trial. *Diab. Care., 23*, 9–14.

Waki, I., Kyo, H., Yasuda, M., & Kimura, M., (1982). Effects of a hypoglycemic component of ginseng radix on insulin biosynthesis in normal and diabetic animals. *J. Pharmacobiodyn., 5*, 547–554.

Walia, K., & Boolchandani, R., (2015). Role of amla in type 2 diabetes mellitus: A review. *Res. J. Recent Sci., 4*, 31–35

Wang, X. Q., Wang, L., Tu, Y. C., & Zhang, Y. C., (2018). Traditional Chinese medicine for refractory nephrotic syndrome: Strategies and promising treatments. *Evid. Based Complement Alternat. Med.*, 8746349.

Wang, Z. Q., Ribnicky, D., Zhang, X. H., Zuberi, A., Raskin, I., Yu, Y., & Cefalu, W. T., (2010). An extract of *Artemisia dracunculus* L. enhances insulin receptor signaling and modulates gene expression in skelmuscle in KK-A(y) mice. *J. Nutr. Biochem., 22*(1), 71–78.

Wang, Z. S., Xiong, F., Xie, X. H., Chen, D., Pan, J. H., & Cheng, L., (2015). Astragaloside IV attenuates proteinuria in streptozotocin induced diabetic nephropathy via the inhibition of endoplasmic reticulum stress. *BMC Nephrol.*, 16, 44.

Wedick, N. M., Pan, A., Cassidy, A., Rimm, E. B., Sampson, L., Rosner, B., Willett, W., Hu, F. B., & Sun, Q., (2012). Dietary flavonoid intakes and risk of type 2 diabetes in US men and women. *Am. J. Clin. Nutr., 95*(4), 925–933.

Wehmeier, U. F., & Piepersberg, W., (2004). Biotechnology and molecular biology of the α-glucosidase inhibitor acarbose. *Appl. Microbiol. Biotechnol., 63*, 613–625.

Wehmeier, U. F., (2003). The biosynthesis and metabolism of acarbose in *Actinoplanes* sp. SE50/110: A progress report. *Biocatal. Biotransformation, 21*, 279–284.

Welihinda, J., Karunanayake, E. H., Sheriff, M. H., & Jayasinghe, K. S., (1986). Effect of *Momordica charantia* on the glucose tolerance in maturity onset diabetes. *J. Ethnopharmacol., 17*(3), 277–282.

World Health Organization, (2016). *Global Report on Diabetes*. Available on the WHO website. http://www.who.int (accessed on 20 May 2020).

Xu, K. Z., Zhu, C., Kim, M. S., Yamahara, J., & Li, Y., (2009). Pomegranate flower ameliorates fatty liver in an animal model of type 2 diabetes and obesity. *J. Ethnopharmacol., 123*, 280–281.

Yamada, K., Hogokowa, M., & Fujimoto, S., (2008a). Effect of corosolic acid on gluconeogenesis in rat liver. *Diabetes Res. Clin. Pract., 80*, 48–55.

Yamada, K., Hosokawa, M., Yamada, C., Watanabe, R., Fujimoto, S., Fujiwara, H., Kunitomo, M., et al., (2008b). Dietary corosolic acid ameliorates obesity and hepatic steatosis in KK-Ay mice. *Biol. Pharm. Bull., 31*(4), 651–655.

Yamakoshi, J., Tokutake, S., Kikuchi, M., Kubota, Y., Konishi, H., & Mitsuoka, T., (2001). Effect of proanthocyanidin-rich extract from grape seeds on human fecal flora and fecal odor. *Microb. Ecol. Health Dis., 13*, 25–31.

Yeh, G. Y., Eisenberg, D. M., Kaptchuk, T. J., & Phillips, R. S., (2003). Systematic review of herbs and dietary supplements for glycemic control in diabetes. *Diab Care., 26*, 1277–1294.

Yibchok-anun, S., Adisakwattana, S., Yao, C. Y., Sangvanich, P., Roengsumran, S., & Hsu, W. H., (2006). Slow acting protein extract from fruit pulp of *Momordica charantia* with insulin secretagogue and insulinomimetic activities. *Biol. Pharm. Bull., 29*(6), 1126–1131.

Yimam, M. S., Brownell, L. A., Jialal, I., Singh, S., & Jia, Q., (2013). Effects of Aloe Vera supplementation in subjects with prediabetes/metabolic syndrome. *Metab. Syndr. Relat. Disord., 11*, 35–40.

Yin, R, V., & Phung, O. J., (2015). Effect of chromium supplementation on glycated hemoglobin and fasting plasma glucose in patients with diabetes mellitus. *Nutr. J., 14*, 14.

Yoon, J. W., Kang, S. M., Vassy, J. L., Shin, H., Lee, Y. H., Ahn, H. Y., Choi, S. H., et al., (2012). Efficacy and safety of ginsam, a vinegar extract from Panax ginseng, in type 2 diabetic patients: Results of a double-blind, placebo-controlled study. *J Diabetes Investig., 3*, 309–317.

Yoshida, M., Vanstone, C. A., Parsons, W. D., Zawistowski, J., & Jones, P. J., (2006). Effect of plant sterols and glucomannan on lipids in individuals with and without type II diabetes. *Eur. J. Clin. Nutr., 60*(4), 529–537.

Zhang, C., Gui, L., Xu, Y., Wu, T., & Liu, D., (2013). Preventive effects of andrographolide on the development of diabetes in autoimmune diabetic NOD mice by inducing immune tolerance. *Int. Immunopharmacol., 16*(4), 451–456

Zhang, H. T., Shi, K., Baskota, A., Zhou, F. L., Chen, Y. X., & Tian, H. M., (2014). Silybin reduces obliterated retinal capillaries in experimental diabetic retinopathy in rats. *Eur. J. Pharmacol., 740*, 233–239.

Zhang, Y. L. S., & Liu, Y. Y., (2007). Effect of panax quinquefolius saponin on insulin sensitivity in patients of coronary heart disease with blood glucose abnormality. *Zhongguo Zhong Xi Yi Jie He Za Zhi., 27*, 1066–1069.

Zhang, Z., Jiang, J., Yu, P., Zeng, X., Larrick, J. W., & Wang, Y., (2009). Hypoglycemic and beta cell protective effects of andrographolide analog for diabetes treatment. *J. Transl. Med., 7*, 62.

Zhu, Y., Dong, Y., Qian, X., Cui, F., Guo, Q., Zhou, X., Wang, Y., Zhang, Y., & Xiong, Z., (2012). Effect of superfine grinding on antidiabetic activity of bitter melon powder. *Int. J. Mol. Sci., 13*(11), 14203–14218.

CHAPTER 11

Marine Drugs: A Source of Medicines for Neuroinflammatory Disorders

ARUNACHALAM MUTHURAMAN,[1,2] NARAHARI RISHITHA,[1] and NALLAPILAI PARAMAKRISHNAN[3]

[1]*Department of Pharmacology, JSS Academy of Higher Education and Research, Mysuru – 570015, Karnataka, India*

[2]*Pharmacology Unit, Faculty of Pharmacy, AIMST University, Semeling, Bedong – 08100, Kedah, Darul Aman, Malaysia, E-mail: arunachalammu@gmail.com*

[3]*Department of Pharmacognosy, JSS Academy of Higher Education and Research, Mysuru – 570015, Karnataka, India*

ABSTRACT

Neuroinflammation is due to the abnormal neurovascular system. In nervous tissue nerve, the blood barrier is altered in pathological conditions of neuro-inflammation which leads to enhancing the complication in peripheral as well as the central neurovascular system. The etiology is due to the activation of endogenous factors such as overexpression of pro-inflammatory and pro-apoptotic mediators; activation of neuroimmune cells like glial, astrocyte, oligodendrocyte, and Schwann cells. In addition, various cellular enzymes, ions, ion channels, and prion proteins are contributing to the pathogenesis of neurovascular disorders. Moreover, the various exogenous factors have also involved the damage of blood-brain and blood-nerve barriers such as smoking, alcohol, and neurotoxic agents. The various conventional medicines are focused to treat the neurovascular system via regulation of ion channels, and enzymes mediated actions. Moreover, the clinical use of conventional medicines is still limited due to the low efficacy and high toxicity. Current drug discovery is focused on the natural source of medicines for neurovascular disorders especially marine drugs. Because marine drugs play an important role in the amelioration of neurovascular disorders via enhancement of the anti-inflammatory mechanism. Therefore, this book chapter focused to explore the marine drugs category and their molecular mechanism for neuroinflammatory disorders such as Alzheimer's disorder (AD), Parkinson's disorders (PD), multiple sclerosis (MS), stroke, vascular dementia (VaD), and neuropathic pain.

11.1 INTRODUCTION

The damage of neuronal tissue is an important hallmark for the progress of neuroinflammatory disorders. The chronic neuroinflammation induces the neurodegenerative process leads to cause the neuroinflammatory disorders like Parkinson disease (PD), Alzheimer's disease (AD), multiple sclerosis (MS), opsoclonus-myoclonus-ataxia syndrome, autoimmune encephalitis, encephalomyelitis, optic neuritis, stroke, transverse myelitis, vascular dementia (VaD), and neuropathic pain (Bhat et al., 2015). The maximum neuroinflammatory disorders are very common in human beings and the rest of the disorders is a rare type of neuroinflammatory disease like opsoclonus-myoclonus-ataxia syndrome (Ratner et al., 2016). Various factors like infection, nutritional deficiency, lifestyle modification, mechanical injury, exposure of neurotoxic agents, pollutants, radiations, drug, and disease-associated factors, and genetic factors are liable for the progress of neuroin-flammation (Lane and Routledge, 1983). In addition, certain physiological abnormalities of neuronal tissue like the sustained opening of excitatory ion channels, neuronal firing, and neuronal excitation also responsible for the initiation of neuroinflammatory signals (Dillman et al., 2009). Further, it also improves the activation of secondary neuroinflammatory messengers like calcium-binding proteins (calmodulin, calpeptin & calsequestrin (Muth-uraman and Kaur, 2016); lipid peroxidation of mitochondrial, nuclear and cellular membranes (Ott et al., 2007); induces the enzymatic and non-enzy-matic factors of inflammation (Kruidenier and Verspaget, 2002); activates the prooxidant and proapoptotic factors; induces the release and synthesis of chemokines and cytokines (Haddad, 2002); expression of abnormal post-translated proteins (prion) (Iwanaga et al., 2014); and fragmentation of mitochondrial and nuclear deoxyribonucleic acid (DNA) (Baek et al., 2011).

The central nervous system (CNS) has specialized cells for the process of neuroinflammatory reactions. Such cells are also known as neuroimmune cells, i.e., glial cells (astrocyte and oligodendrocyte); similarly, the peripheral nervous system (PNS) has Schwann cells (Ruhl, 2005). These cells fight against the pathogens and foreign particles via the process of the host immune system (Finlay and McFadden, 2006). The over-activation of these cells is ready to release and enhance the various neurotoxic proteins which lead to accelerating the neuroinflammatory and neurodegenerative process. Typically, CNS has a privileged site for immunological reaction via communication between the peripheral immune cells (mast cells, macrophage, neutrophils, and lymphocytes), endothelial cells of blood-brain barrier (BBB), neurovascular beds, astrocytes &and oligodendrocyte in different areas of the brain (Biber et al., 2007). Further, the circulatory peripheral immune cells are suppressing the BBB and encounter neuronal misfolded proteins (prions), and glial cells immune-compatibility complex molecules in healthy tissue (Puentes et al., 2016). Whereas, in pathological conditions, these cells are plays a toxic role to the neuronal tissue with widespread interaction of neuroinflammatory cells by the migration of leukocytes via BBB to the brain (Zlokovic, 2008). Therefore, a clear understanding is required about neuroinflammatory disorders with etiology; possible mechanism; drug targets; available medicines (natural, synthetic, and semisynthetic); drawbacks, and drug discovery process of newer medicines. It will be helpful to achieve the better treatment of neuroinflammatory and neurodegenerative disorders of mankind. This book chapter focused to explore marine drugs and their molecular mechanism to treat neuroinflammatory disorders and neurovascular disorders.

11.2   MECHANISM OF NEUROINFLAMMATION

The development of neuroinflammation initiate from exogenous and endogenous factors like toxic metabolites like homocysteine; uremia; autoimmunity reaction like neutrophil recruitment; microbial infections like herpes and human immunodeficiency virus; aging; traumatic brain injury like a head injury; spinal cord injury; air pollutions like biphenyl compounds; alcohol; passive smoking condensate like carbonyl compounds; and heavy metals like iron, copper, lead, and mercury. Even, high fat, cholesterol, and glucose levels are also affecting the neuronal system (Singh et al., 2002). Some of the factors are modifiable (controllable) like smoking, alcohol intake, etc. Some of the factors are not modifiable and it can be controlled with medicines and proper alternative approaches. The mechanism of neuroinflammation can be explained through cellular and molecular aspects (Agarwal et al., 2012). Cellular aspects, endothelial cell, platelet, peripheral, and central immune cells contribute to the development of neuroinflammatory and neurovascular disorders. The activation of platelets is contributed to the release of clotting factors and activates the neutrophils and leukocytes (Iadecola and Anrather, 2011). The activations of neutrophils and leukocytes ready to enter the diapedesis process in the brain through BBB which leads to cause enhance the central neuroinflammatory reactions. The process also called extravasation reactions between vascular and tissue compartments (Ransohoff et al., 2003). In addition, acute a change of cellular ion-exchange homeostasis is induced the cellular dysfunctions via fluid exudation and edema. These reactions also interlinked with extravasation reactions (Pober and Sessa, 2007). Further, the immunological reaction of peripheral, as well as the central process, also contributes to the pathogenesis of neuroinflammation. Innate immune cells of the CNS, i.e., microglial cells enhance the biosynthesis of reactive oxygen species (ROS); inflammatory mediators lead to release the signals for recruitment of peripheral immune cells like neutrophil and leukocytes for neuroinflammatory reactions (Miller et al., 2009).

Out of all cellular reactions, the immunological reactions are key players for neuroinflammatory disorders due to its rapid, sensitive, and sustained reactions. Astrocytes and oligodendrocytes are predominantly engaging in these neuro-immunological reactions (Otmishi et al., 2008). In addition, astrocytes also release the multiple growth factors which make the morphological changes of nerve tissue. This reaction is acute and initial events because once astrocytes reaction is over it will compose to proteoglycan matrix and form the glial scar (Silver and Miller, 2004). The immune cells are communicating via various cell mediators like cytokines and chemokines (Hanisch, 2002). In physiological conditions, cytokines regulate inflammation, cell signaling, and various cellular processes like cell growth and survival (Veldhoven et al., 2006). Chemokines are subset molecules of cytokines and it also regulates the cell functions like the relocation of cells which is attracting immune cells in the place of injury or infection from the circulation of blood, cerebrospinal fluid, and surrounding (adjacent) tissue compartments (Goverman, 2009). The following brain cells are communicating via cytokines and chemokines. Such cells are microglia, astrocytes, endothelial cells oligodendrocytes, and other glial cells (Hanisch, 2002). In addition, chemokines, and cytokines also possess neuromodulatory actions.

In contrast, neuroinflammatory conditions of neuronal tissue release the sustained and abundant quantity of cytokines and chemokines. In this situation, BBB is compromised with immunological reactions leads to accelerating peripheral immune cell recruitment (Heneka et al., 2014). The molecular mediator of cytokines reactions are interleukin-6 (IL-6: responsible for astrogliosis); and interleukin-1 beta (IL-1β: responsible for cytotoxic reaction), and tumor necrosis factor-alpha (TNF-α: responsible for neuroinflammatory reaction). TNF-α has potential neurotoxic and neuro-inflammatory properties (Munoz-Fernandez and Fresno, 1998). Therefore, various immuno-compatibility based medicines are clinically used for neuroinflammatory disorders like immunosuppressants, i.e., tacrolimus, rapamycin. azathioprine, mycophenolate mofetil, and cyclosporine (Villiger and Muller-Ladner, 2017). Marine drugs-based medicines are also targeting to regulate the immunological reaction to treat neuroinflammatory disorders (de Boer and Gaillard, 2007).

CNS immunological reactions are ready to respond to the peripheral immune signals. Astrocytes of endothelial tight junctions are strictly regu-lating the entry of peripheral immune cells via BBB and interstitial space (Shechter et al., 2013). Whereas, immune-compromised BBB endothelial cells are allows recruiting the peripheral immune cells to achieve the adaptive and innate immunological reactions via macrophages, T-cells, and B-cells. These exacerbated reactions induce the neuroinflammatory environment of the CNS leads to cause the neurodegenerative and neuroinflammatory disor-ders (Nguyen et al., 2002). Whereas, the inflammation of PNS due to direct attack of immune cells to the peripheral neuronal tissue. And, PNS immune cells, i.e., Schwann cells alter the nerve-blood barrier function of neuronal (Collins and Hadden, 2017). In addition, CNS, and PNS have the neuronal and non-neuronal (cellular) connection for the regulation and induction of neuroinflammation in both sites. The specialized connection for the inflam-matory reaction of neuronal tissues is the gut-brain axis. The neuroimmune cells and gut microbiota have a connection for the neuroinflammatory reac-tion and progress of neurological disorders (Ordovas-Montanes et al., 2015). The best example of gut-based neuroinflammatory disorders is Parkinson's disease and autism spectral disorders (Mayer et al., 2014).

The molecular mechanism of neuroinflammatory disorders varies based on the category of cells involved, level of mediator released, and duration of process ongoing for neuroinflammatory reaction (Agostinho et al., 2010). In the acute stage, PNS & CNS immune cells and endothelial cells based release of IL and TNF are involved in the neuroinflammatory reaction. In addition, local release of autocoids like prostaglandin, leukotriene, brady-kinins, histamine, and serotonin also contributes to the neuroinflammatory reaction, which leads to alters the ion and fluid exchange process leads to cause the potential neuroinflammation (Kasperska-Zajac et al., 2008). If acute neuroinflammation is prolonged conditions, it activates the multiple cytosolic cellular environments like generation ROS from mitochondrial; accumulation of calcium ion via the release of stored calcium and opening of stored operated calcium ion channels; induces the oxidative enzymatic system; alteration of the phospholipid layer of the membrane via peroxidative process; and modulation of DNA and post-translational proteins including misfolding of expressed proteins leads to cause the neurodegeneration (Rizzuto et al., 2012).

11.3 ROLE OF MARINE DRUGS FOR NEUROINFLAMMATORY DISORDERS

Marine drugs are known to produce multiple pharmacological actions. In addition, some of the marine drugs are also reported that it has potent neuro-protective and prevention of neuroinflammation. The following sections are exploring the role of marine drugs in neuroinflammatory disorders like AD, PD, MS, stroke, VaD, and neuropathic pain.

11.3.1 ROLE OF MARINE DRUGS FOR AD

Alzheimer's disease (AD) is chronic neurodegenerative disorders due to an accumulation of inflammatory markers associated with shrinkage (atrophy) of the brain (Gavett et al., 2011). However, neuroinflammation also occurs in various neurological disorders. Thus, neuroinflammation associated accumulation of beta-amyloid proteins and neurofibrillary tangles are specific hallmarks for AD. The synthesis and catabolism of β-amyloid proteins are cyclic patterns (Parihar and Hemnani, 2004). The neuroimmune cells of CNS, i.e., microglial are contributed to regulates the β-amyloid proteins levels via phagocytosis actions. The cytokine-activated microglial cells fail to produce their phagocytosis ability for the clearance of β-amyloid proteins which leads to forms the β-amyloid plaque and neurofibrillary tangles and it's deposited on the head of neurons leads to produce AD (Aldskogius, 2001). The main inflammatory cytokine, i.e., IL-1β upregulation is identified in AD. In addition, the administration of central acting anti-inflammatory agents ameliorated the progression of AD (Wyss-Coray, 2006). The adminis-tration of conventional nootropic agents and phytomedicines are shown less efficacy & more toxicity (Mukherjee et al., 2008). Nevertheless, the marine drugs like CEP-1347, fascaplysin, ethanolic extract of *Nannochloropsis oceanic* and pyrogallol-phloroglucinol-6,6-bieckol (PPB) attenuated the AD with potential anti-inflammatory and neuroprotective actions (Chen et al., 2014b, Choi et al., 2017, Kang et al., 2013, Manda et al., 2016). In addition, marine drugs are produced the potent anti-inflammatory actions via multiple cellular and molecular mechanism. The collagen-derived peptides of marine sponge (*Chondrosiareniformis*) possess the potent ROS scavenging activity leads to produce the effective wound-healing action (Pozzolini et al., 2018). Other marine sponges, i.e., *Geodia barrette* has potent anti-inflammatory substances. Now, the structure of *Geodia barrette* derived products are identified as 6-bromoindole (Di et al., 2018). Hence, the derivatives of 6-bromoindole may be useful medicines to treat neuroinflammatory disorders like AD. Further, twigs and leaves of *Lumnitzeraracemosa* (marine mangrove plant) possess the potent anti-angiogenic and anti-inflammatory action. Hence, it can be a source for the isolation of newer bioactive molecules to treat neuroinflammatory as well as cancer disorders (Yu et al., 2018). The chronic inflammation causing the multiple disorders via accelerates the expression of various cellular proteins. Recently, marine exopolysaccharide is also identified as an anti-inflammatory agent (Zykwinska et al., 2019). In addition to that, astaxanthin, phlorotannins (*Ecklonia cava*), and zoanthamine alkaloids (*Zoanthus cf. pulchellus*) are identified as marine-derived anti-inflammatory agents (Guillen et al., 2018; Han et al., 2019; Lee et al., 2018).

Some of the marine sponge-derived products inhibit the acetylcholinesterase (AChE) enzyme. This enzyme contributes to regulating the levels of acetylcholine neurotransmitters in the cholinergic nervous system. The effective memory function of the brain is due to the activation of cholinergic nerve impulses with acetylcholine. Therefore, the inhibitors of AChE of marine sponge-derived products are useful for controlling AD. Such agents are 4-acetoxy-plakinamine B (fifty percentage of inhibitory concentration; IC_{50}: 3.75 μM) from *Corticium*Sp.; 2-bromoamphimedine (IC_{50}: 300 μM) from *Petrosian*Sp.; (marine sponge) petrosamine (IC_{50}: 91 μM) from *Petrosian* species. Similarly, the soft coral derived products, i.e., cladidiol (IC_{50}: 67 μM) from *Cladiella sp.;* and marine mollusk derived products, i.e., turbotoxinA (IC_{50}: 67 μM) from *Turbo marmoratus* are also documented to produce the inhibitory action of AChE enzyme. Various enzymatic pathways are explored in the development of AD. The beta-secretase-1 ($BACE_1$) enzyme also plays an important role in the regulation of cholinergic neuron function and controls the β-amyloid proteins. The following marine sponge-derived products with variable concentrations are produced the inhibitory action of beta-secretase-1

(BACE$_1$) enzyme. This enzyme is playing a vital role in the progress of AD. Marine sponge-derived BACE$_1$ inhibitors are xestosaprolD (IC$_{50}$: 93.2 µM) from *Xestospongia* sp.; xestosaprolF (IC$_{50}$: 135 µM); xestosaprolG (IC$_{50}$: 155 µM); xestosaprolH (IC$_{50}$: 82 µM); xestosaprolI (IC$_{50}$: 163 µM); xestosaprolJ (IC$_{50}$: 90 µM); xestosaprolK (IC$_{50}$: 93 µM); xestosaprolL (IC$_{50}$: 98 µM); xestosaprolM (IC$_{50}$: 104 µM); dictyodendrin F(IC$_{50}$: 1.5 µM); dictyodendrin H (IC$_{50}$: 1 µM); dictyodendrin I (IC$_{50}$: 2 µM) from *Xestospongia* sp.; dictyodendrin J (IC$_{50}$: 2 µM) from *Ianthella* sp.; dictazoleA (IC$_{50}$: 135 µM) *Smenospongia cerebriformis*; topsentinolK trisulfate (IC$_{50}$: 1.2 µM) from *Topsentia* sp.; lamellarin O, O1, and O2 (IC$_{50}$: 10 µM) from *Ianthella* sp.; ianthellidoneF (IC$_{50}$: 10 µM) from *Ianthella* sp. (Grosso et al., 2014). Hence, these marine products are useful to treat AD patients.

11.3.2   ROLE OF MARINE DRUGS FOR PD

Parkinson's disease is unique for the chronic inflammatory neurodegenerative disorders along with the alteration of neuromuscular motor unit changes. The symptoms of PD are mostly due to the alteration of dopamine (neurotransmitter) levels in different areas of the brain (Mu et al., 2012). There are four main dopamine pathways are identified in the human brain such pathways are: (1) *Mesocortico-limbic pathway*: it covers the ventral tegmental area (VTA) of the midbrain, ventral striatum (olfactory tubercle and nucleus accumbens). The key function of this pathway is maintaining the reward-related and aversion-related cognition; (2) *Mesocortical pathway*: it covers the VTA, prefrontal cortex. The main function of this pathway is maintaining the executive functions and development of drug addiction behavior; (3) *Nigrostriatal pathway*: it covers the substantia nigra pars compacta (SNc), and dorsal striatum (putamen and caudate nucleus). The key function of this pathway is maintaining the motor function, associative learning, reward-related cognition, and development of drug addiction behavior; and (4) *Tuberoinfundibular pathway*: it covers the arcuate nucleus, hypothalamus, pituitary gland, and hypophyseal portal system. The main function of this pathway is the regulation of prolactin release (Nemeroff, 1986). The major dopaminergic pathway of PD is the nigrostriatal pathway. The degeneration of this pathway showed the neuromotor dysfunctions via the reduction of dopamine levels. The reduction and neurodegeneration of dopaminergic neurons are expected to the acceleration of chronic neuroinflammatory process and alteration of alpha-synuclein (α-Syn) protein aggregation as well as misfolding of α-Synproteins (Kirshner et al., 2012). Furthermore, the development of a "leaky" gut environment, i.e., via dysbiosis of good bacteria and bad bacteria. This "leaky" gut environment induces the neuroinflammatory reactions leads to accelerating the α-Syn misfolding and transfer across the CNS neuron (Engen et al., 2017). Moreover, chronic oxidative stress is also responsible for the action of the neuroinflammatory signal in dopaminergic neurons via the activation of microglial cells (Lull and Block, 2010). The above reaction occurs in Stage 1 PD. In stage 2 PD shown the potential inflammatory reaction in the brainstem and alters the sleep pattern and develops the depression along with impairment of motor functions (Hawkes et al., 2007). In stage 3: inflammatory reactions affect the substantia nigra, dopamine-producing, and secreting cells of the brain leads to show motor dysfunction. Whereas, in stage 4: inflammatory reactions affect all the dopaminergic neurons and produce the neuromotor dysfunction and loss of executive function and memory. The administration of antioxidants and anti-inflammatory substances along with dopamine enhancers reduces the progress of PD (Gao et al., 2003). The administration of phytomedicines and anti-Parkinson drugs are shown less efficacy and more toxicity (Khazdair et al., 2015). However, marine drugs such as CEP-1347, 11-Dehydrosinulariolideare attenuated the PD with anti-inflammatory and neuroprotective actions (Feng et al., 2016; Ning et al., 2018). Some of the marine-derived products are ameliorated with the PD with anti-inflammatory actions. Such agents are xyloketal (Li et al., 2013); 11-dehydrosinulariolide (Feng et al., 2016); *Holothuria scabra* extracts (Chalorak et al., 2018); *Chondrus crispus*; and oligo-porphyrin (Liu et al., 2018). In addition to that, marine drugs have a potential role in axonal transport and amelioration of neurodegenerative disorders via multiple molecular and cellular mechanisms including neuroinflammation (White et al., 2016). Thus, the marine can be an essential medicine for neuroinflammatory disorders.

11.3.3   ROLE OF MARINE DRUGS FOR MS

MS is an important common inflammatory neurological disease. The primary etiological factor of neuroinflammation is due to the action of the CNS immunological system via neuroimmune cells, i.e., astrocytes, and oligodendritic cells (Morrell, 1976). MS is characterized by a chronic process of demyelination and neurodegeneration of brain cells leads to produces the cognitive dysfunction, weakness of limbs, and development of muscular fatigue (Witte et al., 2014). The primary target site of inflammatory cytokines is BBB and also involves the recruitment of peripheral immune cells. The CNS recruited plasma cells and B-cells are readily supports the production of antibodies against the myelin sheath proteins. The degradation of myelin proteins affects the conduction system of CNS and PNS signals (Hemmer et al., 2002). The administration of anti-inflammatory, immunosuppressive drugs, and phytomedicines are producing ameliorative effects in MS disorders (Islam et al., 2008). However, the efficacy &potency of these medicines remain poor. In addition, chronic administration also causes unavoidable toxicity to the biological system (Martins and Fernandes, 2012). The marine drugs like lestaurtinib, enzastaurin, MS14, and dihydro-austrasulfone (DA) ameliorated the MS via immunosuppressive, anti-inflammatory, and neuroprotective actions (Ahmadi et al., 2010; Chen et al., 2014b; Gagalo et al., 2015; Tabatabai et al., 2007). Bryostatin-1 has potential ameliorative action on MS via suppression of matrix metalloprotein and antioxidant actions (Safaeinejad et al., 2018). Similarly, some of the marines drugs are documented to produce the potential neuroprotective action and amelioration of inflammatory brain disorders (Grosso et al., 2014; Tafreshi et al., 2008). The different marine sources are producing the anti-inflammatory and immuno-modulatory action in animals and in humans. From blue starfish (*Linckia laevigata*) derived marine drugs, i.e., linckoside B (IC_{50}: 76 µM); linckoside F (IC_{50}: 30 µM); linckoside G (IC_{50}: <10 µM); linckoside H (IC_{50}: <10 µM); linckoside I (IC_{50}: 40 µM); linckoside J (IC_{50}: 40 µM); linckoside K (IC_{50}: 10 µM); LLG-5 (IC_{50}: 59.3 µM); LLG-3 (IC_{50}: 63.1 µM); and granulatoside A(IC_{50}: 95 µM) are documented to produce the neuroprotection and neuro-immune modulatory actions with different IC_{50} values. Similarly, sea cucumber (*Cucumaria echinata*) derived marine drugs, i.e., CEG-6 (IC_{50}: 43 µM); HLG-3 (IC_{50}: 42 µM); CEG-8 (IC_{50}: 40 µM); CEG-9 (IC_{50}: 35.1 µM); SJG-1 (IC_{50}: 39.1 µM); SJG-2 (IC_{50}: 64.8 µM); and CG-1 (IC_{50}: 43 µM) are also stated as neuroprotection from neuroinflammatory and neuro-immune modulatory actions (Grosso et al., 2014; Safaeinejad et al., 2018). Thus, these agents can be useful to treat MS. However, the detailed studies are essential to the clinical use of this medicine with full efficacy and less toxicity.

11.3.4 ROLE OF MARINE DRUGS FOR STROKE

Stroke is the fourth leading directly to death in the world. Multiple etiological factors are employed in the progress of stroke. The symptoms of stroke are originated by abnormal functions of the neurovascular system (Jean-Louis et al., 2008). The initial events of stroke originate from vascular segments, the narrow or blocking of the cerebral vascular system are making the ischemic environments to the specific area of the brain (Moskowitz et al., 2010). The hypoxic and ischemic events activate the abnormal cellular metabolic activity and activate the inflammatory signals via stimulation of mitochondrial, nuclear, and endoplasmic reticulum in the affected area of the brain (Nathan and Ding, 2010). Myeloid cells, i.e., macrophages, monocytes, eosinophils, basophils, neutrophils, erythrocytes, and megakaryocytes are playing an important role in the management of neuron functions. In addition, T-cells, B-cells, and natural killer cells are also contributed to the regulation of neuronal cell function. Further, the myeloid and lymphoid cells contribute to the formation of dendritic cells (Graf and Enver, 2009). Dendritic cell functions are essential for the maintaining of both adaptive and innate immunity for a healthy brain. Furthermore, dendritic cells have a unique ability to stimulate memory and naïve T-cells which is required to clear the misfolding proteins(Joffre et al., 2012). In a stroke, the chronic neuroinflammation of regional brain tissue is clearing macrophages and migratory dendritic cells which lead to the loss of neuronal regulatory functions. The various conventional anti-inflammatory, anti-thrombotic agents and phytomedicines are showed the ameliorative effect in stroke patients (Reagan-Shaw et al., 2008). Even, it also causes the potential adverse effects due to too complex mechanism and involvement of multiple pathways of neurodegeneration. The marine drugs, i.e., CEP-1347, xyloketalB, and polyphenolic compounds of *Ecklonia cava* are attenuated the stroke symptoms via multiple pharmacological actions like anti-oxidant, anti-inflammatory, immunosuppressive, and neuroprotective actions (Carlsson et al., 2009; Gong et al., 2018; Kim et al., 2012). Xyloketal B is an important potent medicine to treat neuroinflammation (Gong et al., 2018). Further, it also ameliorates the stroke symptoms via inhibition of ROS, toll-like receptor-4, and nuclear factor kappa-B activity (Choi and Choi, 2015; Pan et al., 2017). Currently, various marine drugs are under the preclinical screening stage to explore their improvement of neurovascular functions. Such agents are waixenicin, bryostatin-1, ilimaquinone, 5,8-diepi-ilimaquinone, 4,5-diepi-dactylospongiaquinone, JBIR-59, JBIR-124, conantokin-G, x-conotoxin (Choi and Choi, 2015; Pan et al., 2017; Safaeinejad et al., 2018). Therefore, marine drugs have a promising role in the management of neurovascular disorders including stroke symptoms.

11.3.5 ROLE OF MARINE DRUGS FOR VaD

VaD is also caused by neurovascular inflammation. The primary etiology of neurovascular inflammation is acquired by metabolic waste products like uremic toxin like creatinine, & urea; accumulation of peroxidative products of lipids; formation of carbonyl protein and advanced glycation end products (AGE) in diabetes; and accumulation of toxic amino acid, i.e., homocysteine (Requena et al., 1996). These factors are enhancing the formation of free radicals, activation of neuroglial cells; alteration of mitochondrial function; and breakdown of DNA leads to accelerating the neuroinflammatory process in cerebral endothelial cells and neuronal system (Agostinho et al., 2010). It may subsequent or direct action on the neurovascular system. Further, it also induces the neurodegeneration and misfolding of proteins in the neurovascular system and it causes the loss of memory, and abnormal motor function (Sas et al., 2007). The administration of nootropic agents, anti-oxidant, anti-inflammatory agents, and phytomedicines are able to reduce the symptoms of VaD (Anekonda and Reddy, 2005). However, the potency and efficacy remain poor to treat VaD disorders. Some marine drugs, i.e., CEP-1347, Xyloketal B, polyphenolic compounds of *Ecklonia cava* and pyrogallol-phloroglucinol-6,6-bieckol are ameliorated from the progress of VaD via anti-inflammatory action on the neurovascular system (Carlsson et al., 2009; Gong et al., 2018; Koirala et al., 2017; Oh et al., 2018). Further, some of the marine drugs ameliorate dementia associated with cognitive impairments and neuroinflammation. Such agents are akebiasaponin D, fermented products of *Laminaria japonica* and *Urechis unicinctus*. These agents also possess potent neuroprotective actions and prevention of neurotoxin associated VaD (Reid et al., 2018; Wang et al., 2018; Zhu et al., 2018)

11.3.6 ROLE OF MARINE DRUGS FOR NEUROPATHIC PAIN

Neuropathic pain is also an inflammatory neurological disorder. It may be systemic and regional specific disorders due to the involvement of etiological factors. Some time, mononeuropathy can be making the progress of polyneuropathy pain syndrome (Argoff et al., 2006). The foremost form of neuropathic pain is diabetic neuropathy. In diabetic conditions, the accumulated metabolic waste products like AGE products, carbonylated, and nitrosylated proteins are contributing to the progress of diabetic secondary complications (Pop-Busi et al., 2006). The molecular and cellular mechanisms of neuropathic pain progress are too complex. Neuroinflammation is the primary contributing factor in the development of neuropathic pain (Basbaum et al., 2009). Various categories of medicines are employed in managing the neuropathic pain like non-steroidal anti-inflammatory agents, steroidal drugs; narcotic drugs; anti-epileptic drugs; and anti-depressants drugs are used for the controlling of various neuropathic disorders (Mao and Chen, 2000). The first FDA approved drug for neuropathic pain is carbamazepine for trigeminal neuralgia. Though, chronic administrations of these agents produce the potential adverse effects. Some natural medicines, i.e., herbal origin (phytomedicines) are shown some improvements for the amelioration of neuropathic pain with less toxicity (Brookoff, 2000). However, the clinical usefulness of the medicines remains questionable due to lack of purity, identification of active principle, and molecular (targeted) mechanism (Kelloff et al., 2005). In addition, some of the marine drugs, i.e., enzastaurin, dihydro-austrasulfone, capnellene, lemnalol, flexibilide, and DHA attenuated the neuropathic pain symptoms by regulation of neuronal impulse transmission, anti-inflammatory, and neuroprotective actions(Chen et al., 2014a; Dudek et al., 2008; Jean et al., 2008; Jean et al., 2009; Silva et al., 2017; Wen et al., 2010; Zhang et al., 2018). Further, some newer molecules like frondanol, zoanthamine, fucoxanthin, and 6-bromoindole also possessed the anti-inflammatory and neuroprotective actions (Ahmad et al., 2017; Guillen et al., 2018; Rodriguez-Luna et al., 2018; Subramanya et al., 2018). Various marine products like astaxanthin, α-Conotoxin, μ-Conotoxin, tetrodotoxin, and docosahexaenoic acid attenuate the neuropathic pain symptoms with reduction pro-inflammatory mediators (Castro et al., 2017; Manzhulo et al., 2016; Salas et al., 2015; Sharma et al., 2018; Tosti et al., 2017).

11.4 MARINE DRUGS TREATMENT FOR NEUROINFLAMMATORY DISORDERS

Neuroinflammatory disorders are treated with various conventional medicines and herbal medicines. In addition, certain alternative methods are used to treat neuroinflammatory disorders. However, the chronic administration of these medicines is producing unwanted serious side effects. Occasionally, it also produced life-threatening effects. Due to its more toxicity and poor efficacy, the discovery of newer molecules for controlling the neuroinflammatory disorders is urgently required. The existing approaches and medicines are summarized in Table 11.1. This table also showed the molecular mechanism of conventional, herbal medicines and Miscellaneous approaches for neuroinflammatory action. In addition, Table 11.2 expressed the molecular mechanism of marine drugs for neuroinflammatory action.

TABLE 11.1 Molecular Mechanism of Conventional and Herbal Medicines for Neuroinflammatory Action

Sl. No.	Medicines	Mechanisms	References
Conventional Medicines			
1.	Anatibant	Blocks bradykinin signaling.	Zweckberger and Plesnila, 2009
		Prevents BBB disruption.	
2.	Cyclosporin A	Reduces T-cell counts and activation.	Martinez and Peplow, 2018
3.	Infliximab	Regulates cytokine TNF-α, IL-6, IL-1β, and IFNγ expression.	Dadsetan et al., 2016
4.	Thalidomide	–	Shamim and Laskowski, 2017
5.	Lenalidomide	–	Valera et al., 2015
6.	NMDA antagonist	Inhibit the calcium channel opening.	Vanle et al., 2018
		Decrease the neuronal excitability.	
7.	Erythropoietin	Decreases production of pro-inflammatory cytokines and chemokines.	Ercan et al., 2018
8.	Anakinra	Blocks IL-1 signal transduction.	Cavalli and Dinarello, 2018
9.	Rosuvastatin	Pleiotropic actions on immune regulation.	Lu et al., 2018
10.	Steroids	Inhibit leukocyte activation and infiltration. Modulate cytokine release.	De Nicola et al., 2018; Giatti et al., 2012
11.	Progesterone	Decreases upregulation of IL-1β, TNF, and complement factors 3 and 5. Reduce microglial activation.	De Nicola et al., 2018
12.	Dexanabinol	TNF inhibitor.	Frankola et al., 2011; Hasturk et al., 2018; Shohami et al., 1997

TABLE 11.1 *(Continued)*

Sl. No.	Medicines	Mechanisms	References
Herbal Medicines			
13.	Artemether	Cytokine NF-κB and p38 MAPK signaling	Okorji et al., 2016
14.	Caffeine	G-CSF Granulocytecolony-stimulating factor.	Wadhwa et al., 2018
		Stimulates stem cells to produce granulocytes.	
15.	Capsaicin and its analogs	TRPV1 reduced expression of TNF-α and IL-1β, & ROS.	Motte et al., 2018
16.	Curcumin	PKc activator.	Fu et al., 2018; Kulkarni et al., 2005; Shal et al., 2018; Song et al., 2018; Suk, 2005
17.	Diosgenin, quercetin, ginsenoside Rg3, rosmarinic acid, ginsenoside Rg3, quinic acid, and ligraminol	JNK pathway inhibitor.	
18.	Quercetin, apigenin, rosmarinic acid, curcumin, resveratrol, berberine, 6-shagoal, and huperzine	MEK activator.	
19.	quercetin, berberine	Trk-PI$_3$K activator.	
20.	Apigenin, ginsenoside Rg3, curcumin, resveratrol, berberine, 6-shagoal, and EGCG	CaMK-II/IV activator.	
21.	Quercetin, apigenin, curcumin, and EGCG	MTOR & AKT pathway inhibitor.	
22.	Narigenin, and galantamine	Nrf2 activator.	
23.	Apigenin, curcumin, resveratrol, and berberine	CREB regulator.	
24.	Ginsenoside Rg3, quinin acid, curcumin, resveratrol, berberine, 6-shagoal, huperzine, prosapogeninIII, diosgenin, rosmarinic acid, and ligraminol	Post translational protein inhibitor.	
25.	Galantamine	Protein modifier.	Liu et al., 2018b,

TABLE 11.1 *(Continued)*

Sl. No.	Medicines	Mechanisms	References
Miscellaneous			
26.	IL-4, and IL-10 (immunomodulatory cytokine)	Higher IFNγ and lower IL-10 and IL-4	Eijkelkamp et al., 2016, Latta et al., 2015; Liesz et al., 2014
27.	Amnion-derived cellular cytokine solution	Calcium-binding receptor MPO and ionized calcium-binding adaptor molecule 1 (Iba1) expression.	Deng-Bryant et al., 2015; Steed et al., 2008
28.	Hypertonic saline	Reduces TNF and IL-10. Improves T-cell function.	de Oliveira and Pinto, 2015; Schreibman et al., 2018
29.	Hypothermia	Temperature-dependent activation of a humoral and cellular immune response. Decreased neutrophil accumulation in CNS. Decreased IL-1β via reduction of caspase-1 activity.	Ceulemans et al., 2010, 2011

Abbreviation: AKT, serine/threonine-specific protein kinase; BBB, blood-brain barrier; CaMK, Ca^{2+}/calmodulin (CaM)-dependent protein kinase; CREB, cyclic adenosine monophosphate-response elements binding protein; EGCG, Epigallocatechin-3-gallate; G-CSF, granulocyte colony-stimulating factor; IFNγ, interferon-gamma; IL, interleukin; JNK, c-Jun N-terminal kinase; MEK, mitogen-activated protein kinase; MPO, myeloperoxidase; NF-κB, nuclear factor κB; NMDA, N-methyl-D-aspartate; Nrf2, nuclear factor erythroid 2–related factor 2; PI$_3$K, phosphoinositide 3-kinases; ROS, reactive oxygen species; TNF-α, tumor necrosis factor-α; Trk, tropomyosin-related kinase; and TRPV1, transient receptor potential cation channel subfamily V member 1.

Table 11.2 has been summarized as a possible molecular mechanism of herbal and conventional medicines for neuroinflammatory disorders.

11.5 OPPORTUNITIES FOR MARINE DRUGS FOR NEUROINFLAMMATORY DISORDERS

The present review of the reports documented that; marine drugs have a promising role in controlling the neuroinflammatory reaction. And it also ameliorates the neuroinflammation associated with neurological disorders. The selective marine drugs are attenuated the neuroinflammatory disorders with modification of neuroinflammatory pathways. Such agents are CEP-1347, fascaplysin, ethanolic extract of *Nannochloropsis oceanic* and pyrogallol-phloroglucinol-6,6-bieckol are attenuates the AD (Chen et al., 2014b; Choi et al., 2017; Kang et al., 2013; Manda et al., 2016); PD ameliorated by CEP-1347, 11-Dehydrosinulariolide (Feng et al., 2016; Ning et al., 2018); MS ameliorated by lestaurtinib, enzastaurin, MS14, and Dihydro-austrasulfone (Ahmadi et al., 2010; Chen et al., 2014b; Gagalo et al., 2015; Tabatabai et al., 2007); stroke ameliorated by CEP-1347, Xyloketal B, and polyphenolic compounds of *Ecklonia cava* (Carlsson et al.,2009; Gong et al., 2018; Kim et al.,2012). VaD ameliorated by CEP-1347, Xyloketal B, polyphenolic compounds of *Ecklonia cava* and pyrogallol-phloroglucinol-6,6-bieckol (Carlsson et al., 2009; Gong et al., 2018; Koirala et al., 2017; Oh et al., 2018). Neuropathic pain ameliorated by enzastaurin, dihydro-austrasulfone, capnellene, lemnalol, flexibilide, and DHA (Chen et al., 2014a; Dudek et al., 2008; Jean et al., 2008; Jean et al., 2009; Silva et al., 2017; Wen et al., 2010; Zhang et al., 2018); and neuroinflammation ameliorated by frondanol, zoanthamine, fucoxanthin, and 6-bromoindole (Ahmad et al., 2017; Guillen et al., 2018; Rodriguez-Luna et al., 2018; Subramanya et al., 2018). The summary of a neuroinflammatory mechanism; and molecular mechanism of marine drugs are expressed in Figure 11.1 and Table 11.2, respectively. Even, it has a specialized mechanism for the attenuation of neuroinflammatory reactions. The toxicity level and therapeutic efficacy are better than other natural and conventional medicines.

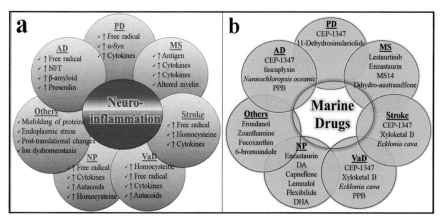

FIGURE 11.1 Summary of neuroinflammatory reactions and list of marine drugs used in neuroinflammatory disorders. Abbreviation: AD, Alzheimer's disorder; PD, Parkinson's disorder; MS, multiple sclerosis; VaD, vascular dementia; NP, neuropathic pain; CEP-1347, semisynthetic derivatives of K-252a (i.e., indolocarbazole); MS14, natural marine herbal drug; DHA, docosahexaenoic acid; PPB, pyrogallol-phloroglucinol-6,6-bieckol; and DA, dihydro-austrasulfone.

TABLE 11.2   Molecular Mechanism of Marine Drugs for the Prevention of Neuroinflammation

Sl. No.	Marine Drugs	Mechanisms	References
1.	Methanol extracts derived from green algae of *Ulva conglobate*	Inhibition of iNOS and COX-2	Kim and Chojnacka, 2015
2.	Ethanol extract of microalgae, i.e., *Nannochloropsis oceanica*	Down-regulation of APP and BACE1 expression	Chen et al., 2014b; Choi et al., 2017
3.	Alginate-derived oligosaccharide derived from brown algae	Inactivation the TLR4-NF-κB signaling pathway	Zhou et al., 2015
4.	Dieckol derived from brown algae of *Ecklonia cava*	Downregulation of ERK, Akt, and NADPH oxidase-mediated pathways	Cui et al., 2015; Lee et al., 2018
5.	Floridoside derived from red algae of *Laurencia undulata*	Inhibition of p38 and ERK	Kim et al., 2013
6.	Phlorofucofuroeckol B derived from brown algae of *Ecklonia stolonifera*	Inhibition of IκB-alpha/NF-κB and Akt/ERK/JNK pathways	Yu et al., 2015
7.	Aurantiamide-acetate derived from a marine fungus of *Aspergillus* sp.	Inhibition of NF-κB, JNK, and p38	Yoon et al., 2014
8.	Citreohybridonol derived from a marine fungus of *Toxicocladosporium* sp.	An inhibitory effect on the NF-κB and p38 pathways	Cho et al., 2016
9.	Sinuleptolide derived from soft coral of *Sinularia kavarattiensis*	Inhibition of IL-1β, IL-6, IL-8, IL-18, and TNF-α	Morretta et al., 2017
10.	Lestaurtinib: Derived K-252a from the marine actinomycete	FLT-3, JAK-2, Trk-A, Trk-B, Trk-C	Obeid et al., 2014
11.	Enzastaurin: Derived from the marine staurosporine	PKCβ, GSK-3β	Kreisl et al., 2010; Ning et al., 2018
12.	CEP-1347: Derived from K-252a	JNKs	Ning et al., 2018; Roux et al., 2002
13.	11-Dehydrosinulariolidederived from soft coral of *Sinulariaflexibilis*	Inhibition of NF-κB The increase of DJ-1 expression	Chen et al., 2016; Ning et al., 2018
14.	DA, a synthetic precursor of austrasulfone of soft coral (*Cladiellaaustralis*)	Inhibition of iNOS and COX-2	Chen et al., 2015; Wen et al., 2010
15.	Capnellenederived from soft coral of *Capnella imbricate*	Inhibition of spinal COX-2	Jean et al., 2009

TABLE 11.2 *(Continued)*

Sl. No.	Marine Drugs	Mechanisms	References
16.	Lemnalol derived from soft coral of *Lemnaliacervicorni*	Inhibition of spinal TNF-α	Jean et al., 2008
17.	Flexibilide derived from soft coral of *Sinulariaflexibilis*	Inhibition of spinal iNOS	Chen et al.,2014a
18.	Xyloketal B derived from the marine fungus of *Xylaria*Sp.	Inhibition of brain caspase-3 and Bax	Gong et al., 2018
19.	DHA-derived from marine fish	Inhibition of spinal p38	Thomas et al., 2015
20.	α-D-glucan derived from *Mytiluscoruscus*	Inhibits the TLR4/NF-κB/ MAPK pathway	Liu et al., 2017
21.	Polyphenols derived from brown algae of *Eckloniacava*	Inhibition of cytosolic calcium	Kim et al.,2012
22.	PPB derived from *Ecklonia cava*. 6-Bromoindole derived from marine sponge *Geodia barrette*. Fucoxanthin (allenic carotenoid) derived from edible brown seaweeds. Zoanthamine derived from *Zoanthus genus*. Frondanol derived fromorange-footed sea cucumber, i.e., *Cucumariafrondosa.* *Isochrysisgalbana* derived from marine microalgae, i.e., *Haptophyta*Sp.	Inhibits the MMP	Guillen et al., 2018; Kang et al., 2011; Oh et al., 2018; Rodriguez-Luna et al., 2018; Rodriguez-Luna et al., 2017; Subramanya et al., 2018
23.	MS14 (natural herbal-marine drug formulation)	Inhibition of spinal lipocalin-2	Ebrahimi Kalan et al., 2014

Abbreviation: APP, amyloid precursor protein; BACE1, beta-secretase 1; Bax, B-cell lymphoma-2-like protein 4; COX-2, cyclooxygenase-2; DA, dihydro-austrasulfone; DHA, Docosahexaenoic acid; DJ-1, Parkinsonism associated deglycase protein 7(PARK7); ERK, extracellular signal-regulated kinase; FLT-3, vascular endothelial growth factor-dependent tyrosine kinase protein; GSK, glycogen synthase kinase; IL, interleukin; iNOS, inducible nitric oxide synthase; IκB, an inhibitor of NF-κB; JAK, Janus kinase; JNK, c-Jun N-terminal kinase; MAPK, mitogen-activated protein kinases; MMP, Matrix metalloproteinase; MS14, natural marine-herbal drug; NADPH, nicotinamide adenine dinucleotide phosphate; NF-κB, nuclear factor κB; PKC, protein kinase C; PPB, pyrogallol-phloroglucinol-6,6-bieckol; TLR4, toll-like receptor 4; TNF-α, tumor necrosis factor-α; and Trk, tropomyosin-related kinase.

Table 11.2 has been summarized as a possible mechanism of marine drugs for neuroinflammatory disorders.

11.6 FUTURE DIRECTION

Based on data availability, marine drugs are beneficial for neurological disorders. And it has ample scope to treats chronic inflammatory and neuro-infectious disorders. However, more extensive and detailed information with preclinical data is required before starting to practice in clinical trials as well as treatments to a human being. Because, still lack preclinical data like toxicity profile including the toxic level of dose, toxicokinetic, toxico-dynamic reactions; and interaction with biological tiny molecules, nutrients, and other medicaments. Similarly, pharmacological details with adverse drug effects and allergic effects also need to investigate safe usage to treat neuroinflammatory disorders. This book chapter helps to choose the best medicines for the specific neuroinflammatory disorder and makes the aware-ness for the screening of marine drugs for human neurological disorders.

11.7 CONCLUSIONS

Hence, marine drugs may be used as neuroprotective agents for neuroinflam-mation and inflammation associated with neurological disorders. However, the extensive preclinical research documents are required for the safe usage of marine drugs in neuroinflammatory disorders in a human being.

ACKNOWLEDGMENTS

The authors are thankful toJSS College of Pharmacy, JSS Academy of Higher Education and Research, Mysuru, India; and Faculty of Pharmacy, AIMST University, Semeling, Malaysia for support to make this book chapter.

KEYWORDS

- cognitive disorder
- drug discovery
- neuroimmune system

- **neuroinflammation**
- **neuropathic pain**
- **neurovascular disorders**

REFERENCES

Agarwal, A., Aponte-Mellado, A., Premkumar, B. J., Shaman, A., & Gupta, S., (2012). The effects of oxidative stress on female reproduction: A review. *Reprod. Biol. Endocrinol., 10*, 49.

Agostinho, P., Cunha, R. A., & Oliveira, C., (2010). Neuroinflammation, oxidative stress and the pathogenesis of Alzheimer's disease. *Curr. Pharm. Des., 16*(25), 2766–2778.

Ahmad, T. B., Rudd, D., Smith, J., Kotiw, M., Mouatt, P., Seymour, L. M., Liu, L., & Benkendorff, K., (2017). Anti-Inflammatory activity and structure-activity relationships of brominated indoles from a marine mollusc. *Mar. Drugs., 15*(5).

Ahmadi, A., Habibi, G., & Farrokhnia, M., (2010). MS14, an Iranian herbal-marine compound for the treatment of multiple sclerosis. *Chin. J. Integr Med., 16*(3), 270–271.

Aldskogius, H., (2001). Regulation of microglia-potential new drug targets in the CNS. *Expert Opin. Ther. Targets., 5*(6), 655–668.

Anekonda, T. S., & Reddy, P. H., (2005). Can herbs provide a new generation of drugs for treating Alzheimer's disease? Brain research. *Brain. Res. Rev., 50*(2), 361–376.

Argoff, C. E., Cole, B. E., Fishbain, D. A., & Irving, G. A., (2006). Diabetic peripheral neuropathic pain: Clinical and quality-of-life issues. *Mayo Clinic Proceedings, 81*(4), S3–S11.

Baek, H. J., Lee, M. Y., Lee, H., & Min, M. S., (2011). Mitochondrial DNA data unveil highly divergent populations within the genus *Hynobius* (Caudata: Hynobiidae) in South Korea. *Mol. Cells., 31*(2), 105–112.

Basbaum, A. I., Bautista, D. M., Scherrer, G., & Julius, D., (2009). Cellular and molecular mechanisms of pain. *Cell, 139*(2), 267–284.

Bhat, A. H., Dar, K. B., Anees, S., Zargar, M. A., Masood, A., Sofi, M. A., & Ganie, S. A., (2015). Oxidative stress, mitochondrial dysfunction and neurodegenerative diseases; a mechanistic insight. *Biomed. Pharmacother., 74*, 101–110.

Biber, K., Neumann, H., Inoue, K., & Boddeke, H. W., (2007). Neuronal 'on' and 'off' signals control microglia. *Trends Neurosci., 30*(11), 596–602.

Brookoff, D., (2000). Chronic pain: 1. a new disease? *Hosp. Pract., 35*(7), 45–52, 59.

Carlsson, Y., Leverin, A. L., Hedtjarn, M., Wang, X., Mallard, C., & Hagberg, H., (2009). Role of mixed lineage kinase inhibition in neonatal hypoxia-ischemia. *Dev. Neurosci., 31*(5), 420–426.

Castro, J., Harrington, A. M., Garcia-Caraballo, S., Maddern, J., Grundy, L., Zhang, J., Page, G., et al., (2017). Alpha-conotoxin Vc1.1 inhibits human dorsal root ganglion neuron excitability and mouse colonic nociception via GABAB receptors. *Gut, 66*(6), 1083–1094.

Cavalli, G., & Dinarello, C. A., (2018). Anakinra therapy for non-cancer inflammatory diseases. *Front Pharmacol., 9*, 1157.

Ceulemans, A. G., Zgavc, T., Kooijman, R., Hachimi-Idrissi, S., Sarre, S., & Michotte, Y., (2010). The dual role of the neuroinflammatory response after ischemic stroke: Modulatory effects of hypothermia. *J Neuroinflammation., 7*, 74.

Ceulemans, A. G., Zgavc, T., Kooijman, R., Hachimi-Idrissi, S., Sarre, S., & Michotte, Y., (2011). Mild hypothermia causes differential, time-dependent changes in cytokine expression and gliosis following endothelin-1-induced transient focal cerebral ischemia. *J Neuroinflammation., 8*, 60.

Chalorak, P., Jattujan, P., Nobsathian, S., Poomtong, T., Sobhon, P., & Meemon, K., (2018). *Holothuriascabra* extracts exhibit anti-Parkinson potential in *C. elegans*: A model for anti-Parkinson testing. *Nutr. Neuro Sci., 21*(6), 427–438.

Chen, C. H., Chen, N. F., Feng, C. W., Cheng, S. Y., Hung, H. C., Tsui, K. H., Hsu, C. H., et al., (2016). A coral-derived compound improves functional recovery after spinal cord injury through its antiapoptotic and anti-inflammatory effects. *Mar Drugs, 14*(9).

Chen, N. F., Huang, S. Y., Lu, C. H., Chen, C. L., Feng, C. W., Chen, C. H., Hung, H. C., et al., (2014). Flexibilide obtained from cultured soft coral has anti-neuroinflammatory and analgesic effects through the upregulation of spinal transforming growth factor-beta-1 in neuropathic rats. *Mar. Drugs., 12*(7), 3792–3817.

Chen, S. C., Chien, Y. C., Pan, C. H., Sheu, J. H., Chen, C. Y., & Wu, C. H., (2014). Inhibitory effect of dihydro-austrasulfone alcohol on the migration of human non-small cell lung carcinoma A549 cells and the antitumor effect on a Lewis lung carcinoma-bearing tumor model in C57BL/6J mice. *Mar. Drugs., 12*(1), 196–213.

Chen, Y. C., Wen, Z. H., Lee, Y. H., Chen, C. L., Hung, H. C., Chen, C. H., Chen, W. F., & Tsai, M. C., (2015). Dihydroaustrasulfone alcohol inhibits PDGF-induced proliferation and migration of human aortic smooth muscle cells through inhibition of the cell cycle. *Mar, Drugs, 13*(4), 2390–2406.

Cho, K. H., Kim, D. C., Yoon, C. S., Ko, W. M., Lee, S. J., Sohn, J. H., Jang, J. H., Ahn, J. S., Kim, Y. C., & Oh, H., (2016). Anti-neuroinflammatory effects of citreohybridonol involving TLR4-MyD88-mediated inhibition of NF-small ka, CyrillicB and MAPK signaling pathways in lipopolysaccharide-stimulated BV2 cells. *Neurochem Int., 95*, 55–62.

Choi, D. Y., & Choi, H., (2015). Natural products from marine organisms with neuroprotective activity in the experimental models of Alzheimer's disease, Parkinson's disease and ischemic brain stroke: Their molecular targets and action mechanisms. *Arch. Pharm. Res., 38*(2), 139–170.

Choi, J. Y., Hwang, C. J., Lee, H. P., Kim, H. S., Han, S. B., & Hong, J. T., (2017). Inhibitory effect of ethanol extract of *Nannochloropsis oceanica* on lipopolysaccharide-induced neuroinflammation, oxidative stress, amyloidogenesis, and memory impairment. *Oncotarget., 8*(28), 45517–45530.

Collins, M. P., & Hadden, R. D., (2017). The nonsystemic vasculitic neuropathies. *Nat. Rev. Neurol., 13*(5), 302–316.

Cui, Y., Park, J. Y., Wu, J., Lee, J. H., Yang, Y. S., Kang, M. S., Jung, S. C., et al., (2015). Dieckol attenuates microglia-mediated neuronal cell death via ERK, Akt and NADPH oxidase-mediated pathways. *Korean J. Physiol. Pharmacol., 19*(3), 219–228.

Dadsetan, S., Balzano, T., Forteza, J., Cabrera-Pastor, A., Taoro-Gonzalez, L., Hernandez-Rabaza, V., et al., (2016). Reducing peripheral inflammation with infliximab reduces neuroinflammation and improves cognition in rats with hepatic encephalopathy. *Front Mol. Neurosci., 9*, 106.

De Boer, A. G., & Gaillard, P. J., (2007). Drug targeting to the brain. *Annu. Rev. Pharmacol. Toxicol., 47*, 323–355.

De Nicola, A. F., Garay, L. I., Meyer, M., Guennoun, R., Sitruk-Ware, R., Schumacher, M., & Gonzalezm D, M. C., (2018). Neurosteroidogenesis and progesterone anti-inflammatory/neuroprotective effects. *J. Neuroendocrinol., 30*(2).

De Oliveira, M. F., & Pinto, F. C., (2015). Hypertonic saline: A brief overview of hemodynamic response and anti-inflammatory properties in head injury. *Neural Regen. Res., 10*(12), 1938–1939.

Deng-Bryant, Y., Readnower, R. D., Leung, L. Y., Cunningham, T. L., Shear, D. A., & Tortella, F. C., (2015). Treatment with amnion-derived cellular cytokine solution (ACCS) induces persistent motor improvement and ameliorates neuroinflammation in a rat model of penetrating ballistic-like brain injury. *Restor. Neurol. Neurosci., 33*(2), 189–203.

Di, X., Rouger, C., Hardardottir, I., Freysdottir, J., Molinski, T. F., Tasdemir, D., & Omarsdottir, S., (2018). 6-Bromoindole derivatives from the icelandic marine sponge geodiabarretti: Isolation and anti-inflammatory activity. *Mar. Drugs., 16*(11).

Dillman, J. F., Phillips, C. S., Kniffin, D. M., Tompkins, C. P., Hamilton, T. A., & Kan, R. K., (2009). Gene expression profiling of rat hippocampus following exposure to the acetylcholinesterase inhibitor soman. *Chem. Res. Toxico., 22*(4), 633–638.

Dudek, A. Z., Zwolak, P., Jasinski, P., Terai, K., Gallus, N. J., Ericson, M. E., & Farassati, F., (2008). Protein kinase C-beta inhibitor enzastaurin (LY317615.HCl) enhances radiation control of murine breast cancer in an orthotopic model of bone metastasis. *Invest New Drugs, 26*(1), 13–24.

Ebrahimi, K. A., Soleimani, R. J., Kafami, L., Mohamadnezhad, D., Khaki, A. A., & Mohammadi, R. A., (2014). MS14, a marine herbal medicine, an immunosuppressive drug in experimental autoimmune encephalomyelitis. *Iran Red Crescent Med. J., 16*(7), e16956.

Eijkelkamp, N., Steen-Louws, C., Hartgring, S. A., Willemen, H. L., Prado, J., Lafeber, F. P., Heijnen, C. J., Hack, C. E., Van, R. J. A., & Kavelaars, A., (2016). IL4-10 fusion protein is a novel drug to treat persistent inflammatory pain. *J. Neurosci., 36*(28), 7353–7363.

Engen, P. A., Dodiya, H. B., Naqib, A., Forsyth, C. B., Green, S. J., Voigt, R. M., Kordower, J. H., Mutlu, E. A., Shannon, K. M., & Keshavarzian, A., (2017). The potential role of gut-derived inflammation in multiple system atrophy. *J. Parkinsons Dis., 7*(2), 331–346.

Ercan, I., Tufekci, K. U., Karaca, E., Genc, S., & Genc, K., (2018). Peptide derivatives of erythropoietin in the treatment of neuroinflammation and neurodegeneration. *Adv. Protein. Chem. Struct. Biol., 112*, 309–357.

Feng, C. W., Hung, H. C., Huang, S. Y., Chen, C. H., Chen, Y. R., Chen, C. Y., Yang, S. N., et al., (2016). Neuroprotective effect of the marine-derived compound 11-dehydrosinulariolide through DJ-1-related pathway in *in vitro* and *in vivo* models of Parkinson's disease. *Mar. Drugs., 14*(10).

Finlay, B. B., & McFadden, G., (2006). Anti-immunology: Evasion of the host immune system by bacterial and viral pathogens. *Cell, 124*(4), 767–782.

Frankola, K. A., Greig, N. H., Luo, W., & Tweedie, D., (2011). Targeting TNF-alpha to elucidate and ameliorate neuroinflammation in neurodegenerative diseases. *CNS Neurol. Disord. Drug Targets, 10*(3), 391–403.

Fu, Y., Yang, J., Wang, X., Yang, P., Zhao, Y., Li, K., Chen, Y., & Xu, Y., (2018). Herbal compounds play a role in neuroprotection through the inhibition of microglial activation. *J. Immunol. Res.*, 9348046.

Gagalo, I., Rusiecka, I., & Kocic, I., (2015). Tyrosine kinase inhibitor as a new therapy for ischemic stroke and other neurologic diseases: Is there any hope for a better outcome? *Curr. Neuropharmacol., 13*(6), 836–844.

Gao, H. M., Liu, B., Zhang, W., & Hong, J. S., (2003). Novel anti-inflammatory therapy for Parkinson's disease. *Trends Pharmacol. Sci., 24*(8), 395–401.

Gavett, B. E., Stern, R. A., & McKee, A. C., (2011). Chronic traumatic encephalopathy: A potential late effect of sport-related concussive and subconcussive head trauma. *Clin. Sports Med., 30*(1), 179–188.

Giatti, S., Boraso, M., Melcangi, R. C., & Viviani, B., (2012). Neuroactive steroids, their metabolites, and neuroinflammation. *J. Mol. Endocrinol., 49*(3), R125-R134.

Gong, H., Luo, Z., Chen, W., Feng, Z. P., Wang, G. L., & Sun, H. S., (2018). Marine compound xyloketal B as a potential drug development target for neuroprotection. *Mar. Drugs., 16*(12).

Goverman, J., (2009). Autoimmune T cell responses in the central nervous system. *Nat. Rev Immunol., 9*(6), 393–407.

Graf, T., & Enver, T., (2009). Forcing cells to change lineages. *Nature, 462*(7273), 587–594.

Grosso, C., Valentao, P., Ferreres, F., & Andrade, P. B., (2014). Bioactive marine drugs and marine biomaterials for brain diseases. *Mar. Drugs., 12*(5), 2539–2589.

Guillen, P. O., Gegunde, S., Jaramillo, K. B., Alfonso, A., Calabro, K., Alonso, E., Rodriguez, J., Botana, L. M., & Thomas, O. P., (2018). Zoanthamine alkaloids from the Zoantharian Zoanthus cf. pulchellus and their effects in neuroinflammation. *Mar. Drugs., 16*(7).

Haddad, J. J., (2002). Cytokines and related receptor-mediated signaling pathways. *Biochem. Biophys. Res. Commun., 297*(4), 700–713.

Han, J. H., Lee, Y. S., Im, J. H., Ham, Y. W., Lee, H. P., Han, S. B., & Hong, J. T., (2019). Astaxanthin ameliorates lipopolysaccharide-induced neuroinflammation, oxidative stress, and memory dysfunction through inactivation of the signal transducer and activator of transcription 3 pathway. *Mar. Drugs., 17*(2).

Hanisch, U. K., (2002). Microglia as a source and target of cytokines. *Glia., 40*(2), 140–155.

Hasturk, A. E., Gokce, E. C., Yilmaz, E. R., Horasanli, B., Evirgen, O., Hayirli, N., Gokturk, H., Erguder, I., & Can, B., (2018). Therapeutic evaluation of tumor necrosis factor-alpha antagonist etanercept against traumatic brain injury in rats: Ultrastructural, pathological, and biochemical analyses. *Asian J. Neurosurg., 13*(4), 1018–1025.

Hawkes, C. H., Del, T. K., & Braak, H., (2007). Parkinson's disease: A dual-hit hypothesis. *Neuropathol. Appl. Neurobiol., 33*(6), 599–614.

Hemmer, B., Archelos, J. J., & Hartung, H. P., (2002). New concepts in the immunopathogenesis of multiple sclerosis. *Nat. Rev. Neurosci., 3*(4), 291–301.

Heneka, M. T., Kummer, M. P., & Latz, E., (2014). Innate immune activation in neurodegenerative disease. *Nat. Rev. Immunol., 14*(7), 463–477.

Iadecola, C., & Anrather, J., (2011). The immunology of stroke: From mechanisms to translation. *Nature Med., 17*(7), 796–808.

Islam, M. S., Murata, T., Fujisawa, M., Nagasaka, R., Ushio, H., Bari, A. M., Hori, M., & Ozaki, H., (2008). Anti-inflammatory effects of phytosterylferulates in colitis induced by dextran sulphate sodium in mice. *Br. J. Pharmacol., 154*(4), 812–824.

Iwanaga, M., Tsukui, K., Uchiyama, K., Katsuma, S., Imanishi, S., & Kawasaki, H., (2014). Expression of recombinant proteins by BEVS in a macula-like virus-free silkworm cell line. *J. Invertebr. Pathol., 123*, 34–37.

Jean, Y. H., Chen, W. F., Duh, C. Y., Huang, S. Y., Hsu, C. H., Lin, C. S., Sung, C. S., Chen, I. M., & Wen, Z. H., (2008). Inducible nitric oxide synthase and cyclooxygenase-2 participate in anti-inflammatory and analgesic effects of the natural marine compound lemnalol from Formosan soft coral *Lemnalia cervicorni*. *Eur. J. Pharmacol., 578*(2–3), 323–331.

Jean, Y. H., Chen, W. F., Sung, C. S., Duh, C. Y., Huang, S. Y., Lin, C. S., Tai, M. H., Tzeng, S. F., & Wen, Z. H., (2009). Capnellene, a natural marine compound derived from soft coral, attenuates chronic constriction injury-induced neuropathic pain in rats. *Br. J. Pharmacol., 158*(3), 713–725.

Jean-Louis, G., Zizi, F., Clark, L. T., Brown, C. D., & McFarlane, S. I., (2008). Obstructive sleep apnea and cardiovascular disease: Role of the metabolic syndrome and its components. *JCSM, 4*(3), 261–272.

Joffre, O. P., Segura, E., Savina, A., & Amigorena, S., (2012). Cross-presentation by dendritic cells. *Nat. Rev. Immunol., 12*(8), 557–569.

Kang, I. J., Jang, B. G., In, S., Choi, B., Kim, M., & Kim, M. J., (2013). Phlorotannin-rich *Ecklonia cava* reduces the production of beta-amyloid by modulating alpha- and gamma-secretase expression and activity. *Neurotoxicology, 34*, 16–24.

Kang, S. M., Lee, S. H., Heo, S. J., Kim, K. N., & Jeon, Y. J., (2011). Evaluation of antioxidant properties of a new compound, pyrogallol-phloroglucinol-6,6'-bieckol isolated from brown algae, *Ecklonia cava*. *Nutr. Res. Pract., 5*(6), 495–502.

Kasperska-Zajac, A., Brzoza, Z., & Rogala, B., (2008). Platelet-activating factor (PAF): A review of its role in asthma and clinical efficacy of PAF antagonists in the disease therapy. *Recent Pat. Inflamm. Allergy Drug Discov., 2*(1), 72–76.

Kelloff, G. J., Krohn, K. A., Larson, S. M., Weissleder, R., Mankoff, D. A., Hoffman, J. M., Link, J. M., et al., (2005). The progress and promise of molecular imaging probes in oncologic drug development. *Clin. Cancer Res., 11*(22), 7967–7985.

Khazdair, M. R., Boskabady, M. H., Hosseini, M., Rezaee, R., & Tsatsakis, A. M., (2015). The effects of *Crocus sativus* (saffron), and its constituents on nervous system: A review. *Avicenna J. Phytomed., 5*(5), 376–391.

Kim, J. H., Lee, N. S., Jeong, Y. G., Lee, J. H., Kim, E. J., & Han, S. Y., (2012). Protective efficacy of an *Ecklonia cava* extract used to treat transient focal ischemia of the rat brain. *Anat. Cell Biol., 45*(2), 103–113.

Kim, M., Li, Y. X., Dewapriya, P., Ryu, B., & Kim, S. K., (2013). Floridoside suppresses pro-inflammatory responses by blocking MAPK signaling in activated microglia. *BMB Rep., 46*(8), 398–403.

Kim, S. K., & Chojnacka, K., (2015). *Marine Algae Extracts: Processes, Products, and Applications*. John Wiley & Sons, New Jersey.

Kirshner, M., Galron, R., Frenkel, D., Mandelbaum, G., Shiloh, Y., Wang, Z. Q., & Barzilai, A., (2012). Malfunctioning DNA damage response (DDR) leads to the degeneration of nigro-striatal pathway in mouse brain. *J. Mol Neurosci., 46*(3), 554–568.

Koirala, P., Jung, H. A., & Choi, J. S., (2017). Recent advances in pharmacological research on Ecklonia species: A review. *Arch Pharm. Res., 40*(9), 981–1005.

Kreisl, T. N., Kotliarova, S., Butman, J. A., Albert, P. S., Kim, L., Musib, L., Thornton, D., & Fine, H. A., (2010). A phase I/II trial of enzastaurin in patients with recurrent high-grade gliomas. *Neuro Oncol., 12*(2), 181–189.

Kruidenier, L., & Verspaget, H. W., (2002). Review article: Oxidative stress as a pathogenic factor in inflammatory bowel disease-radicals or ridiculous? *Aliment Pharmacol. Ther., 16*(12), 1997–2015.

Kulkarni, A. P., Kellaway, L. A., & Kotwal, G. J., (2005). Herbal complement inhibitors in the treatment of neuroinflammation: Future strategy for neuroprotection. *Ann. NY Acad. Sci., 1056*, 413–429.

Lane, R. J., & Routledge, P. A., (1983). Drug-induced neurological disorders. *Drugs, 26*(2), 124–147.

Latta, C. H., Sudduth, T. L., Weekman, E. M., Brothers, H. M., Abner, E. L., Popa, G. J., Mendenhall, M. D., et al., (2015). Determining the role of IL-4 induced neuroinflammation in microglial activity and amyloid-beta using BV2 microglial cells and APP/PS1 transgenic mice. *J. Neuroinflammation., 12*, 41.

Lee, S., Youn, K., Kim, D. H., Ahn, M. R., Yoon, E., Kim, O. Y., & Jun, M., (2018). Anti-neuroinflammatory property of phlorotannins from *Ecklonia cava* on Abeta25–35-induced damage in PC12 cells. *Mar. Drugs., 17*(1).

Li, S., Shen, C., Guo, W., Zhang, X., Liu, S., Liang, F., Xu, Z., Pei, Z., Song, H., Qiu, L., Lin, Y., & Pang, J., (2013). Synthesis and neuroprotective action of xyloketal derivatives in Parkinson's disease models. *Mar. Drugs., 11*(12), 5159–5189.

Liesz, A., Bauer, A., Hoheisel, J. D., & Veltkamp, R., (2014). Intracerebral interleukin-10 injection modulates post-ischemic neuroinflammation: An experimental microarray study. *Neurosci. Lett.*, 579, 18–23.

Liu, F., Zhang, X., Li, Y., Chen, Q., Zhu, X., Mei, L., Song, X., et al., (2017). Anti-inflammatory effects of a *Mytiluscoruscus* alpha-d-glucan (MP-A) in activated macrophage cells via TLR4/NF-kappaB/MAPK pathway inhibition. *Mar. Drugs., 15*(9).

Liu, Y., Geng, L., Zhang, J., Wang, J., Zhang, Q., & Duan, D., (2018). Oligo-porphyran ameliorates neurobehavioral deficits in Parkinsonian mice by regulating the PI3K/Akt/Bcl-2 pathway. *Mar. Drugs., 16*(3).

Liu, Y., Zhang, Y., Zheng, X., Fang, T., Yang, X., Luo, X., Guo, A., Newell, K. A., Huang, X. F., & Yu, Y., (2018). Galantamine improves cognition, hippocampal inflammation, and synaptic plasticity impairments induced by lipopolysaccharide in mice. *J. Neuroinflammation., 15*(1), 112.

Lu, D., Liu, Y., Mai, H., Zang, J., Shen, L., Zhang, Y., & Xu, A., (2018). Rosuvastatin reduces neuroinflammation in the hemorrhagic transformation after rt-PA treatment in a mouse model of experimental stroke. *Front Cell Neurosci., 12*, 225.

Lull, M. E., & Block, M. L., (2010). Microglial activation and chronic neurodegeneration. *Neurotherapeutics., 7*(4), 354–65.

Manda, S., Sharma, S., Wani, A., Joshi, P., Kumar, V., Guru, S. K., Bharate, S. S., Bhushan, S., Vishwakarma, R. A., Kumar, A., & Bharate, S. B., (2016). Discovery of a marine-derived bis-indole alkaloid fascaplysin, as a new class of potent P-glycoprotein inducer and establishment of its structure-activity relationship. *Eur. J. Med. Chem., 107*, 1–11.

Manzhulo, I. V., Ogurtsova, O. S., Kipryushina, Y. O., Latyshev, N. A., Kasyanov, S. P., Dyuizen, I. V., & Tyrtyshnaia, A. A., (2016). Neuron-astrocyte interactions in spinal cord dorsal horn in neuropathic pain development and docosahexaenoic acid therapy. *J Neuroimmunol., 298*, 90–97.

Mao, J., & Chen, L. L., (2000). Systemic lidocaine for neuropathic pain relief. *Pain, 87*(1), 7–17.

Martinez, B., & Peplow, P. V., (2018). Neuroprotection by immunomodulatory agents in animal models of Parkinson's disease. *Neural Regen. Res., 13*(9), 1493–1506.

Martins, S., & Fernandes, L., (2012). Delirium in elderly people: A review. *Front Neurol.*, 3, 101.

Mayer, E. A., Knight, R., Mazmanian, S. K., Cryan, J. F., & Tillisch, K., (2014). Gut microbes and the brain: Paradigm shift in neuroscience. *J. Neurosci., 34*(46), 15490–15496.

Miller, A. H., Maletic, V., & Raison, C. L., (2009). Inflammation and its discontents: The role of cytokines in the pathophysiology of major depression. *Biol. Psychiatry., 65*(9), 732–741.

Morrell, R. M., (1976). Immunopathology of the nervous system. *Adv. Exp. Med. Biol., 73*, 121–146.

Morretta, E., Esposito, R., Festa, C., Riccio, R., Casapullo, A., & Monti, M. C., (2017). Discovering the biological target of 5-epi-sinuleptolide using a combination of poteomic approaches. *Mar. Drugs., 15*(10).

Moskowitz, M. A., Lo, E. H., & Iadecola, C., (2010). The science of stroke: Mechanisms in search of treatments. *Neuron, 67*(2), 181–198.

Motte, J., Ambrosius, B., Gruter, T., Bachir, H., Sgodzai, M., Pedreiturria, X., Pitarokoili, K., & Gold, R., (2018). Capsaicin-enriched diet ameliorates autoimmune neuritis in rats. *J. Neuroinflammation, 15*(1), 122.

Mu, L., Sobotka, S., Chen, J., Su, H., Sanders, I., Adler, C. H., Shill, H. A., Caviness, J. N., Samanta, J. E., & Beach, T. G., (2012). Altered pharyngeal muscles in Parkinson disease. *J. Neuropathol. Exp. Neurol., 71*(6), 520–530.

Mukherjee, P. K., Kumar, V., Kumar, N. S., & Heinrich, M., (2008). The Ayurvedic medicine *Clitoriaternatea* from traditional use to scientific assessment. *J. Ethnopharmacol., 120*(3), 291–301.

Munoz-Fernandez, M. A., & Fresno, M., (1998). The role of tumor necrosis factor, interleukin 6, interferon-gamma, and inducible nitric oxide synthase in the development and pathology of the nervous system. *Prog. Neurobiol., 56*(3), 307–340.

Muthuraman, A., & Kaur, P., (2016). Renin-angiotensin-aldosterone system: A current drug target for the management of neuropathic pain. *Curr. Drug Targets, 17*(2), 178–195.

Nathan, C., & Ding, A., (2010). Nonresolving inflammation. *Cell, 140*(6), 871–882.

Nemeroff, C. B., (1986). The interaction of neurotensin with dopaminergic pathways in the central nervous system: Basic neurobiology and implications for the pathogenesis and treatment of schizophrenia. *Psychoneuroendocrinology., 11*(1), 15–37.

Nguyen, M. D., Julien, J. P., & Rivest, S., (2002). Innate immunity: The missing link in neuroprotection and neurodegeneration? *Nat. Rev. Neurosci., 3*(3), 216–227.

Ning, C., Wang, H. D., Gao, R., Chang, Y. C., Hu, F., Meng, X., & Huang, S. Y., (2018). Marine-derived protein kinase inhibitors for neuroinflammatory diseases. *Biomed. Eng. Online, 17*(1), 46.

Obeid, M., Rosenberg, E. C., Klein, P. M., & Jensen, F. E., (2014). Lestaurtinib (CEP-701) attenuates "second hit" kainic acid-induced seizures following early life hypoxic seizures. *Epilepsy Res., 108*(4), 806–810.

Oh, S., Son, M., Lee, H. S., Kim, H. S., Jeon, Y. J., & Byun, K., (2018). Protective effect of pyrogallol-phloroglucinol-6,6-bieckol from Ecklonia cava on monocyte-associated vascular dysfunction. *Mar. Drugs, 16*(11).

Okorji, U. P., Velagapudi, R., El-Bakoush, A., Fiebich, B. L., & Olajide, O. A., (2016). Antimalarial drug artemether inhibits neuroinflammation in BV2 microglia through Nrf2-dependent mechanisms. *Mol. Neurobiol., 53*(9), 6426–6443.

Ordovas-Montanes, J., Rakoff-Nahoum, S., Huang, S., Riol-Blanco, L., Barreiro, O., & Von, A. U. H., (2015). The regulation of immunological processes by peripheral neurons in homeostasis and disease. *Trends Immunol., 36*(10), 578–604.

Otmishi, P., Gordon, J., El-Oshar, S., Li, H., Guardiola, J., Saad, M., Proctor, M., & Yu, J., (2008). Neuroimmune interaction in inflammatory diseases. *Clin. Med. Circ. Respirat. Pulm. Med., 2*, 35–44.

Ott, M., Gogvadze, V., Orrenius, S., & Zhivotovsky, B., (2007). Mitochondria, oxidative stress and cell death. *Apoptosis, 12*(5), 913–922.

Pan, N., Lu, L. Y., Li, M., Wang, G. H., Sun, F. Y., Sun, H. S., Wen, X. J., et al., (2017). Xyloketal B alleviates cerebral infarction and neurologic deficits in a mouse stroke model by suppressing the ROS/TLR4/NF-κB inflammatory signaling pathway. *Acta Pharmacol. Sin., 38*(9), 1236–1247.

Parihar, M. S., & Hemnani, T., (2004). Alzheimer's disease pathogenesis and therapeutic interventions. *J. Clin. Neurosci., 11*(5), 456–467.

Pober, J. S., & Sessa, W. C., (2007). Evolving functions of endothelial cells in inflammation. *Nat. Rev. Immunol., 7*(10), 803–815.

Pop-Busui, R., Sima, A., & Stevens, M., (2006). Diabetic neuropathy and oxidative stress. *Diabetes Metab. Res. Rev., 22*(4), 257–273.

Pozzolini, M., Millo, E., Oliveri, C., Mirata, S., Salis, A., Damonte, G., Arkel, M., & Scarfi, S., (2018). Elicited ROS scavenging activity, photoprotective, and wound-healing properties of collagen-derived peptides from the marine sponge *Chondrosiareniformis*. *Mar. Drugs., 16*(12).

Puentes, F., Malaspina, A., Van, N. J. M., & Amor, S., (2016). Non-neuronal cells in ALS: Role of glial, immune cells and blood-CNS barriers. *Brain Pathol., 26*(2), 248–257.

Ransohoff, R. M., Kivisakk, P., & Kidd, G., (2003). Three or more routes for leukocyte migration into the central nervous system. *Nat. Rev. Immunol., 3*(7), 569–581.

Ratner, N., Brodeur, G. M., Dale, R. C., & Schor, N. F., (2016). The "neuro" of neuroblastoma: Neuroblastoma as a neurodevelopmental disorder. *Ann. Neurol., 80*(1), 13–23.

Reagan-Shaw, S., Nihal, M., & Ahmad, N., (2008). Dose translation from animal to human studies revisited. *FASEB J., 22*(3), 659–661.

Reid, S. N. S., Ryu, J. K., Kim, Y., & Jeon, B. H., (2018). GABA-enriched fermented *Laminaria japonica* improves cognitive impairment and neuroplasticity in scopolamine- and ethanol-induced dementia model mice. *Nutr. Res. Pract., 12*(3), 199–207.

Requena, J. R., Fu, M. X., Ahmed, M. U., Jenkins, A. J., Lyons, T. J., & Thorpe, S. R., (1996). Lipoxidation products as biomarkers of oxidative damage to proteins during lipid peroxidation reactions. *Nephrol. Dial Transplant, 11*(5), 48–53.

Rizzuto, R., De Stefani, D., Raffaello, A., & Mammucari, C., (2012). Mitochondria as sensors and regulators of calcium signaling. *Nat. Rev. Mol. Cell Biol., 13*(9), 566–578.

Rodriguez-Luna, A., Avila-Roman, J., Gonzalez-Rodriguez, M. L., Cozar, M. J., Rabasco, A. M., Motilva, V., & Talero, E., (2018). Fucoxanthin-containing cream prevents epidermal hyperplasia and UVB-induced skin erythema in mice. *Mar. Drugs., 16*(10).

Rodriguez-Luna, A., Talero, E., Terencio, M. D. C., Gonzalez-Rodriguez, M. L., Rabasco, A. M., De Los Reyes, C., Motilva, V., & Avila-Roman, J., (2017). Topical application of glycolipids from *Isochrysis galbana* prevents epidermal hyperplasia in mice. *Mar. Drugs., 16*(1).

Roux, P. P., Dorval, G., Boudreau, M., Angers-Loustau, A., Morris, S. J., Makkerh, J., & Barker, P. A., (2002). K252a and CEP1347 are neuroprotective compounds that inhibit mixed-lineage kinase-3 and induce activation of Akt and ERK. *J. Biol. Chem., 277*(51), 49473–49480.

Ruhl, A., (2005). Glial cells in the gut. *Neurogastroenterol. Motil., 17*(6), 777–790.

Safaeinejad, F., Bahrami, S., Redl, H., & Niknejad, H., (2018). Inhibition of inflammation, suppression of matrix metalloproteinases, induction of neurogenesis, and antioxidant property make bryostatin-1 a therapeutic choice for multiple sclerosis. *Front Pharmacol., 9*, 625.

Salas, M. M., McIntyre, M. K., Petz, L. N., Korz, W., Wong, D., & Clifford, J. L., (2015). Tetrodotoxin suppresses thermal hyperalgesia and mechanical allodynia in a rat full thickness thermal injury pain model. *Neurosci. Lett., 607*, 108–113.

Sas, K., Robotka, H., Toldi, J., & Vecsei, L., (2007). Mitochondria, metabolic disturbances, oxidative stress and the kynurenine system, with focus on neurodegenerative disorders. *J. Neurol. Sci., 257*(1–2), 221–239.

Schreibman, D. L., Hong, C. M., Keledjian, K., Ivanova, S., Tsymbalyuk, S., Gerzanich, V., & Simard, J. M., (2018). Mannitol and hypertonic saline reduce swelling and modulate inflammatory markers in a rat model of intracerebral hemorrhage. *Neurocrit. Care, 29*(2), 253–263.

Shal, B., Ding, W., Ali, H., Kim, Y. S., & Khan, S., (2018). Anti-neuroinflammatory potential of natural products in attenuation of Alzheimer's disease. *Front Pharmacol., 9*, 548.

Shamim, D., & Laskowski, M., (2017). Inhibition of inflammation mediated through the tumor necrosis factor alpha biochemical pathway can lead to favorable outcomes in Alzheimer's disease. *J. Cent. Nerv. Syst. Dis., 9*, 1179573517722512.

Sharma, K., Sharma, D., Sharma, M., Sharma, N., Bidve, P., Prajapati, N., Kalia, K., & Tiwari, V., (2018). Astaxanthin ameliorates behavioral and biochemical alterations in *in-vitro* and in-vivo model of neuropathic pain. *Neurosci. Lett., 674*, 162–170.

Shechter, R., London, A., & Schwartz, M., (2013). Orchestrated leukocyte recruitment to immune-privileged sites: Absolute barriers versus educational gates. *Nat. Rev Immunol., 13*(3), 206–218.

Shohami, E., Gallily, R., Mechoulam, R., Bass, R., & Ben-Hur, T., (1997). Cytokine production in the brain following closed head injury: Dexanabinol (HU-211) is a novel TNF-alpha inhibitor and an effective neuroprotectant. *J. Neuroimmunol., 72*(2), 169–177.

Silva, R. V., Oliveira, J. T., Santos, B. L. R., Dias, F. C., Martinez, A. M. B., Lima, C. K. F., & Miranda, A. L. P., (2017). Long-chain omega-3 fatty acids supplementation accelerates nerve regeneration and prevents neuropathic pain behavior in mice. *Front Pharmacol., 8*, 723.

Silver, J., & Miller, J. H., (2004). Regeneration beyond the glial scar. *Nat. Rev. Neurosci., 5*(2), 146–156.

Singh, R. B., Mengi, S. A., Xu, Y. J., Arneja, A. S., & Dhalla, N. S., (2002). Pathogenesis of atherosclerosis: A multifactorial process. *Exp. Clin. Cardiol., 7*(1), 40–53.

Song, D. H., Kim, G. J., Lee, K. J., Shin, J. S., Kim, D. H., Park, B. J., & An, J. H., (2018). Mitigation effects of a novel herbal medicine, hepad, on neuroinflammation, neuroapoptosis, and neuro-oxidation. *Molecules, 23*(11).

Steed, D. L., Trumpower, C., Duffy, D., Smith, C., Marshall, V., Rupp, R., & Robson, M., (2008). Amnion-derived cellular cytokine solution: A physiological combination of cytokines for wound healing. *Eplasty., 8*, e18.

Subramanya, S. B., Chandran, S., Almarzooqi, S., Raj, V., Al-Zahmi, A. S., Al-Katheeri, R. A., Al-Zadjali, S. A., et al., (2018). Frondanol, a nutraceutical extract from *Cucumaria frondosa*, attenuates colonic inflammation in a DSS-induced colitis model in mice. *Mar. Drugs, 16*(5).

Suk, K., (2005). Regulation of neuroinflammation by herbal medicine and its implications for neurodegenerative diseases. A focus on traditional medicines and flavonoids. *Neurosignals., 14*(1–2), 23–33.

Tabatabai, G., Weller, M., Bamberg, M., Tatagiba, M., & Wick, W., (2007). Preclinical and clinical applications of enzastaurin in the treatment of malignant gliomas. *Aktuelle Neurologie., 34*(S2), V145.

Tafreshi, A. P., Ahmadi, A., Ghaffarpur, M., Mostafavi, H., Rezaeizadeh, H., Minaie, B., Faghihzadeh, S., & Naseri, M., (2008). An Iranian herbal-marine medicine, MS14, ameliorates experimental allergic encephalomyelitis. *Phytother. Res., 22*(8), 1083–1086.

Thomas, J., Thomas, C. J., Radcliffe, J., & Itsiopoulos, C., (2015). Omega-3 fatty acids in early prevention of inflammatory neurodegenerative disease: A focus on Alzheimer's disease. *Biomed. Res. Int.*, 172801.

Tosti, E., Boni, R., & Gallo, A., (2017). Micro-conotoxins modulating sodium currents in pain perception and transmission: A therapeutic potential. *Mar. Drugs, 15*(10).

Valera, E., Mante, M., Anderson, S., Rockenstein, E., & Masliah, E., (2015). Lenalidomide reduces microglial activation and behavioral deficits in a transgenic model of Parkinson's disease. *J. Neuroinflammation., 12*, 93.

Vanle, B., Olcott, W., Jimenez, J., Bashmi, L., Danovitch, I., & IsHak, W. W., (2018). NMDA antagonists for treating the non-motor symptoms in Parkinson's disease. *Transl. Psychiatry., 8*(1), 117.

Veldhoen, M., Hocking, R. J., Atkins, C. J., Locksley, R. M., & Stockinger, B., (2006). TGFbeta in the context of an inflammatory cytokine milieu supports de novo differentiation of IL-17-producing T cells. *Immunity, 24*(2), 179–189.

Villiger, P. M., & Muller-Ladner, U., (2017). Tapering and termination of immunosuppressive therapy. *Z Rheumatol., 76*(1), 6–7.

Wadhwa, M., Chauhan, G., Roy, K., Sahu, S., Deep, S., Jain, V., Kishore, K., et al., (2018). Caffeine and modafinil ameliorate the neuroinflammation and anxious behavior in rats during sleep deprivation by inhibiting the microglia activation. *Front Cell Neurosci., 12*, 49.

Wang, Y., Shen, J., Yang, X., Jin, Y., Yang, Z., Wang, R., Zhang, F., & Linhardt, R. J., (2018). Akebiasaponin D reverses corticosterone hypersecretion in an Alzheimer's disease rat model. *Biomed. Pharmacother., 107*, 219–225.

Wen, Z. H., Chao, C. H., Wu, M. H., & Sheu, J. H., (2010). A neuroprotective sulfone of marine origin and the *in vivo* anti-inflammatory activity of an analog. *Eur. J. Med. Chem., 45*(12), 5998–6004.

White, J. A., Banerjee, R., & Gunawardena, S., (2016). Axonal transport and neurodegeneration: How marine drugs can be used for the development of therapeutics. *Mar. Drugs., 14*(5).

Witte, M. E., Mahad, D. J., Lassmann, H., & Van, H. J., (2014). Mitochondrial dysfunction contributes to neurodegeneration in multiple sclerosis. *Trends in Molecular Medicine, 20*(3), 179–187.

Wyss-Coray, T., (2006). Inflammation in Alzheimer disease: Driving force, bystander, or beneficial response? *Nature Med., 12*(9), 1005–1015.

Yoon, C. S., Kim, D. C., Lee, D. S., Kim, K. S., Ko, W., Sohn, J. H., Yim, J. H., Kim, Y. C., & Oh, H., (2014). Anti-neuroinflammatory effect of aurantiamide acetate from the marine fungus *Aspergillus* sp. SF-5921: Inhibition of NF-kappaB and MAPK pathways in lipopolysaccharide-induced mouse BV2 microglial cells. *Int. Immunopharmacol., 23*(2), 568–574.

Yu, D. K., Lee, B., Kwon, M., Yoon, N., Shin, T., Kim, N. G., Choi, J. S., & Kim, H. R., (2015). Phlorofucofuroeckol B suppresses inflammatory responses by down-regulating nuclear factor kappaB activation via Akt, ERK, and JNK in LPS-stimulated microglial cells. *Int. Immunopharmacol., 28*(2), 1068–1075.

Yu, S. Y., Wang, S. W., Hwang, T. L., Wei, B. L., Su, C. J., Chang, F. R., & Cheng, Y. B., (2018). Components from the leaves and twigs of mangrove *Lumnitzera racemosa* with anti-angiogenic and anti-inflammatory effects. *Mar. Drugs., 16*(11).

Zhang, L., Terrando, N., Xu, Z. Z., Bang, S., Jordt, S. E., Maixner, W., Serhan, C. N., & Ji, R. R., (2018). Distinct analgesic actions of DHA and DHA-derived specialized pro-resolving mediators on post-operative pain after bone fracture in mice. *Front Pharmacol., 9*, 412.

Zhou, R., Shi, X. Y., Bi, D. C., Fang, W. S., Wei, G. B., & Xu, X., (2015). Alginate-derived oligosaccharide inhibits neuroinflammation and promotes microglial phagocytosis of beta-amyloid. *Mar. Drugs, 13*(9), 5828–5846.

Zhu, Y. Z., Liu, J. W., Wang, X., Jeong, I. H., Ahn, Y. J., & Zhang, C. J., (2018). Anti-BACE1 and antimicrobial activities of steroidal compounds isolated from marine *Urechisunicinctus. Mar. Drugs., 16*(3).

Zlokovic, B. V., (2008). The blood-brain barrier in health and chronic neurodegenerative disorders. *Neuron, 57*(2), 178–201.

Zweckberger, K., & Plesnila, N., (2009). Anatibant, a selective non-peptide bradykinin B2 receptor antagonist, reduces intracranial hypertension and histopathological damage after experimental traumatic brain injury. *Neurosci. Lett., 454*(2), 115–117.

Zykwinska, A., Marquis, M., Godin, M., Marchand, L., Sinquin, C., Garnier, C., Jonchere, C., et al., (2019). Microcarriers Based on glycosaminoglycan-like marine exopolysaccharide for tgf-beta1 long-term protection. *Mar. Drugs., 17*(1).

Index